Relativistic Dynamical Quantum Mechanics

With Applications In FOUR Dimensions

K. M. Eason, PhD

First Edition 1.0

March 2022

ISBN: 978-0-578-57845-3

Library of Congress Control Number: 2021906236

Front cover image by: Symphonious Publishing House
Book design by: Symphonious Publishing House
Published by: Symphonious Publishing House
Website: symphoniouspub.com
Contact: info@symphoniouspub.com

First printing, March 2022

Dedications

To Asa J. Davis, PhD and Albert Einstein, PhD. Even in the afterlife, I hope you can find some pleasure in these fruits of your labors.

Acknowledgements

The author gratefully acknowledges invaluable revision contributions from Professor Simon Greaves of Tohoku University in Sendai, Japan.

Preface To The First Edition

Full 4D Relativistic Quantum Mechanics introduces the fourth dimension formally, where this dimension is shown to contain valuable time-dependence information, otherwise, missing in quantum mechanics. Taking an obvious queue from the theory of relativity, a new interpretation of so-called *space-time* implies that *the fourth dimension is not real-valued*, but purely imaginary. This strange, yet simple interpretation ensures the independence of four dimensions and enables a relativistic formulation of quantum mechanics that extends time-dependent phenomena in quantum mechanics. Consequently, it also leads to many new descriptions of elementary particles as functions of time, including, for example, transitions between quantum states.

This text logically introduces the special theory of relativity, which leads to a formal definition of the fourth dimension. This provides the foundation to develop a generalization of the Dirac equation that fully includes this imaginary fourth dimension. We'll see that this leads to a host of time-dependent phenomena, including spin-dependent transient phenomena applicable *e.g.* in spintronics applications. Perennial well-known heuristic problems in quantum mechanics are also generalized further. For example, the propagating plane-wave gains a time-dependent amplitude as well-known with electromagnetic waves, and the harmonic oscillator generalizes to the *damped* harmonic oscillator, etc.

In order to cover the content effectively, the text begins in Chapter 1, with the special theory of relativity, including some demonstrative experiments of the past that could only be explained by the theory of relativity. This topic sets the stage for a formal definition of the fourth dimension, enabling its inclusion into a quantum mechanics formulation.

The text also includes a fresh look at superconductors using a purely relativistic quantum mechanical treatment with higher order corrections, with support from the fourth dimension. New information emerges in the application which includes how spin-orbit interaction can give rise to well-known observations in superconductors. Also, additional quantum operators are shown to provide the origins of type I and type II superconductors. All of these topics are relevant today in a number of technologies on the horizon.

Our intent is that the reader finds effective coverage of the information presented on the fourth dimension, and with sufficient fidelity to capture it's splendor, richness, and usefulness. Just as important, may the reader absorb the information sufficiently in order to apply the content to either solve new problems and/or take the subject further. Lastly, the author thanks you truly for your interest in this content. May it serve you and the universe in some significant way.

Sincerely,

K. Makai Eason, PhD

Contents

Relevant Physical Constants In SI Units

Quantity	Symbol	Value
Avogadro constant	N_a	6.0221415×10^{23} $\mathrm{mol^{-1}}$
Bohr radius	a_0	$0.5291772108 \times 10^{-10}$ m
Bohr magneton	μ_B	$9.27400949 \times 10^{-24}$ $\mathrm{JT^{-1}}$
Boltzmann's constant	k_B	$1.3806505 \times 10^{-23}$ $\mathrm{J \cdot K^{-1}}$
Compton wavelength	λ_C	$2.426310238 \times 10^{-12}$ m
elementary charge	e	$1.60217662 \times 10^{-19}$ C
electron g-factor	g	2.0023193043718
electron mass	m_e	$9.1093826 \times 10^{-31}$ kg
flux quantum	Φ_0	$2.06783372 \times 10^{-15}\mathrm{Tesla \cdot m^2}$
gyromagnetic ratio	γ	$8.7941001136 \times 10^{-10}$ $\mathrm{HzT^{-1}}$
light speed in vacuum	c	$299,792,458$ $\mathrm{ms^{-1}}$
Planck constant	h	$6.6260693 \times 10^{-34}$ $\mathrm{J \cdot s}$
Planck reduced constant	$\hbar$	$1.05457168 \times 10^{-34}$ $\mathrm{J \cdot s}$
proton mass	m_p	$1.67262171 \times 10^{-27}$ kg
quanta of resistance	R_q	$25.8k\Omega$
Rydberg constant	R_d	$109,737,31.568525$ $\mathrm{m^{-1}}$
universal gas constant	R	$8.3144\mathrm{JK^{-1}mol^{-1}}$
vacuum permittivity	ϵ_0	$8.854187817 \times 10^{-12}$ $\mathrm{Fm^{-1}}$
vacuum permeability	μ_0	$4 \times \pi \times 10^{-7}$ $\mathrm{N \cdot A^{-2}}$

CHAPTER 1

A Very Special Theory Of Relativity

Numbers are indeed things. However, they are especially unique *things* because their properties are inherited from all other things! This may sound simple (or not), but this fact is profound in its own way. Consider this: the reason $1 + 1 = 2$ is because if we hold up one digit or finger, then add one more, we then have what we call 2 fingers up. That number 2 corresponds to what is observed. It is *real* both literally and mathematically. In other words, the equation $1 + 1 = 2$ mirrors what is observed. Extending the idea of mirroring the real world further, we find even more of this in properties of more complex 'numbers' like vector spaces. Similar to scalars, they inherit properties from our physical n–dimensional world. Another example, and central to this text, is the so-called *imaginary unit* $i = \sqrt{-1}$. Scientists know too well that we cannot get very far trying to describe our physical world *without* the peculiar i. This is very noticeable when systems have to undergo some kind of physical change periodically with time, or dynamically. A cogent example is given by electromagnetic waves known as the so-called Maxwell equations. But, does i also mirror anything from the physical world? Amazingly, the answer is a resounding *yes*. Here, we will be uncovering the origins of this indispensable unit i. But first, to set the stage for what is to come, let's revisit an old newspaper article raising an important question for our consideration.

On May 04, 1935, the New York Times (NYT) published an article which leaked comments from private discussions between three physicists. One of the physicists was none other than Albert Einstein. The other two were Boris Podolsky and Nathan Rosen. At this time, the subject of quantum physics was still young, and not all physicists shared the same understanding or interpretations. In some instances, this lead to a somewhat contentious climate stemming from trying to grasp the shear awkwardness revealed by quantum physics, as well as the theory of relativity. These new, yet, non-intuitive subjects motivated scientists to tread cautiously in their public interpretations of physical phenomena. The NYT article concerned a quantum mechanics thought experiment known today as the *EPR Paradox*. The EPR Paradox is a thought experiment between Einstein, Podolsky and Rosen (EPR) implying what some perceived as *uncomfortable* versions of a physical reality. Ultimately, some were enticed to conclude that quantum mechanics was *incomplete*. An example of something that steered some in this direction was the fact that quantum mechanics did not allow for an elementary particle to have definite position and momentum of arbitrary precision at the same time. This was a challenging notion to wrap the mind around. However,

EINSTEIN ATTACKS QUANTUM THEORY

Scientist and Two Colleagues Find It Is Not 'Complete' Even Though 'Correct.'

SEE FULLER ONE POSSIBLE

Believe a Whole Description of 'the Physical Reality' Can Be Provided Eventually.

Figure 1.1: Newspaper article published in the NYT May 04, 1935, relating to the EPR Paradox.

today, we can say that quantum mechanics, in fact, *can not be complete without this strange fact*. It translates to what is known formally as the *uncertainty principle*. It is a statement about the delicate nature of the wave-like properties possessed by elementary particles. Descriptions like this lie wholly within the grasp of quantum mechanics and the theory of relativity.

Still, the question of whether quantum mechanics is complete or not, as a subject or body of knowledge, is still very much relevant. In the spirit of knowledge and exploration, let's hope *not*. In fact, without a doubt, it is *not*. In the words of Einstein, Podolsky, and Rosen (though expressed in a different context):

"...of what is formulated, it is correct, however, it is not complete."

A clear example of this has been the historic inability of quantum mechanics to describe the dynamical behavior of an electron spin in a magnetic field. When a magnetic field is introduced, the electron spin (or magnetic moment) undergoes a unique motion known as *precession*, accompanied by eventual polar rotation causing anti-alignment (alignment) with the magnetic field[1]. The simplest of experiments evidencing the polar rotation, in particular, can be found in magnetic saturation curves (hysteresis curves) which overwhelmingly demonstrate this behavior of the collective spin angular momentum. An example of a *hysteresis curve* is shown in Fig. 1.2.

Recently, however, fundamental quantum mechanics *has* been able to predict this kind of precessional + polar rotational behavior naturally for the electron spin, as well as, orbital angular momentum[2]. We will unveil this *duo-dynamical*[3] behavior, as well, in an alternative manner. We'll achieve this result using *four-dimensional quantum mechanics*. Spoiler alert...we'll find that it agrees with the alternative derivation using *relativistic* electromagnetic laws, and more importantly, with real physical magnetic systems.

There is a subtle leap in a quantum theory that includes such a physical description. An electron spin, described this way, is an example of

[1]We'll see later that this motion also includes a third relaxation degree of freedom along the spin axis

[2]**2019 Quantum Mechanics With A SPIN On It (book), Eason.**

[3]more generally, spin has a *trio-dynamical* as there are up to three degree of motion.

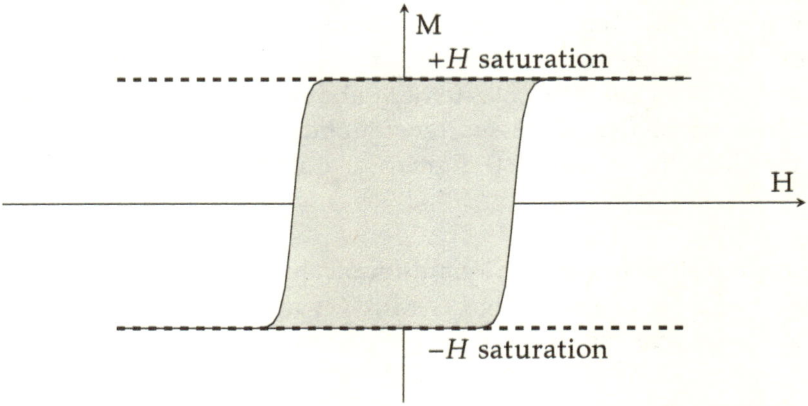

Figure 1.2: Hysteresis loop illustrating the response of a localized electrons in magnetic material, resulting from a phenomenon known as spin-orbit coupling as exchange interaction. To achieve reversal, the applied magnetic field (x-axis) *must* overcome the internal effective (spin-orbit + exchange) field, given approximately by the crossing value along the H (or x) axis.

a *dynamical transition between equilibrium quantum states.* Quantum mechanics having the ability to predict both equilibrium quantum states as well as natural dynamical transitions between them represents a step *forward*. As we will see shortly, obtaining such a quantum mechanics formalism *requires the fourth dimension,* which depends on relativity. In *four-dimensional quantum mechanics* which may equivalently be called *relativistic dynamical quantum mechanics,* quantum transitions naturally emerge. Thus, we have some useful content to unpack here, and we'll be doing this precisely, in the early coming pages.

Before digging in, let take up the question as to why anyone should take the time to learn more about quantum mechanics. If you are already familiar with some parts of the subject, this will be an opportunity to expand what you know in a useful way to understand how to describe the dynamics of quantum mechanical systems accurately. Much of the results developed here should be useful in many ways. It is also laid out in such a way to understand *how* to use it.

To begin, we'll need to have precise discussions of concepts like dimensions and their connections with the physical description of the universe. As mentioned, it is important to understand that *numbers are physical* because they mirror the properties of the physical world, from a vector space to a complex scalar. This becomes even more apparent when connecting the fourth dimension.

Additionally, quantum mechanics is arguably one of the most accurate and far-reaching branches of physics available today. Taking this into consideration, it is probably not a bad idea to learn some of it. It can help us better grasp the properties of our physical universe. So, how can we pull this off for the reader while introducing some of the above topics giving rise to four dimensional quantum mechanics? In our aim towards these objectives, you will find the following:

1. Using relativity along with mathematics, this text provides a formal derivation of the fourth dimension.

2. Using this fourth dimension, the text uniquely develops a general formulation of four-dimensional quantum mechanics consistent with *relativistic* quantum mechanics *i.e.* Dirac's equation.

3. The text establishes that four-dimensional quantum mechanics generalizes to necessitate *non-Hermitian* operators (having complex eigenvalues). In so many words, the assumption of *strictly* Hermitian operators are demonstrated as unnecessarily restrictive.

4. The text also demonstrates it's consequences and utility in applications, including revisiting perennial pedagogical concepts in quantum mechanics such as plane waves, the harmonic oscillator, *etc.* Both generalize to possess decaying amplitudes.

5. In presenting these topics, the reader will also find complete derivations of nearly all results to provide a reliable pedagogical scaffold in the construction of all conclusions.

As many have found, quantum mechanics is *not* an easy subject to learn (properly), no matter how you slice it. This is largely due to the levels of mathematical rigor (and general lack of mathematical experience with too many students) that is required. Moreover, we are going to be adding a new layer by introducing the fourth dimension. Regardless, in learning any topics in physics, know that the root language of the universe is *math*. Sooner or later, this must be reconciled for any serious student of science. In our opinion, this should not be intimidating, but rather inviting. Know that if persistence and diligence is within you, then you *will* find success in understanding the subject properly[4] and penetrating any barriers to insights you may have. In so far as our role in helping you achieve this, we have organized the content of this book for aiming to not only extend your knowledge of the subject, but guide, step by step, through the unfolding of everything including four-dimensional relativistic quantum mechanics, as well as applications of it. So, please stick with it. You got this.

With that out of the way, it is probably fitting to begin our discussions at the apex of scientific knowledge existing *c*.1905, when Einstein (and others) published a strange, yet undeniably correct body of work known as the *special theory of relativity*. From here, we can build something useful towards 4D quantum mechanics. Let's get started, shall we?

1.1 Lorentz Factor and Einstein's Special Theory

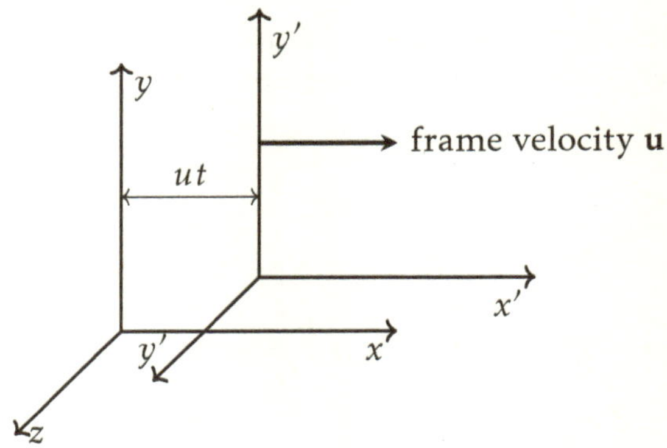

Figure 1.3: Illustration two inertial reference frames: a reference frame $\mathbf{r}'$ moving at uniform velocity $\mathbf{u}$, relative to a *stationary* frame $\mathbf{r}$.

What does relativity even mean, anyway, in physics, anyway? *Ordinary* relativity predates the 1905 works of Einstein *et. al.* Relativity is tied to the formal relations of a system's coordinates as seen between two distinct reference frames. For example, if I observed the motion of a star from California, could someone looking at the same star from New York or Beijing take my info as truth? To take this question one step

[4]this goes for any subject!

further, what about the physical laws, since they all depend on space and time? Questions like these become thorny quickly. So, a good place to start is with a description of the physical coordinates characterizing space and time. Relativity is tied to these questions. In its simplest expression, it translates to a relationship between the three dimensional *position* $\mathbf{r} = [x, y, z]^T$ and time t of a point attached to an object described in two *different inertial reference frames*[5]. With such frames, the velocity, denoted by $\mathbf{u}$, (or velocity of one frame with respect to the other) is often taken as *constant*. Fig. 1.3 illustrates examples of two frames in this context.

One frame $\mathbf{r}$ is taken to be at rest, so designated as the *origin*, while the other frame $\mathbf{r}' = [x', y', z']^T$ is moving with *uniform* velocity $\mathbf{u}$ *relative* to the frame described by $\mathbf{r}$. Common intuition based on everyday experience leads us to relate the coordinates between the two frames as

$$\mathbf{r}' = \mathbf{r} - \mathbf{u}t \quad \text{or} \quad \mathbf{r} = \mathbf{r}' + \mathbf{u}t \tag{1.1a}$$

$$t = t' \tag{1.1b}$$

Eqs. (1.1) relate *four* variables and is known as the *Galilean transformation* equations. They relate the position $\mathbf{r}$ and time t of an object in one frame to those in the other. If the relative velocity $\mathbf{u}$ is *constant* in the coordinate of time t, they are *linear* equations in the four coordinates. Moreover, a signature feature of the Galilean transformation is that *time is identical between the two frames.* Einstein conveyed a very similar concept using the idea of synchronicity of clocks[6]. Under this condition, we say *time is invariant in the Galilean frame.* Then, it follows that time-differentiation of Eqs. (1.1a) and (1.1b) gives the analogous velocity relations as

$$\mathbf{v}' = \mathbf{v} - \mathbf{u} \tag{1.2a}$$

$$\mathbf{v} = \mathbf{v}' + \mathbf{u} \tag{1.2b}$$

These relations are called *relative velocities* because Eq. (1.2a) expresses the velocity *relative* to $\mathbf{v}$, while Eq. (1.2b) is the velocity relative to $\mathbf{v}'$

Towards the end of the 19th century, after success was found measuring the speed of light, as well as other properties of light, new experiments

[5]inertial here means both frames have at most steady motion

[6]clock synchronicity means both clocks begin at precisely the same instant *and* remain ticking in lockstep.

were being contrived where some of them started to reveal contradictory behavior to Eqs. (1.2). For example, in 1851, French physicist Hippolyte Fizeau (1819-1896) performed experiments measuring characteristics of the speed of light in flowing water. It turned out that in order to understand Fizeau's empirical results concerning light speeds in different mediums, *special relativity was needed.*[7] Much later, c.1886, Michelson and Morley refined Fizeau's experiments and they extended experimental efforts beyond a water medium. They are known for attempting to find evidence of motion of a background *luminiferous aether*, which was a widely held concept by many physicists at the time. It was postulated as the medium in which electromagnetic waves propagate (*e.g.* light). The Michelson-Morley experiment on this, however, could not detect such evidence. In *light* of new observations like these, in particular, revealing contradictions to Eqs. (1.2), physicists began working on theories that might explain these observations. From those efforts, rising to the surface, the *Lorentz factor* eventually bubbled up.

1.1.1 The Lorentz Factor

One physicist's contributions that would prove a fruitful seed was that of Hendrik Lorentz (1853-1928). As mentioned, an impetus came from experimental observations suggesting the *speed of light was capped* or had a finite maximum. If one measured the speed of light in a given medium, then did a similar observation while moving the same light-carrying medium, one would find that light *always* measures less than a certain finite value regardless of how fast you *move* the light-carrying medium. Mathematically speaking, we say that the principle of vector superposition (*aka* linear combinations) of the velocities *alone* was insufficient for velocities in mediums carrying light. Because such behavior was at odds with Eqs. (1.2), it was recognized that, perhaps, our understanding of space and time could be improved. The beginning of that *improved* understanding came when Lorentz generalized the linear Galilean transformation in Eq. (1.1) to one that is *nonlinear*. Let's work out how this is possible.

If positions x, y, and z could be different between two inertial frames, *depending on the frame velocity*, perhaps the *real* time coordinates might also be related differently, *e.g.* as functions of the moving frame velocity. To generalize Eqs. (1.1), let us consider only a single spatial coordinate

[7] We'll discuss the details of this experiment shortly

with time (for simplicity) x and x'. Let us now write the following generalization:

$$x' = (x - ut)\gamma_r \tag{1.3a}$$

$$x = (x' + ut')\gamma_r \tag{1.3b}$$

We have two equations in two unknowns, namely γ_r and t'. Once we find γ_r, we can find the relation between t and t'. Let us find γ_r by first computing $x'x$, which gives

$$x'x = \gamma_r^2(x - ut)(x' + ut') = \gamma_r^2(xx' - u^2tt' + xut' - x'ut) \tag{1.4}$$

Here is where the maximum light speed enters the picture; An explicit assumption made by Lorentz was to take the *speed of light as constant* in both frames. We can imagine an experiment where the maximum speed of light is measured in one frame (say $\mathbf{r}, t$), then we introduce another reference frame with relative motion to it. In experiments, no frame could be found that would produce light faster than the well-known value $c \approx 3 \times 10^8$ m/s. Today, this is taken as one of the *postulates of relativity*:

Ⓡ *First Postulate of Relativity*: There does not exist a physical reference frame where the measurement of the speed of light exceeds $c \approx 3 \times 10^8$ m/s. It follows that since *all matter* travels at speeds less than that of light, *all matter* as well as light must be limited to the max speed of light denoted by c, being the *ultimate terminal velocity*.

Taking the above postulate into consideration, if x and x' are taken as the position of the light to be measured, then it follows that

$$x = ct \quad \text{and} \quad x' = ct' \tag{1.5}$$

The above statement implicitly assumes *only* initial synchronization at time $t = 0.$[8]. Dividing both sides of Eq. (1.4) by $x'x$, then substituting Eq. (1.5) gives

$$1 = \gamma_r^2\left(1 - \frac{u^2tt'}{x'x} + \frac{xut'}{x'x} - \frac{x'ut}{x'x}\right)$$

[8]but without lockstep

$$= \gamma_r^2\left(1 - \frac{u^2}{c^2} + \frac{u}{c} - \frac{u}{c}\right)$$

$$= \gamma_r^2\left(1 - \frac{u^2}{c^2}\right)$$

Solving for γ_r, we obtain the *Lorentz factor* given by

Lorentz Factor

$$\gamma_r = \frac{1}{\sqrt{1 - \frac{u^2}{c^2}}} \tag{1.6}$$

Eq. (1.6) was derived by Lorentz *c*.1895.

Using γ_r, we can determine t' by using an alternative form of xx', by multiplying Eq. (1.3a) by x, so we have

$$xx' = (x^2 - utx)\gamma_r = (c^2t^2 - utx)\gamma_r \tag{1.7}$$

Using the *postulate* relation $x = ct$, it follows that

$$c^2t't = (x^2 - utx)\gamma_r = (c^2t^2 - utx)\gamma_r \tag{1.8}$$

Dividing both sides by c^2t, we obtain

$$t' = \left(t - \frac{ux}{c^2}\right)\gamma_r \tag{1.9}$$

Analogously, multiplying Eq. (1.3b) by $x' = ct'$ and dividing by c^2t', we obtain the following for t:

$$t = \left(t' + \frac{ux'}{c^2}\right)\gamma_r \tag{1.10}$$

These results not only contain a generalization of the relations between time and space coordinates, but maintains a consistent form of symmetry, as well. Notice, also, that no assumptions, whatsoever, were made regarding the uniformness of **u**. In other words, these results allow **u** to be nonuniform in time.

From the above results, Lorentz gave us the following four relations between x, x', t, and t', known today as the *Lorentz transformation*:

Lorentz Transformation Equations

$$\mathbf{r}' = \frac{\mathbf{r} - \mathbf{u}t}{\sqrt{1 - \frac{u^2}{c^2}}} \quad \text{and} \quad \mathbf{r} = \frac{\mathbf{r}' + \mathbf{u}t'}{\sqrt{1 - \frac{u^2}{c^2}}} \tag{1.11a}$$

$$t' = \frac{t - \frac{\mathbf{u}\cdot\mathbf{r}}{c^2}}{\sqrt{1 - \frac{u^2}{c^2}}} \quad \text{and} \quad t = \frac{t' + \frac{\mathbf{u}\cdot\mathbf{r}'}{c^2}}{\sqrt{1 - \frac{u^2}{c^2}}} \tag{1.11b}$$

Fig. 1.4 illustrates γ_r as a function of the parameter $\beta = u/c$.

While, for simplicity we have only used x for the spatial coordinate, the general vector result given in Eq. (1.11) is found by replacing $x' \to \mathbf{r}'$ and u with its vector form $\mathbf{u} = [u_x, u_y, u_z]^T$. The profound implications of the Lorentz transformations would not begin to reveal themselves until around ten years later, with the works of people like Einstein.

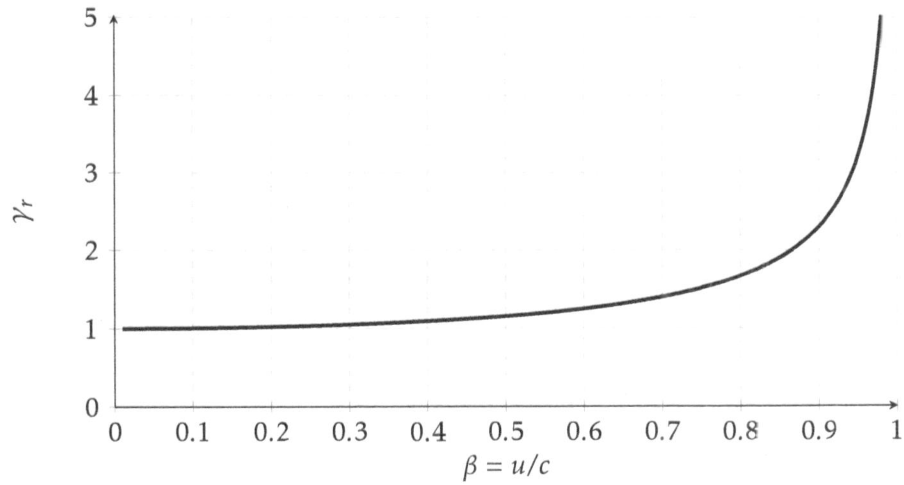

Figure 1.4: Lorentz factor γ_r as a function of the parameter $\beta = u/c$, where u if the relative frame velocity and $c = 3 \times 10^8$ m/s is the maximum velocity of light. γ_r has a minimum value of 1 approaching ∞ as $u \to \infty$. Notice that γ_r satisfies the relation $\gamma_r \geq 1$.

1.1.2 Lorentz Factor Enables Special Theory of Relativity

Einstein examined the implications of the Lorentz transformation on general physical systems, since $\mathbf{r} = [x, y, z]^T$ and t are usually involved in any physical law. One of the first relations Einstein considered was the corresponding generalization of relative velocities between two *Lorentzian* frames, *i.e.* the analog of Eqs. (1.2).

To obtain the analog of Eqs. (1.2a) and (1.2b), we can consider the difference between any two points (t_1, x_1) and (t_2, x_2) *in the same Lorentzian frame*. Using the Lorentz transformation under the *assumption* of a uniform frame velocity $\mathbf{u}$, the differences (Δ) in both frames, respectively, are given by

$$\Delta x' = (\Delta x - u\Delta t)\gamma_r \quad \text{and} \quad \Delta x = (\Delta x + u\Delta t')\gamma_r \tag{1.12a}$$

$$\Delta t' = \left(\Delta t - \frac{u\Delta x}{c^2}\right)\gamma_r \quad \text{and} \quad \Delta t = \left(\Delta t' + \frac{u\Delta x'}{c^2}\right)\gamma_r \tag{1.12b}$$

Because u is taken as constant, γ_r is also *constant*. The ratios $\Delta x'/\Delta t'$ and $\Delta x/\Delta t$ are then given by

$$\frac{\Delta x'}{\Delta t'} = \frac{\Delta x - u\Delta t}{\Delta t - \frac{u\Delta x}{c^2}}$$

$$\frac{\Delta x}{\Delta t} = \frac{\Delta x' + u\Delta t'}{\Delta t' + \frac{u\Delta x'}{c^2}}$$

Taking the limits as $\Delta \to 0$, respectively leads to a result for the relativistic velocities given by

Velocity Vector Addition Formulae (constant frame velocity)

$$\mathbf{v}' = \frac{\mathbf{v} - \mathbf{u}}{1 - \frac{\mathbf{u}\cdot\mathbf{v}}{c^2}} \tag{1.13a}$$

$$\mathbf{v} = \frac{\mathbf{v}' + \mathbf{u}}{1 + \frac{\mathbf{u}\cdot\mathbf{v}'}{c^2}} \tag{1.13b}$$

These relations are known as the *velocity addition formulae*, describing the relative velocity of a system in frame $\mathbf{r}'$ with respect to $\mathbf{r}$. By comparing Eq. (1.13a) and (1.13b) to Eqs. (1.2a) and (1.2b), we can see that the former approximates to the latter when $uv/c^2 << 1$, *or* when body motion is perpendicular to the frame's motion. Shortly, when we discuss Fizeau's experiments with light in moving water, we'll see how these conditions are satisfied, and thus, Eqs. (1.13) become practical. Eqs. (1.13) are a generalization of the relative velocity equations given by Eqs. (1.2). Let's continue examining some of the consequences of the Lorentz transformation.

1.2 Length Contraction and Time Dilation

The Lorentz transformation turns out to have surprisingly deep consequences for *all* physical systems. There are even effects on very fundamental concepts such as our notions of *length* and *elapsed time*. Let's now look at these concepts a little closer through a Lorenzian lens.

Consider two points in space described by two *different Lorentzian* frames. Let us call the pair of points (x_1, x_2) in the stationary frame and (x_1', x_2') in the moving frame, respectively. The Lorentz transformation relates these coordinates as

$$x_1' = \gamma_r(x_1 - vt_1) \qquad x_2' = \gamma_r(x_2 - vt_2) \tag{1.14a}$$

$$t_1' = \gamma_r\left(t_1 - \frac{vx_1}{c^2}\right) \qquad t_2' = \gamma_r\left(t_2 - \frac{vx_2}{c^2}\right) \tag{1.14b}$$

$$t_1 = \gamma_r\left(t_1' + \frac{vx_1'}{c^2}\right) \qquad t_2 = \gamma_r\left(t_2' + \frac{vx_2'}{c^2}\right) \tag{1.14c}$$

Using Eq. (1.14), we can evaluate the quantity $L' = x_2' - x_1'$ observed in the moving frame, but in terms of $L = x_2 - x_1$, where L is observed in the stationary frame. L' can be written as

$$L' = x_2' - x_1' = \gamma_r(x_2 - vt_2) - \gamma_r(x_1 - vt_1) \tag{1.15}$$

$$= \gamma_r(x_2 - x_1) - \gamma_r v(t_2 - t_1) \tag{1.16}$$

Substituting t_1 and t_2 from Eq. (1.14c), we have

$$L' = x_2' - x_1' = \gamma_r(x_2 - x_1) - \gamma_r v\left[\gamma_r\left(t_2' + \frac{vx_2'}{c^2}\right) - \gamma_r\left(t_1' + \frac{vx_1'}{c^2}\right)\right] \tag{1.17a}$$

$$= \gamma_r(x_2 - x_1) - \gamma_r^2 \cdot \frac{v^2}{c^2}(x_2' - x_1') - v\gamma_r^2(t_2' - t_1') \tag{1.17b}$$

Since t'_2 and t'_1 correspond to the times that x'_2 and x'_1 are measured in the frame $\mathbf{r}'$, we can set $t'_2 = t'_1$, which just means they are measured *simultaneously*.[9] Then, the last term vanishes, and we have

$$x'_2 - x'_1 + \gamma_r^2 \cdot \frac{v^2}{c^2}(x'_2 - x'_1) = \left(1 + \gamma_r^2 \cdot \frac{v^2}{c^2}\right)(x'_2 - x'_1) = \gamma_r(x_2 - x_1) \tag{1.18}$$

The term on the RHS in parentheses with the Lorentz factor can be replaced by the relation

$$1 + \gamma_r^2 \cdot \frac{v^2}{c^2} = \gamma_r^2$$

Substituting this relation into Eq. (1.18), we have

$$\gamma_r^2(x'_2 - x'_1) = \gamma_r(x_2 - x_1) \Rightarrow x'_2 - x'_1 = \frac{x_2 - x_1}{\gamma_r} \tag{1.19}$$

or

Relativistic Length Contraction Relation

$$L' = \frac{L}{\gamma_r} \tag{1.20}$$

As we observed in Fig. 1.4, $\gamma_r \geq 1$, so Eq. (1.20) implies that measuring length in a moving Lorentzian frame $\mathbf{r}'$ yields *smaller* measurements than one would measure in the stationary frame $\mathbf{r}$. This phenomenon is known as *length contraction*. Lorentz also proved this result for a body using purely electrodynamics, or the Maxwell equations. That, however, is a much more complex analysis. In the end-of-chapter problems, we'll have you derive[10] the corresponding elapsed time relation which is given by

Relativistic Elapsed-Time Dilation

$$\Delta t' = \gamma_r \Delta t \tag{1.21}$$

[9] simultaneity leads to the most precise length measurement

[10] follow a similar logic as with length contraction, but consider an analog of simultaneity for space, *i.e.* $x'_2 = x'_1$

Eq. (1.21) is known as *time dilation* because, opposite to length measurements, measurements of elapsed time in the moving Lorentzian frame are *larger* than those of the stationary frame. t is the time measured using a clock in the static frame, which for practical purposes, can be regarded as a laboratory frame (*e.g.* earth). Thus, t' is measured with a clock in the moving frame.

Therefore, the Lorentz transformation begins to reveal distinct effects on the behavior of physical systems. Indeed, this story continues as we explore more implications on physical systems. Let's recap what we have so far. From the Galilean transformation, we saw that time between two frames are related by $t = t'$. We say that time is *invariant*. In a Lorentzian frame, we found a more general result where $t \neq t'$. Yet, the result converges to the Galilean transformation when $u/c << 1$. Generally, in the Lorentzian frame, $t = f(t', x')$ just as with spatial coordinates where $x = g(t', x')$, *etc*. Thus, unlike in the Galilean frame, for any coordinate in a Lorentzian frame, *any coordinate more generally relates to the four coordinates of the other frame.*

This *fourness* arising from the Lorentzian transformation is sometimes referred to as *space-time*. However, this concept becomes problematic for a physical n-dimensional space, not only from the standpoint of unit-consistency, but also in it's inability to satisfy the properties of a dimensional space, which is a vector space.[11] It turns out that such difficulties can be extricated if we formalize a fourth dimension by extending the fundamental properties of vector spaces to the Lorentz transformation. This just means for us to allow consistent rules of a 3D vector space extended to 4D. Like Lorentz did, we only need to approach the problem as a generalization to what is already known to be correct. When we do, we *can* find a representation of the fourth dimension.

1.3 The Fourth Dimension Restores Conservation

Length contraction also has mathematical implications such as an inability for a 3D vector to satisfy conservation in a 3D Euclidean norm

[11]Since all three spatial coordinates x, y, z necessarily vary with time, space-time coordinates do not mutually satisfy the property of independent coordinates.

between Lorentzian frames. More precisely, only an approximate conservation becomes possible when $\gamma_r \approx 1$, but still inexact.

It turns out that we can *restore* the conservation of the Euclidean norm between Lorentzian frames only if we include the fourth dimension. Then, the approximation becomes exact. Let us layout a few rules to be expected for the fourth dimension x_4 to ensure consistency with the first three x_1, x_2, and x_3. Of course, *3D* is a more familiar concept to our everyday experience. The Lorentz transformation, however, has revealed to us that *four* coordinates must be used to more generally relate a physical system in one Lorentzian frame to another. With Lorentz, it is described using the four coordinates x, y, z, and t.

However, as already mentioned, a rather inexorable characteristic of *space-time* is that x, y, and z necessarily *depend* on time t. In fact, it is not possible to create a physical change, say, in x, without a corresponding change in time t. Because of this one-to-one relationship with time, *it cannot be a physical dimension since there cannot be a change in the other three without a change in time.* If this were *not* the case, then a physical system could move along any dimension *without* a change in time. This would imply the possibility of *infinite velocities.* This, of course, contradicts everything we know about physical systems, which all demonstrate some finite terminal speed, including light itself, which has the maximum terminal speed $c \approx 3 \times 10^8$m/s. Indeed, all physical evidence ever observed suggests there is an absolute upper bound on the 3D speed known as the speed of light in vacuum, or c. From this, we may deduce that *any* physical dimension must satisfy the following basic requirements:

1. Any dimensional coordinate of a physical n-dimensional space must be posable in an equivalent form having *consistent physical units e.g.* all can be posed in the unit of space, or equivalently, a unit of time, *etc.* where all forms equally satisfy the n-D vector space properties.

2. Related to the one-to-oneness of 3D space with time between frames, the fourth dimension of the Lorentzian frame must also be in proportion to time (as with x, y, z, and t) to be valid in a frame having motion. This is compulsory because all physical systems, in fact, have motion. Even if they do not have *net* motion (*e.g.* in their expectation value), the system is composed of smaller 'parts'

that do. In other words, motion is ubiquitous in the universe necessitating this property for any physical dimension.

3. There must exist a proper vector space norm that is conserved in the vector space. Since the norm for 1, 2, and 3 dimensions is well known, there is no reason to expect a different norm when the fourth dimension is included. In other words, we may expect an inner-product exists that is *conserved* between Lorentzian frames just as a 3D self inner-product is conserved.

Taking these points into consideration along with the Lorentz transformation, let us write a vector of *four* principle dimensions x_1, x_2, x_3 and x_4 or a *four-vector* as

$$\mathbf{S}\,[x_1,\ x_2,\ x_3,\ x_4]^T = [x,\ y,\ z,\ 4(t)]^T \tag{1.22}$$

Each component represents a physical dimension where $4(t)$ is a function of time representing the fourth. In line with property 2, let four$(t) = \kappa t$. We can determine κ by using the *invariance* of the *inner product norm* in a four-dimensional space[12]. For convenience only, we consider *one* of the three dimensions x, y, z since the extension to include all three dimensions is straightforward. The inner-product, in this case, becomes

$$\mathbf{S}\cdot\mathbf{S} = \kappa^2 t^2 + x^2 = \kappa^2\left(t' + \frac{ux'}{c^2}\right)^2 \gamma_r^2 + (x' + ut')^2\,\gamma_r^2 \tag{1.23}$$

This must equal to the inner-product in the other Lorentzian frame r', given by

$$\mathbf{S}'\cdot\mathbf{S}' = \kappa^2 t'^2 + x'^2 \tag{1.24}$$

Squaring the terms of Eq. (1.23) gives

$$\mathbf{S}\cdot\mathbf{S} = \left(\kappa^2 t'^2 + \frac{2t'ux'}{c^2} 2t'ux' + \frac{u^2 x'^2}{c^4} + x'^2 + 2x'ut' + u^2 t'^2\right)\gamma_r^2$$

We can rewrite the above as

$$\mathbf{S}\cdot\mathbf{S} = = \kappa^2 t'^2\left(\frac{1+\frac{u^2}{\kappa^2}}{1-\frac{u^2}{c^2}}\right) + x'^2\left(\frac{1+\frac{\kappa^2 u^2}{c^4}}{1-\frac{u^2}{c^2}}\right) + 2t'ux'\left(\frac{1+\frac{\kappa^2}{c^2}}{1-\frac{u^2}{c^2}}\right) \tag{1.25}$$

[12] If we can achieve exactness of the Euclidean norm between frames under these assumptions, then our result is valid.

Eq. (1.25) must equal $\kappa^2 t'^2 + x'^2 = \mathbf{S}' \cdot \mathbf{S}'$. For this to be true, the first and second terms in parentheses on the RHS must be unity, while the third/last term must vanish. All three of these conditions are satisfied if and only if $1 + \kappa^2/c^2 = 0$ or

$$\kappa^2 = -c^2 \Rightarrow \kappa = \pm ic \tag{1.26}$$

where $i = \sqrt{-1}$. One can check that this value makes all three coefficients 1, 1, and 0 respectively since $\kappa^2 = -c^2$. So we find the fourth dimension is purely imaginary. We then have a valid *four-vector* $\mathbf{S} = \mathbf{S}_\ell$ with homogeneous units of length given by

Spatial Four-Vector

$$\mathbf{S}_\ell = [x,\ y,\ z,\ ict]^T \tag{1.27}$$

Because $4(t)$ is imaginary, adding the fourth dimension means that a 4D vector belongs to a *complex* vector space. We may loosely think of this extension from 3D to 4D as analogous to how a complex scalar is a generalization of a real scalar. *This also provides insight into the origin of complex numbers in the laws of physics.*

One must also keep in mind that as in the case of three dimensions, there are *two* possible vectors satisfying the norm, given by the positive (+) or *rightward* and the negative (−) or *leftward*. The 4D inner-product is, therefore, identical to the so-called *space-time interval* given by

$$\mathbf{S} \cdot \mathbf{S} = x^2 + y^2 + z^2 - c^2 t^2 \tag{1.28}$$

In this form, all four dimensions have units of *length* defining a *four-dimensional space,* but *not* a space-time. As expressed in consideration 1, we could alternatively define the complex four-vector in *units of time* by defining $\mathbf{S} = \mathbf{S}_t$ as

$$\mathbf{S}_t \equiv \left[\frac{x}{ic},\ \frac{y}{ic},\ \frac{z}{ic},\ t\right]^T \tag{1.29}$$

or $[x/c,\ y/c,\ z/c,\ it]^T$. This ability ensures the laws of physics described in a Lorentian frame remain unchanged, allowing proper operations such as normalizing of units or non-dimensionalizing.

Let us find an explicit relation between the more general 4D vectors and the Lorentz factor γ_r. For this, consider the vector of differences $\Delta\mathbf{S}$ given by

$$\Delta\mathbf{S}_\ell = [\Delta x,\ \Delta y,\ \Delta z,\ ic\Delta t]^T \tag{1.30}$$

Dividing both sides by $dx_4 = ic\Delta t$ leads to

$$\frac{\Delta\mathbf{S}_\ell}{\Delta x_4} = \frac{1}{ic}\frac{\Delta\mathbf{S}_\ell}{\Delta t} = \left[1,\ \frac{\Delta x}{ic\Delta t},\ \frac{\Delta y}{ic\Delta t},\ \frac{\Delta z}{ic\Delta t}\right]^T \tag{1.31}$$

Taking the inner-product of Eq. (1.31) then gives

$$-\frac{1}{c^2}\frac{\Delta\mathbf{S}_\ell}{\Delta t}\cdot\frac{\Delta\mathbf{S}}{\Delta t} = 1 - \frac{1}{c^2}\left(\frac{\Delta x}{\Delta t}^2 + \frac{\Delta y}{\Delta t}^2 + \frac{\Delta z}{\Delta t}^2\right) \tag{1.32}$$

Evaluating the limit as $\Delta t \to 0$ yields the result

$$-\frac{1}{c^2}\frac{\partial\mathbf{S}_\ell}{\partial t}\cdot\frac{\partial\mathbf{S}_\ell}{\partial t} = 1 - \frac{1}{c^2}\left(v_x^2 + v_y^2 + v_z^2\right) = 1 - \frac{\mathbf{v}\cdot\mathbf{v}}{c^2} = \frac{1}{\gamma_r^2} \tag{1.33}$$

We find that the Lorentz factor is directly related to the *derivative of the four-vector* with respect to the 4th dimension $x_4 = ict$[13]. Since we know the RHS of Eq. (1.33) is strictly positive when $v < c$, the LHS must also be positive under the same conditions. This leads to the inequality

$$\frac{\partial\mathbf{S}_\ell}{\partial t}\cdot\frac{\partial\mathbf{S}_\ell}{\partial t} = v^2 - c^2 \leq 0 \quad (v \leq c) \tag{1.34}$$

The inner-product of $\partial\mathbf{S}_\ell/\partial t$, thus, *always* yields a *negative* real number. Such properties are possible because the 4D vector space contain an *imaginary* or complex dimension. Eq. (1.34) also informs that $v = 0$ corresponds to $\dot{\mathbf{S}_\ell}^2 = -c^2$, while $v = c$ corresponds to $\dot{\mathbf{S}_\ell}^2 = 0$ suggesting that v^2 has one-to-oneness with $\dot{\mathbf{S}}^2$.

Let us now consider the implied form of the *unit vector* for our fourth dimension.

1.3.1 The Unit Vector of the 4th Dimension

In three dimensions, we are more familiar with unit vectors. For example, in a 3D Cartesian coordinate system, the set of unit vectors each

[13] equivalently, the derivative with respect to time

have amplitude *unity*. They may be written as

$$x \Rightarrow \mathbf{e}_x = \frac{\mathbf{x}}{|x|}, \quad y \Rightarrow \mathbf{e}_y = \frac{\mathbf{y}}{|y|}, \quad z \Rightarrow \mathbf{e}_z = \frac{\mathbf{z}}{|z|} \tag{1.35}$$

The vectors $\mathbf{x}$, $\mathbf{y}$, and $\mathbf{z}$ are the vectors along the direction of each dimension, having length x, y, z, respectively. We want to establish an analogous vector for the fourth dimension. First, consider the spatial component described by the Lorentz transformation, given by

$$x' = (x - u_x t)\gamma_r \tag{1.36}$$

This can also be written as

$$x' = (x - \mathbf{u} \cdot \mathbf{e}_x t)\,\gamma_r \tag{1.37}$$

where $\mathbf{e}_x$ is the unit vector along the direction of x. For the second and third dimensions, y and z, we similarly have

$$y' = \left(y - \mathbf{u} \cdot \mathbf{e}_y t\right)\gamma_r \qquad z' = (z - \mathbf{u} \cdot \mathbf{e}_z t)\,\gamma_r \tag{1.38}$$

Recall that the Lorentz transformation gives, for the time coordinate,

$$t' = \left(t - \frac{\mathbf{u} \cdot \mathbf{r}}{c^2} \cdot t\right)\gamma_r \tag{1.39}$$

If we multiply Eq. (1.39) by ic, we have

$$ict' = \left(ict - \frac{ic\mathbf{u} \cdot \mathbf{r}}{c^2}\right)\gamma_r \tag{1.40}$$

$$= \left(ict - \frac{\mathbf{u} \cdot i\mathbf{r}}{c}\right)\gamma_r \tag{1.41}$$

Recall that when we derived the Lorentz transformation using a 1D model, we used the relation $x = ct$. In the 3D case, this generalizes to $r = |\mathbf{r}| = ct$. Using this relation along with the unit vector along $\mathbf{r}$, for $x_4 = ict$, we then get

$$ict' = \left(ict - \frac{\mathbf{u} \cdot ict\mathbf{e}_r}{c}\right)\gamma_r \tag{1.42}$$

$$= (ict - (\mathbf{u} \cdot i\mathbf{e}_r)t)\,\gamma_r \tag{1.43}$$

Comparing this result to Eq. (1.37), we find the unit vector along the *fourth* dimension to be

Fourth Dimension Unit Vector

$$\mathbf{e}_4 \equiv i\mathbf{e}_r \tag{1.44}$$

From this, we can also write a general form for the transformation equations valid for any of the four dimensions as

4D Reference Frame Transformation Equations

$$x_i' = (x_i - \mathbf{u}\cdot\mathbf{e}_i)\cdot\gamma_r \quad \text{and} \quad x_i = (x_i + \mathbf{u}\cdot\mathbf{e}_i)\cdot\gamma_r \qquad i = 1,2,3,4 \tag{1.45}$$

In this form, note that the frame velocity vector **u** is generally *composed of four components*, not three.

So, we have so far established a direct connection between the 4D vector space and the Lorentz factor γ_r, as well as a form of the 4D unit vector, which is along an imaginary radial axis.

To prepare for the application of four dimensions to quantum mechanics, it is still helpful to review well-known relativistic results based explicitly on the Lorentz factor. So now, let us continue and revisit the implications of the special theory of relativity on first order variables such as momentum, and other important variables which are relevant to the energy of a physical system.

1.4 Relativistic Momentum

Momentum is a fundamental concept in physics. It is more general than velocity because in quantum mechanical systems, both massless (*e.g.* photons) and mass systems (*e.g.* electrons) have well-defined momenta. Recall that in a non-relativistic frame, the *momentum* is related to the non-relativistic force **F** as

$$\mathbf{F} = \frac{d\mathbf{p}}{dt} \Rightarrow d\mathbf{p} = \mathbf{F}dt \tag{1.46}$$

If we define the differential of energy dE as

$$\Delta E = \int \mathbf{F} \cdot d\mathbf{x} \tag{1.47}$$

then, we have

$$\Delta E = \int \mathbf{F} \cdot \frac{d\mathbf{x}}{dt} dt = \int \mathbf{v} \cdot \mathbf{F} dt \tag{1.48}$$

This also implies that the power dE/dt is given by

$$\frac{dE}{dt} = \mathbf{v} \cdot \mathbf{F} \tag{1.49}$$

Since $\mathbf{F} = d\mathbf{p}/dt$, Eq. (1.49) is equivalent to

$$dE = \mathbf{v} \cdot d\mathbf{p} \tag{1.50}$$

Since momentum is more general, it makes sense that its changes give rise to changes in energy. In order to evaluate the energy, we would need both $\mathbf{v}$ and $\mathbf{p}$. One way around this is to assume a form for $\mathbf{p}(\mathbf{v})$ which means $\mathbf{p}$ is taken as an explicit function of velocity $\mathbf{v}$. This is possible since γ_r is only a function of v^2. Recall that the momentum in the Galilean frame is defined as

$$p_x \equiv m_0 \times \lim_{\Delta t \to 0} \frac{\Delta x}{\Delta t} = m_0 \frac{dx}{dt} = m_0 v_x \tag{1.51}$$

However, in Eq. (1.51), the derivative is taken with respect to t, which is *invariant only in a Galilean frame*. We know it is *not* invariant in the Lorentzian frame. If we can identify an invariant unit of time in the Lorentzian frame, then evaluating the relativistic momentum in terms of velocity becomes business as usual, meaning that we can mimic Eq. (1.51) by just replacing t with the Lorentz invariant time $t*$ which can be found. We have already laid the ground work to be able to find $t*$ because we know that by including the fourth dimension, the inner-product is invariant, and can also be expressed in either units of space or time. This fact can be exploited here. Similar to the analysis in the previous section, we can use Eqs. (1.31) and (1.32) to write

$$\frac{\Delta \mathbf{S}}{ic} \cdot \frac{\Delta \mathbf{S}}{ic} = -\frac{\Delta x^2 + \Delta y^2 + \Delta z^2}{c^2} + \Delta t^2 \tag{1.52}$$

$$= \Delta t^2 \left(1 - \frac{\Delta x^2 + \Delta y^2 + \Delta z^2}{\Delta t^2 c^2}\right) \tag{1.53}$$

$$= \Delta t^2 \left(1 - \frac{1}{c^2}\frac{\Delta \mathbf{r}}{\Delta t} \cdot \frac{\Delta \mathbf{r}}{\Delta t}\right) \tag{1.54}$$

Since the term on the RHS in parentheses is dimensionless, this equation has units of time *squared*. It is also an *invariant* quantity in any Lorentzian frame, having inherited this property from the conserved 4D inner-product. Provided $v < c$, we can define an invariant change in time $\Delta t*$ as

$$\Delta t^* \equiv \Delta t \sqrt{1 - \frac{1}{c^2}\frac{\Delta \mathbf{r}}{\Delta t} \cdot \frac{\Delta \mathbf{r}}{\Delta t}} \tag{1.55}$$

Taking the limit as $\Delta t \to 0$, we obtain a relation for the invariant time *differential* given by

4D-to-3D Time Differential Relation

$$dt^* = \frac{dt}{\gamma_r} \tag{1.56}$$

Eq. (1.56) can be used in evaluating derivatives with respect to time for a body in a Lorentzian frame. t is the ordinary *measured* value of time while t^* is the invariant time. To use Eq. (1.56), all we need to do is replace dt by dt^* using the same approach from Eq. (1.51). This leads to a relativistic momentum given by

$$p_x = m_0 \times \lim_{\Delta t^* \to 0} \frac{\Delta x}{\Delta t^*} = m_0 \frac{dx}{dt*} = m_0 \gamma_r \frac{dx}{dt}$$

For 3D motion, the *relativistic momentum* becomes

Relativistic Momentum

$$\mathbf{p} = \frac{m_0 \mathbf{v}}{\sqrt{1 - \frac{v^2}{c^2}}} \tag{1.57}$$

Notice that the *relativistic momentum* also converges to the *nonrelativistic* momentum ($m_0 v$) when $v/c << 1$ or $\gamma_r \to 1$.

The relativistic momentum given by Eq. (1.57) can be used to find higher order relativistic variables that depend on momentum, e.g. energy. So, let's use the above results to determine the relativistic *kinetic* energy.

1.5 Relativistic Kinetic Energy

As with momentum, *energy* is also fundamental to physical systems. Often, it is the energy that is used to enable deeper levels of analysis. Here, we turn our attention to the energy of motion, known as *kinetic energy*.

1.5.1 Kinetic Energy Using Invariant Time

To obtain the *relativistic* energy, we can use an approach analogous to how we determined the relativistic momentum. We just need to use the familiar expressions and replace dt by dt^*. Then, let us first modify Eq. (1.46) by defining the *relativistic force* as

$$\mathbf{F} = \frac{d\mathbf{p}}{dt^*} \tag{1.58}$$

$\mathbf{p}$ is the *relativistic momentum* given by Eq. (1.57). If the relativistic energy is found using the invariant unit of time t^*, the change in kinetic energy becomes

$$\Delta K = \int \mathbf{F} \cdot d\mathbf{x} = \int \frac{d\mathbf{p}}{dt*} \cdot d\mathbf{x} = \int \gamma_r \mathbf{v} \cdot d\mathbf{p} \tag{1.59}$$

$d\mathbf{p}$ can be found as

$$d\mathbf{p} = m_0 \mathbf{v} d\gamma_r + m_0 \gamma_r d\mathbf{v}$$

Here, we are accounting for the dependence of the relativistic momentum $\mathbf{p}$ on $\gamma_r(v)$. Evaluating $d\gamma_r$, we have

$$d\gamma_r = \frac{d\gamma_r}{dv_x} \cdot dv_x + \frac{d\gamma_r}{dv_y} \cdot dv_y + \frac{d\gamma_r}{dv_z} \cdot dv_z \tag{1.60a}$$

$$= -\frac{1}{2}\left(1 - \frac{v^2}{c^2}\right)^{-3/2} \cdot -\frac{2}{c^2} \cdot \left(v_x dv_x + v_y dv_y + v_z dv_z\right) \tag{1.60b}$$

$$= \frac{\gamma_r^3}{c^2} \mathbf{v} \cdot d\mathbf{v} \tag{1.60c}$$

With this result, $d\mathbf{p}$ becomes

$$d\mathbf{p} = m_0 \mathbf{v} d\gamma_r + m_0 \gamma_r d\mathbf{v} = m_0 \left[\frac{\gamma_r^3}{c^2} (\mathbf{v} \cdot d\mathbf{v})\mathbf{v} + \gamma_r d\mathbf{v}\right] \tag{1.61}$$

Now, we can write the change in relativistic kinetic energy as

$$\Delta K = \int \gamma_r \mathbf{v} \cdot d\mathbf{p} = m_0 \int \gamma_r \left[\gamma_r + \gamma_r^3 \left(\frac{v^2}{c^2} \right) \right] \mathbf{v} \cdot d\mathbf{v} \tag{1.62}$$

$$= m_0 \int \gamma_r^2 \left[1 + \gamma_r^2 \left(\frac{v^2}{c^2} \right) \right] \mathbf{v} \cdot d\mathbf{v} \tag{1.63}$$

The term in parentheses can be rewritten from the following relation

$$\gamma_r^2 = \frac{1}{1 - v^2/c^2} \Rightarrow \gamma_r^2 \left(1 - \frac{v^2}{c^2} \right) = 1 \Rightarrow 1 + \gamma_r^2 \left(\frac{v^2}{c^2} \right) = \gamma_r^2 \tag{1.64}$$

Substituting this result, the change in kinetic energy becomes

$$\Delta K = m_0 \int \gamma_r^4 \mathbf{v} \cdot d\mathbf{v} = m_0 \int \left[\frac{1}{1 - v^2/c^2} \right]^{\frac{4}{2}} \mathbf{v} \cdot d\mathbf{v} \tag{1.65}$$

To evaluate the integral, we can use substitution of $c^2 u = v^2 \rightarrow c^2 du/2 = v dv$ along with the relation

$$\frac{d}{du}(1-u)^{-1} = -(1-u)^{-2}$$

Using the above integrals, we find that the relativistic kinetic energy depending on the *invariant* time is given by

$$\Delta K = \frac{m_0}{2} \frac{c^2}{[1 - v^2/c^2]} \bigg|_{v_i}^{v_f} = \frac{m_0}{2} \Delta \gamma_r^2 c^2 = \frac{\gamma_r}{2} \Delta m c^2 \tag{1.66}$$

In Eq. (1.66), we have used the conventional definition for the *relativistic mass m*, given by

$$m = \gamma_r m_0 \tag{1.67}$$

From Eq. (1.66), the *relativistic kinetic energy K* becomes

Relativistic Kinetic Energy (invariant time)

$$K = \frac{\gamma_r m c^2}{2} \tag{1.68}$$

Since $\gamma_r \geq 1$, the *absolute lowest value* of K is *not* zero, but a positive quantity given by

$$K \geq \frac{m_0 c^2}{2} \tag{1.69}$$

As mentioned, this indicates that motion is truly ubiquitous, and even an observed *motionless* system is somehow created through a combination of systems in motion. Such a property inevitably requires the concept of waves, which we'll begin in Chap. 2.

In the *non*relativistic limit, the kinetic energy has a factor of $\gamma_r/2 \to 1/2$. This is consistent with the non-relativistic kinetic energy of $(1/2)mv^2$. However, so was Einstein's result. For a more comprehensive comparison, let us also work out how one obtains $K = E = mc^2$, so a more in-depth comparison can be made. In short, Einstein's result is possible if the *ordinary Galilean time t* is used in the force, rather than the invariant time t^*.

1.5.2 Obtaining $E = mc^2$

If we define the force using the Galilean time t, then we have

$$\Delta K = \int \mathbf{F} \cdot d\mathbf{x} = \int \frac{d\mathbf{p}}{dt} \cdot d\mathbf{x} = \int \mathbf{v} \cdot d\mathbf{p} \tag{1.70}$$

As we did above, accounting for the dependence of $\mathbf{p}$ on $\gamma_r(v)$, we found that $d\gamma_r$ is given by

$$d\gamma_r = \frac{\gamma_r^3}{c^2} \mathbf{v} \cdot d\mathbf{v} \tag{1.71}$$

so $d\mathbf{p}$ now becomes

$$d\mathbf{p} = m_0 \mathbf{v} d\gamma_r + m_0 \gamma_r d\mathbf{v} = m_0 \left[\frac{\gamma_r^3}{c^2} (\mathbf{v} \cdot d\mathbf{v}) \mathbf{v} + \gamma_r d\mathbf{v} \right] \tag{1.72}$$

Then, the change in kinetic energy can be expressed as

$$\Delta K = m_0 \int \left[\frac{\gamma_r^3 v^2}{c^2} (\mathbf{v} \cdot d\mathbf{v}) + \gamma_r \mathbf{v} \cdot d\mathbf{v} \right] \tag{1.73a}$$

$$= m_0 \int \left[\gamma_r + \gamma_r^3 \left(\frac{v^2}{c^2} \right) \right] \mathbf{v} \cdot d\mathbf{v} \tag{1.73b}$$

$$= m_0 \int \gamma_r \left[1 + \gamma_r^2 \left(\frac{v^2}{c^2} \right) \right] \mathbf{v} \cdot d\mathbf{v} \tag{1.73c}$$

Substituting Eq. (**??**), we have

$$\Delta K = m_0 \int \gamma_r^3 \mathbf{v} \cdot d\mathbf{v} = m_0 \int \left[\frac{1}{1 - v^2/c^2} \right]^{\frac{3}{2}} \mathbf{v} \cdot d\mathbf{v} \tag{1.74}$$

Similar to the above analysis, evaluating the integral in Eq. (1.74) can be done by using substitution, where $c^2 u = v^2 \rightarrow (c^2/2)du = v \cdot dv$, which gives

$$\Delta K = \frac{m_0 c^2}{2} \int \left[\frac{1}{1 - u} \right]^{\frac{3}{2}} du \tag{1.75}$$

We can, again, use the relation

$$\frac{d}{du}(1-u)^{-1/2} = -\frac{1}{2}(1-u)^{-3/2}$$

so Eq. (1.75) evaluates to

$$\Delta K = \left. \frac{m_0 c^2}{[1 - v^2/c^2]^{1/2}} \right|_{v_i}^{v_f} = \Delta \gamma_r m_0 c^2 \tag{1.76}$$

According to Eq. (1.76), the kinetic energy K is now given by the familiar result

$$K = \gamma_r m_0 c^2 = mc^2 \tag{1.77}$$

Let us compare the expansions of both results given by $K = \gamma_r mc^2/2$ and $K = mc^2$.

1.5.3 Comparing Relativistic Kinetic Energies

Consider the *Taylor Series expansion* of the kinetic energy given by Eq. (1.68) around the point $x_0 = v^2 = 0$. The expansion becomes

$$K(x) = K(x_0) + \frac{1}{1!}\frac{dK}{dx} \cdot x + \frac{1}{2!}\frac{d^2K}{dx^2} \cdot x^2 + \frac{1}{3!}\frac{d^3K}{dx^3} \cdot x^3 + \cdots \tag{1.78}$$

Letting $K = \gamma_r mc^2/2$ and defining $K_0 = m_0 c^2/2$, expanding around the point $x_0 = 0$ gives

$$K(x) = K_0 \gamma_r^2 \tag{1.79a}$$

$$= \frac{m_0c^2}{2}\left[1 + \frac{1}{1!}\frac{d}{dx_0}\left(1 - \frac{x_0}{c^2}\right)^{-1} \cdot x + \frac{1}{2!}\frac{d^2}{dx_0^2}\left(1 - \frac{x_0}{c^2}\right)^{-1} \cdot x^2 + ...\right] \tag{1.79b}$$

The derivative terms, starting with the first, are as follows:

$$\frac{d}{dx_0}\left(1 - \frac{x_0}{c^2}\right)^{-1} = -1 \cdot \left(1 - \frac{x_0}{c^2}\right)^{-2} \cdot \left(-\frac{1}{c^2}\right) = +\frac{1}{c^2} \tag{1.80}$$

Evaluating the second derivative at $x_0 = 0$, we have

$$\frac{d^2}{dx_0^2}\left(1 - \frac{x_0}{c^2}\right)^{-1} = \frac{2}{c^4}\left(1 - \frac{x_0}{c^2}\right)^{-3} = \frac{2}{c^4} \tag{1.81}$$

Substituting the evaluated derivatives along with $x = v^2$ into Eq. (1.79) gives

$$K(x) = \frac{m_0c^2}{2}\left[1 + \frac{x}{c^2} + \frac{x^2}{c^4} + ...\right] \tag{1.82}$$

$$= \underline{\frac{m_0c^2}{2} + \frac{1}{2}m_0v^2 + \frac{1}{2}\frac{m_0v^4}{c^2} + \cdots} \tag{1.83}$$

This is the result for $K = \gamma_r mc^2/2$. The analogous expansion for $K = mc^2$ is given by

$$K(x) = m_0c^2\left[1 + \frac{x}{c^2} + \frac{3x^2}{8c^4} + ...\right] \tag{1.84}$$

$$= \underline{m_0c^2 + \frac{1}{2}m_0v^2 + m_0\frac{3v^4}{8c^2} + \cdots} \tag{1.85}$$

Notice how the second terms in both expansions are identical. This term is actually the *non-relativistic kinetic energy*. However, this is teh only common term. Terms to the left of the common term are smaller in the corrected form, while those to the right are larger in comparison to the expansion of $K = mc^2$. Both lead to a consistent general form for the *nonrelativistic approximation* in conditions of small v^2/c^2 (where higher order terms become negligible) having the approximate form

Nonrelativistic Kinetic Energy

$$K(v) - K_0 \approx \frac{1}{2}m_0v^2 \tag{1.86}$$

Fig. 1.5 plots both γ_r versus $\gamma_r^2/2$, which correspond to both forms of the relativistic kinetic energy we have discussed.

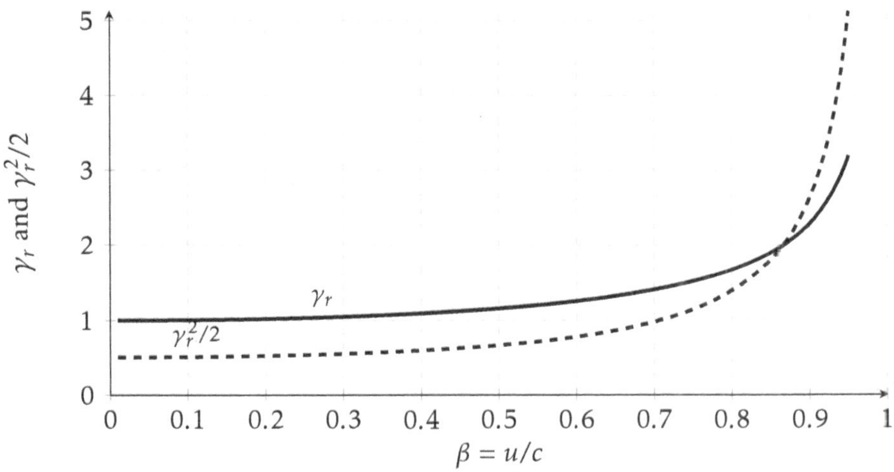

Figure 1.5: Lorentz factor γ_r *vs.* the corrected form $\gamma_r^2/2$ as a function of the parameter $\beta = u/c$. u if the relative frame velocity and $c = 3 \times 10^8$ m/s is the maximum velocity of light. There is a single crossing at $\gamma_r = 2$, where they are equal, however, differ everywhere else. To the right of the crossover, $\gamma_r^2/2$ becomes larger.

In the limit of $\beta = u/c \to 0$, the corrected energy, being based on the *invariant time*, converges to 1/2 instead of 1. Therefore, at low speeds, the corrected kinetic energy is lower than what is predicted by $K = mc^2$, while at higher speeds, the corrected energy contributions are larger. This has important implications, for example, because contributions like spin-orbit interactions (to be introduced in the next Chapter) turn out to be larger than expected. The relativistic momentum is more useful than velocity because it makes life easier in determining more forms of kinetic energy including *massless* kinetic energy of a photon. For example, photon energy is given by pc. For quantum mechanics, it is convenient to express the kinetic energy in terms of the relativistic momentum. Let's do this now.

1.5.4 Momentum-Dependent Form Of Kinetic Energy

Eq. (1.77) can be cast into a form depending explicitly on the relativistic momentum $\mathbf{p}$ by manipulating the quantity p^2c^2 as follows:

$$p^2c^2 = \frac{m_0^2 v^2 c^2}{1 - v^2/c^2} \tag{1.87a}$$

$$= \frac{m_0^2 (v^2/c^2) c^4}{1 - v^2/c^2} \tag{1.87b}$$

$$= \frac{m_0^2 (v^2/c^2) c^4 - m_0^2 c^4 + m_0^2 c^4}{1 - v^2/c^2} \tag{1.87c}$$

$$= \frac{m_0^2 c^4 [v^2/c^2 - 1]}{1 - v^2/c^2} + \frac{m_0^2 c^4}{1 - v^2/c^2} \tag{1.87d}$$

$$= -m_0^2 c^4 + m^2 c^4 \tag{1.87e}$$

Since $K = m\gamma_r c^2/2 = m_0 \gamma_r^2 c^2/2$, the second term on the RHS can be used to express *squared relativistic kinetic energy* K^2 as

$$K^2 = \frac{\gamma_r^2}{4} \cdot \left(p^2c^2 + m_0^2 c^4\right) \tag{1.88}$$

Thus, the *relativistic kinetic energy* K can alternatively be written as

Relativistic Kinetic Energy (Momentum-Dependent)

$$K = \frac{\gamma_r}{2} \cdot \sqrt{p^2c^2 + m_0^2 c^4} \tag{1.89}$$

If we consider the limiting case when $m_0 = 0$, which defines a *massless* system, Eq. (1.89) reduces to

Relativistic Kinetic Energy (Massless System)

$$K = \frac{\gamma_r p c}{2} \tag{1.90}$$

Eq. (1.89) implies an *alternative* definition for a *relativistic mass m* if we define it using a kinetic energy of the form

$$K = \frac{1}{2}mc^2 \tag{1.91}$$

Equating (1.91) and (1.89) leads to

$$\frac{1}{2}mc^2 = \frac{c^2}{2}\gamma_r\sqrt{m_0^2 + \frac{p^2}{c^2}} = \frac{m_0c^2}{2}\gamma_r\sqrt{1 + \frac{p^2}{m_0^2c^2}} \tag{1.92}$$

From Eq. (1.92), we find a dimensionless relativistic mass m/m_0 given by

$$\frac{m}{m_0} = \gamma_r\sqrt{1 + \frac{p^2}{m_0^2c^2}} \tag{1.93}$$

Eq. (1.93) has a simpler form which can be found by substituting the relativistic momentum $p = m_0\gamma_r v$. When this is done, we obtain the result

$$\frac{m}{m_0} = \gamma_r^2 \tag{1.94}$$

According to Eq. (1.33), *mass* is also related to the 4D velocity inner product. What we traditionally think of as *mass*, in the relativistic picture, *is not independent of motion*, but depends on momentum/velocity.

Fig. 1.6 plots relativistic mass (m/m_0) given by Eq. (1.93). Since the kinetic energy is proportional to the relativistic mass, which depends on the speed, it is the relativistic mass function $m(v)$ that determines the behavior of the kinetic energy. From the Taylor Series expansion, we can see that it is second order in v *only* in approximation, provided this term dominates. However, generally, it is not strictly quadratic.

Now that we have an understanding of the special theory of relativity, let's consider how humans began to be convinced in the correctness of it. After all, any physical law should be observable somehow in the physical world. So, let's address some of the observations validating this peculiar theory.

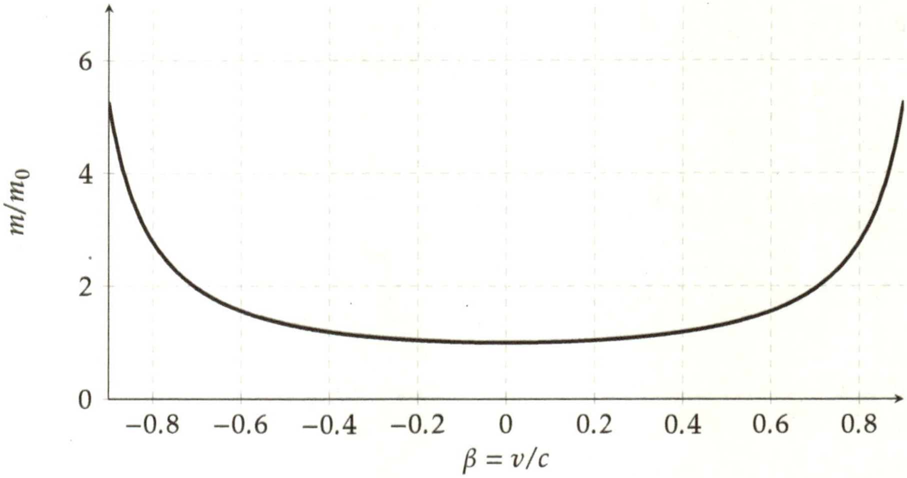

Figure 1.6: The relativistic mass m as a function of $\beta = v/c$. As in the case of kinetic energy, mass possesses a symmetric dependence on speed v.

1.6 Experiments Explained ONLY By Special Relativity

The most convincing evidence to help establish the correctness of any physical theory comes by comparing predictions from the theory with experimental observations *under the same conditions assumed by the theory*. So, we now consider a couple of unique historic experiments that required the special theory of relativity to correctly explain.

1.6.1 Light Speed In Moving Water

Around 1851, a set of experiments measuring speeds of light were carried out by French physicist Armand Hippolyte Louis Fizeau (1819-1896). Fizeau took measurements of the speed of light in *flowing water*. He was already a pioneer in measuring the velocity of light in air.

For Fizeau's experiments, let v be the velocity of the light in the laboratory frame, so taken as the stationary frame. The water's frame is moving at velocity $\mathbf{u}$, which is also measured with respect to the rest frame of the laboratory. Let v' be the velocity of the light as measured relative to the water. Conventional thinking at the time lead Fizeau to expect something in accordance with Eq. (1.2b). However, Fizeau found that it did not agree completely. Instead, he found that it follows an *empirical* relation given by

$$v = v' + u\left(1 - \frac{1}{n'^2}\right) \tag{1.95}$$

$n' = c/v'$ is the *index of refraction* based on measuring light speed in *stationary* water or relative to water at constant velocity. The term in parentheses was regarded as a *drag coefficient* and it lead to a considerable *reduction* in the observed velocity of light in moving water. The index of refraction of water is $n' \sim 1.33$. Substituting n' using an average water flow speed of $u = 1\text{m/s}$, with $c \approx 3 \times 10^8\text{m/s}$, one finds the measured speed becomes

$$\begin{aligned} v &= v' + u\left(1 - \frac{1}{n'^2}\right) \\ &= v' + u\left(1 - \frac{1}{1.33^2}\right) \\ &= v' + 0.434u < v' + u \end{aligned}$$

Instead of adding the *full* speed of the water u, one only needed to add ~43% of it. We'll see shortly that as long as $u << c$, the *relativistic* velocity leads exactly to Fizeau's empirical result describing the coordinates of light. Using $v' = c/n'$, we can also write

$$v = \frac{c}{n'} + u\left(1 - \frac{1}{n'^2}\right) \tag{1.96}$$

Since $n' \geq 1$, the first term satisfies the condition $c/n' \leq c$. Moreover, as n' decreases towards unity, corresponding to higher speeds for v', the second term contributes less and less. This renders the experiments incapable of observing speeds greater than c. In the limiting case $n' = 1 \rightarrow v = c$, and *never greater*. As an empirical law, we desire a theory to not only predict this relation, but provide some insight as to whether the observations made are legitimate or not, or perhaps some errors may be lurking somewhere.

The relativistic result given by Eq. (1.13) (velocity addition formula) turns out to be a generalization of Fizeau's empirical relation given by Eq. (1.96). To see this, we can write Eq. (1.13b) as follows:

$$v = \frac{v' + u}{1 + \frac{uv'}{c^2}}$$

$$= \frac{(v'+u)\left(1-\frac{uv'}{c^2}\right)}{\left(1+\frac{uv'}{c^2}\right)\left(1-\frac{uv'}{c^2}\right)}$$

$$= \frac{v' - \frac{uv'^2}{c^2} + u - \frac{u^2v'}{c^2}}{\left(1+\frac{uv'}{c^2}\right)\left(1-\frac{uv'}{c^2}\right)}$$

$$= \frac{v'\left(1-\frac{u^2v'^2}{c^4}\right) + u\left(\frac{uv'^3}{c^4} - \frac{v'^2}{c^2} + 1 - \frac{uv'}{c^2}\right)}{\left(1+\frac{uv'}{c^2}\right)\left(1-\frac{uv'}{c^2}\right)}$$

$$= v' + u\frac{\left[\frac{uv'}{c^2}\left(1-\frac{v'^2}{c^2}\right) - \left(1-\frac{v'^2}{c^2}\right)\right]}{\left(1+\frac{uv'}{c^2}\right)\left(1-\frac{uv'}{c^2}\right)}$$

$$= v' + u\frac{\left[1-\frac{v'^2}{c^2}\right]}{1+\frac{uv'}{c^2}}$$

Using $v'^2/c^2 = 1/n'^2$, the above result reduces to Fizeau's empirical formula in the approximation $1 + uv'/c^2 \sim 1$, where we have

$$v \approx v' + u\left(1 - \frac{1}{n'^2}\right)$$

As mentioned, v is measured in the laboratory frame. The above result is in agreement Eq. (1.95). This experiment deals, particularly, with measurements of light in moving frames, and effectively illustrates a consistency with the theory of relativity, as well as how the measured speed of light is limited. Remarkably, light does not appear to able to exceed the value of $c \approx 3 \times 10^8$ m/s.

Next, let us consider another experiment. This one involves interactions between x-rays and electrons bound to nuclei in a metal. It serves as yet another cogent example of experimental observations that required the special theory of relativity to predict and provide a clear explanation of the observations.

1.6.2 Compton Scattering

In 1920, experimental physicist Arthur Holly Compton (1892-1962) investigated the scattering of monochromatic (single wavelength) x-rays from various materials. x-rays are a specific part of the electromagnetic radiation spectrum, having a wavelength λ on the order of 10^{-10} m, or an angstrom. Since this is the length scale of subatomic particles, x-rays can interact with electrons of an atom. Thus, they are also known as ionizing radiation.

They can be produced using something called of a cathode ray tube (CRT). Inside the CRT, electrons are accelerated from a cathode (a negatively charged terminal) at high speeds towards the positively charged anode. As they hit the metal anode, they release energy. Most of the kinetic energy converts to heat, but a few percent of the electrons' energy is converted into x-rays. Around 1922, Compton found that a quantum of x-rays that was scattered by the valence electrons in a metal acquired *longer wavelengths*, or reduced frequency. Thus, the scattered x-ray possessed less energy than the incoming x-ray. Just as with a classical system, the difference was transferred to the scattered electron. This phenomenon is known as the *Compton effect* or *Compton scattering*, between x-ray quanta and electrons. The measured increase in wavelength $\Delta\lambda = \lambda_f - \lambda_i$ was found to follow the relation

$$\lambda_f - \lambda_i = \frac{2h}{m_0 c} \times \sin^2\left(\frac{\theta}{2}\right) \tag{1.97}$$

λ_i is the initial wavelength before scattering; λ_f is the wavelength after scattering. h is a constant known as Planck's constant; m_0 is the electron's rest mass ; c is the speed of light; and θ is the scattering angle. An example of Compton scattering is illustrated in Fig. 1.7.

Eq. (1.97) can be derived using conservation of momentum along with conservation of relativistic kinetic energy. Before we go on, we must point out that this analysis uses a form for the momentum that we have not introduced. It is known as the de Broglie relation, and it will be derived in the next Chapter. For now, we will only use the result which says that the relativistic momentum p can be expressed as

$$p = \frac{\hbar}{\lambda} \qquad \text{(de Broglie relation)}$$

$\hbar = h/2\pi$ is called Planck's reduced constant and λ is the x-ray wavelength. This relation along with a relativistic energy allowed Compton to derive a solution that agreed with his experimental observations shown in Fig. 1.8. Let's see how.

The energy for both mass and massless systems (c.Eq. 1.90) and momentum conservation along x and y components of the momentum leads to

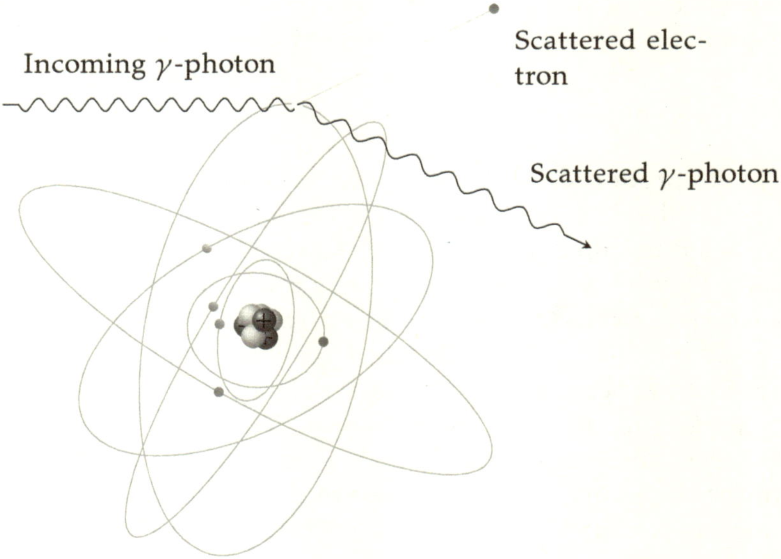

Figure 1.7: Illustration of Compton scattering. The γ-ray penetrates the metal and scatters with outer atomic electrons. The scattering angle is determined by knowing the incoming angle, and the angle after scattering.

the following three equations:

$$\frac{h}{\lambda_i} = \frac{h}{\lambda_f}\cos\theta + p_e\cos\phi \quad \text{(x-components)} \tag{1.98a}$$

$$0 = \frac{h}{\lambda_f}\sin\theta - p_e\sin\phi \quad \text{(y-components)} \tag{1.98b}$$

$$\frac{hc}{\lambda_i} + K_{e,i} = \frac{hc}{\lambda_f} + K_{e,f} \quad \text{(energy conservation)} \tag{1.98c}$$

$K_{e,i}$ and $K_{e,f}$ are the initial and final (after scattering) kinetic energies of the electron in the valence shell of the metallic atom. To express this explicitly, we can use the *relativistic* kinetic energy from Eq. (1.68), or

$$K = \frac{\gamma_r mc^2}{2}$$

For an electron occupying a quantum state in the atom, while it may have zero net velocity $\langle v \rangle = 0$, it does *not* have zero squared velocity $\langle v^2 \rangle > 0$. For this reason, $\gamma \neq 1$. Also, their mean energies, depending on $\langle v^2 \rangle$, are likely to be raised further from the x-ray bombardment. Thus,

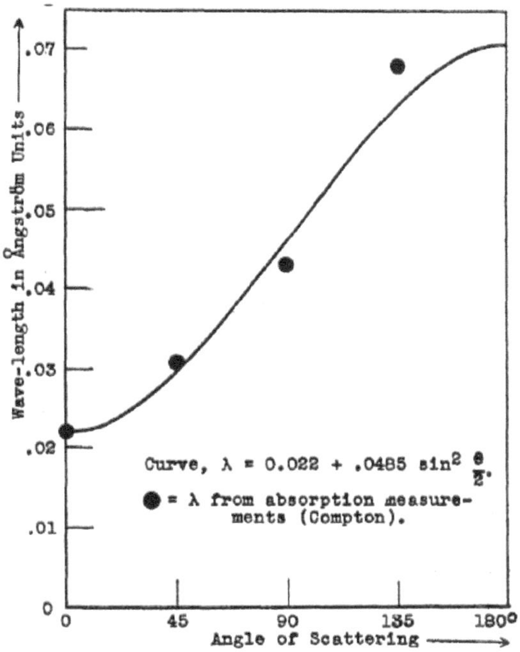

Figure 1.8: Comparison of Compton's measured/observed increase in wavelength compared to theory (Eq. 1.97), adapted from his famous 1923 paper in The Physical Review, A Quantum Theory Of The Scattering of Xrays By Light Elements. Compton, who discovered the effect carrying out measurements also gave us a theory describing it based on relativity.

we *cannot* take the nonrelativistic limit $m_0c^2/2$ for the electron. In short, the electron will possess a larger energy. But, if we take a larger value of $K = m_0c^2$, it corresponds to

$$\frac{\gamma^2 m_0 c^2}{2} = m_0c^2 \Rightarrow \gamma_r = \sqrt{2} \tag{1.99}$$

This value of γ_r corresponds to

$$\frac{v^2}{c^2} = \frac{1}{2}$$

for the electron in the metal. With this, Eq. (1.98c) becomes

$$\frac{hc}{\lambda_i} + m_0c^2 = \frac{hc}{\lambda_f} + \sqrt{m_0^2c^4 + p_e^2c^2} \quad \text{(energy conservation)} \tag{1.100}$$

Note that we have arrived at the same equations used by Compton, however, the rationale behind the energy balance equation differs based on the form of the relativistic energy. Still, we have three equations in the three unknowns λ_f, p_e, and ϕ. p_e is the momentum of the scattering electron.

Using the y-component of momentum given by Eq. (1.98b), we can write $\phi(\lambda_f, p_e)$ as

$$\sin\phi = \frac{h}{p_e\lambda_f}\sin\theta \Rightarrow \cos^2\phi = 1 - \frac{h^2}{p_e^2\lambda_f^2}\sin^2\theta \tag{1.101}$$

Substitution into the x-component in Eq. (1.98a) leads to

$$\frac{h}{\lambda_i} = \frac{h}{\lambda_f}\cos\theta + p_e\sqrt{1 - \frac{h^2}{p_e^2\lambda_f^2}\sin^2\theta} \tag{1.102}$$

$$= \frac{h}{\lambda_f}\cos\theta + \sqrt{p_e^2 - \frac{h^2}{\lambda_f^2}\sin^2\theta} \tag{1.103}$$

Bringing the cosine term to the LHS and solving for p_e^2 (electron momentum after scattering, as it is assumed to have zero momentum initially) gives

$$p_e^2 = \frac{h^2}{\lambda_i^2} - \frac{2h^2}{\lambda_i\lambda_f}\cos\theta + \frac{h^2}{\lambda_f^2} \tag{1.104}$$

Eq. (1.104) can be substituted into the energy conservation equation of Eq. (1.100) along with $c = \lambda f$. Then, we have

$$\frac{hc}{\lambda_i} + m_0c^2 = \frac{hc}{\lambda_f} + \sqrt{m_0^2c^4 + c^2\left(\frac{h^2}{\lambda_i^2} - \frac{2h^2}{\lambda_i\lambda_f}\cos\theta + \frac{h^2}{\lambda_f^2}\right)} \tag{1.105}$$

Then, leaving the radical term on the RHS, and arranging everything else on the LHS, followed by squaring both sides, we then have

$$\left[\frac{2hc}{\lambda_i} - \frac{2hc}{\lambda_f} + m_0c^2\right]^2 = m_0^2c^4 + c^2\left(\frac{h^2}{\lambda_i^2} - \frac{2h^2}{\lambda_f\lambda_f}\cos\theta + \frac{h^2}{\lambda_f^2}\right) \tag{1.106}$$

In squaring the LHS, there are nine terms total. However, three of them are identical to three of the four terms on the RHS (leaving only the cosine term). Thus, these all cancel from both sides. Then, the remaining six on the LHS are twice of three unique terms (hence the factor of two on the LHS). The result reduces to

$$\frac{2hc^3m_0}{\lambda_i} - \frac{2hc^3m_0}{\lambda_f} - \frac{2h^2c^2}{\lambda_i\lambda_f} = -\frac{2h^2c^2}{\lambda_i\lambda_f}\cos\theta \tag{1.107}$$

This can be rearranged to give

$$\frac{1}{\lambda_i} - \frac{1}{\lambda_f} = -\frac{h}{\lambda_i\lambda_f m_0 c}(1-\cos\theta) \tag{1.108}$$

Multiplying both sides by $\lambda_i\lambda_f$ gives the final result for the wavelength shift as

Compton Scattering Wavelength Change

$$\lambda_f - \lambda_i = \frac{h}{m_0 c}(1-\cos\theta) \tag{1.109}$$

The maximum amplitude-wavelength h/m_0c on the RHS is known as the *Compton wavelength*. To obtain the identical form used by Compton, we can use the trigonometric identity $2\sin^2(\theta/2) = 1-\cos\theta$. Substitution into Eq. (1.109) gives

$$\lambda_f - \lambda_i = \frac{2h}{m_0 c}\sin^2\left(\frac{\theta}{2}\right) \tag{1.110}$$

So, we again find with this experiment, the *special theory of relativity* is needed to predict the observations correctly. More and more observations began accumulating in the 19th century, making it difficult to ignore the unique capability of the *special theory of relativity* in describing physical behavior of systems. It simply could not be ignored any longer.

Velocities are not always constant in magnitude nor direction. This means that forces and torques may be involved giving rise to dynamical behavior. Up to now, we've examined energy, however, we also want to take a look at the implications of the Lorentz factor on forces and torques. The torque, for example, plays an important role in magnetic systems which possess angular momentum. So, before closing out the chapter, let us also examine force and torque relations resulting from the Lorentzian transformation.

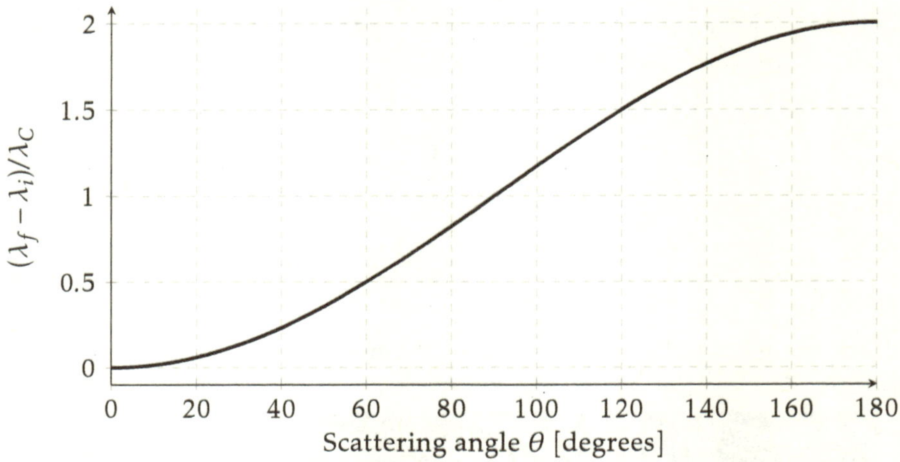

Figure 1.9: Photon wavelength change $\lambda_f - \lambda_i$ relative to the Compton wavelength $\lambda_C = h/m_0c$ as a function of the scattering angle θ.

1.7 The Relativistic Force and Torque Laws

If the velocity vector $\mathbf{v}$ is *non-constant*, then $\mathbf{v}\cdot\mathbf{v} = v^2$ is also *non-constant*. In such conditions, we can account for this in the Lorentz factor γ_r, which is a function of v^2. Taking the time-derivative of the momentum, we have

$$\frac{d\mathbf{p}}{dt^*} = \gamma_r \frac{d}{dt}(m_0\gamma_r\mathbf{v}) \tag{1.111a}$$

$$= m_0\gamma_r\frac{d\gamma_r}{dt}\mathbf{v} + m_0\gamma_r^2\frac{d\mathbf{v}}{dt} \tag{1.111b}$$

Since $\gamma_r = 1/\sqrt{1 - v^2/c^2}$, as we've already seen, using the relation $d(v^2)/dt = d(\mathbf{v}\cdot\mathbf{v})/dt = 2\mathbf{v}\cdot d\mathbf{v}/dt$, $d\gamma_r/dt$ becomes

$$\frac{d\gamma_r}{dt} = \frac{d}{dt}\left(1 - \frac{v^2}{c^2}\right)^{-\frac{1}{2}} \tag{1.112a}$$

$$= -\frac{1}{2}\left(1 - \frac{v^2}{c^2}\right)^{-\frac{3}{2}}\left(-\frac{2}{c^2}\mathbf{v}\cdot\frac{d\mathbf{v}}{dt}\right) \tag{1.112b}$$

$$= \frac{\gamma_r^3}{c^2}\left(\mathbf{v}\cdot\frac{d\mathbf{v}}{dt}\right) \tag{1.112c}$$

Substitution into Eq. (1.111) gives

$$\frac{d\mathbf{p}}{dt^*} = \frac{m_0\gamma_r^4}{c^2}\left(\mathbf{v}\cdot\frac{d\mathbf{v}}{dt}\right)\mathbf{v} + m_0\gamma_r^2\frac{d\mathbf{v}}{dt} \tag{1.113}$$

Since $dt^* = dt/\gamma_r$, Eq. (1.113) is equivalent to

$$\frac{d\mathbf{p}}{dt} = \frac{m_0\gamma_r^3}{c^2}\left(\mathbf{v}\cdot\frac{d\mathbf{v}}{dt}\right)\mathbf{v} + m_0\gamma_r\frac{d\mathbf{v}}{dt} \tag{1.114}$$

Since t is what is directly measured in a laboratory frame, for example, Eq. (1.114) becomes useful in practice. Therefore, the *relativistic force law* can be defined by $d\mathbf{p}/dt$ where $\mathbf{F}$ becomes

Relativistic Force Law

$$\mathbf{F} = m_0\gamma_r\frac{d\mathbf{v}}{dt} + \frac{m_0\gamma_r^3}{c^2}\left(\mathbf{v}\cdot\frac{d\mathbf{v}}{dt}\right)\mathbf{v} \tag{1.115}$$

In the relativistic picture, the force *picks up* an additional contribution along the axis of the velocity vector $\mathbf{v}$. This term is only zero when v^2 is constant. This can be seen from the fact that

$$\frac{d}{dt}\left(v^2\right) = \frac{d}{dt}(\mathbf{v}\cdot\mathbf{v}) = 2\left(\mathbf{v}\cdot\frac{d\mathbf{v}}{dt}\right)$$

Thus, if $v^2 \neq$ constant, $d\mathbf{v}/dt$ has a nonzero component along the axis of $\mathbf{v}$, in addition to the familiar contribution along the acceleration direction. Even in perfect circular motion, it is well-known that there is a component of acceleration proportional to the velocity vector.[14]

The force law given by Eq. (1.115) also suggests that the relationship between force $\mathbf{F}$ and $\frac{d\mathbf{v}}{dt}$ can be expressed using a tensor, or matrix relation. Let's determine the tensor form, as well.

The time derivative of momentum $d\mathbf{p}/dt$ can be expressed another way by making use of a vector algebra identity given by

$$\mathbf{A}\times\mathbf{B}\times\mathbf{C} = (\mathbf{A}\cdot\mathbf{C})\mathbf{B} - (\mathbf{A}\cdot\mathbf{B})\mathbf{C} \tag{1.116}$$

[14] e.g. for an explicit proof of this statement, see Section 7.1 in Quantum Mechanics With A SPIN On It (Eason 2022, Second Edition)

Applying this identity to the term $\left(\mathbf{v}\cdot\frac{d\mathbf{v}}{dt}\right)\mathbf{v}$, we have

$$\left(\mathbf{v}\cdot\frac{d\mathbf{v}}{dt}\right)\mathbf{v} = -\mathbf{v}\times\frac{d\mathbf{v}}{dt}\times\mathbf{v}+v^2\frac{d\mathbf{v}}{dt} \tag{1.117a}$$

$$= \mathbf{v}\times\mathbf{v}\times\frac{d\mathbf{v}}{dt}+v^2\frac{d\mathbf{v}}{dt} \tag{1.117b}$$

Substitution of this result into Eq. (1.115) leads to

$$\frac{d\mathbf{p}}{dt} = \frac{m_0\gamma_r^3}{c^2}\left(\mathbf{v}\times\mathbf{v}\times\frac{d\mathbf{v}}{dt}+v^2\frac{d\mathbf{v}}{dt}\right)+m_0\gamma_r\frac{d\mathbf{v}}{dt} \tag{1.118a}$$

$$= m_0\gamma_r\left[\frac{\gamma_r^2}{c^2}\mathbf{v}\times\mathbf{v}\times\frac{d\mathbf{v}}{dt}+\left(1+\gamma_r^2\frac{v^2}{c^2}\right)\frac{d\mathbf{v}}{dt}\right] \tag{1.118b}$$

$$= \frac{m_0\gamma_r^3}{c^2}\mathbf{v}\times\mathbf{v}\times\frac{d\mathbf{v}}{dt}+m_0\gamma_r^3\frac{d\mathbf{v}}{dt} \tag{1.118c}$$

In the above, we have used a relation we have seen already, namely

$$1+\gamma_r^2\frac{v^2}{c^2} = \gamma_r^2$$

Then, $d\mathbf{p}/dt$ or the force $\mathbf{F}$ becomes

Relativistic Force Law-Form II

$$\mathbf{F} = m_0\gamma_r^3\left[\frac{d\mathbf{v}}{dt}+\frac{1}{c^2}\mathbf{v}\times\mathbf{v}\times\frac{d\mathbf{v}}{dt}\right] \tag{1.119}$$

It is the *second* term on the RHS of Eq. (1.119), involving the vector cross product, which gives rise to a *tensor* form. Including the second term not only modifies the magnitude of the force $\mathbf{F}$, but it *also modifies the orientation such that the force is generally non-parallel to* $d\mathbf{v}/dt$. To re-write Eq. (1.119), we can use the skew matrix representation for the cross product, given by

$$\mathbf{A}\times\mathbf{B} = \begin{bmatrix} 0 & -A_z & A_y \\ A_z & 0 & -A_x \\ -A_y & A_x & 0 \end{bmatrix}\begin{bmatrix} B_x \\ B_y \\ B_z \end{bmatrix} \tag{1.120}$$

Using the above equation, it is straightforward to show that $\mathbf{A}\times\mathbf{A}\times\mathbf{B}$ is

$$\mathbf{A}\times\mathbf{A}\times\mathbf{B} = \begin{bmatrix} -A_y^2 - A_z^2 & A_xA_y & A_xA_z \\ A_xA_y & -A_x^2 - A_z^2 & A_yA_z \\ A_xA_z & A_yA_z & -A_x^2 - A_y^2 \end{bmatrix}\begin{bmatrix} B_x \\ B_y \\ B_z \end{bmatrix} \tag{1.121}$$

Then, using the above matrix relations, a tensor relation for the force **F** becomes

Relativistic Force-Acceleration Tensor Relation

$$\frac{d\mathbf{p}}{dt} = m_0\gamma_r^3 \begin{bmatrix} 1 - \frac{v_y^2+v_z^2}{c^2} & \frac{v_xv_y}{c^2} & \frac{v_xv_z}{c^2} \\ \frac{v_yv_x}{c^2} & 1 - \frac{v_x^2+v_z^2}{c^2} & \frac{v_yv_z}{c^2} \\ \frac{v_zv_x}{c^2} & \frac{v_zv_y}{c^2} & 1 - \frac{v_x^2+v_y^2}{c^2} \end{bmatrix} \frac{d\mathbf{v}}{dt} \tag{1.122}$$

In the limit of $v^2/c^2 \to 0$, the force tensor approaches the 3×3 identity matrix $\mathbf{I}(3)$. However, as v^2/c^2 increases, the diagonal elements necessarily become smaller, while the off-diagonal terms become larger, indicating rotations in $d\mathbf{p}/dt$ relative to $d\mathbf{v}/dt$.

1.7.1 Relativistic Torque

The force can also be used to determine the corresponding time-derivative of the orbital angular momentum **L**. If we use the definition $\mathbf{L} = \mathbf{r}\times\mathbf{p}$, then it follows that

$$\frac{d\mathbf{L}}{dt^*} = \gamma_r\frac{d\mathbf{L}}{dt} = \gamma_r\left[\frac{d}{dt}(\mathbf{r}\times\mathbf{p})\right] = \gamma_r\left[\frac{d\mathbf{r}}{dt}\times\mathbf{p} + \mathbf{r}\times\frac{d\mathbf{p}}{dt}\right] \tag{1.123}$$

Recall that the velocity vector $\mathbf{v} = d\mathbf{r}/dt$ and $\mathbf{p} = m_0\gamma_r\mathbf{v}$. Consequently, the first term vanishes since it is a cross-product between two *parallel* vectors. Then,

$$\frac{d\mathbf{L}}{dt} = \mathbf{r}\times\frac{d\mathbf{p}}{dt} \tag{1.124}$$

Thus, we only need to substitute Eq. (1.114) for $d\mathbf{p}/dt$ to obtain $d\mathbf{L}/dt$. Doing this, we have

$$\frac{d\mathbf{L}}{dt} = \mathbf{r}\times\frac{d\mathbf{p}}{dt} \tag{1.125a}$$

$$= \mathbf{r} \times m_0 \gamma_r \left[\frac{\gamma_r^2}{c^2} \left(\mathbf{v} \cdot \frac{d\mathbf{v}}{dt} \right) \mathbf{v} + \frac{d\mathbf{v}}{dt} \right] \tag{1.125b}$$

$$= \frac{m_0 \gamma_r^3}{c^2} \left(\mathbf{v} \cdot \frac{d\mathbf{v}}{dt} \right) \mathbf{r} \times \mathbf{v} + m_0 \gamma_r \mathbf{r} \times \frac{d\mathbf{v}}{dt} \tag{1.125c}$$

Since the orbital angular momentum $\mathbf{L} = \mathbf{r} \times \mathbf{p}$ and $\mathbf{p} = m_0 \gamma_r \mathbf{v}$, the first term in the last line is proportional to $\mathbf{L}$. Substitution then gives

Relativistic Time-Derivative Of Orbital Angular Momentum

$$\frac{d\mathbf{L}}{dt} = m_0 \gamma_r \mathbf{r} \times \frac{d\mathbf{v}}{dt} + \underbrace{\frac{\gamma_r^3}{c^2} \left(\mathbf{v} \cdot \frac{d\mathbf{v}}{dt} \right) \mathbf{L}}_{\text{longitudinal motion}} \tag{1.126}$$

Of particular interest in Eq. (1.126) is the second term on the RHS, also being a relativistic contribution. It means that the time-dependent variation of the orbital angular momentum **L** also has a component *along the direction of* **L**, which is known as *longitudinal relaxation*.The existence of longitudinal relaxation in several magnetic systems is a foregone conclusion, both at the macroscopic level and down to the atomic level. Thus, relativity reveals itself again!

Summarizing, we find that applying the special theory of relativity leads to generalizations of some familiar laws of physics, particularly forces **F** and torques **T**. Any of these laws can be *projected* down to the scale of quantum mechanics, as we'll see shortly.

1.8 Chapter Summary

In this chapter, we have introduced the *special theory of relativity*, starting with the Lorentz factor and its concomitant transformation. We then considered several consequences from the Lorentz transformation. In parallel, we also determined an explicit form of the fourth dimension. We examined the relationship between it and the Lorentz factor. These concepts were also used to obtain momentum, energy, force, and torque, in a relativistic framework.

These results turn out to be crucial if we wish to project down to the fundamental length scale of quantum mechanics. For this, we must introduce the concept of waves. We'll start this introduction in the next chapter. Then, we'll see even more clearly just how *special* this theory from Lorentz, Einstein, *etc.* truly is.

Table 1.1 summarizes some key equations from this chapter.

Table 1.1: Chapter 1 Summary Equations.

Name	Equation
Lorentz factor	$\gamma_r = \frac{1}{\sqrt{1-\frac{u^2}{c^2}}}$
Length Contraction	$L' = \frac{L}{\gamma_r}$
Spatial 4D-Vector	$\mathbf{S}_\ell = [x,\ y,\ z,\ ict]^T$
Relativistic Momentum	$\mathbf{p} = \frac{m_0\mathbf{v}}{\sqrt{1-\frac{v^2}{c^2}}}$
Rel. Kinetic Energy*	$K = \frac{\gamma_r mc^2}{2}$
Compton Scattering	$\lambda_f - \lambda_i = \frac{h}{m_0 c}(1-\cos\theta)$
Relativistic Force	$\mathbf{F} = m_0\gamma_r\frac{d\mathbf{v}}{dt} + \frac{m_0\gamma_r^3}{c^2}\left(\mathbf{v}\cdot\frac{d\mathbf{v}}{dt}\right)\mathbf{v}$
Relativistic Torque	$\frac{d\mathbf{L}}{dt} = m_0\gamma_r\mathbf{r}\times\frac{d\mathbf{v}}{dt} + \frac{\gamma_r^2}{c^2}\left(\mathbf{v}\cdot\frac{d\mathbf{v}}{dt}\right)\mathbf{L}$

1.9 Chapter Problems

Problem 1.1 Calculate the Lorentz factor

$$\gamma_r = \frac{1}{\sqrt{1-\frac{u^2}{c^2}}}$$

for a particle frame having speeds $u = 0.5c, 0.75c$ and c. Then, calculate the same for the cases $u = 2c, 3c$, and $4c$. Discuss the pattern of the numbers, and whether or not there exists a mathematical limitation for calculating γ_r.

Problem 1.2 In the Lorentz transformation given by

$$\mathbf{r}' = (\mathbf{r} - \mathbf{u}t)\gamma_r \quad \text{and} \quad (\mathbf{r}' + \mathbf{u}t')\gamma_r$$

$$t' = \left(t - \frac{\mathbf{u} \cdot \mathbf{r}}{c^2}\right)\gamma_r \quad \text{and} \quad t = \left(t' + \frac{\mathbf{u} \cdot \mathbf{r}'}{c^2}\right)$$

we use the symbol u to denote the speed of one of the reference frames, measured relative to a fixed frame. How can this be applied to an electron bound to an atom? Which frame should be the fixed frame and which should be the other reference frame in the Lorentz transformation equations?

Problem 1.3 For a particle frame having speeds $u = 0.5c, 0.75c$, and $0.95c$, calculate the relative position r and time t using Lorentz transformation given in Prob. 1.2. Compare them to the corresponding results from a Galilean frame described by

$$\mathbf{r} = \mathbf{r}' + \mathbf{u}t' \quad , \quad t = t'$$

Problem 1.4 In this chapter, we determined the square of the 4D velocity as

$$\frac{\partial \mathbf{S}_\ell}{\partial t} \cdot \frac{\partial \mathbf{S}_\ell}{\partial t} = v^2 - c^2$$

Calculate values for c/γ_r and $\mathbf{S}_\ell$ using velocities $v = 0, .5c, c$, and $1.5c$. How do they compare?

Problem 1.5 Using the Lorentz transformation, show that

$$\Delta t' = \gamma_r \Delta t$$

This result is known is *time dilation.*

Hint: Use the method used to derive Eq. (1.20).

Problem 1.6 Show that the crossover point when $\gamma_r = \gamma_r^2/2$ is given by

$$\frac{v}{c} = \sqrt{\frac{3}{4}}$$

Problem 1.7 Using the relativistic momentum

$$p_r = m_0\gamma_r v = m_0 v_r$$

Derive an expression for the relativistic kinetic energy $E_r = p_r^2/2m_0$ using the relativistic momentum and the non-relativistic form E_0 by taking the limit $\gamma_r = 1$. Calculate the values of both energies for $v = c$.

Then, show that

$$E_r - E_0 = \frac{1}{2m_0c^2}\left(\frac{v^2}{c^2 - v^2}\right)$$

The figure illustrates the two energies as a function of velocity or

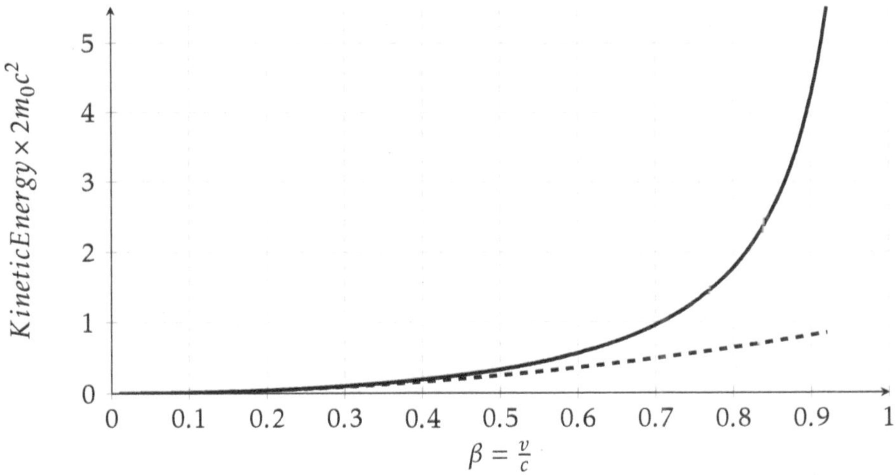

$\beta = v/c$. Notice that for a finite range of velocities $v \in (0, c)$ or $\beta \in (0, 1)$ the relativistic energy has *boundless* energy, while the non-relativistic has finite energy. In quantum mechanics, energies are boundless so that relativistic is more consistent, especially at higher energies.

Problem 1.8 Earlier, we obtained a relativistic torque equation given by

$$\frac{d\mathbf{L}}{dt} = m_0\mathbf{r} \times \frac{d\mathbf{v}}{dt} + \underbrace{\frac{\gamma_r^2}{c^2}\left(\mathbf{v} \cdot \frac{d\mathbf{v}}{dt}\right)\mathbf{L}}_{\text{longitudinal motion}}$$

Let us consider the longitudinal term only, given by

$$\frac{d\mathbf{L}_{long}}{dt} = \frac{\gamma_r^2}{c^2}\left(\mathbf{v}\cdot\frac{d\mathbf{v}}{dt}\right)\mathbf{L} = \frac{1}{c^2 - v^2}\left(\mathbf{v}\cdot\frac{d\mathbf{v}}{dt}\right)\mathbf{L}$$

Consider an electron traveling in a perfect circular orbit, having a instantaneous velocity v of

$$\mathbf{v} = \mathbf{r} \times \omega$$

In the case of constant frequency ω along z, then

$$\frac{d\mathbf{v}}{dt} = \frac{d\mathbf{v}}{dt} \times \omega$$

In an (r, θ, z) coordinate system, what is the direction of $\frac{d\mathbf{v}}{dt}$? Is it along r, θ, or z?

Once you have your answer, then consider an instaneous velocity given by

$$\mathbf{v}_f = \mathbf{v}_i + \frac{d\mathbf{v}}{dt}\delta t \Rightarrow \Delta\mathbf{v} = \frac{d\mathbf{v}}{dt}\delta t$$

Based on this, along which axis will $\Delta\mathbf{v}$ be? Additionally, discuss which conditions will $\frac{d\mathbf{L}_{long}}{dt}$ be zero, and under which conditions will it be nonzero?

1.10 Suggested Readings & References

[1] A. Einstein, *On the Electrodynamics of Moving Bodies*, Annalen der Physik (1905).

[2] K. Eason, *Quantum Mechanics With A SPIN On It*, Sym. Pub (2019).

[3] A. H. Compton, *A Quantum Theory of the Scattering of X-rays by the Light Elements*, The Physical Review, **21**, No. 5 (1923).

CHAPTER 2

Including the 4th Dimension

We can use the *special theory of relativity*, and the *four-dimensional vector* formalized in Chapter 1 to project useful ideas or concepts down to quantum mechanical systems having wave-particle duality. This will be a focus of this chapter...taking such concepts and developing a quantum mechanical framework with them. From this endeavor, we find more macroscopic equations can be obtained, including the relativistic electromagnetic laws.[1] To assist us in developing some intuition for a four-dimensional quantum mechanical wave, we start with a discussion of a special geometrical structure that aids in providing intuition, namely, the *helix*. From here, we can begin to bring together the old and the new, or the big, and the small.

A *helical wave* can be imagined as being "constrained" to a helical structure. For the moment, let us not ponder so much the 'why' for it is like some mathematical solutions...simply the trick that leads to a worthy conclusion. With this, we can develop a helical wave that allows for a natural progression into quantum mechanical wave functions and their *operators*.[2] Ultimately, we will arrive at a type of helical structure known

[1]This statement ordinarily is not obvious. However, we shall find that it becomes more so after the fourth dimension is included.

[2]These operators are purely mathematical that are applied to q.m. wave functions to

as a *modulated helix*. It is a helix whose *radius is generally non-uniform* along its axis. In our case, it is due to a time-dependence. So, let's begin, specifically, with a discussion of the *unit* helix.

2.1 The Helical Wave Structure

The coil structure shown in Fig. 2.1 is an example of a *unit* helix. As a unit, it has a radius of unity ($r = 1$). Also imagine a *corpuscle* or a localized *particle*-like distribution constrained in its motion to the path along the helix.

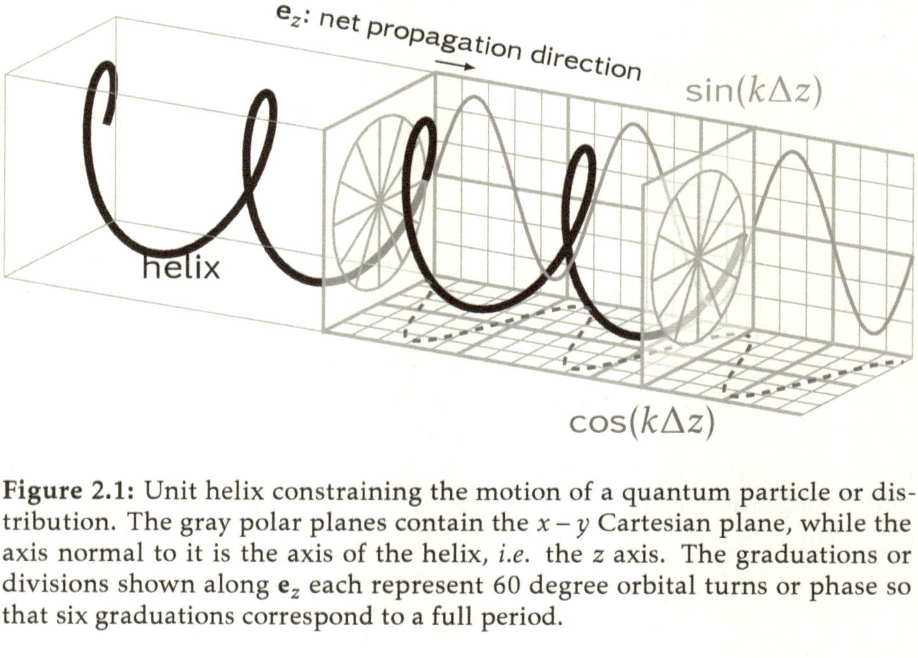

Figure 2.1: Unit helix constraining the motion of a quantum particle or distribution. The gray polar planes contain the $x-y$ Cartesian plane, while the axis normal to it is the axis of the helix, *i.e.* the z axis. The graduations or divisions shown along $\mathbf{e}_z$ each represent 60 degree orbital turns or phase so that six graduations correspond to a full period.

Let us now consider a point or position in three dimensions in motion along the *unit* helix (*i.e.* $r = 1$). The 3D position can be written as

$$\mathbf{r}(x,y,z) = \cos(\delta\phi)\widehat{\mathbf{e}}_x + \sin(\delta\phi)\widehat{\mathbf{e}}_y + z\widehat{\mathbf{e}}_z = \cos(kz)\widehat{\mathbf{e}}_x + \sin(kz)\widehat{\mathbf{e}}_y + (vt)\widehat{\mathbf{e}}_z \quad (2.1)$$

The scalar k is called the *wavenumber* parameter to be derived shortly. Here, it is associated with the z–axis. $\widehat{\mathbf{e}}_x$, $\widehat{\mathbf{e}}_y$ and $\widehat{\mathbf{e}}_z$ are *unit* vectors along

yield useful information.

those respective directions. The coordinates x and y lie in the polar planes resembling dart-boards shown in Fig. 2.1. As mentioned, the z-axis is parallel to the axis of the helix. Owing to the helical structure, the particle's constrained motion is both along the z-axis, while simultaneously *rotating* about the z-axis.

Because $\int \cos x dx = 0$ and $\int \sin x dx = 0$ over any period of the motion, the net or average propagation direction is along the z axis, or along the axis of the helix. However, though the net motion is along this axis, the linear motion along z is also *accompanied by some form of angular momentum* imparted by curvature of the helical structure. The vertical polar planes (dart boards) in Fig. 2.1 can be used to *visualize* the rotational phase $\Delta\phi$ of the point around the propagation axis. As the motion proceeds, there can be up to two components of angular momentum: first, the orbital contribution about the axis and second, the internal spinning arising from rotation along the track. The physical scenario parallels doing a 'donut' in a fast car which is in a circular pattern, as it has two angular momentum contributions: first, about the center of the circle, as well as the wheels spinning around the axle.[3] The $y - z$ plane can be used to observe *only* the periodic behavior projected onto the y-axis. This component is described by the sine function (vertical curve shown in red). Likewise, the $x - z$ plane provides the same for the x-axis periodic motion, described by the cosine function (bottom curve shown in blue).

The distance along the propagation direction that corresponds to *one complete cycle of motion* along the helix is defined as one *wavelength* λ. Now, imagine that you had a stopwatch, and could sit on the wave while in motion to measure the time it takes to travel one wavelength λ. This interval or unit period of time to travel one wavelength is called the *wave period* $\Delta t = T_p$. It follows from these two definitions that the *linear or mean wave velocity* **v** *along the propagation direction* becomes

$$v = \frac{\Delta z}{\Delta t} = \frac{\lambda}{T_p} = \lambda f \Rightarrow \Delta t = \frac{\Delta z}{\lambda f} \tag{2.2}$$

Then, if we know the properties of the wave like the wavelength λ and period T_p (along with elapsed time), we do not need to write the component of motion along z which is $v = v_z$ or $\Delta z = vt$.[4] Therefore, *two* of the three coordinates along with the wave parameter completely describe the *mean* observable motion.

[3] Not that I've ever done this.

[4] This is true under the condition of constant velocity.

Because the particle is also rotating about the propagation direction z, there is an associated rotation $\Delta\phi$. Thus, there is also a rate $\omega = d\phi/dt$ which can be visualized on the vertical circular phase planes. A full rotational period of 2π corresponds to the same elapsed unit of time T_p, so ω is given by

$$\omega = \frac{2\pi}{T_p} = 2\pi f \tag{2.3}$$

Using Eqs. (2.2) and (2.3), the rotational phase $\Delta\phi$ becomes

$$\Delta\phi = k\Delta z = \omega\Delta t \tag{2.4a}$$

$$= 2\pi f \left[\frac{\Delta z}{\lambda f}\right] \tag{2.4b}$$

$$= \frac{2\pi\Delta z}{\lambda} \tag{2.4c}$$

From this result, we can define the *wavenumber* k associated with the motion along the propagation direction as

Wavenumber

$$k = \frac{2\pi}{\lambda} \tag{2.5}$$

This inverse length parameter, or *wavenumber*, provides the phase information using only the linear displacement along z. Let us consider an equivalent representation of the same information, however, using *complex functions* instead.

2.1.1 Complex Representation And Wave Functions

The periodic motion of a point constrained to a unit helix was described above using *real* Cartesian coordinates along e_x, e_y, and e_z. The coefficient of each unit vector is a *real number*, while the unit vector has amplitude $|e_i| = 1$. However, there are many alternative representations of the same position. For example, we can consider a form where we replace the unit vector e_y, satisfying the condition $|e_y| = 1$, with a purely imaginary unit i also satisfying $|i| = 1$. Instead of $R_y(z) = \sin kz \cdot \widehat{\mathbf{e}}_y$, let us write

$$y(z) = \sin(kz) \cdot i \tag{2.6}$$

where $i = \sqrt{-1}$. The complex scalar given by $x(z)\mathbf{e}_x + y(z)$ using Eq. (2.6) is also an equivalent representation including two dimensions because a complex scalar has *two* independent degrees of freedom: One is given by the real; the other by the imaginary part, *i.e.* y. *All* the same information can be preserved. We'll see an explicit example of this shortly. But, this parallel can be exploited to represent any system composed (partially or wholly) of a two-degree-of-freedom system, *i.e.*

$$[\widehat{\mathbf{e}}_x, \widehat{\mathbf{e}}_y] \Leftrightarrow (1, i)$$

For constant mean motion[5], since the component of motion along z in helical motion is redundant (if ω/speed and λ are known), *the motion along a helix can be represented wholly by a 2D complex system.*

The coordinates of 2D space also have associated structure which can be evaluated using a suitable *norm* operation.[6] For example, we know that the *length* of a 2D real vector $[a \quad b]^T$ is given by

$$\mathbf{A} = \begin{bmatrix} a \\ b \end{bmatrix} \Rightarrow |\mathbf{A}|^2 = \mathbf{A}^T \cdot \mathbf{A} = a \cdot a + b \cdot b = a^2 + b^2 \tag{2.7}$$

Since $i = \sqrt{-1}$, if we evaluate the same operator using the *complex* representation, we get

$$\mathbf{A} = \begin{bmatrix} a \\ bi \end{bmatrix} \Rightarrow |\mathbf{A}|^2 = \mathbf{A}^T \cdot \mathbf{A} = a \cdot a + ib \cdot ib = a^2 - b^2 \tag{2.8}$$

Since $a^2 - b^2 \neq a^2 + b^2$, in order to represent the *same* real-valued quantity $a^2 + b^2$ *i.e.* preserve the information, we must, instead, use $\mathbf{A}$ and it's complex conjugate A^* $(a - ib)$ (obtained by multiplying the imaginary component by -1), so we then have

$$\mathbf{A} = \begin{bmatrix} a \\ bi \end{bmatrix} \Rightarrow |\mathbf{A}|^2 = [\mathbf{A}^*]^T \cdot \mathbf{A} = \mathbf{A}^\dagger \mathbf{A} = a \cdot a - ib \cdot ib = a^2 + b^2 \tag{2.9}$$

---

[5] Constant motion here means $|v|$ is constant along the axis of the helix.

[6] A norm is an operation that takes the elements of a dimensional space and map to a single scalar.

Thus, using the complex conjugate ensures conservation of information between both spaces, a 2D real space and a 2D complex space. The corresponding norms are

$$\mathbf{A}^T\mathbf{A} \Rightarrow \mathbf{A}^\dagger\mathbf{A} \tag{2.10}$$

The transpose-conjugate notation used in Eq. (2.10) is a more concise notation known as *dagger* (†). It is a generalization of the ordinary inner product for vectors that contain complex number elements.

Then, an alternative form for the position vector $\mathbf{r}$, using a *complex* function for the position, becomes

$$\mathbf{r} = \cos(kz) + i\sin(kz) = e^{ikz} \tag{2.11}$$

Then the square of the magnitude of the 2D unit vector is given by

$$\begin{aligned}\mathbf{r}^2 = \mathbf{r}^\dagger\mathbf{r} &= \begin{bmatrix}\cos(kz) & -i\sin(kz)\end{bmatrix}\cdot\begin{bmatrix}\cos(kz)\\ i\sin(kz)\end{bmatrix}\\ &= \cos^2(kz) + i\cos(kz)\sin(kz) - i\sin(kz)\cos(kz) + \sin^2(kz)\\ &= \cos^2(kz) + \sin^2(kz) = 1\end{aligned}$$

The complex form for 2D spatial information turns out to be very useful in quantum mechanics. For example, in conditions where the *radius of the helix is infinitesimally small* in comparison to the propagation length-scale, it's more convenient to use fixed wavelength and frequency values, rather than tiny distances changing with time, *etc.*

In addition to the vector magnitude, the more general 3D form of the wavenumber k, known as the *wave-vector* $\mathbf{k}$ (wavenumber $k = |\mathbf{k}|$) is along the direction of motion. With this picture, it is straightforward to *relax* the *unit* helix radius assumption to allow for any radius R_0, so we have

$$R(x,y,z) = R_0 e^{ikz} \Rightarrow R(x,y,z) = R_0 e^{i\mathbf{k}\cdot\mathbf{r}} \tag{2.12}$$

In this form,Eq. (2.12) describes a general *distribution* in space whose collective motion is constrained to a helix, *i.e.* a distributed *tiny cloud* described by a set of points in motion.

As mentioned, as long as the wavelength λ and frequency $\omega = 2\pi f$ are known, the mean motion along z is also known, provided the linear velocity is constant.

With the above, taking the limit as $\lambda \to \infty$ leads to

$$\lim_{\lambda\to\infty} R(x,y,z) = R_0 e^{i\Delta z/\infty} = R_0 \cdot 1 = R_0 \tag{2.13}$$

We find that the position function tends to lose its complex scalar form and converges to a real-valued constant. This *implies a static position* in space. This result is a problem as it *contradicts the special theory of relativity*, which suggests motion is ubiquitous or $(dR/dt)^2 = v^2 \neq 0$ (c. Eq. 1.69). In quantum mechanics, only zero *net* motion is achievable through the multitude of possibilities from its constituents possessing nonzero kinetic energy. The atom in a solid phase material is a cogent example.

A remedy to this result can be found with the inclusion of the fourth dimension because Eq. (2.12) only relies on three coordinates x, y, and z. Let's see how the fourth dimension can restore consistency with relativity.

2.1.2 Recovering Motion With A Non-uniform Helix

Recall that the *wavevector* **k** contains information about the mean change in position $\langle \Delta R \rangle$. Since all position *changes* are accompanied by corresponding changes in time, this information contains time-dependent information. Bridging the explicit time dependence to **k** can be done by introducing the fourth dimension x_4 into the wave description. Let us extend Eq. (2.12) to 4D by writing

$$S(x_1, x_2, x_3, x_4) = S_0 e^{i\left(k_x x_1 + k_y x_2 + k_z x_3 + k_4 x_4\right)} \tag{2.14}$$

If we set $k_4 x_4 = k_4 \cdot (ict) = -\omega_4 t$, Eq. (2.14) takes the form

$$S(x,y,z,t) = S_0 e^{i(\mathbf{k}\cdot\mathbf{r} - \omega_4 t)} \tag{2.15}$$

where the frequency $\omega_4 = ik_4 c \neq \omega$, so we have[7]

$$S(x,y,z,t) = R_0 e^{-k_4 ct} e^{i\mathbf{k}\cdot\mathbf{r}} = R_0(t) e^{i\mathbf{k}\cdot\mathbf{r}} \tag{2.16}$$

This form is equivalent to a propagating wave whose distribution amplitude varies with time. Note that the time-dependent information, specifically, has been injected by the fourth dimension. Eq. (2.16) is

[7] recall ω from Eq. (2.4) relates directly to k which is tied to the first *three* dimensions.

also a helix, however, it has non-uniform radius along its propagation direction due to time dependence. Now, in the limiting case of $\lambda \to \infty$, the amplitude is generally *non-constant*. Thus, by adding the fourth dimension, we have recovered motion in the limiting case because $d(R^\dagger R)/dt \neq 0$.

Fig. 2.2 illustrates an example of a *non-uniform* helical wave structure based on *four* dimensions. The complex-valued description of motion

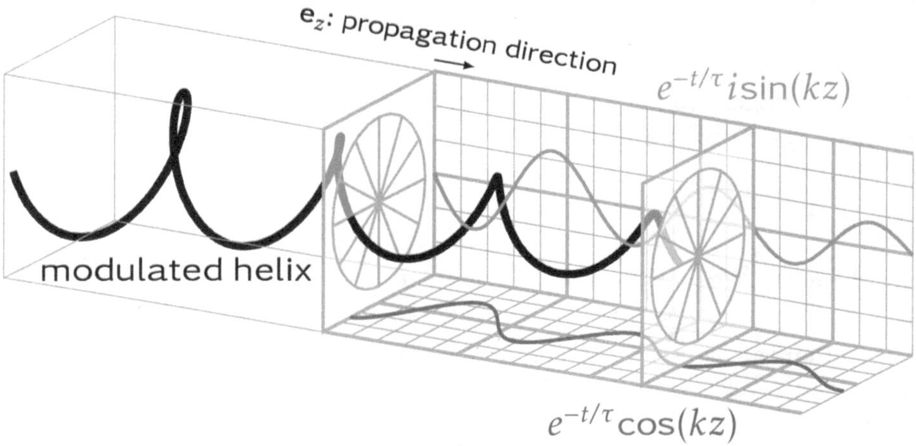

Figure 2.2: Example of a nonuniform helix. With motion along the axis, the helix can alter its radial amplitude in time because all physical dimensions also vary in time.

associated with *a modulated helix represents a four-dimensional wave*. It is a generalization of a 3D one, and thus, preserves and consistently extends all the same properties from 3D.

It turns out that there are specific *mathematical operations* that can be applied in order to *extract* useful physical-system information. To take this further, we must develop an *indirect* description of complex functions like those given in Eq. (2.16). We can do this by utilizing concepts from calculus. The results can be used in quantum mechanics to describe physical systems in-line with the theory of relativity. The resulting form leads us to the so-called *quantum mechanical wave function*.

2.1.3 Properties Of Quantum Mechanical Wave Functions

We have found a general description of motion along a helical structure using a complex exponential function. The position function, though a complex scalar, still maintains physical units of length, *e.g.* meters. Since the waves are generally distributed in space, we can also use the concept of *density* ρ. Let's use the *strict volume density* ρ defined at any point of 3D space $R(x, y, z)$ where the units are purely inverse spatial dimensions. For example, a one dimensional (1D) density ρ has units of m^{-1}; in 2D it has units of m^{-2}; and in 3D units m^{-3} *etc.* This form allows determination of *any* other volume density quantity such as mass density ρ_m, charge density ρ_c, *etc.*

One objective here is to establish an explicit relationship between the position function given by Eq. (2.16), the strict density ρ, and the *wave function*, which is conventionally denoted by $\psi(x, y, z)$. To do this, let's revisit a fundamental concept in calculus known as the *mean* or *average* of a function of a single variable. In one dimension, a length L or equivalently, a difference in 1D spatial positions $L = \Delta x$ can be written as

$$\Delta x = \int_{x_i}^{x_f} dx \tag{2.17}$$

Dividing both sides by Δx, we get the result

$$1 = \frac{1}{\Delta x}\int_{x_i}^{x_f} dx = \frac{1}{\Delta x}\int_{x_i}^{x_f} 1(x)dx \tag{2.18}$$

In Eq. (2.18), $1(x)$ is the *constant function* equal to unity at every value of x. When expressed this way, it is more obvious that we may interpret Eq. (2.18) as the *average value* of the function $1(x)$. Eq. (2.18) is an example of a *probability distribution* $P(x)$ which must satisfy the integral condition

$$1 = \int_{x_i}^{x_f} P(x) \cdot 1(x)dx \Rightarrow P(x) = \frac{1}{\Delta x} \tag{2.19}$$

Eq. (2.19) is also an example of a *continuous uniform distribution*. As long as the limits of integration are *finite*, Eq. (2.19) is valid. In describing any physical body of presumably finite size, the body measures a length $\Delta x = L$, and therefore, this condition is always satisfied in practice.

The probability distribution $P(x)$ is more useful to determine the *average* or *mean* value of a *non-uniform* integrable function $f(x)$ as

$$\text{average of } f(x) = \frac{1}{\Delta x}\int_{x_i}^{x_f} f(x)dx = \int_{x_i}^{x_f} P(x)\cdot f(x)dx \tag{2.20}$$

$f(x)$ replaces $1(x)$ in Eq. (2.18) and is related to the *Mean Value Theorem* of calculus and can be *generalized to functions of any number of variables.* For example, if the function $f(x,y)$ depends on two variables x and y, then Eq. (2.20) becomes

$$\text{average of } f(x,y) = \frac{1}{\Delta x\Delta y}\int_{x_i}^{x_f}\int_{y_i}^{y_f} f(x,y)dxdy \tag{2.21}$$

$$= \int_{x_i}^{x_f}\int_{y_i}^{y_f} P(x,y)\cdot f(x,y)dxdy \tag{2.22}$$

Alternatively, if x and y are *spatial* variables, Eq. (2.21) can be written as

$$\text{average of } f(x,y) = \frac{1}{\Delta A}\int_{x_i}^{x_f}\int_{y_i}^{y_f} f(x,y)dxdy \tag{2.23}$$

ΔA is the bounded area containing the distributed points $f(x,y)$.

Likewise, if the function $f(x,y,z)$ depends on three variables x, y, and z, then we have

$$\text{average of } f(x,y,z) = \frac{1}{\Delta x\Delta y\Delta z}\int_{x_i}^{x_f}\int_{y_i}^{y_f}\int_{z_i}^{z_f} f(x,y,z)dxdydz \tag{2.24}$$

$$= \int_{x_i}^{x_f}\int_{y_i}^{y_f}\int_{z_i}^{z_f} P(x,y,z)\cdot f(x,y,z)dxdydz \tag{2.25}$$

or

$$\text{average of } f(x,y,z) = \frac{1}{\Delta V}\int_{x_i}^{x_f}\int_{y_i}^{y_f}\int_{z_i}^{z_f} f(x,y,z)dV \tag{2.26}$$

Thus, *the explicit distribution function as well as its physical units depend on the number of variables or dimensions.* If all variables are taken as spatial coordinates, then $1/\Delta x$ is inverse length *e.g.* m^{-1}, $1/\Delta x\Delta y$ is inverse area *e.g.* m^{-2}, while $1/\Delta x\Delta y\Delta z$ is inverse volume *e.g.* m^{-3}. These quantities are the units of *strict* volume densities, in which we are interested.

The uniform probability distributions for up to three *spatial* variables can summarized as

$$1\text{D} \Rightarrow P(x) = \frac{1}{\Delta x} \quad \text{(inverse length)} \tag{2.27a}$$

$$2\text{D} \Rightarrow P(x,y) = \frac{1}{\Delta x \Delta y} = \frac{1}{\Delta A} \quad \text{(inverse area)} \tag{2.27b}$$

$$3\text{D} \Rightarrow P(x,y,z) = \frac{1}{\Delta x \Delta y \Delta z} = \frac{1}{\Delta V} \quad \text{(inverse volume)} \tag{2.27c}$$

How can this be related directly to the *distributions* characterized by helical wave motion? One difficulty is that if x is real-valued, Δx (and therefore $1/\Delta x$) is *real-valued*, while we have established the distributed function $S(x, y, z, t)$ generally as a *complex scalar* function.

To extricate ourselves from this dilemma, let us *define* the following complex conjugate-pair of functions ψ and ψ^* as

$$\psi \equiv \frac{N}{\sqrt{S}} \quad \text{and} \quad \psi^* \equiv \frac{N^*}{\sqrt{S^*}} \tag{2.28a}$$

Let us call N the *normalization constant*. From this, it follows that

$$\psi^* \psi = \frac{|N|^2}{|S|^2} \tag{2.29}$$

Since S represents points distributed in space over some *finite* extent, Eq. (2.29) can also be integrated over that same finite extent. One useful purpose for the normalization constant/scalar N is to ensure that

$$\int \psi^* \psi dV = \int \frac{|N|^2}{|S|} dV = 1 \tag{2.30}$$

When Eq. (2.30) is satisfied, the wave function ψ is said to be *normalized*. Moreover, the *normalized function is a probability distribution* that can be used to determine mean values of quantities varying over x_1, x_2, and x_3. Although the definitions only include 3D spatial coordinates, the mean values can still vary independently in time. As we will see, this is the subtle fact that enables the power of the fourth dimension.

Let's look at a simple example of finding a value of N that guarantees Eq. (2.30) is satisfied.

Exercise 2.1 Let ψ and ψ^* be given by

$$\psi \equiv \frac{N}{\sqrt{x}} \quad \text{and} \quad \psi^* = \psi \equiv \frac{N}{\sqrt{x}}$$

Using the result that $\int dx/x = \ln x$ to solve Eq. (2.30) leads to

$$\int \psi^* \psi dx = \int \frac{|N|^2}{x} dx = |N|^2 \ln\left(x_i/x_f\right) = 1$$

Solving for N, we get

$$|N| = \frac{1}{\sqrt{\ln\left(x_i/x_f\right)}}$$

Because there is a real distribution over some finite length, it is certain that $x_f \neq x_i$, guaranteeing a nonzero value for the log term, and a finite value for N. Thus, Eq. (2.30) is satisfied. Then, ψ, by definition, is also *normalized*. ■

As evidenced in Example 2.1, the parameter $|N|$ carries information about the spatial extent. Since quantum eigentates are usually quantized depending on some *quantum number*, $|N|$ is often quantized so that *it may not take on arbitrary values*. Still, for any finite distribution over space, it is always possible to find the value of N for any wave function ψ so that

Normalization Condition

$$1 = \int_V |\psi|^2 dV \tag{2.31}$$

As mentioned, when a quantum mechanical wave function satisfies Eq. (2.31), it said to be *normalized*. In this form, $|\psi|^2$ is also guaranteed to be a *probability distribution* which can be used to compute useful expectation values of spatially-dependent parameters, even if they are time-dependent.

So, we have a suitable definition for the wave function ψ that can satisfy the properties of a probability distribution as given by Eq. (2.31). At the same time, it effectively contains information relating to motion along a helical structure. Before we continue building up a description of quantum mechanical wave functions, let us examine the macroscropic scale to consider some cases where *wave equations* are known to appear. Eventually, we will make the connection between both scales.

2.2 Classical Wave Equations

Let us now discuss a special and relevant form of equation that appears in various areas of macroscopic physics. For example, they appear in mechanics, electromagnetics, heat transfer, as well as other areas. Let us revisit some of these classical wave equations, starting with mechanics.

2.2.1 Mechanical Wave In A String

Consider a string ordinarily pulled in tension along the x direction and pinned on its extremities. At any point other than the endpoints, the string can be *displaced* vertically. For example, imagine we initially displace the string vertically, say with a finger, from a taught equilibrium position as shown in Fig. 2.3.

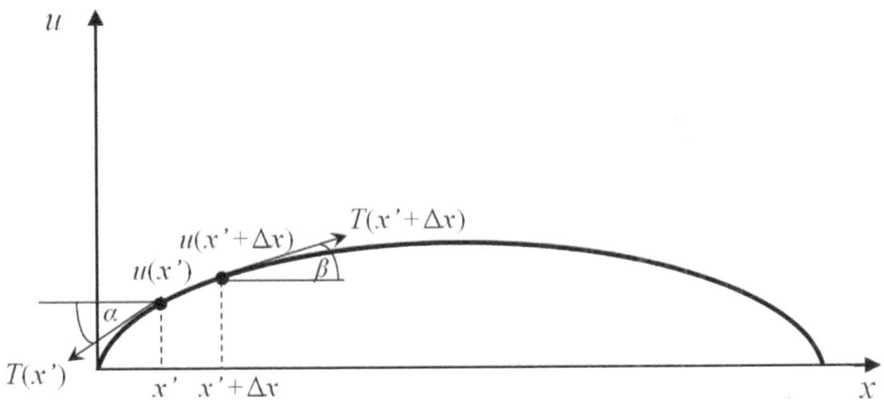

Figure 2.3: Illustration if a vibrating string with associated tensions T and vertical displacement $u(x)$. As shown, tension T is always tangential for a string of general structure.

Since the string is displaced *vertically*, the net forces along the horizontal direction must continue to sum to zero (since there is no net force in this direction). If at any point x, the string forms an angle θ with the horizontal, the x and y components of tension are given by

$$T_x = T\cos\theta \quad and \quad T_y = T\sin\theta \tag{2.32}$$

At two different locations corresponding to ends of a differential element of the string Δx at x and $x + \Delta x$, let us denote their corresponding angles by $\theta(x) = \alpha$ and $\theta(x + \Delta x) = \beta$, respectively. Then, for the net horizontal forces, we have

$$T(x' + \Delta x)\cos\beta - T(x')\cos\alpha = 0 \Rightarrow$$

$$T(x' + \Delta x)\cos\beta = T(x')\cos\alpha \tag{2.33}$$

Initially, the string is displaced or perturbed vertically while there are also constraints on the endpoints which always counter forces along the x axis. However, before the perturbation, there is an existing condition in the string, of a *horizontal equilibrium tension* T_∞ from the endpoint constraints creating it. Since it is assumed that the displacement is strictly vertical, there is no motion along the horizontal direction ($a_x = 0$). Then, the component of tension along x remains constant and is given by the x-component of the non-equilibrium tension T. But, this is initially equal to T_∞. So, for the force balance along the x–axis, we also have

$$T(x' + \Delta x)\cos\beta = T(x')\cos\alpha = T_\infty \tag{2.34}$$

Again, T_∞ is the horizontal equilibrium tension *without* any perturbation in the string. It is induced solely by the forces from the endpoint constraints. [8]

Similarly, for the vertical forces, being the direction in which there is force *and* motion, we have

$$T(x' + \Delta x)\sin\beta - T(x')\sin\alpha = ma_y = \rho_x \Delta x \frac{\partial^2 u}{\partial t^2} \tag{2.35}$$

$\rho_x = m/\Delta x$ is the *linear* mass density. We can divide Eq. (2.35) by the equilibrium tension amplitude T_∞ so we have

$$\frac{T(x' + \Delta x)\sin\beta}{T_x(x' + \Delta x)} - \frac{T(x')\sin\alpha}{T_x(x' + \Delta x)} = \frac{\rho_x \Delta x}{T_\infty}\frac{\partial^2 u}{\partial t^2} \tag{2.36}$$

[8] A string can only support tension, and not compression.

or

$$\frac{T(x'+\Delta x)\sin\beta}{T(x'+\Delta x)\cos\beta} - \frac{T(x')\sin\alpha}{T(x')\cos\alpha} = \tan\beta - \tan\alpha = \frac{\rho_x \Delta x}{T_\infty}\frac{\partial^2 u}{\partial t^2} \tag{2.37}$$

Eq. (2.37) can be expressed in another form. To do this, note that the element under consideration forms a triangle whose horizontal length is Δx while the vertical height is $\Delta u = u(x'+\Delta x) - u(x')$. It follows that the local tangents become

$$\tan\beta = \frac{\sin\beta}{\cos\beta} = \frac{\Delta u}{\Delta x}(x'+\Delta x) \tag{2.38}$$

$$\tan\alpha = \frac{\sin\alpha}{\cos\alpha} = \frac{\Delta u}{\Delta x}(x') \tag{2.39}$$

Using this result, then dividing by Δx, we have

$$\frac{\tan\beta - \tan\alpha}{\Delta x} = \frac{\frac{\Delta u}{\Delta x}(x'+\Delta x) - \frac{\Delta u}{\Delta x}(x')}{\Delta x} = \frac{\rho_x}{T_\infty}\frac{\partial^2 u}{\partial t^2} \tag{2.40}$$

Taking the limit of Eq. (2.40) as $\Delta x \to 0$, we obtain an example of a classical 1D wave equation given by

1D Classical Wave Equation

$$\frac{\partial^2 u}{\partial x^2} = \frac{\rho_x}{T_\infty}\frac{\partial^2 u}{\partial t^2} \tag{2.41}$$

If we examine the physical units of the coefficient given by ρ_x/T_∞, we find that

$$\text{units of}\left(\frac{\rho_x}{T_\infty}\right) = \frac{kg}{m} \div \frac{kg-m}{s^2} = \frac{s^2}{m^2} \tag{2.42}$$

This unit is the inverse of the units of velocity squared or $1/v_x^2$. This velocity turns out to be the speed of the wave propagating along the string (x–axis).

It is a remarkable fact that the solution to Eq. (2.41) has a general form given by

1D Wave Solution Type I

$$u(x,t) = A \cdot f(x - v_x t) + B \cdot g(x + v_x t) \tag{2.43}$$

It is not even necessary to write an explicit function dependence on x nor t that satisfies the wave equation. Any shape or distribution $(f(x), g(x))$ function pair will still satisfy the wave equation as long as it can be expressed in the form of Eq. (2.43).

Let us check that Eq. (2.43) is the solution to the 1D wave equation of Eq. (2.42). Let $\beta_\pm = x \pm v_x t$ so that Eq. (2.43) becomes

$$u(\beta_-, \beta_+) = A \cdot f(\beta_-) + B \cdot g(\beta_+) \tag{2.44}$$

Evaluating the first derivative of u with respect to x, we get

$$\frac{\partial u}{\partial x} = A \cdot \frac{\partial f}{\partial \beta_-}\frac{\partial \beta_-}{\partial x} + B \cdot \frac{\partial g}{\partial \beta_+}\frac{\partial \beta_+}{\partial x} \tag{2.45a}$$

$$= A \cdot \frac{\partial f}{\partial \beta_-} + B \cdot \frac{\partial g}{\partial \beta_+} \tag{2.45b}$$

We have used the fact that $\partial\beta_\pm/\partial x = 1$. Using this result in the second derivative, we then have

$$\frac{\partial^2 u}{\partial x^2} = \frac{\partial}{\partial x}\frac{\partial u}{\partial x} \tag{2.46a}$$

$$= A \cdot \frac{\partial^2 f}{\partial \beta_-^2} + B \cdot \frac{\partial^2 g}{\partial \beta_+^2} \tag{2.46b}$$

This result corresponds to the LHS of Eq. (2.41). Now, let us get the first derivative with respect to time t, where we have

$$\frac{\partial u}{\partial t} = A \cdot \frac{\partial f}{\partial \beta_-}\frac{\partial \beta_-}{\partial t} + B \cdot \frac{\partial g}{\partial \beta_+}\frac{\partial \beta_+}{\partial t} \tag{2.47a}$$

$$= A \cdot \frac{\partial f}{\partial \beta_-}(-v_x) + B \cdot \frac{\partial g}{\partial \beta_+}(+v_x) \tag{2.47b}$$

$$= v_x\left(B \cdot \frac{\partial g}{\partial \beta_+} - A \cdot \frac{\partial f}{\partial \beta_-}\right) \tag{2.47c}$$

Taking the derivative again, taking v_x independent of time, and making use of the relations $\partial\beta_\pm/\partial t = \pm v_x$, we have

$$\frac{\partial^2 u}{\partial t^2} = \frac{\partial}{\partial t}\frac{\partial u}{\partial t} \tag{2.48a}$$

$$= v_x \frac{\partial}{\partial t}\left(B \cdot \frac{\partial g}{\partial \beta_+} - A \cdot \frac{\partial f}{\partial \beta_-}\right) \tag{2.48b}$$

$$= v_x \left(B \cdot \frac{\partial^2 g}{\partial \beta_+^2}\frac{\partial \beta_+}{\partial t} - A \cdot \frac{\partial^2 f}{\partial \beta_-^2}\frac{\partial \beta_-}{\partial t}\right) \tag{2.48c}$$

$$= v_x \left(B \cdot \frac{\partial^2 g}{\partial \beta_+^2}(v_x) - A \cdot \frac{\partial^2 f}{\partial \beta_-^2}(-v_x)\right) \tag{2.48d}$$

$$= v_x^2 \left(B \cdot \frac{\partial^2 g}{\partial \beta_+^2} + A \cdot \frac{\partial^2 f}{\partial \beta_-^2}\right) \tag{2.48e}$$

Multiplying by $1/v_x^2$ leads to Eqs. (2.46). Thus, we find that Eq. (2.43) *is* the solution to Eq. (2.41).

For our purposes, there is a more specific form of the solution given by Eq. (2.43). Let us now consider a solution of the form

$$u_2(x,t) = A \cdot f(ikx - i\omega t) + B \cdot g(-ikx + i\omega t) \tag{2.49a}$$

$$= A \cdot f(\beta_-) + B \cdot g(\beta_+) \tag{2.49b}$$

Taking the spatial derivative of Eq. (2.49a) leads to

$$\frac{\partial u_2}{\partial x} = Aik\frac{\partial f}{\partial \beta_-} - Bik\frac{\partial g}{\partial \beta_+} \tag{2.50a}$$

$$= ik\left(A\frac{\partial f}{\partial \beta_-} - B\frac{\partial g}{\partial \beta_+}\right) \tag{2.50b}$$

The second derivative then becomes

$$\frac{\partial}{\partial x}\frac{\partial u_2}{\partial x} = ik\frac{\partial}{\partial x}\left(A\frac{\partial f}{\partial \beta_-} - B\frac{\partial g}{\partial \beta_+}\right) \tag{2.51a}$$

$$= ik\left(Aik\frac{\partial^2 f}{\partial \beta_-^2} - B(-ik)\frac{\partial^2 g}{\partial \beta_+^2}\right) \tag{2.51b}$$

$$= -k^2\left(A\frac{\partial^2 f}{\partial \beta_-^2} + B\frac{\partial^2 g}{\partial \beta_+^2}\right) \tag{2.51c}$$

An analogous evaluation with respect to time t gives for the first derivative

$$\frac{\partial u_2}{\partial t} = -Ai\omega\frac{\partial f}{\partial \beta_-} + Bi\omega\frac{\partial g}{\partial \beta_+} \tag{2.52a}$$

$$= i\omega\left(-A\frac{\partial f}{\partial \beta_-} + B\frac{\partial g}{\partial \beta_+}\right) \tag{2.52b}$$

The second derivative with respect to time then becomes

$$\frac{\partial}{\partial t}\frac{\partial u_2}{\partial t} = i\omega\frac{\partial}{\partial t}\left(-A(-i\omega)\frac{\partial^2 f}{\partial \beta_-^2} + Bi\omega\frac{\partial^2 g}{\partial \beta_+^2}\right) \tag{2.53a}$$

$$= -\omega^2\left(A\frac{\partial^2 f}{\partial \beta_-^2} + B\frac{\partial^2 g}{\partial \beta_+^2}\right) \tag{2.53b}$$

By comparing Eq. (2.53) to Eq. (2.51a), it follows that

$$\frac{\partial^2 u_2}{\partial x^2} = -k^2\left(A\frac{\partial^2 u_2}{\partial \beta_-^2} + B\frac{\partial^2 u_2}{\partial \beta_+^2}\right) = \frac{-k^2}{-\omega^2}\frac{\partial^2 u_2}{\partial t^2} \tag{2.54a}$$

$$= \frac{k^2}{\omega^2}\frac{\partial^2 u_2}{\partial t^2} \tag{2.54b}$$

Therefore, u_2 is also a solution to the wave equation if and only if

$$v_x^2 = \frac{\omega^2}{k^2} \tag{2.55}$$

Note that there are no restrictions on k and ω. The relation holds if they are real or complex numbers. So, we have a *second type* of solution to the wave equation given by

Wave Equation Solution Type II

$$u_2(x,t) = A\cdot f(ikx - i\omega t) + B\cdot g(-ikx + i\omega t) \text{ if } v^2 = \frac{\omega^2}{k^2} \tag{2.56}$$

Extending the wave equation to higher dimensions is straightforward. If the displacement u is still constrained at its extremities, we can replace the linear string with a 2D sheet in the $x-y$-plane, for example. Then, Eq. (2.36) becomes

$$\frac{T(x'+\Delta x, y'+\Delta y)\sin\beta}{T_x(x'+\Delta x, y'+\Delta y)} - \frac{T(x',y')\sin\alpha}{T_x(x'+\Delta x, y'+\Delta y)} = \frac{\rho_a \Delta x \Delta y}{T_\infty}\frac{\partial^2 u}{\partial t^2} \tag{2.57}$$

Basically, the x–direction becomes an arbitrary direction of tension in the 2D plane and the mass generalizes to $m = \rho_A \Delta x \Delta y = \rho_A \Delta A$. Then, the 2D analysis leads to

2D Wave Equation

$$\frac{\partial^2 u}{\partial x^2} + \frac{\partial^2 u}{\partial y^2} = \frac{\rho_A}{T_\infty} \frac{\partial^2 u}{\partial t^2} \tag{2.58}$$

ρ_A is now the *areal* density instead of the linear density. Likewise, the 3D wave equation becomes

3D Wave Equation

$$\frac{\partial^2 u}{\partial x^2} + \frac{\partial^2 u}{\partial y^2} + \frac{\partial^2 u}{\partial z^2} = \frac{\rho_V}{T_\infty} \frac{\partial^2 u}{\partial t^2} \tag{2.59}$$

ρ_V is the volume density. From the 3D form, a compact representation using the Laplacian scalar operator is given by

$$\nabla^2 u = \frac{\rho_v}{T_\infty} \frac{\partial^2 u}{\partial t^2} = \frac{1}{v^2} \frac{\partial^2 u}{\partial t^2} \tag{2.60}$$

where v is the propagation speed of the wave as it propagates along the string. With this understanding, Eq. (2.60) is a familiar equation and so is it's solution. Note, however, that any homogeneous solution can be added to the above general wave solution to provide a *complete* solution to a specific problem.

2.2.2 Electromagnetic Waves From Maxwell Equations

Another area of physics where the wave equation appears is in electromagnetism, which is described by the Maxwell Equations. We can use two of the equations for electromagnetism, namely, Faraday's and Ampére-Maxwell's law. Both are given by

$$\nabla \times \mathbf{H} = \mathbf{J} + \frac{\partial \mathbf{D}}{\partial t} \qquad \text{(Maxwell-Ampére Law)} \tag{2.61}$$

$$\nabla \times \mathbf{E} = -\frac{\partial \mathbf{B}}{\partial t} \qquad \text{(Faraday's Law)} \tag{2.62}$$

H is the magnetic field; **J** is the electrical current density; **D** is known as the electric displacement field. To expose an explicit wave-like equation, let us assume the following relations between magnetic terms **H** and **B**, and electric terms **E** and **D** respectively, namely

$$\mathbf{D} = \epsilon \mathbf{E} \tag{2.63}$$

$$\mathbf{B} = \mu \mathbf{H} \tag{2.64}$$

ϵ is the *electrical permittivity*. In *vacuum*, it has a measured value in SI units, of

$$\epsilon_0 = 8.854187816 \times 10^{-12} \mathrm{C}^2 \cdot \mathrm{J}^{-1} \cdot \mathrm{m}^{-1} \tag{2.65}$$

C denotes units of Coulomb, being the SI unit of charge. J denotes Joules, the SI unit of energy, and m for meters, the SI unit of length. Similarly, *Faraday's law* applied in *vacuum* leads to

$$\nabla \times \mathbf{E} = -\mu_0 \frac{\partial \mathbf{H}}{\partial t} \tag{2.66}$$

μ_0 is the *magnetic permeability in vacuum* whose measured value is approximately

$$\mu_0 = 4 \times \pi \times 10^{-7} \mathrm{N} \cdot \mathrm{A}^{-2} \tag{2.67}$$

In Eq. (2.67), N is the SI force unit which is the Newton and A is the electrical current unit of Ampére. By taking the curl of Eq. (2.66), then substituting the Ampére-Maxwell's law on the right-hand side (RHS), we have

$$\nabla \times \nabla \times \mathbf{E} = -\epsilon_0 \mu_0 \frac{\partial^2 \mathbf{E}}{\partial t^2} \tag{2.68}$$

It can be shown that if the solution is of the form

$$u = A \cdot f(x - iv't) + B \cdot g(x + iv't) \tag{2.69}$$

then the solution will *not* have steady amplitude, but will generally vary in time. This provides the possibility of decaying amplitudes, which is a known behavior in electromagnetic waves propagating in a medium. However, there is also the limiting case of an electromagnetic wave propagating in *vacuum*. It is in this limiting case, that we see wave propagation without decay, etc. For the limiting case, let us express the electromagnetic equations using a well-known vector identity given by

$$\nabla \times \nabla \times \mathbf{E} = \nabla(\nabla \cdot \mathbf{E}) - \nabla^2 \mathbf{E} \tag{2.70}$$

For an electromagnetic wave in a space of *zero charge density*, it follows from Gauss's law (a static law of electromagnetism) that $\nabla \cdot \mathbf{E} = 0$. Thus, we have the result

$$\nabla^2 \mathbf{E} = \epsilon_0 \mu_0 \frac{\partial^2 \mathbf{E}}{\partial t^2} = \frac{1}{c^2} \frac{\partial^2 \mathbf{E}}{\partial t^2} \tag{2.71}$$

From our earlier discussion, $c = 1/\sqrt{\epsilon_0 \mu_0}$ then becomes the speed of an electromagnetic wave in a chargeless space *i.e.* vacuum. So, we find Eq. (2.71) has identical form to Eq. (2.60).

There is another macroscopic wave equation we can consider, involving heat.

2.2.3 Heat Conduction As A Wave Equation

Another example of a classical wave equation shows up in the thermodynamic process of heat conduction. Heat conduction is a well-known mode of heat transfer whenever there exists a difference in temperature across space, or length. It is described by combining Fourier's empirical law with the First Law of Thermodynamics to express a single spatio-temporal equation for the temperature T. Fourier's law is given by

$$q = -k \cdot \frac{\partial T}{\partial x} \qquad \text{Fourier's Emperical Law} \tag{2.72}$$

q is the rate of heat transfer *per unit area* and k is the conducting material's *heat transfer coefficient*. This relation says that heat flows *more* across larger temperature gradients, just like water flows more across larger pressure gradients. The negative sign ensures that a *positive* flow is from a higher temperature toward a lower.

The second equation we need follows from application of the first law of thermodynamics (an energy balance) to a *closed* differential element, which leads to an equation given by[9]

$$\nabla \cdot q = \rho c_h \cdot \frac{\partial T}{\partial t} \qquad \text{First Law of Thermodynamics} \tag{2.73}$$

[9] in thermodynamics, a *closed* system is a system where no mass flows in or out of the system.

ρ is the mass density of the conducting material in which the heat is propagating. c_h is called the *heat capacity* of the material.[10] Substituting Fourier's law for q leads to an equation given by

$$\frac{\partial}{\partial x}\left(-k\frac{\partial T}{\partial x}\right) = \rho c_h \cdot \frac{\partial T}{\partial t} \qquad \text{1D Heat Conduction Equation} \tag{2.74}$$

Also, taking k as uniform in space, we can write

$$\frac{\partial^2 T}{\partial x^2} = -\frac{\rho c_h}{k} \cdot \frac{\partial T}{\partial t} \tag{2.75}$$

Eq. (2.75) is starting to resemble the wave equation, however, *there is a difference*; Notice the time derivative on the RHS is *first* order, not second order as we have seen in the wave equation.

It turns out that we *are* looking at a wave equation, however, this fact is buried in an alternative form that appears for strictly one-directional waves.[11] To see this clearly, we can first solve Eq. (2.75) to obtain it's solution, then find out what the equivalent equation looks like in terms of the second order derivative. So, we must first solve Eq. (2.75).

Fortunately, it can be solved easily using a method known as the *method of separation of variables*. Using this method, we first assume a solution of function factors given by

$$T = T_x(x) \cdot T_t(t) \tag{2.76}$$

Substitution of Eq. (2.76) into Eq. (2.75) leads to

$$\frac{d^2 T_x}{dx^2} \cdot T_t = -\frac{\rho c_h}{k} \cdot T_x \cdot \frac{dT_t}{dt} \tag{2.77}$$

Diving both sides by the solution $T_x \cdot T_t$ gives

$$\cdot\frac{1}{T_x}\frac{d^2 T_x}{dx^2} = -\frac{\rho c_h}{k} \cdot \frac{1}{T_t}\frac{dT_t}{dt} \tag{2.78}$$

Eq. (2.78) suggests that a function of time t on the RHS equals to a function of x on the LHS. Because these are taken as independent variables, this equation can only be true if both sides equal a constant $1/\lambda^2$. Then, we have

$$\frac{dT_t}{dt} = -\frac{k}{\rho c_h \lambda^2} T_t(t) \tag{2.79a}$$

[10] a measure of the material's energy change per unit change in temperature

[11] Recall that heat flows only from high temperature to low, naturally.

$$\frac{d^2T_x}{dx^2} = \frac{1}{\lambda^2} \cdot T_x \tag{2.79b}$$

Let us write the solution for the spatial component, which is given by

$$T_x(x) = Ae^{i(x/\lambda)} + Be^{-i(x/\lambda)} \tag{2.80}$$

where λ has units of length. Substituting Eq. (2.80) into Eq. (2.79b), one can verify that it is a solution.

Then, for the time component, the solution becomes

$$T_t(t) = T_t(t=0) \cdot e^{-(k/\rho c_h \lambda^2)t} = T_0 \cdot e^{-t/\tau} \qquad \tau = \frac{\rho c_h \lambda^2}{k} \tag{2.81}$$

Substitution into Eq. (2.79a) confirms it is a solution. Then, the *total* solution for $T(x,t)$ becomes

$$T(x,t) = T_t(t=0) \cdot e^{-\frac{k}{\rho c_h \lambda^2} \cdot t} \left(Ae^{i(x/\lambda)} + Be^{-i(x/\lambda)}\right) \tag{2.82}$$

It follows from Eq. (2.81) that

$$\frac{dT_t}{dt} = T_t(t=0) \cdot \left(\frac{-k}{\rho c_h \lambda^2}\right) e^{-(k/\rho c_h \lambda^2)t} = \left(\frac{-k}{\rho c_h \lambda^2}\right) \cdot T_t(t) \tag{2.83}$$

Differentiating this relation again with respect to time t, we get

$$\frac{d^2T_t}{dt^2} = \left(\frac{-k}{\rho c_h \lambda^2}\right) \frac{dT_t}{dt} \Rightarrow \frac{dT_t}{dt} = -\left(\frac{\rho c_h \lambda^2}{k}\right) \cdot \frac{d^2T_t}{dt^2} = -\tau \frac{d^2T_t}{dt^2} \tag{2.84}$$

The relation given by Eq. (2.84) can be used to write *an explicit* wave equation for heat conduction. Substitution into Eq. (2.75) then gives

$$\frac{\partial^2 T}{\partial x^2} = -\frac{\rho c_h}{k} \cdot \frac{\partial T}{\partial t} = -\frac{\rho c_h}{k} \cdot T_x \cdot \frac{dT_t}{dt} \tag{2.85}$$

Substituting Eq. (2.84), we have

$$\frac{\partial^2 T}{\partial x^2} = \frac{\rho c_h}{k} \cdot T_x \cdot \tau \cdot \frac{d^2T_t}{dt^2} = \frac{\rho c_h \tau}{k} \frac{\partial^2 T}{\partial t^2} \tag{2.86a}$$

$$= \frac{\rho^2 c_h^2 \lambda^2}{k^2} \frac{\partial^2 T}{\partial t^2} \tag{2.86b}$$

This can be written as

$$\frac{\partial^2 T}{\partial x^2} = \frac{\tau^2}{\lambda^2} \frac{\partial^2 T}{\partial t^2} = \frac{1}{v^2} \frac{\partial^2 T}{\partial t^2} \tag{2.87}$$

From the above results, the *speed* of the heat-wave becomes

$$v = \frac{\lambda}{\tau} \tag{2.88}$$

or

Heat Wave Speed

$$v = \frac{k}{\rho c_h \lambda} \tag{2.89}$$

In practice, to measure characteristic heat wave speed, one must to characterize or measure the length scales λ over which the temperature varies. Once λ is known, with the additional material parameters, v can be determined.

The purpose of this exercise is to illustrate the pervasive nature of waves in physics. As we see in the case of heat conduction, they are even lurking in less obvious forms. Please note that we have only obtained this form to reveal the presence of an explicit wave equation, obtain the conduction heat-wave speed, *etc.*, however, this form with second order time derivatives should *never* be used in practice with heat conduction. Indeed, the conduction equation is appropriate.

We are beginning to notice that the wave equation is common in a variety of physical systems. Additionally, a compulsory characteristic of quantum mechanics is to demonstrate consistency with macroscopic systems. To enable this, we must discuss the important concept of quantum mechanical *operators*. From here, we can bridge the gap between the wide variety of wave equations found in macroscopic systems and the wave functions of quantum mechanics.

2.3 Momentum And Momentum Operators

One of the first operators of interest in quantum mechanics relates to momentum. It is a central topic in physics, so why not start with its quantum mechanical operator. We'll review its form in 3D, then go on to generalize it to four-dimensions.

From Eq. (2.15), we can see a resemblance in form to the solution for an electromagnetic wave described by the Maxwell equations. An *electric field* **E** plane-wave solution has a general form given by

$$\mathbf{E}(x, y, z, t) = \mathbf{E}_0 e^{i(kz - \omega t)} \tag{2.90}$$

In electromagnetics, one assumes ω to be real-valued. Consequently, k *generally* acquires an imaginary component to be able to correctly describe electromagnetic systems. For example, imaginary components in k are necessary to predict the phenomenon of *evanescent waves,* which involves a wave form at an interface. When one of the materials at the interface is metallic, the interfacial wave is known as a surface plasmon polariton (SPP).

In quantum mechanics, there is an important *postulated* equation known as the *de Broglie relation.* It describes the momentum p of a fundamential particle as

$$\underline{p = \frac{h}{\lambda}} = \frac{h}{2\pi} \cdot \frac{2\pi}{\lambda} = \hbar k \tag{2.91}$$

h is known as *Planck's constant* having an empirical value of

$$h = 6.6260693 \times 10^{-34} \quad \text{J} \cdot \text{s} \tag{2.92}$$

Given the form of the wave function ψ, we can use the de Broglie relation to introduce h into ψ. If the wave function is given by

$$\psi(x_1, x_2, x_3) = \psi_0 e^{i\mathbf{k}\cdot\mathbf{r}} = \psi_0 e^{i(k_1 x_1 + k_2 x_2 + k_3 x_3)} \tag{2.93}$$

then, using the de Broglie relation, we have

$$\psi(x_1, x_2, x_3) = \psi_0 e^{i\mathbf{k}\cdot\mathbf{r}} = \psi_0 e^{\frac{i}{\hbar}\mathbf{p}\cdot\mathbf{r}} = \psi_0 e^{\frac{i}{\hbar}\left(p_x x + p_y y + p_z z\right)} \tag{2.94}$$

In a quantum mechanical wave, the coefficient ψ_0 *can only depend on coordinates perpendicular to the motion.* To reflect this, let us assume motion along x_3, so $\psi_0 = \psi_0(x_1, x_2)$, and we have

$$\psi(x_1, x_2, x_3) = \psi(x, y, z) = \psi_0(x, y) e^{\frac{i}{\hbar}(p_z z)} \tag{2.95}$$

In this form, we can begin to see that it is possible to apply certain mathematical operations to Eq. (2.95) to extract momentum p_z information. For example, taking the derivative with respect to z gives

$$\frac{\partial \psi}{\partial z} = \frac{i}{\hbar} p_z \psi_0(x, y) e^{\frac{i}{\hbar}(p_z z)} = \frac{i}{\hbar} p_z \psi \tag{2.96}$$

or

$$-i\hbar \frac{\partial \psi}{\partial z} = p_z \psi \tag{2.97}$$

Eq. (2.97) says that if we take the derivative of the wave function along the direction of motion, then multiply it by $i\hbar$, then we can get the scalar quantity $p_z\psi$. Eq. (2.97) is an example of an *eigenoperator equation*, which is a type of eigenvalue equation. Consequently, the solution to Eq. (2.97) is the called the *eigenfunction* of the quantum mechanical momentum.

If the momentum **p** is taken generally along an arbitrary 3D direction $\mathbf{r} = [x, y, z]^T$, then there are components along each direction described by *eigenoperator* equations given by

$$-i\hbar\frac{\partial\psi}{\partial x} = p_x\psi \quad , \quad -i\hbar\frac{\partial\psi}{\partial y} = p_y\psi \quad , \quad -i\hbar\frac{\partial\psi}{\partial z} = p_z\psi \tag{2.98}$$

The above can be expressed for each component, generally, as

$$-i\hbar\frac{\partial\psi}{\partial x_i} = p_i\psi \tag{2.99}$$

Using Eq. (2.99), we can then express the *fourth* component of the momentum as

$$-i\hbar\frac{\partial\psi}{\partial x_i} = -i\hbar\frac{\partial\psi}{\partial(ict)} = -\frac{\hbar}{c}\frac{\partial\psi}{\partial t} = p_4\psi \tag{2.100}$$

Adding the fourth dimension in this way to the wave function ψ then gives

$$\psi = \psi_0 e^{i(\mathbf{k}\cdot\mathbf{S})} = \psi_0 e^{\frac{i}{\hbar}(\mathbf{p}\cdot\mathbf{S})} = \psi_0 e^{\frac{i}{\hbar}(p_1x_1+p_2x_2+p_3x_3+p_4x_4)} \tag{2.101}$$

k_1 is the component of the wave vector along the direction of $x = x_1$, or the first dimension. Likewise, k_2 is the component of the wave vector along the direction of $y = x_2$, or the second dimension; and k_3 is the component of the wave vector along the direction of $z = x_3$, or the third dimension; k_4 is the corresponding component of the wave vector along the fourth dimension. **S** is the four-dimensional position vector $\mathbf{S} = [x_1, x_2, x_3, x_4]^T$.

With the fourth dimension, the momentum operators can be written generally as

$$\widehat{\mathbf{p}} = -i\hbar\begin{bmatrix}\frac{\partial}{\partial x_1} & \frac{\partial}{\partial x_2} & \frac{\partial}{\partial x_3} & \frac{\partial}{\partial x_4}\end{bmatrix}^T \tag{2.102}$$

Substituting the fourth dimension $x_4 = ict$, we have the *4D momentum vector operator* given by

4D Momentum Eigenoperator

$$\widehat{\mathbf{p}} = -i\hbar \begin{bmatrix} \frac{\partial}{\partial x} & \frac{\partial}{\partial y} & \frac{\partial}{\partial z} & \frac{1}{ic}\frac{\partial}{\partial t} \end{bmatrix}^T \tag{2.103}$$

The 4D momentum can be used to find an additional contribution to the macroscopic relativistic kinetic energy arising from including the fourth dimension.

Let's now look at energy operators and how momentum can be used with an energy operator.

2.4 Kinetic Energy Eigen-operator Equation

When we see the operator, $\widehat{p}^2$, the interpretation is to apply the operator to the wave function *twice*. With the addition of the 4th dimension, everything is *unchanged* for the first three dimensions. There is just an additional term which involves time t. As we have seen, the effective result is to introduce an additional time dependence into the eigenoperator equation. Now, let us return to the relativistic kinetic energy and *unpack* the contributions that are present. Earlier, we found that the *relativistic kinetic energy* K to be

$$K = \frac{\gamma_r}{2} \cdot \sqrt{p^2c^2 + m_0^2c^4} \tag{2.104}$$

Eq. (2.104) was found by solving

$$\Delta K = \int \mathbf{F} \cdot d\mathbf{r} = \int \frac{d\mathbf{p}}{dt*} \cdot d\mathbf{r} = \int \gamma_r \mathbf{v} \cdot d\mathbf{p} \tag{2.105}$$

Let us revisit the derivation now including the the fourth dimension. For this, we must *evaluate a 4D integral* where we have

$$\Delta K = \int \frac{d\mathbf{p}}{dt*} \cdot d\mathbf{r} + \int \frac{dp_4}{dt*} dx_4 \tag{2.106}$$

The first integral on the RHS contains the contributions from the first three dimensions. In Sec. 1.5.1, it is shown that Eq. (1.68) can also be written as Eq. (2.104).

Let us evaluate the additional contribution from the fourth dimension, denoted by K_4. This gives

$$\Delta K_4 = \int \frac{dp_4}{dt*} \cdot dx_4 = \int \gamma_r \frac{dp_4}{dt} \cdot d(ict) \tag{2.107a}$$

$$= ic \int \gamma_r \frac{dp_4}{dt} \cdot dt \tag{2.107b}$$

$$= ic \int \gamma_r dp_4 \tag{2.107c}$$

$$= \underline{ic\gamma_r \Delta p_4} \tag{2.107d}$$

In Eq. (2.107), γ_r has been pulled out of the integral because γ_r *only depends on the first three dimensions* and their time correlations. Thus, the additional contribution from the fourth dimension is given by

4th Dimension Kinetic Energy Contribution

$$K_4 = ic\gamma_r p_4 \tag{2.108}$$

Eq. (2.108) and Eq. (2.99) can be used to find the quantum mechanical operator for this contribution, as

$$\widehat{K_4} = ic\widehat{p_4} = ic\left(-i\hbar\frac{\partial}{\partial x_4}\right) = ic\left(-i\hbar\frac{\partial}{ic\partial t}\right) \tag{2.109a}$$

$$= -i\hbar\frac{\partial}{\partial t} \tag{2.109b}$$

The operator for the fourth dimension's kinetic energy contribution is then given by

4th Dimension Kinetic Energy Operator

$$\widehat{K_4} = -i\hbar\gamma_r\frac{\partial}{\partial t} \tag{2.110}$$

Eq. (2.110) may look familiar if you have dealt with energy operators in quantum mechanics. It is usually defined as the *total energy operator*.

However, the 4D energy operator contains both real and imaginary energy contributions. It can be expressed as

Total Energy Operator

$$i\hbar\frac{\partial}{\partial t} = \widehat{E}_R - \widehat{E}_I \tag{2.111}$$

$\widehat{E}_R$ is the quantum mechanical operator that yields the real component of the energy E_R, from the wave function ψ; $\widehat{E}_I$ yields the imaginary component of energy E_I. The use of the minus sign with the imaginary component in Eq. (2.111) ensures we are consistent with a wave function of the form $\psi \propto e^{-i(kx-\omega t)}$.

The corresponding total energy eigenoperator equation is given by

$$i\hbar\frac{\partial\psi}{\partial t} = \left[\widehat{E}_R - \widehat{E}_I\right]\psi = (E_R - iE_I)\psi \tag{2.112}$$

In addition to the kinetic energy, there may also be a potential energy V. Then, the *total relativistic energy* $E = K + V$ is given by

4D Relativistic Kinetic Energy

$$E = \frac{\gamma_r}{2}\sqrt{p^2c^2 + m_0^2c^4} + ic\gamma_r p_4 + V \tag{2.113}$$

Eq. (2.113) can be used to find a corresponding general quantum mechanical energy operator useful in describing electrons having both charge and spin. With the introduction of spin, we must allow for occupation of either a single quantum state or the superposition of multiple quantum states. We'll focus on the simplest case of 2 states.

In the next section, we discuss how the scalar energy can be expressed in its equivalent multi-quantum state form to include descriptions of superposition.

2.4.1 Including Superposition

In order to include wave-like properties like *superposition*, we do not have to reinvent the wheel. This step was introduced by Dirac in working out his relativistic quantum mechanical operator. Likewise, we can introduce it by using higher order state vectors to describe the system energy in an eigenoperator equation.[12] To begin, let us find the solutions for matrices σ_1, σ_2 satisfying the following kinetic energy equation:

$$E = \frac{1}{2}\sqrt{p^2c^2 + m_0^2c^4} = \sigma_1 pc + \sigma_2 m_0 c^2 \tag{2.114}$$

As a single equation with two unknowns, it follows that σ_1, σ_2 must be matrices of atleast size 2×2 because no solution exists if they are 1×1 scalars. It is this subtle step that injects the *superpositional potential* into the system. Squaring both sides of Eq. (2.114), we have

$$\frac{1}{4}\left(p^2c^2 + m_0^2c^4\right) = \left(\sigma_1 pc + \sigma_2 m_0 c^2\right)\left(\sigma_1 pc + \sigma_2 m_0 c^2\right) \tag{2.115a}$$

$$= \sigma_1^2 p^2 c^2 + \sigma_1\sigma_2 pcm_0c^2 + \sigma_2\sigma_1 pcm_0c^2 + \sigma_2^2 m_0^2 c^4 \tag{2.115b}$$

$$= \sigma_1^2 p^2 c^2 + \sigma_2^2 m_0^2 c^4 + (\sigma_1\sigma_2 + \sigma_2\sigma_1) pcm_0c^2 \tag{2.115c}$$

Equating left and right hand sides of Eq. (2.115), we have σ_1 and σ_2 must satisfy the following conditions:

$$(\sigma_1)^2 = (\sigma_2)^2 = \frac{1}{4} \tag{2.116a}$$

$$\sigma_1\sigma_2 + \sigma_2\sigma_1 = 0 \tag{2.116b}$$

The conditions above are satisfied minimally by a well-known set of 2×2 matrices known as *spin matrices*. Note, however, that spin matrices are also useful in problems excluding electron spin. Also, we have an additional factor of 1/2 owing to the correction to the relativistic kinetic energy. Therefore, the spin matrices in our case, satisfying the conditions of Eq. (2.116) are given by

[12] This essentially means to wave function becomes a vector of wave functions *i.e.* of order higher than 1.

Principle Spin Matrices

$$\widehat{\sigma_x} = \begin{bmatrix} 0 & \frac{1}{2} \\ \frac{1}{2} & 0 \end{bmatrix} \quad \widehat{\sigma_y} = \begin{bmatrix} 0 & -\frac{i}{2} \\ \frac{i}{2} & 0 \end{bmatrix} \quad \widehat{\sigma_z} = \begin{bmatrix} \frac{1}{2} & 0 \\ 0 & -\frac{1}{2} \end{bmatrix} \tag{2.117}$$

The eigenvalues for each spin matrix are $\lambda_\sigma = \pm 1/2$. Since we are interested in describing a *four-dimensional space*, it is natural to choose matrices as 4×4 in size, as Dirac did in contriving the Dirac equation. These larger matrices can depend on the 2×2 spin matrices of Eq. (2.117) so they still satisfy the conditions of Eq. (2.116). As is known, Dirac's choices of 4×4 matrices *do* satisfy the conditions of Eq. (2.116).[13] There are three 4×4 matrices given in terms of the 2×2 principle spin matrices σ_i. The x-component is defined as

$$\widehat{\Sigma}_x = \begin{bmatrix} 0 & \widehat{\sigma_x} \\ \widehat{\sigma_x} & 0 \end{bmatrix} = \begin{bmatrix} 0 & 0 & 0 & \frac{1}{2} \\ 0 & 0 & \frac{1}{2} & 0 \\ 0 & \frac{1}{2} & 0 & 0 \\ \frac{1}{2} & 0 & 0 & 0 \end{bmatrix} \tag{2.118a}$$

The y-component is defined as

$$\widehat{\Sigma}_y = \begin{bmatrix} 0 & \widehat{\sigma_y} \\ \widehat{\sigma_y} & 0 \end{bmatrix} = \begin{bmatrix} 0 & 0 & 0 & \frac{i}{2} \\ 0 & 0 & -\frac{i}{2} & 0 \\ 0 & \frac{i}{2} & 0 & 0 \\ -\frac{i}{2} & 0 & 0 & 0 \end{bmatrix} \tag{2.118b}$$

And the z-component $\widehat{\Sigma}_z$ is defined as

$$\widehat{\Sigma}_z = \begin{bmatrix} 0 & \widehat{\sigma_z} \\ \widehat{\sigma_z} & 0 \end{bmatrix} = \begin{bmatrix} 0 & 0 & \frac{1}{2} & 0 \\ 0 & 0 & 0 & -\frac{1}{2} \\ \frac{1}{2} & 0 & 0 & 0 \\ 0 & -\frac{1}{2} & 0 & 0 \end{bmatrix} \tag{2.118c}$$

The fourth 4×4 matrix $\widehat{\Sigma}_0$ is defined as

$$\widehat{\Sigma}_0 = \begin{bmatrix} \mathbf{I}(2) & 0 \\ 0 & -\mathbf{I}(2) \end{bmatrix} = \begin{bmatrix} \frac{1}{2} & 0 & 0 & 0 \\ 0 & \frac{1}{2} & 0 & 0 \\ 0 & 0 & -\frac{1}{2} & 0 \\ 0 & 0 & 0 & -\frac{1}{2} \end{bmatrix} \tag{2.118d}$$

[13] However, Dirac equations had 1 instead of 1/4.

Substituting $\sigma_i \rightarrow \Sigma_i$ in the linear higher order form of the kinetic energy, while replacing the fourth dimension contribution by its operator given in Eq. (2.110), we can write a 4D quantum mechanics energy operator. Recall the definition for the total energy operator given by

$$i\hbar\frac{\partial}{\partial t}\psi = \left[\widehat{E}_R + \widehat{E}_I\right]\psi \tag{2.119}$$

If all the operators we have considered contribute to $\widehat{E}_R$, then we can write a *total 4D energy operator* for Eq. (2.113) as

4D Total Energy Operator

$$\left[\mathbf{I}(4)\cdot\left(-i\hbar\frac{\partial}{\partial t}\right) + c\cdot\widehat{\Sigma}\cdot(\mathbf{p}+e\mathbf{A}) + \widehat{\Sigma}_0 m_0 c^2 + \mathbf{I}(4)V\right]\psi = iE_I\psi \tag{2.120}$$

where $\Sigma = [\Sigma_x \quad \Sigma_y \quad \Sigma_z]^T$. In general, it is a non-Hermitian Hamiltonian. The 4D energy operator given by Eq. (2.120) ultimately contains a description of dynamical transitions between quantum states. This will be evident when we begin to apply the 4D operator along with the Eq. (2.119).

We are also ready to further unpack the relativistic kinetic energy operator. Then, we'll consider some explicit operators therein to apply to problems, *etc.*

2.5 Contributions To Relativistic Energy Operator

The richness of the relativistic energy may not be obvious in initial appearance. However, we can dig into the relativistic energy using mathematical tools to reveal more plainly just what kind of energy contributions are inside. Recall the form of the *3D* relativistic energy (ignoring the fourth dimension for now) given by

$$\left[\mathbf{I}(4)V + \frac{c\gamma_r}{2}\widehat{\Sigma}'\cdot(\widehat{\mathbf{p}}+e\mathbf{A}) + \widehat{\Sigma}_0'\frac{m_0\gamma_r c^2}{2}\right]\psi = \overline{E}\psi \tag{2.121}$$

The quantum state eigenvector ψ is a 4×1 vector containing eigenfunctions ψ_i, or

$$\psi = \begin{bmatrix} \psi_1 \\ \psi_2 \\ \psi_3 \\ \psi_4 \end{bmatrix}$$

We have pulled out the factor of 1/2 from $\widehat{\Sigma}$ so that $\widehat{\Sigma}' = 2 \cdot \widehat{\Sigma}$. The reason for this is just for convenience in allowing us to readily see how things scale in comparison to the Dirac form. Then, Eq. (2.121) can be rearranged to give

$$\left[(E - V)\mathbf{I}(4) - \frac{c\gamma_r}{2}\widehat{\Sigma}' \cdot (\widehat{\mathbf{p}} + e\mathbf{A}) - \widehat{\Sigma}_0' \frac{m_0 \gamma_r c^2}{2} \right] \psi = 0 \tag{2.122}$$

Substituting the Σ' 4×4 matrices leads to two sets of equations for the first pair (ψ_1, ψ_2) and second pair of wave functions (ψ_3, ψ_4), respectively, where we have

$$\left(E - V - \frac{m_0 \gamma_r c^2}{2} \right) \begin{bmatrix} \psi_1 \\ \psi_2 \end{bmatrix} - \frac{c\gamma_r}{2} \widehat{\boldsymbol{\sigma}}' \cdot (\widehat{\mathbf{p}} + e\mathbf{A}) \begin{bmatrix} \psi_3 \\ \psi_4 \end{bmatrix} = 0 \tag{2.123a}$$

$$-\frac{c\gamma_r}{2} \widehat{\boldsymbol{\sigma}}' \cdot (\widehat{\mathbf{p}} + e\mathbf{A}) \begin{bmatrix} \psi_1 \\ \psi_2 \end{bmatrix} + \left(E - V + \frac{m_0 \gamma_r c^2}{2} \right) \begin{bmatrix} \psi_3 \\ \psi_4 \end{bmatrix} = 0 \tag{2.123b}$$

In Eqs. (2.123a) and (2.123b), note that $\widehat{\boldsymbol{\sigma}}'$ is the vector of 2×2 principle spin matrices having eigenvalues of 1. Eq. (2.123b) can be used to solve for ψ_3 and ψ_4 where we have

$$\begin{bmatrix} \psi_3 \\ \psi_4 \end{bmatrix} = \frac{c\gamma_r}{2} \cdot \frac{1}{E - V + \frac{\gamma_r m_0 c^2}{2}} \widehat{\boldsymbol{\sigma}}' \cdot (\widehat{\mathbf{p}} + e\mathbf{A}) \begin{bmatrix} \psi_1 \\ \psi_2 \end{bmatrix}$$

Substitution of this result into Eq. (2.123a) leads to

$$\left(E - V - \frac{m_0 \gamma_r c^2}{2} \right) \begin{bmatrix} \psi_1 \\ \psi_2 \end{bmatrix} - \cdots$$

$$\widehat{\boldsymbol{\sigma}}' \cdot (\widehat{\mathbf{p}} + e\mathbf{A}) \frac{\gamma_r^2 c^2/4}{E - V + \frac{m_0 \gamma_r c^2}{2}} \widehat{\boldsymbol{\sigma}}' \cdot (\widehat{\mathbf{p}} + e\mathbf{A}) \begin{bmatrix} \psi_1 \\ \psi_2 \end{bmatrix} = 0 \tag{2.124}$$

Now, let $\Gamma = m_0 \gamma_r c^2/2$. This *almost* leads to an eigenvalue equation for the *shifted* energy $E_S = E - \Gamma$ since Eq. (2.124) may be written as follows:

$$V\begin{bmatrix}\psi_1\\ \psi_2\end{bmatrix}+[\widehat{\sigma}'\cdot(\widehat{\mathbf{p}}+e\mathbf{A})]\frac{\gamma_r^2c^2/4}{E-V+\Gamma}[\widehat{\sigma}'\cdot(\widehat{\mathbf{p}}+e\mathbf{A})]\begin{bmatrix}\psi_1\\ \psi_2\end{bmatrix}$$
$$=(E-\Gamma)\begin{bmatrix}\psi_1\\ \psi_2\end{bmatrix}=\overline{E}_S\begin{bmatrix}\psi_1\\ \psi_2\end{bmatrix}\quad(2.125)$$

The sandwiched scalar ratio term on the LHS complicates things because it depends on the potential energy V as well as the energy E. So, to move forward, an approximation can be made using a *generalized binomial series expansion*. The binomial expansion of the function $f(x)=(1+x)^k$ is given by

$$(1+x)^k=1+kx+\frac{k(k-1)}{2!}x^2+\frac{k(k-1)(k-2)}{3!}x^3+\cdots\frac{k(k-1)\cdots(k-n+1)}{n!}x^n+\cdots$$

To use this series expansion, we can write the sandwiched term in the form $(1+x)^k$ by rearranging as follows:

$$\frac{\gamma_r^2c^2/4}{E-V+\Gamma}=\frac{\gamma_r^2c^2}{4\Gamma}\frac{1}{1+\frac{E-V}{\Gamma}}$$
$$=\frac{\gamma_r}{2m_0}\left[1+\frac{E-V}{\Gamma}\right]^{-1}$$

We have used the fact that $4\Gamma=4\times\gamma_r m_0c^2/2=2m_0\gamma_r c^2$. Applying the binomial expansion, keeping only the first two terms with $x=(E-V)/\Gamma$ and $k=-1$, we have

$$\frac{\gamma_r}{2m_0}\left[1+\frac{E-V}{\Gamma}\right]^{-1}\approx\frac{\gamma_r}{2m_0}-\frac{\gamma_r(E-V)}{2m_0\Gamma}$$

Substitution of this approximation into Eq. (2.125), then using the relation $\gamma_r/\Gamma=2/m_0c^2$ gives

$$(E_S-V)\begin{bmatrix}\psi_1\\ \psi_2\end{bmatrix}\approx\widehat{\sigma}'\cdot(\widehat{\mathbf{p}}+e\mathbf{A})\gamma_r\left[\frac{1}{2m_0}-\frac{E-V}{m_0^2c^2}\right]\widehat{\sigma}'\cdot(\widehat{\mathbf{p}}+e\mathbf{A})\begin{bmatrix}\psi_1\\ \psi_2\end{bmatrix}$$
$$=\frac{\gamma_r[\widehat{\sigma}'\cdot(\widehat{\mathbf{p}}+e\mathbf{A})]^2}{2m_0}-\widehat{\sigma}'\cdot(\widehat{\mathbf{p}}+e\mathbf{A})\gamma_r\left[\frac{E-V}{m_0^2c^2}\right]\widehat{\sigma}'\cdot(\widehat{\mathbf{p}}+e\mathbf{A})\begin{bmatrix}\psi_1\\ \psi_2\end{bmatrix}$$

The term $[\widehat{\sigma}'\cdot(\widehat{\mathbf{p}}+e\mathbf{A})]^2$ can be replaced using an identity involving the vector of principle spin matrices $\widehat{\sigma}'$ given by

$$[\widehat{\sigma}'\cdot(\widehat{\mathbf{p}}+e\mathbf{A})]^2=(\widehat{\mathbf{p}}+e\mathbf{A})^2+e\hbar\widehat{\sigma}'\cdot\mathbf{B}\quad(2.126)$$

where $\mathbf{B}$ is the magnetic field related to the magnetic vector potential $\mathbf{A}$ via $\mathbf{B}=\nabla\times\mathbf{A}$. The appearance of γ_r may appear to complicate things, however, this factor can be lumped into the wave function ψ.[14] Thus, the lowest order expansion for Eq. (2.126) gives

$$E_S\begin{bmatrix}\psi_1\\ \psi_2\end{bmatrix}=\left[V+\frac{(\widehat{\mathbf{p}}+e\mathbf{A})^2}{2m_0}+\left(\frac{e\hbar}{2m_0}\right)\widehat{\boldsymbol{\sigma}}'\cdot\mathbf{B}-\widehat{\boldsymbol{\sigma}}'\cdot(\widehat{\mathbf{p}}+e\mathbf{A})\left[\frac{E-V}{m_0^2c^2}\right]\widehat{\boldsymbol{\sigma}}'\cdot(\widehat{\mathbf{p}}+e\mathbf{A})\right]\begin{bmatrix}\psi_1\\ \psi_2\end{bmatrix} \tag{2.127}$$

We can recognize some of the terms in Eq. (2.127). Particularly, the second and third terms on the RHS are the *non-relativistic kinetic energy* and a magnetic energy contribution known as the *anomalous Zeeman energy*.

Kinetic Energy+Anomalous Zeeman Energy

$$\widehat{E}_K+\widehat{E}_Z=\frac{(\widehat{\mathbf{p}}+e\mathbf{A})^2}{2m_0}+\left(\frac{e\hbar}{2m_0}\right)\widehat{\boldsymbol{\sigma}}'\cdot\mathbf{B} \tag{2.128}$$

To add to these energy contributions, let's turn our attention back to the last term of Eq. (2.127). It is unique to relativistic quantum mechanics. To pry into this term, we can do some mathematical rearrangements using the *commutation relations* of these operators. If this concept is new to the reader, you can bookmark here, then go check out the chapter appendix Sec. B.11 on this topic, then come back.

To obtain some of the terms of interest, let us carry out some manipulation involving operators, as follows:

$$(E-V)\widehat{\boldsymbol{\sigma}}'\cdot(\widehat{\mathbf{p}}+e\mathbf{A})=\widehat{\boldsymbol{\sigma}}'\cdot(\widehat{\mathbf{p}}+e\mathbf{A})(E-V)+[E-V,\widehat{\boldsymbol{\sigma}}'\cdot(\widehat{\mathbf{p}}+e\mathbf{A})]$$

We have used the commutation operator definition where $[\widehat{A},\widehat{B}]=\widehat{AB}-\widehat{BA}$, or $\widehat{AB}=\widehat{BA}+[\widehat{A},\widehat{B}]$. In the first term, $E-V$ is the kinetic energy. Generally, there are many terms included in it. One of the largest contributions is given by $\widehat{\mathbf{P}}^2/2m_0$. For this contribution, we have

$$(E-V)\widehat{\boldsymbol{\sigma}}'\cdot(\widehat{\mathbf{p}}+e\mathbf{A})=\widehat{\boldsymbol{\sigma}}'\cdot(\widehat{\mathbf{p}}+e\mathbf{A})\frac{(\widehat{\mathbf{p}}+e\mathbf{A})^2}{2m_0}+[E-V,\widehat{\boldsymbol{\sigma}}\cdot(\widehat{\mathbf{p}}+e\mathbf{A})]$$

For the second term, we can use the fact that $\widehat{\boldsymbol{\sigma}}'$ is a vector of constant 2×2 matrices, so they commute with $\widehat{\mathbf{P}}$. Then, we can write the following

[14]This point is discussed further in Sec. 2.7.

relations:

$$\begin{aligned}[E - V, \widehat{\sigma}' \cdot (\widehat{\mathbf{p}} + e\mathbf{A})] &= (E - V)\widehat{\sigma}' \cdot (\widehat{\mathbf{p}} + e\mathbf{A}) - \widehat{\sigma}' \cdot (\widehat{\mathbf{p}} + e\mathbf{A})(E - V) \\ &= (E - V)(\widehat{\mathbf{p}} + e\mathbf{A}) \cdot \widehat{\sigma}' - \widehat{\sigma} \cdot (\widehat{\mathbf{p}} + e\mathbf{A})(E - V) \\ &= E(\widehat{\mathbf{p}} + e\mathbf{A}) \cdot \widehat{\sigma}' - \widehat{\sigma}' \cdot (\widehat{\mathbf{p}} + e\mathbf{A})E - V(\widehat{\mathbf{p}} + e\mathbf{A}) \cdot \widehat{\sigma}' + \widehat{\sigma}' \cdot (\widehat{\mathbf{p}} + e\mathbf{A})V \\ &= \widehat{\sigma}' \cdot (\widehat{\mathbf{p}} + e\mathbf{A})V - V(\widehat{\mathbf{p}} + e\mathbf{A}) \cdot \widehat{\sigma}' \\ &= \widehat{\sigma}' \cdot [\widehat{\mathbf{p}} + e\mathbf{A}, V]\end{aligned}$$

We have used the fact that E is a scalar constant. Using the kinetic energy operator along with this result, we obtain

$$(E - V)\widehat{\sigma}' \cdot (\widehat{\mathbf{p}} + e\mathbf{A}) = \widehat{\sigma}' \cdot (\widehat{\mathbf{p}} + e\mathbf{A})\frac{(\widehat{\mathbf{p}} + e\mathbf{A})^2}{2m_0} + \widehat{\sigma}' \cdot [\widehat{\mathbf{p}} + e\mathbf{A}, V]$$

Bringing this back into our Hamiltonian term of interest leads to

$$-\widehat{\sigma}' \cdot (\widehat{\mathbf{p}} + e\mathbf{A})\left[\frac{E - V}{m_0^2 c^2}\right]\widehat{\sigma}' \cdot (\widehat{\mathbf{p}} + e\mathbf{A}) \tag{2.129a}$$

$$= -\frac{\widehat{\sigma}' \cdot (\widehat{\mathbf{p}} + e\mathbf{A})}{m_0^2 c^2}\left[\widehat{\sigma}' \cdot (\widehat{\mathbf{p}} + e\mathbf{A})\frac{(\widehat{\mathbf{p}} + e\mathbf{A})^2}{2m_0} + \widehat{\sigma}' \cdot [\widehat{\mathbf{p}} + e\mathbf{A}, V]\right] \tag{2.129b}$$

$$= -[\widehat{\sigma} \cdot (\widehat{\mathbf{p}} + e\mathbf{A})]^2 \frac{(\widehat{\mathbf{p}} + e\mathbf{A})^2}{2m_0^3 c^2} - \frac{\widehat{\sigma}' \cdot (\widehat{\mathbf{p}} + e\mathbf{A})\widehat{\sigma}' \cdot [\widehat{\mathbf{p}} + e\mathbf{A}, V]}{m_0^2 c^2} \tag{2.129c}$$

We can make use of another identity for the spin operators, given by

$$(\mathbf{a} \cdot \widehat{\sigma}')(\mathbf{b} \cdot \widehat{\sigma}') = (\mathbf{a} \cdot \mathbf{b}) + i(\mathbf{a} \times \mathbf{b}) \cdot \widehat{\sigma}' \tag{2.130}$$

In applying this, we take from the first term $\mathbf{a} = \mathbf{b} = \widehat{\mathbf{p}} + e\mathbf{A}$, while from the second term $\mathbf{a} = \widehat{\mathbf{p}} + e\mathbf{A}$ but $\mathbf{b} = [\widehat{\mathbf{p}} + e\mathbf{A}, V]$. Applying this to both terms in the last line of Eq. (2.129), we then have

$$-\widehat{\sigma}' \cdot (\widehat{\mathbf{p}} + e\mathbf{A})\left[\frac{E - V}{m_0^2 c^2}\right]\widehat{\sigma}' \cdot (\widehat{\mathbf{p}} + e\mathbf{A})$$

$$= -\frac{(\widehat{\mathbf{p}} + e\mathbf{A})^4}{2m_0^3 c^2} - \frac{(\widehat{\mathbf{p}} + e\mathbf{A}) \cdot [\widehat{\mathbf{p}} + e\mathbf{A}, V]}{m_0^2 c^2} - i\frac{\widehat{\sigma}' \cdot (\widehat{\mathbf{p}} + e\mathbf{A}) \times [\widehat{\mathbf{p}} + e\mathbf{A}, V]}{m_0^2 c^2}$$

Then, using the commutation relation $[\widehat{\mathbf{p}} + e\mathbf{A}, V] = -i\hbar\nabla V$ (c.Eq. B.195), we have

$$-\widehat{\sigma}' \cdot (\widehat{\mathbf{p}} + e\mathbf{A}) \left[\frac{E - V}{m_0^2 c^2} \right] \widehat{\sigma}' \cdot (\widehat{\mathbf{p}} + e\mathbf{A})$$

$$= -\frac{(\widehat{\mathbf{p}} + e\mathbf{A})^4}{2m_0^3 c^2} + \frac{i\hbar}{m_0^2 c^2}(\widehat{\mathbf{p}} + e\mathbf{A}) \cdot \nabla V - \frac{\hbar}{m_0^2 c^2} \widehat{\sigma}' \cdot (\widehat{\mathbf{p}} + e\mathbf{A}) \times \nabla V \tag{2.131}$$

The first term is a higher order correction to the kinetic energy (*e.g.* Eq. 2.127). We'll have more to say about this term in Chap. 7, which is a discussion of *superconductivity*. The second term is new, and of the form $\mathbf{P} \cdot \mathbf{E}$, having a *complex coefficient*.[15] However, it is the last term that we are after, which gives

General Spin-Orbit Interaction

$$\widehat{H}_{SO} = -\frac{\hbar}{m_0^2 c^2} \widehat{\sigma}' \cdot (\widehat{\mathbf{p}} + e\mathbf{A}) \times \nabla V \tag{2.132}$$

Eq. (2.132) is the *spin-orbit interaction*. Note that using the original Dirac equation as a starting point, the result becomes

$$\widehat{H}_{SO} = -\frac{\hbar}{4m_0^2 c^2} \widehat{\sigma} \cdot (\widehat{\mathbf{p}} + e\mathbf{A}) \times \nabla V \quad \text{(without correction)} \tag{2.133}$$

As can be seen, there is a difference in a factor of 4, where the corrected form of Eq. (2.132) is the *larger*.

In the spirit of demonstrating the many energy contributions contained in the relativistic energy, there is one additional term we must discuss. This contribution is contained in the term $(\widehat{\mathbf{p}} + e\mathbf{A})^2/2m_C$ of Eq. (2.128).

2.6 Magnetic Orbital Energy Of Free Charged Particle

Adding the magnetic vector potential $\mathbf{A}$ to the momentum as $\widehat{\mathbf{P}} = \widehat{\mathbf{p}} + e\mathbf{A}$, we enable the Hamiltonian to include the description of forces on a

[15]The operator is still Hermitian because its first order in $\widehat{\mathbf{p}}$, which carries a factor of $-i\hbar$.

charged mass particle in a *static* electric and magnetic field, where $\mathbf{B} = \nabla \times \mathbf{A}$.

Consider the following conditions where an electron is *initially* moving along the z axis of a cylindrical structure. Additionally, let this electron be exposed to an external magnetic field $\mathbf{B}$ pointing also along the cylindrical axis (z-axis), as illustrated in Fig. 2.4. We anticipate

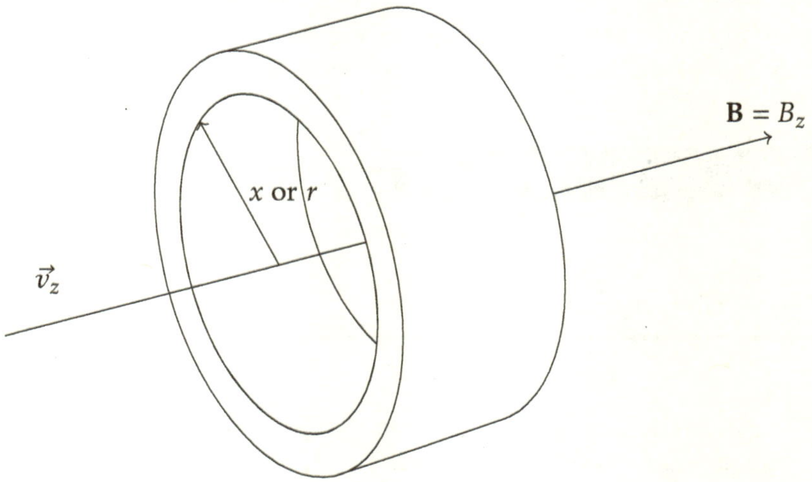

Figure 2.4: Illustration of an electron traveling along the longitudinal axis, or z axis, while confined within the cylindrical structure.

that the cylindrical structure imposes boundary conditions on any wave function. This fact will be useful when we solve for the eigensolutions of the quantum mechanical operator we are about to find. Note that the magnetic field is assumed to be uniform so the magnetic vector potential $\mathbf{A}$ is linear in the spatial variables x, y, and z and independent of time t. Given the relationship between $\mathbf{B}$ and $\mathbf{A}$ where $B = [0, 0, B_z] = [0, 0, \partial A_y/\partial x - \partial A_x/\partial y]$, it can be shown that the following vector potential yields a magnetic field along the $z-axis$:

$$\mathbf{A} = \frac{1}{2}[-B_z \cdot y, B_z \cdot x, 0]^T \tag{2.134}$$

Eq. (2.134) is known as a *symmetric gauge* because it expresses a symmetry in the x and y coordinates. The energy operator (or Hamiltonian) we'll consider only includes the kinetic energy term under conditions *with* a magnetic field. Then, the energy operator is given by

$$\widehat{H} = \frac{(\widehat{\mathbf{p}} + e\mathbf{A})^2}{2m_0}\psi = E_\lambda \psi \tag{2.135}$$

Expanding the square in Eq. (2.135) leads to

$$\frac{1}{2m_0}\left[\widehat{\mathbf{p}}^2 + e\widehat{\mathbf{p}} \cdot \mathbf{A} + e\mathbf{A} \cdot \widehat{\mathbf{p}} + e^2\mathbf{A}^2\right]\psi = E_\lambda \psi \tag{2.136}$$

$\widehat{\mathbf{p}}^2$ is just $-\hbar^2\nabla^2\psi$. Let us consider the term $e\widehat{\mathbf{p}} \cdot \mathbf{A}$. Since all operators are applied to the wave function ψ, we can make use of a vector-calculus identity for the *gradient of a (vector times a scalar)*.[16] Applying it to the second term of Eq. (2.136) leads to

$$e\widehat{\mathbf{p}} \cdot (\mathbf{A}\psi) = e\left[(\widehat{\mathbf{p}} \cdot \mathbf{A})\psi + \mathbf{A} \cdot (\widehat{\mathbf{p}}\psi)\right]$$

The first term on the RHS is proportional to the divergence of **A**. Because of the form of **A** given by Eq. (2.134), an inner product with the momentum operator (or divergence) is identically zero so we have

$$\frac{1}{2m_0}\left[\widehat{\mathbf{p}}^2 + 2e\mathbf{A} \cdot \widehat{\mathbf{p}} + e^2\mathbf{A}^2\right]\psi = E_\lambda \psi \tag{2.137}$$

Substituting the momentum operator into Eq. (2.137) then gives

$$E_\lambda \psi = \frac{(\widehat{\mathbf{p}} + e\mathbf{A})^2}{2m_0}\psi \tag{2.138a}$$

$$= \frac{1}{2m_0}\left[\widehat{\mathbf{p}}^2 + 2e\left(-i\hbar B_z \frac{y}{2}\frac{\partial \psi}{\partial x} + i\hbar B_z \frac{x}{2}\frac{\partial \psi}{\partial y}\right) + \frac{e^2 B_z^2}{4}\left(x^2 + y^2\right)\right]\psi \tag{2.138b}$$

$$= \frac{1}{2m_0}\left[\widehat{\mathbf{p}}^2 + e\left(-i\hbar B_z y \frac{\partial \psi}{\partial x} + i\hbar B_z x \frac{\partial \psi}{\partial y}\right) + \frac{e^2 B_z^2}{4}\left(x^2 + y^2\right)\right]\psi \tag{2.138c}$$

Now, consider the following operator:

$$-\mathbf{r} \times \widehat{\mathbf{p}} = \begin{bmatrix} i & j & k \\ x & y & z \\ -i\hbar\frac{\partial}{\partial x} & -i\hbar\frac{\partial}{\partial y} & -i\hbar\frac{\partial}{\partial z} \end{bmatrix} = -\begin{bmatrix} -i\hbar y\frac{\partial}{\partial z} + i\hbar z\frac{\partial}{\partial y} \\ i\hbar x\frac{\partial}{\partial z} - i\hbar z\frac{\partial}{\partial x} \\ -i\hbar x\frac{\partial}{\partial y} + i\hbar y\frac{\partial}{\partial x} \end{bmatrix} \tag{2.139}$$

---

[16] it is the generalization of the product rule of differentiation

Then, taking $-\mathbf{r}\times\widehat{\mathbf{p}}\cdot\mathbf{B}$ gives

$$-\mathbf{r}\times\widehat{\mathbf{p}}\cdot\mathbf{B} = -\widehat{\mathbf{L}}\cdot\mathbf{B} = \left(-i\hbar B_z y\frac{\partial}{\partial x} + i\hbar B_z x\frac{\partial}{\partial y}\right) \tag{2.140}$$

$\widehat{\mathbf{L}}$, of course, is the *orbital angular momentum* operator. Comparing Eqs. (2.138a) and (2.140), it can be seen that the component of of Eq. (2.140) along the z direction (the magnetic field B direction) is identical to the term in Eq. (2.138a).

Then, using $r^2 = x^2 + y^2$ we can write Eq. (2.138a) as

$$\frac{(\widehat{\mathbf{p}}+e\mathbf{A})^2}{2m_0}\psi = \left[\frac{\widehat{\mathbf{p}}^2}{2m_0} - \frac{e\hbar}{2m_0}\cdot\widehat{\mathbf{L}}\cdot\mathbf{B} + \frac{e^2B_z^2r^2}{8m_0}\right]\psi = E_\lambda\psi \tag{2.141}$$

The above is intended to illustrate that the quantum mechanical operator $(\widehat{\mathbf{p}}+e\mathbf{A})^2/2m_0$ also *contains orbital angular momentum* energy created by an external magnetic field **B**. Next, let's find the exact solution to Eq. (2.141).

2.6.1 Wave Function Solutions

With rotational motion in mind, it is better to use a coordinate system that describes rotation more naturally. To obtain the solution to Eq. (2.141), a cylindrical coordinate system (r,φ,z) is most useful.

Using the *symmetric gauge* given by $\mathbf{A} = (1/2)[-B\cdot y, B\cdot x, 0]^T$, we can use the relations $y = r\sin\varphi$ and $x = r\cos\varphi$ in the cylindrical coordinate system. Also, since the azimuthal unit vector in the cylindrical coordinate system is given by $\varphi = -\sin\varphi\widehat{\mathbf{x}} + \cos\varphi\widehat{\mathbf{y}}$, the magnetic vector potential **A** is oriented azimuthally, or circles around the cylindrical z–axis. $\widehat{\mathbf{x}}$ and $\widehat{\mathbf{y}}$ are unit vectors. Then, using the above relations in the symmetric gauge, we can also write the magnetic vector potential **A** as

$$\mathbf{A} = [0 \quad rB_z/2 \quad 0]^T \tag{2.142}$$

In *cylindrical* coordinates, the momentum operator is given by

$$\widehat{\mathbf{p}} = -i\hbar\nabla = -i\hbar\left[\frac{\partial}{\partial r} \quad \frac{1}{r}\frac{\partial}{\partial\varphi} \quad \frac{\partial}{\partial z}\right]^T$$

Then, the Eq. (2.141) in cylindrical coordinates becomes

$$E_\lambda \psi = \frac{1}{2m_0}\left[-\hbar^2\nabla^2\psi - ier\hbar B\mathbf{e}_\varphi \cdot \nabla\psi + \frac{e^2B^2}{4}r^2\psi\right] \tag{2.143a}$$

$$= \frac{1}{2m_0}\left[-\hbar^2\left(\frac{1}{r}\frac{\partial}{\partial r}\left(r\frac{\partial\psi}{\partial r}\right) + \frac{1}{r^2}\frac{\partial^2\psi}{\partial\varphi^2} + \frac{\partial^2\psi}{\partial z^2}\right) - ie\hbar B\frac{\partial\psi}{\partial\varphi}\right. \tag{2.143b}$$

$$\left. + \frac{e^2B^2}{4}r^2\psi\right] \tag{2.143c}$$

Let ψ be *separable*, or given by

$$\psi = R(r)\Phi(\varphi)Z(z) \quad \text{and} \quad E_\lambda = E_r + E_\varphi + E_z \tag{2.144}$$

Substituting Eq. (2.144) into Eq. (2.143b), we can readily separate into distinct equations for z and the combination (r,φ). Then, for the z dependence, $Z(z)$ is described by

$$-\frac{\hbar^2}{2m_0}\frac{d^2Z}{dz^2} = E_zZ(z) \Rightarrow \frac{d^2Z}{dz^2} + k_z^2Z = 0 \tag{2.145}$$

The z dependent solution to Eq. (2.145) is given by

Z(z) Solution for ψ

$$Z(z) = Z_0e^{ik_zz} \quad \text{where} \quad k_z^2 = \frac{2m_0E_z}{\hbar^2} \tag{2.146}$$

One can verify Eq. (2.146) is a solution to Eq. (2.145).

The remaining part of Eq. (2.143b) depends on r and φ in the following equation describing functions $R(r)$ and $\Phi(\varphi)$:

$$\frac{1}{2m_0}\left[-\hbar^2\left(\Phi\frac{1}{r}\frac{\partial}{\partial r}\left(r\frac{\partial R}{\partial r}\right) + R\frac{1}{r^2}\frac{\partial^2\Phi}{\partial\varphi^2}\right) - ie\hbar B_zR\frac{\partial\Phi}{\partial\varphi} + \frac{e^2B_z^2}{4}r^2R\Phi\right] = (E_r + E_\varphi)R\Phi \tag{2.147}$$

We can still use *separability* for Eq. (2.147), however, there is a trick we must exploit.

Particularly, because it can be shown that if $\Phi(\varphi) = C_\varphi e^{ik_\varphi r\varphi}$, then both terms in Eq. (2.147) depending on φ sum to zero, which also means $E_\varphi = 0$, or there is no contribution to energy from this degree of freedom. To determine the value of k_φ, we sum all the terms depending on φ, which gives

$$-\frac{\hbar^2}{2m_0}\left(\frac{1}{r^2}\frac{\partial^2\Phi}{\partial\varphi^2}\right) - i\frac{e\hbar B_z}{2m_0}\frac{\partial\Phi}{\partial\varphi} = E_\varphi\Phi \tag{2.148}$$

Let the solution be written as $\Phi(\varphi) = C_\varphi e^{ik_\varphi r\varphi}$. Evaluating the derivatives of Φ leads to

$$\frac{\partial\Phi}{\partial\varphi} = C_\varphi ik_\varphi re^{ik_\varphi r\varphi} = ik_\varphi r\Phi \tag{2.149}$$

$$\frac{\partial^2\Phi}{\partial\varphi^2} = C_\varphi(ik_\varphi r)^2 e^{ik_\varphi r\varphi} = -k_\varphi^2 r^2\Phi \tag{2.150}$$

Then, substitution of these results into the LHS of Eq. (7.51) gives

$$-\frac{\hbar^2}{2m_0}\left(\frac{1}{r^2}\frac{\partial^2\Phi}{\partial\varphi^2}\right) - i\frac{e\hbar B_z}{2m_0}\frac{\partial\Phi}{\partial\varphi} = -\frac{\hbar^2}{2m_0}\frac{1}{r^2}\left(-k_\varphi^2 r^2\Phi\right) - i\frac{e\hbar B_z}{2m_0}\left(ik_\varphi r\Phi\right) \tag{2.151a}$$

$$= \left[\frac{\hbar^2 k_\varphi^2}{2m_0} + \frac{e\hbar B_z k_\varphi r}{2m_0}\right]\Phi \tag{2.151b}$$

We want the value of k_φ that makes Eq. (2.151) vanish satisfying the condition

$$0 = \hbar k_\varphi + eB_z r$$

Solving for k_φ, we have

Azimuthal Wavenumber k_φ

$$k_\varphi = -\frac{eB_z r}{\hbar} \tag{2.152}$$

Since the function $\Phi(\varphi) = \Phi(r,\varphi) = C_\varphi e^{ik_\varphi r\varphi}$, we also have for the azimuthal solution $\Phi(r,\varphi)$

Azimuthal Solution $\Phi(\varphi)$

$$\Phi(r,\varphi) = C_\varphi e^{-i\frac{eB_z r^2}{\hbar}\varphi} \tag{2.153}$$

The momentum in the azimuthal direction is given by

$$p_\varphi = \hbar k_\varphi = -eB_z r \tag{2.154}$$

Equivalently, for a mass particle m_0 where $p \approx m_0 v$, we have

$$v_\varphi = \frac{p_\varphi}{m_0} = -\frac{eB_z r}{m_0} = -\omega_0 r \tag{2.155}$$

From Eq. (2.155), we can define a rotational frequency ω_0 as

Cyclotron Frequency ω_0

$$\omega_0 = \frac{eB_z}{m_0} \tag{2.156}$$

Eq. (2.156) is known as the *cyclotron frequency*. It is the rotational frequency of a mass m_0 having charge strength e, moving in the direction of φ due to the axial field of strength B_z. We'll use this parameter again shortly when we determine the quantized energy levels for the radial equation.

C_φ is called the *normalization factor* which can be found using the *normalization condition*. This is done by evaluating the following in cylindrical coordinates

$$\begin{aligned}
1 &= \int_0^{2\pi} C_\varphi^* C_\varphi e^{i\frac{e}{\hbar}Br^2\varphi} e^{-i\frac{e}{\hbar}Br^2\varphi} d\varphi \\
&= |C_\varphi|^2 \int_0^{2\pi} d\varphi \\
&= |C_\varphi|^2 \cdot 2\pi
\end{aligned}$$

This gives the **azimuthal normalization factor** as

Normalization Factor For $\Phi(\varphi)$

$$|C_{\varphi}| = \frac{1}{\sqrt{2\pi}} \tag{2.157}$$

We now have solutions $Z(z)$ and $\Phi(r,\varphi)$, so only $R(r)$ remains. The remaining terms, being only radial, are given by[17]

$$\frac{1}{2m_0}\left[-\hbar^2\frac{1}{r}\frac{d}{dr}\left(r\frac{dR}{dr}\right)+\frac{e^2B_z^2}{4}r^2R\right] = E_rR \tag{2.158}$$

Solving for the radial part $R(r)$ is the more challenging of the separated coordinate equations. To solve Eq. (2.158), we must express it in a different (but equivalent) form. Using this different form, it becomes amenable to a powerful method for solving equations like this, known as the *Method of Frobenius*. . The Method of Frobenius finds a solution in the form of an infinite series expansion, and is nontrivial. Therefore, it what follows, we shall include all the steps keeping in mind that it is all to allow us to apply this method to find an exact solution to Eq. (2.158).

First, expanding the derivative term and rearranging, we have

$$\frac{d^2R}{dr^2}+\frac{1}{r}\frac{dR}{dr}+\left(\frac{2m_0E_r}{\hbar^2}-\frac{e^2B_z^2}{4\hbar^2}r^2\right)R = 0 \tag{2.159}$$

Let us express Eq. (2.159) in terms of an alternative ζ defined by the relation

$$\sqrt{\zeta} = \sqrt{\frac{eB_z}{2\hbar}}r = \nu r \Rightarrow \zeta = \nu^2r^2 \tag{2.160}$$

From the definition of ζ, the following useful relations can be obtained:

$$d\zeta = 2\nu^2 r dr \quad \text{or} \quad \frac{d\zeta}{dr} = 2\nu^2 r$$

$$\frac{dR}{dr} = \frac{dR}{d\zeta}\frac{d\zeta}{dr} = 2\nu^2 r\frac{dR}{d\zeta}$$

$$\frac{d^2R}{dr^2} = \frac{d}{dr}\left(2\nu^2 r\frac{dR}{d\zeta}\right) = 2\nu^2\frac{dR}{d\zeta}+2\nu^2 r\frac{d\zeta}{dr}\frac{d}{d\zeta}\left(\frac{dR}{d\zeta}\right)$$

[17] since R is only a function of r, partial derivatives become total derivatives.

$$= 2\nu^2 \frac{dR}{d\zeta} + 4\nu^4 r^2 \frac{d^2R}{d\zeta^2} = 2\nu^2 \frac{dR}{d\zeta} + 4\nu^2 \zeta \frac{d^2R}{d\zeta^2}$$

Substitution of these relations into Eq. (2.159) leads to the following equivalent differential equation in terms of ζ:

$$4\nu^2 \zeta \frac{d^2R}{d\zeta^2} + 4\nu^2 \frac{dR}{d\zeta} + \left(\frac{2m_0 E_r}{\hbar^2} - \nu^2 \zeta \right) R(\zeta) = 0 \tag{2.161}$$

or

$$4\zeta \frac{d^2R}{d\zeta^2} + 4\frac{dR}{d\zeta} + \left(\frac{2m_0 E_r}{\nu^2 \hbar^2} - \zeta \right) R(\zeta) = 0 \tag{2.162}$$

Next, let us write the wave function R in the form $R = R_\infty(\zeta) R_1(\zeta)$, where $R_\infty(\zeta)$ is an *asymptotic* solution we will find first. For this, consider the asymptotic limit $\zeta \to \infty$, where Eq. (2.162) becomes

$$4\frac{d^2R_\infty}{d\zeta^2} - R_\infty(\zeta) = 0 \Rightarrow \frac{d^2R_\infty}{d\zeta^2} - \frac{1}{4} R_\infty(\zeta) = 0 \tag{2.163}$$

Eq. (2.163) has a solution given by

$$R_\infty(\zeta) = C_\infty e^{-\frac{\zeta}{2}} + D_\infty e^{+\frac{\zeta}{2}} = C_\infty e^{-\sqrt{\frac{eB}{8\hbar}} r^2} \tag{2.164}$$

The result follows because the other mathematical solution $R_\infty(\zeta) = D_\infty e^{+\frac{\zeta}{2}}$ goes to infinity as $\zeta \to \infty$. This does not satisfy the boundary condition where $\psi(r = \infty) = 0$. Therefore, the coefficient $D_\infty = 0$. Because $R_\infty(\zeta)$ is now in the form of a *Gaussian*, it turns out that we can obtain an equation purely in terms of R_1. If $R = R_\infty R_1$, we have the following derivatives of $R(\zeta)$ in terms of R_1:

$$\frac{dR}{d\zeta} = \frac{d}{d\zeta}(R_\infty R_1) = \left[\frac{dR_1}{d\zeta} - \frac{1}{2} R_1 \right] e^{-\frac{1}{2}\zeta}$$

$$\frac{d^2R}{d\zeta^2} = \left[\left(\frac{d^2R_1}{d\zeta^2} - \frac{1}{2}\frac{dR_1}{d\zeta} \right) + \left(\frac{dR_1}{d\zeta} - \frac{1}{2} R_1 \right) \left(-\frac{1}{2} \right) \right] e^{-\frac{1}{2}\zeta}$$

Using these relations in Eq. (2.162), we obtain the following equation for R_1:

$$4\zeta \left[\left(\frac{d^2R_1}{d\zeta^2} - \frac{1}{2}\frac{dR_1}{d\zeta} \right) + \left(\frac{dR_1}{d\zeta} - \frac{1}{2} R_1 \right) \left(-\frac{1}{2} \right) \right] e^{-\frac{1}{2}\zeta} + 4\left[\frac{dR_1}{d\zeta} - \frac{1}{2} R_1 \right] e^{-\frac{1}{2}\zeta} + \left[\frac{2m_0 E_r}{\nu^2 \hbar^2} - \frac{1}{2}\zeta \right] R_1 e^{-\frac{1}{2}\zeta} = 0 \tag{2.165}$$

We can cancel the Gaussian factors and combine respective terms as

$$4\zeta\frac{d^2R_1}{d\zeta^2}+4(1-\zeta)\frac{dR_1}{d\zeta}+\left[\frac{2m_0E_r}{\nu^2\hbar^2}-2\right]R_1=0$$

or

$$\zeta\frac{d^2R_1}{d\zeta^2}+(1-\zeta)\frac{dR_1}{d\zeta}-\left[\frac{2-\frac{2m_0E_r}{\nu^2\hbar^2}}{4}\right]R_1=0 \tag{2.166}$$

Defining Ω as

$$\Omega=\frac{2m_0E_r}{\nu^2\hbar^2} \tag{2.167}$$

We have

$$\zeta\frac{d^2R_1}{d\zeta^2}+(1-\zeta)\frac{dR_1}{d\zeta}-\left(\frac{2-\Omega}{4}\right)R_1=0 \tag{2.168}$$

Eq. (2.168) is in the form of the *confluent hypergeometric equation*(CHE) which is given by

$$x\frac{d^2y}{dx^2}+(c-x)\frac{dy}{dx}-ay(x)=0$$

The solution to this equation is given by the *confluent hypergeometric series* (CHS).[18] The CHS is given by

$$\mathrm{CHS}(\zeta;a,c)={}_1F_1(\zeta;a,c)=1+\frac{1}{1!}\frac{a}{c}\zeta+\frac{1}{2!}\frac{a(a+1)}{c(c+1)}\zeta^2+$$

$$\frac{1}{3!}\frac{a(a+1)(a+2)}{c(c+1)(c+2)}\zeta^3+\frac{1}{4!}\frac{a(a+1)(a+2)(a+3)}{c(c+1)(c+2)(c+3)}\zeta^4+$$

$$\frac{1}{5!}\frac{a(a+1)(a+2)(a+3)(a+4)}{c(c+1)(c+2)(c+3)(c+4)}\zeta^5+$$

$$\frac{1}{6!}\frac{a(a+1)(a+2)(a+3)(a+4)(a+5)}{c(c+1)(c+2)(c+3)(c+4)(c+5)}\zeta^6+\ldots$$

From Eq. , we see that

$$c=1\quad\text{and}\quad a=\frac{2-\Omega}{4}=\frac{2-\frac{2m_0E_r}{\hbar^2\nu^2}}{4} \tag{2.169}$$

[18]for a detailed derivation, see Appendix Chap. A

For a valid quantum mechanical wave function solution, it must also satisfy the boundary conditions. As already mentioned, this means that $\psi(r=\infty)=0$. For this to happen, it turns out that the infinite series ${}_1F_1(\zeta;a,c)$ has to be truncated dictating that the parameter a only takes negative integer values, or

$$a=\frac{2-\Omega}{4}=-n \tag{2.170}$$

This condition on a also determines the allowed values for E_r because

$$\frac{2-\frac{2m_0E_r}{\hbar^2\nu^2}}{4}=-n\Rightarrow-\frac{2m_0E_r}{\hbar^2\nu^2}=-4n-2 \tag{2.171}$$

Substitution of $\nu^2=eB_z/2\hbar$, and solving for E_r, we have

$$E_r=\hbar\omega_0\left(n+\frac{1}{2}\right) \tag{2.172}$$

or

Radial Eigenenergy

$$E_r=\frac{\hbar eB_z}{m}\left(n+\frac{1}{2}\right) \tag{2.173}$$

where $n=0,1,2,3,...$ ω_0 is the cyclotron frequency given by Eq. (2.156).

Finally, we have arrived at a *complete* solution to Eq. (2.141) given by

Electron Kinetic Energy Equation Eigensolution

$$\psi(r,\varphi,z)=C_ne^{-\frac{eB_z}{8\hbar}r^2}e^{i\frac{eB_zr^2}{\hbar}\varphi}{}_1F_1\left(\frac{eB_z}{2\hbar}\cdot r^2;-n,1\right)e^{ik_zz} \tag{2.174}$$

C_n is the normalization constant. C_n can be obtained by evaluating the integral

$$1=\int\psi^*\psi dV \tag{2.175}$$

$$= \int |C_n|^2 e^{-\frac{eB_z}{2\hbar}r^2} {}_1F_1\left(\frac{eB_z}{2\hbar}r^2;-n,1\right) {}_1F_1\left(\frac{eB_z}{2\hbar}r^2;-n,1\right) r dr d\varphi dz \tag{2.176}$$

To rewrite the integral of Eq. (2.175) in a form that enables use of a lookup table, we can express in terms of $\zeta = eB_z r^2/2\hbar = \nu^2 r^2$. Then $d\zeta = 2\nu^2 r dr$, and the integral can be written as

$$1 = \frac{1}{2\nu^2}\int |C_n|^2 e^{-\zeta} {}_1F_1(\zeta;-n,1) {}_1F_1(\zeta;-n,1) d\zeta d\varphi dz \tag{2.177}$$

Eq. (2.177) has a form that can be found in *Factorization Method In Quantum Mechanics*,[19] where we have

$$1 = \frac{1}{2\nu^2}\left(|C_n|^2 2\pi L_z \frac{n!\Gamma(1)^2}{\Gamma(n+1)}\right) \tag{2.178}$$

$\Gamma()$ is the *Gamma* function, L_z is the thickness of the domain along the z direction. Using $\Gamma(1) = 1$, and solving for $|C_n|$, we have a normalization constant given by

Normalization Constant

$$|C_n| = \sqrt{\frac{eB\Gamma(n+1)}{2\pi\hbar n! L_z}} \tag{2.179}$$

Some examples of the wave function solution from Eq. (2.174) are shown in Table 2.1 for a variety of quantum numbers n and magnetic field strengths.

[19] *Factorization Method In Quantum Mechanics* [Shi-Hai Dong, 2007]

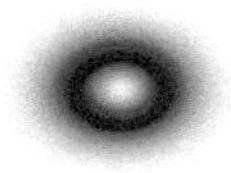

(a) $n = 0, B = 0.5T$

(b) $n = 0, B = 1.5T$

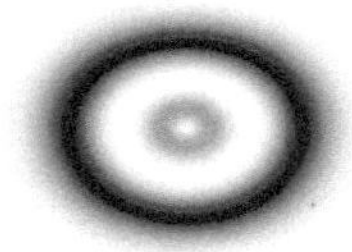

(c) $n = 1, B = 0.5T$

(d) $n = 1, B = 1.5T$

(e) $n = 2, B = 0.5T$

(f) $n = 2, B = 1.5T$

(g) $n = 3, B = 0.5T$ (h) $n = 3, B = 1.5T$

Table 2.1: n–vortices for different values of $n = 0, 1, 2, 3$ and external magnetic field $B = 0.5, 1.5$T. As pure radial functions, they have azimuthal symmetry in the x and y plane. Note that the probability density functions are *not* normalized here. All figures shown are on the same length scale spanning ± 200nm, or 400nm.

All plots shown illustrate distributions on the same length scale and range spanning ± 200nm. From the solutions, the electron is predominantly confined to an orbital exterior ring. The orbital motion about the magnetic field inside a confined space is known as the *cyclotron orbit*, and these orbits are *quantized*.

The general effect of the magnetic field strength is to further confine the orbital ring radially. These *cyclotron orbits* illustrate the motion of an electron when exposed to a magnetic field, with or without propagation along the z–axis. In the case of nonzero propagation along z ($k_z \neq 0$), the motion becomes *spiral*. A cyclotron orbit, therefore, is generally a spiral of motion of the electron as it moves along the cylindrical axis.

2.7 Wave Functions Contain Lorentz Factor

In quantum mechanics, one rarely explicitly includes the Lorentz factor γ_r. Given the 4D eigenenergy, it turns out that it is not necessary and we can still obtain correct eigensolutions. To see why, recall that we obtained a 4D kinetic energy given by

$$E = \frac{\gamma_r}{2}\sqrt{p^2c^2 + m_0^2c^4} + ic\gamma_r p_4 = E_0\gamma_r \quad (\gamma_r \geq 1) \tag{2.180}$$

Because quantum mechanics solves for eigenfunctions, from solving equations of the form

$$\widehat{O}\psi = E_0\psi \tag{2.181}$$

Right-multiplying by γ_r, we have

$$\widehat{O}\psi \cdot \gamma_r = E_0\psi \cdot \gamma_r \tag{2.182}$$

In the case that γ_r is constant, then

$$\widehat{O}\psi \cdot \gamma_r = \widehat{O}(\psi \cdot \gamma_r) \tag{2.183}$$

However, since γ_r is independent of the three dimensions x, y, z, Eq. (2.183) still holds, so

$$\widehat{O}\psi' = E_0'\psi' \tag{2.184}$$

Thus, the solution for the ground state E_0, as well as, any other quantum states having energy $E_\lambda \geq E_0$ are all described by equivalent eigenequations. Thus, γ_r *can* be omitted in relativistic 4D quantum mechanics.

2.8 Chapter Summary

In this chapter, we have focused on the inclusion of the fourth dimension in the description of quantum mechanical systems. To aid in visualizing 4D physics, we started with a geometrical construct known as a helix constructed effectively in both three and four dimensions. We then went on to connect the concept of the wave function, extended to four dimensions. This included a discussion of 4D momentum, ultimately leading the way to a general 4D Hamiltonian similar in form to the Dirac equation. We then looked at a variety of contributions to the quantum mechanical relativistic kinetic energy, that part being limited to the three dimensions.

Another aim has been to demonstrate that the relativistic kinetic energy contains a multitude of contributions. However, the above eigensolution we considered only provided an example of a wave function depending only on three dimensions. Therefore, we want to continue along the path of analyzing solutions, however, we will need to add the fourth dimension, as well.

Table 2.2 summarizes some key equations from this chapter.

Table 2.2: Chapter 1 Summary Equations.

Name	Equation
Wave number	$k = \frac{2\pi}{\lambda}$
Normalization Condition	$\int_V \lvert\psi\rvert^2 dV = 1$
1D Classical Wave Equation	$\frac{\partial^2 u}{\partial x^2} = \frac{\rho_x}{T_\infty}\frac{\partial^2 u}{\partial t^2}$
Wave Eq Sol (Type II):	$u = Af(ikx - i\omega t) + Bg(-ikx + i\omega t)$
Type II Condition	$v^2 = \frac{\omega^2}{k^2}$
4D Momentum Operator	$\widehat{\mathbf{p}} = -i\hbar\begin{bmatrix}\frac{\partial}{\partial x} & \frac{\partial}{\partial y} & \frac{\partial}{\partial z} & \frac{1}{ic}\frac{\partial}{\partial t}\end{bmatrix}^T$
4th Dimension KE	$K_4 = ic\gamma_r p_4$
4D Kinetic Energy	$E = \frac{\gamma_r}{2}\sqrt{p^2c^2 + m_0^2c^4} + ic\gamma_r p_4 + V$
4D Energy Operator ($\gamma_r = 1$)	$\mathbf{I}(4)\cdot\left(-i\hbar\frac{\partial}{\partial t}\right) + c\cdot\widehat{\Sigma}\cdot(\mathbf{p}+e\mathbf{A}) + \widehat{\Sigma}_0 m_0 c^2 + \mathbf{I}(4)V$

2.9 Chapter Problems

Problem 2.1 Consider a helix described by the relation

$$\mathbf{r} = \cos(\delta\phi)\widehat{\mathbf{e}}_x + \sin(\delta\phi)\widehat{\mathbf{e}}_y$$

where $\delta\phi = k\Delta z$. *a.* Calculate the change in position Δz of a fictitious particle whose position is constrained to a helix and has experienced a phase change of 9π. Assume a helical wave length of $\lambda = 1$nm.

b. If the elapsed time $\Delta t = 1$ns, calculate the frequency or rate of rotation ω while in motion along the helical axis.

Problem 2.2 The component of momentum for the 4th dimension $x_4 = ict$ is

$$\widehat{p_4}\psi = -i\hbar \frac{\partial \psi}{\partial x_4}$$

Solve the eigenvalue equation for $\psi = A \cdot f(t)$ (*i.e.* find $f(t)$) for this component given by

$$\widehat{p_4}\psi = \lambda\psi$$

We also found the so-called *normalization condition* for any quantum mechanical wave function given by

$$1 = \int \psi^* \psi dV$$

Can the normalization condition be used with ψ to solve for A? Woud the result be time-dependent and if so, is $A(t)$ a valid wave function amplitude for ψ?

Problem 2.3 Consider the 1D wave function $\psi(x)$ given by

$$\psi(x) = \frac{\alpha}{\sqrt{x^2 + 1}}$$

Find α by evaluating the normalization condition

$$1 = \int \psi^* \psi dx dA = A \int \psi^* \psi dx$$

Then, show that as $x_f - x_i \Rightarrow \infty$, $\alpha \Rightarrow \sqrt{\pi}$.

Note: This last part indicates that even as the body length grows, α still approaches a *finite limit*.

Problem 2.4 Consider the wave function ψ given by

$$\psi(x) = \frac{\alpha}{x^n}$$

Use and evaluate the normalization condition to find a and show that

$$\alpha^2 = \frac{2(n+1)}{x_f^{2n+2} - x_i^{2n+2}}$$

For which values of n is the above solution invalid? Can α still be found even if/when the above is invalid?

Note: In QM, when polynomials are involved in quantum states, the order of the polynomial corresponds to the quantum state/number.

Problem 2.5 The electromagnetic wave equations can be expressed as

$$\nabla \times \mathbf{E} = -\frac{\partial \mathbf{B}}{\partial t}$$

and

$$\nabla \times \mathbf{B} = \mu \mathbf{J} + \mu \frac{\partial \mathbf{D}}{\partial t}$$

The above equations were used to find the wave equation

$$\nabla^2 \mathbf{E} = \frac{1}{c^2} \frac{\partial \mathbf{E}}{\partial t}$$

Let $\mathbf{E} = \mathbf{E}_0 e^{i(kx-\omega t)}$ and $\mathbf{B} = \mathbf{B}_0 e^{i(kx-\omega t)}$. Show that

$$\mathbf{k} \times \mathbf{E} = \omega \mathbf{B}$$

$$\mathbf{k} \times \mathbf{B} = -\mu \mathbf{J} - \mu \omega \mathbf{D}$$

Use the above relations to show that

$$c^2 = \frac{\omega^2}{k^2}$$

Problem 2.6 In this chapter, we found the radial eigenenergy for the kinetic energy of an electron in a magnetic field to be

$$E_r = \frac{\hbar e B_z}{m} \left(n + \frac{1}{2} \right)$$

Calculate the ground state energy $n = 0$, as well as the quantum state energy for $n = 1, n = 2$, and $n = 3$ in a magnetic field of 1 Tesla.

Compare the energy it takes to reach E_3 from E_2, E_2 from E_1, and E_1 from E_0 (ground state).

Problem 2.7 Write out the explicit wave functions (ignoring the normalization factor)

$$\psi(r,\varphi,z) = e^{-\frac{eB_z}{8\hbar}r^2} e^{i\frac{eB_z r^2}{\hbar}\varphi} {}_1F_1\left(\frac{eB_z}{2\hbar}\cdot r^2; -n, 1\right) e^{ik_z z}$$

for an electron in a magnetic field of 1 Tesla for the first five quantum states $n = 0, 1, 2, 3$ and 4 (assume $k_z = 0$). Use a computing language of your choice (*e.g.* python) to plot the wave functions.

Discuss your observations of the behavior of the wave functions as n increases.

2.10 Suggested Readings & References

[1] P. A. M. Dirac, *The Quantum Theory Of The Electron*, **117**, pp. 610-624 (1928)(*note*:available from Proceedings of the Royal Society of London. Series A, Containing Papers of a Mathematical and Physical Character. www.jstor.org).

[2] L. Liu, et. al., *Current-Induced Switching of Perpendicularly Magnetized Magnetic Layers Using Spin Torque from the Spin Hall Effect*, PRL 109, 096602 (2012).

[3] K. Eason, *Quantum Mechanics With A SPIN On It*, Sym. Pub (2019).

[4] Shi-Hai Dong *Factorization Method In Quantum Mechanics* (2007)

Chapter 3

Revisiting Fundamental Problems In 4D

Now that we have worked out a proper quantum mechanical operator for the 4D system, we can revisit some well-known problems in quantum mechanics and apply the 4D operator we obtained in Eq. (2.120). This will be the focus of this chapter, demonstrating how the 4D Hamiltonian can be applied to real problems and used to find exact solutions.

We'll analyze three problems in 4D, starting from simple to more complex. We'll find that all of the obtained eigenfunction generalize from their 3D (x, y, z) solutions to include *real* time-dependence. This will enable, for example, generalizations of the time-dependence of quantities such as expectation value. So, let us begin.

3.1 Free Particle/Plane Wave In Linear Motion

For a free particle restricted to 1D motion, Eq. (2.120) simplifies to

$$\left[-\frac{\hbar^2}{2m_0}\frac{\partial^2}{\partial x^2} - i\hbar\frac{\partial}{\partial t}\right]\psi = (iE_I)\psi \tag{3.1}$$

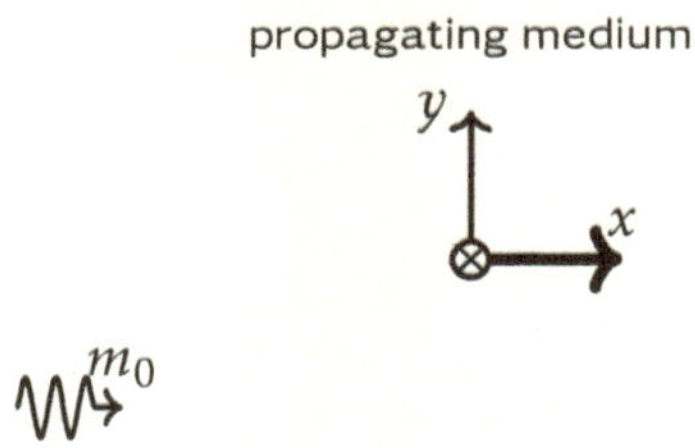

Figure 3.1: Illustration of a fundamental particle traveling in a single direction.

We also require the total energy operator given by[1]

$$i\hbar\frac{\partial\psi}{\partial t} = (E_R - iE_I)\psi \tag{3.2}$$

To solve the above eigen-operator equations, take the wave function $\psi(x,t)$ to be *separable*, so ψ can be written as

$$\psi(x,t) = \psi_x(x)\cdot\psi_t(t) \tag{3.3}$$

Substitution of Eq. (3.3) into Eq. (3.1), we have

$$-\frac{\hbar^2}{2m_0}\frac{d^2\psi_x}{dx^2}\cdot\psi_t(t) - i\hbar\cdot\psi_x(x)\cdot\frac{d\psi_t}{dt} = iE_I\psi_x(x)\cdot\psi_t(t) \tag{3.4}$$

Dividing both sides by $\psi = \psi_x\cdot\psi_t$ leads to

$$-\frac{\hbar^2}{2m_0}\frac{1}{\psi_x}\frac{d^2\psi_x}{dx^2} - i\hbar\frac{1}{\psi_t}\frac{d\psi_t}{dt} = iE_I \tag{3.5}$$

Since the RHS is a complex constant scalar, while the LHS depends on two terms depending respectively on independent variables x and t, it follows that each respective term must be equal to a constant (*i.e.* E_R and $E_R - iE_I$). So, we can write the following two equations:

$$-\frac{\hbar^2}{2m_0}\frac{1}{\psi_x}\frac{d^2\psi_x}{dx^2} = E_R \tag{3.6a}$$

$$i\hbar\frac{1}{\psi_x}\frac{d\psi_t}{dt} = E_R - iE_I \tag{3.6b}$$

or

$$\frac{d^2\psi_x}{dx^2} = -\frac{2m_0E_R}{\hbar^2}\psi_x \tag{3.7a}$$

[1]The negative imaginary component is assumed in order to be consistent with the form $e^{i(kx-\omega t)}$

$$\frac{d\psi_t}{dt} = \frac{E_R - iE_I}{\hbar}\psi_t \tag{3.7b}$$

The solutions to Eq. (3.7) are given by

$$\psi_x(x) = Ae^{+i\frac{\sqrt{2m_0E_R}}{\hbar}x} + Be^{-i\frac{\sqrt{2m_0E_R}}{\hbar}x} \tag{3.8a}$$

$$\psi_t(t) = \psi(0)e^{-i\frac{E_R}{\hbar}t}e^{-\frac{E_I}{\hbar}t} \tag{3.8b}$$

The total solution $\psi = \psi_x \cdot \psi_t$ is then given by

$$\psi(x,t) = \psi(0)e^{-i\frac{E_R}{\hbar}t}e^{-\frac{E_I}{\hbar}t}\left[Ae^{+i\frac{\sqrt{2m_0E_R}}{\hbar}x} + Be^{-i\frac{\sqrt{2m_0E_R}}{\hbar}x}\right] \tag{3.9}$$

If $E_I = 0$, then we get the time-independent 1D solution

$$\psi(x,t) = Ae^{-i\frac{E_R}{\hbar}t}\left[e^{+i\frac{\sqrt{2m_0E_R}}{\hbar}x} + Be^{-i\frac{\sqrt{2m_0E_R}}{\hbar}x}\right] \tag{3.10}$$

In this 4D example, the plane wave solution generalizes to a plane wave *with decaying amplitude*. Notice that the result corresponds to motion restricted to a single axis.

Given a time dependence, this solution ensures the wave function amplitude vanishes at $x = \infty$. The condition can change, however, in problems involving quantum state superposition. In this case, we find that single dimensional motion carries a decaying amplitude. Let's consider another problem in the 4D frame, known as the *harmonic oscillator problem*.

3.2 The Damped Harmonic Oscillator

The *harmonic oscillator* is also a well-known problem in quantum mechanics. Surprisingly, as simple as its description is, it's quantum mechanical solution is nontrivial due to spatial dependence. As with the plane wave problem, only the *x − coordinate* is considered from the first three dimensions. However, the 4D Hamiltonian for a harmonic oscillator must also include a *potential energy* due to a connection to a spring (literally or figuratively). This energy is given by

$$V = \frac{1}{2}kx^2 \tag{3.11}$$

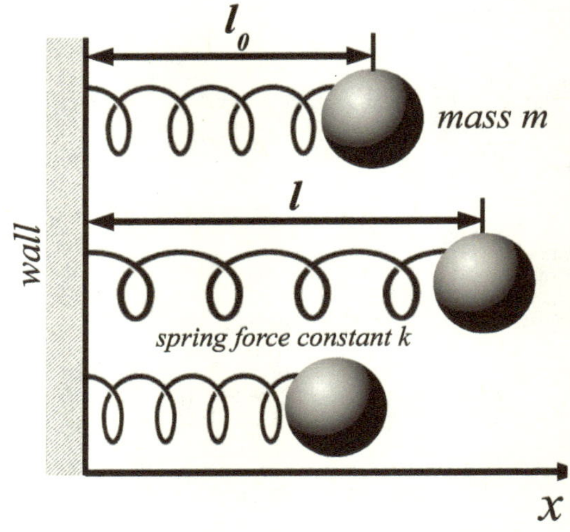

Figure 3.2: A *classical* harmonic oscillator consisting of a mass m attached to a spring (figuratively or literally). It is shown at different amounts of displacement. In a state of no motion, there is an equilibrium position l_0 at which the system desires to be. Displacing the mass produces a force that counters the displacement $x = l - l_0$.

k is called the spring stiffness constant. If we define a harmonic oscillator *natural frequency* ω_0 as[2]

$$\omega_0 = \sqrt{\frac{k}{m}} \tag{3.12}$$

then, the potential energy for a harmonic oscillator can also be written as

$$V = \frac{1}{2} m\omega_0^2 x^2 \tag{3.13}$$

The 4D Hamiltonian then becomes

$$\left[\frac{\widehat{\mathbf{p}}^2}{2m} + V(x) - i\hbar\frac{\partial}{\partial t}\right]\psi = iE_I\psi \tag{3.14}$$

or

$$\left[-\frac{\hbar^2}{2m}\frac{\partial^2}{\partial x^2} + \frac{1}{2}m\omega_0^2 x^2 - i\hbar\frac{\partial}{\partial t}\right]\psi = iE_I\psi \tag{3.15}$$

[2]The next two relations are well-known for the classical harmonic oscillator, so we do not derive them here.

Again, we take the wave function $\psi(x,t)$ to be separable so that ψ can be written as

$$\psi(x,t) = \psi_x(x) \cdot \psi_t(t) \tag{3.16}$$

Substitution of Eq. (3.16) into Eq. (3.15), we have

$$-\frac{\hbar^2}{2m}\frac{d^2\psi_x}{dx^2} \cdot \psi_t(t) + \frac{1}{2}m\omega_0^2 x^2 \psi_x \cdot \psi_t \cdots$$
$$-i\hbar \cdot \psi_x(x) \cdot \frac{d\psi_t}{dt} = iE_I \psi_x(x) \cdot \psi_t(t) \tag{3.17}$$

Dividing both sides by $\psi = \psi_x \cdot \psi_t$, we have

$$-\frac{\hbar^2}{2m}\frac{1}{\psi_x}\frac{d^2\psi_x}{dx^2} + \frac{1}{2}m\omega_0^2 x^2 - i\hbar\frac{1}{\psi_t}\frac{d\psi_t}{dt} = iE_I \tag{3.18}$$

Just as we found in the plane wave problem in the previous example, this leads to two independent equations since the LHS has two terms depending on respective independent variables. This leads to following pair of equations to be solved:

$$-\frac{\hbar^2}{2m}\frac{d^2\psi_x}{dx^2} + \frac{1}{2}m\omega_0^2 x^2 \psi_x = E_R \psi_x \tag{3.19a}$$

$$\frac{d\psi_t}{dt} = \frac{E_R - iE_I}{i\hbar}\psi_t \tag{3.19b}$$

For the *time-dependent* part, we know from the previous example that the solution is given by

$$\psi_t(t) = \psi(0)e^{-i\frac{E_R}{\hbar}t}e^{-\frac{E_I}{\hbar}t} \tag{3.20}$$

Substituting Eq. (3.20) into Eq. (3.19b) verifies that it is a solution.

In this problem, the time-dependent solution, in particular, contains a *new* addition to the harmonic oscillator eigensolution. Consequently, Eq. (3.20) leads to a more general solution describing a *damped* harmonic oscillator with a position expectation value $\langle x \rangle(t)$ given by

$$\langle x \rangle(t) = x_0 e^{-\frac{E_I}{2\hbar}t}\cos(\omega_0 t) \tag{3.21}$$

This result will be worked out in detail below. However, in what remains to be solved, though it is nontrivial, it is identical to the spatial solution *without* the 4D contribution. However, we include the derivation of the spatial solution so that it can be used to determine the expectation value given by Eq. (3.21) . If you are already familiar with it, feel free to skip ahead.

3.2.1 Spatial Dependence Solvable By Frobenius Method

Eq. (3.19a) can be solved for $\psi_x(x)$ by using the same method we deployed for the cyclotron problem in Sec. 2.6. It is known as the *Method of Frobenius* or *Frobenius method.* The method solves a certain class of second order ordinary differential equations by assuming that the solution takes the form of an infinite series. Substituting the infinite series into the differential equation of interest, and setting up appropriate conditions allow determination of all parameters associated with the series. In our case, we must further impose boundary conditions of the wave function ψ_x, as well. Using this method, the complete spatial solution ψ_x can be obtained. Its combination with Eq. (3.20) provides the complete solution.

First, let us write Eq. (3.19a) as

$$\frac{d^2\psi_x}{dx^2} + \frac{2m}{\hbar^2}\left[E_R - \frac{1}{2}m\omega_0^2x^2\right]\psi_x = 0 \tag{3.22}$$

As we did in Sec. 2.6, we introduce some useful parameters to both simplify the notation as well as modify its form to be more susceptible to the *Frobenious Method.* Let the dimensionless parameter ζ, replacing x, be defined as

$$\zeta = \alpha x \Rightarrow d\zeta = \alpha dx \tag{3.23}$$

We'll define the parameter α shortly. From this definition, the first derivative of ψ_x becomes

$$\frac{d\psi_x}{dx} = \frac{d\psi_x}{d\zeta}\frac{d\zeta}{dx} = \alpha\frac{d\psi_x}{d\zeta}$$

Continuing differentiation, the second derivative becomes

$$\frac{d^2\psi_x}{dx^2} = \frac{d\zeta}{dx}\frac{d}{d\zeta}\left(\alpha\frac{d\psi_x}{d\zeta}\right) = \alpha^2\frac{d^2\psi_x}{d\zeta^2}$$

Substituting of the first and second derivatives into Eq. (3.22), we have an equivalent equation in terms of ζ, given by

$$\alpha^2\frac{d^2\psi_x}{d\zeta^2} + \frac{2m}{\hbar^2}\left[E_R - \frac{1}{2\alpha^2}m\omega^2\zeta^2\right]\psi_x(\zeta) = 0 \tag{3.24}$$

or

$$\frac{d^2\psi_x}{d\zeta^2} + \left[\frac{2mE_R}{\alpha^2\hbar^2} - \frac{m^2\omega^2}{\alpha^4\hbar^2}\zeta^2\right]\psi_x(\zeta) = 0 \tag{3.25}$$

This form can be used to define the parameter α. Let α be defined as that value that makes the coefficient for ζ^2 (last term in square brackets) equal unity or 1. To satisfy this condition, let

$$\alpha^4 = \frac{m^2\omega^2}{\hbar^2}\zeta^2 \Rightarrow \underline{\alpha = \sqrt{\frac{m\omega}{\hbar}}} \tag{3.26}$$

To check that $\zeta = \alpha x$ is *dimensionless*, or that α has units of inverse length, we can check the following:

$$\begin{aligned} \alpha = \sqrt{\frac{m\omega}{\hbar}} \Rightarrow \text{units}(\alpha) &= \sqrt{\frac{\text{kg}\cdot\text{rad}}{\text{s}\cdot\text{J}\cdot\text{s}}} \\ &= \sqrt{\frac{\text{kg}\cdot\text{rad}\cdot\text{s}^2}{\text{s}^2\cdot\text{kg}\cdot\text{m}^2}} \\ &= \sqrt{\frac{1}{\text{m}^2}} = \frac{1}{\text{m}} = \text{m}^{-1} \end{aligned}$$

With this, Eq. (3.25) simplifies to

$$\frac{d^2\psi_x}{d\zeta^2} + \left[\frac{2E_R}{\hbar\omega} - \zeta^2\right]\psi_x(\zeta) = 0 \tag{3.27}$$

Let us define another *dimensionless* parameter Ω as

$$\underline{\Omega = 2E_R/\hbar\omega_0} \tag{3.28}$$

In SI units, the units of $\hbar\omega_0$ =J-s·s^{-1} =J (Joules), consistent with the energy units of E_R. Eq. (3.27) now becomes

$$\frac{d^2\psi_x}{d\zeta^2} + \left[\Omega - \zeta^2\right]\psi_x(\zeta) = 0 \tag{3.29}$$

We want to transform Eq. (3.29) into the form of the Confluent Hypergeometric Equation (CHE), given by

$$x\frac{d^2y}{dx^2} + (c - x)\cdot\frac{dy}{dx} - ay(x) = 0 \tag{3.30}$$

We know how to obtain the solution to Eq. (3.30). So, if we get Eq. (3.29) into the form of Eq. (3.30), then we can identify the necessary parameters a and c as the parameters in the solution.

For this objective, we must make a few more transformations. Let us also define the parameter γ as

$$\gamma = \zeta^2 \rightarrow \frac{d\gamma}{d\zeta} = 2\zeta \tag{3.31}$$

Using γ, the first derivative $d\psi_x/d\zeta$ becomes

$$\frac{d\psi_x}{d\zeta} = \frac{d\psi_x}{d\gamma}\frac{d\gamma}{d\zeta} = \frac{d\psi_x}{d\gamma}2\zeta$$

For the second derivative, we have

$$\frac{d^2\psi_x}{d\zeta^2} = \frac{d}{d\zeta}\left[\frac{d\psi_x}{d\gamma}2\zeta\right] = \frac{d}{d\gamma}\left[\frac{d\psi_x}{d\gamma}2\zeta\right]\frac{d\gamma}{d\zeta} \tag{3.32}$$

This gives

$$\frac{d^2\psi_x}{d\zeta^2} = 4\zeta^2\frac{d^2\psi_x}{d\gamma^2} + 2\frac{d\psi_x}{d\gamma} = 4\gamma\frac{d^2\psi_x}{d\gamma^2} + 2\frac{d\psi_x}{d\gamma} \tag{3.33}$$

Substituting the result in Eq. (3.33) and dividing by 4γ, leads to

$$\frac{d^2\psi_x}{d\gamma^2} + \frac{1}{2\gamma}\frac{d\psi_x}{d\gamma} + \left[\frac{\Omega}{4\gamma} - \frac{1}{4}\right]\psi_x(\gamma) = 0 \tag{3.34}$$

or, if we multiply by 4γ, we have

$$4\gamma\frac{d^2\psi_x}{d\gamma^2} + 2\frac{d\psi_x}{d\gamma} + [\Omega - \gamma]\psi_x(\gamma) = 0 \tag{3.35}$$

To obtain an asymptotic solution, we may take the limit as $\gamma \rightarrow \infty$. Then, Eq. (3.34) becomes

$$\frac{d^2\psi_x}{d\gamma^2} - \frac{1}{4}\psi_x(\gamma) = 0 \tag{3.36}$$

Eq. (3.36) has a general solution given by

$$\psi_{x\infty} = Ce^{-\frac{1}{2}\gamma} + De^{+\frac{1}{2}\gamma} \tag{3.37}$$

C and D are integration constants independent of γ. Further consideration of the required boundary condition that must be satisfied, namely

$$\psi_x(\infty) = 0 \tag{3.38}$$

It follows that $D = 0$, otherwise ψ would not satisfy Eq. (3.38). Thus,

$$\psi_{x\infty} = Ce^{-\frac{1}{2}\gamma} \tag{3.39}$$

For now, let $C = 1$ for convenience, and set $\psi_x(\gamma) = \psi_{x0} \cdot \psi_{x\infty} = \psi_{x\infty}(\gamma)e^{-\frac{1}{2}\gamma}$. If we are able to obtain the solution under this assumption, then, we will be done. With ψ_x in this form, the *first* derivative becomes

$$\frac{d\psi_x}{d\gamma} = \left[\frac{d\psi_{x0}}{d\gamma} - \frac{1}{2}\psi_{x0}\right]e^{-\frac{1}{2}\gamma} \tag{3.40}$$

Using Eq. (3.40) to find the *second* derivative, we have

$$\frac{d^2\psi_x}{d\gamma^2} = \left[\frac{d^2\psi_{x0}}{d\gamma^2} - \frac{d\psi_{x0}}{d\gamma} + \frac{1}{4}\psi_{x0}\right]e^{-\frac{1}{2}\gamma} \tag{3.41}$$

Now we see that all terms have a factor $e^{-\frac{1}{2}\gamma}$, so this just factors out. Substituting these results into Eq. (3.35) gives

$$4\gamma\left[\frac{d^2\psi_{x0}}{d\gamma^2} - \frac{d\psi_{x0}}{d\gamma} + \frac{1}{4}\psi_{x0}\right] + 2\left[\frac{d\psi_{x0}}{d\gamma} - \frac{1}{2}\psi_{x0}\right] + [\Omega - \gamma]\psi_{x0} = 0 \tag{3.42}$$

This result can be rearranged into the following *Confluent Hypergeometric differential equation*:

Oscillator Confluent Hypergeometric Equation

$$\gamma\frac{d^2\psi_{x0}}{d\gamma^2} + \left(\frac{1}{2} - \gamma\right)\frac{d\psi_{x0}}{d\gamma} - \left(\frac{1-\Omega}{4}\right)\psi_{x0} = 0 \tag{3.43}$$

Note that our equation also has been divided by 4. The dimensionless parameter Ω is given by Eq. (3.28). By comparing Eq. (3.30) and Eq. (3.43), we find that the parameters of the Confluent Hypergeometric function are given by

$$a = \frac{1-\Omega}{4} \quad \text{and} \quad c = \frac{1}{2} \tag{3.44}$$

Summarizing what we have established so far, we have the complete time dependent solution factor $T(t)$. For the spatially dependent part ψ_x, we have found that the exact solution is in the form $X = \psi_{x0}(\gamma)e^{-\frac{1}{2}\gamma}$ where ψ_{x0} is described by the ordinary differential equation given in Eq. (3.43). Thus, all that remains is to solve for $\psi_{x0}(\gamma)$.

3.2.2 Application of Frobenious Method

We can now find the solution to Eq. (3.30) because the CHE can be solved using the *method of Frobenius,* which yields a series in terms of the two parameters a and c given by Eq. (3.44). Assume that $\psi_{x0}(\gamma)$ takes the form of an infinite power series given by

$$\psi_{x0}(\gamma) = \sum_{r=0}^{\infty} d_r \gamma^{p+r} \tag{3.45a}$$

The parameter p is introduced intentionally to ensure that by taking the derivatives, it ends up in the coefficients of the equation. With Eq. (3.45a), the first derivative of $\psi_{x0}(\gamma)$ becomes

$$\frac{d\psi_{x0}}{d\gamma} = \sum_{r=0}^{\infty} (p+r) d_r \gamma^{p+r-1} \tag{3.45b}$$

Continuing on, the second derivative is given by

$$\frac{d^2\psi_{x0}}{d\gamma^2} = \sum_{r=0}^{\infty} (p+r-1)(p+r) d_r \gamma^{p+r-2} \tag{3.45c}$$

Substituting Eqs. (3.45a) - (3.45c) into Eq. (3.43), we have

$$\gamma\left[\sum_{r=0}^{\infty}(p+r-1)(p+r)d_r\gamma^{p+r-2}\right] + (c-\gamma)\left[\sum_{r=0}^{\infty}(p+r)d_r\gamma^{p+r-1}\right] - a\left[\sum_{r=0}^{\infty} d_r\gamma^{p+r}\right] = 0 \tag{3.46}$$

Multiplying γ, we get

$$\left[\sum_{r=0}^{\infty}(p+r-1)(p+r)d_r\gamma^{p+r-1}\right] + c\left[\sum_{r=0}^{\infty}(p+r)d_r\gamma^{p+r-1}\right] - \left[\sum_{r=0}^{\infty}(p+r)d_r\gamma^{p+r}\right] - a\left[\sum_{r=0}^{\infty} d_r\gamma^{p+r}\right] = 0 \tag{3.47}$$

Some of the terms in Eq. 3.47 have the same exponent powers so they can be combined. In fact, there are only two unique terms in the above result. The combining of terms actually reduces Eq. (3.47) to

$$\sum_{r=0}^{\infty} d_r(p+r+c-1)(p+r)\gamma^{p+r-1} + \sum_{r=0}^{\infty} d_r(p+r+a)\gamma^{p+r} = 0 \tag{3.48}$$

The solution remains the same if we multiply Eq. (3.48) by γ^{1-p}. This gives

$$\sum_{r=0}^{\infty} d_r(p+r+c-1)(p+r)\gamma^{r} - \sum_{r=0}^{\infty} d_r(p+r+a)\gamma^{r+1} = 0$$

or

$$\sum_{r=0}^{\infty} d_r(p+r+c-1)(p+r)\gamma^{r} = \sum_{r=0}^{\infty} d_r(p+r+a)\gamma^{r+1} \tag{3.49}$$

If Eq. (3.49) is to be valid *for any value of* γ, all the coefficients on the left must be equal to the corresponding coefficients on the right having the same exponent factor, *etc*. The power or exponent on the RHS is $r+1$, while on the LHS, it is r. This means that the *very first* term on the left ($r=0$) has no counterpart on the RHS. In other words, this term must be equal to zero, or that

Indicial Equation

$$d_0 p(p+c-1) = 0 \tag{3.50}$$

Eq. (3.50) if true if and only if

$$p = 1-c \quad \text{or} \quad p = 0 \tag{3.51}$$

Eq. (3.50) is known as the *indicial equation*. It determines the allowed values of parameter p defining the two independent solutions to the second order ordinary differential equation. We must find *both* solutions corresponding to each value of p, given by Eq. (3.51).

Let us write out the first few terms of the series from both sides. This gives

$$d_0(p+c-1)(p) + d_1(p+c)(p+1)\gamma + d_2(p+c+1)(p+2)\gamma^2 + \cdots =$$

$$d_0(p+a)\gamma + d_1(p+a+1)\gamma^2 + d_2(p+a+2)\gamma^3 + \cdots \quad (3.52)$$

The d_0 coefficient on the RHS equates with (has the same exponent) as the d_1 coefficient on the LHS; the d_1 term on the RHS equates with the d_2 term, and so on. By equating the coefficients of equal powers, we have the relation

$$(p+r-1+a)\cdot d_{r-1} = (p+r+c-1)\cdot(p+r)\cdot d_r$$

or

Recurrence Relation

$$\frac{d_r}{d_{r-1}} = \frac{(p+r-1+a)}{(p+r)(p+r+c-1)} \quad (3.53)$$

Eq. (3.53) is known as the *recurrence relation* or *recurrence ratio*. It defines an *iterative* sequence that determines all the coefficients for the infinite power series solution. We only need to determine d_0. Fortunately for us, its a simple matter because wave function solutions are normalized (*i.e.* must satisfy $\int_{-\infty}^{\infty} |\psi_x|^2 dx = 1$). So, we can assume $d_0 = 1$. Then, for both independent solutions, all coefficients d_r can be computed sequentially, or in a recurring manner by using Eq. (3.53). This provides for *two* independent solutions for $\psi_{x0}(\gamma)$.

Let us denote the independent solution depending on parameter p by $S_p(\gamma)$. Using Eq. (3.53), each of the independent solutions is given by

$$S_p(\gamma;a,c) = 1 + d_0\frac{(p+a)}{(p+1)(p+c)}\gamma + d_1\frac{(p+a+1)}{(p+2)(p+c+1)}\gamma^2 +$$

$$d_2\frac{(p+a+2)}{(p+3)(p+c+2)}\gamma^3 + d_3\frac{(p+a+3)}{(p+4)(p+c+3)}\gamma^4 +$$

$$d_4\frac{(p+a+4)}{(p+5)(p+c+4)}\gamma^5 + d_5\frac{(p+a+5)}{(p+6)(p+c+5)}\gamma^6 + \ldots \quad (3.54)$$

Since we know the values for p (given by Eq. (3.51)) from the *indicial equation*, we can write S_p in an explicit form for the *first* independent solution S_0, where $p = 0$. This solution becomes

$$S_0(\gamma;a,c) = 1 + d_0\frac{a}{1(c)}\gamma + d_1\frac{(a+1)}{2(c+1)}\gamma^2 +$$
$$d_2\frac{(a+2)}{3(c+2)}\gamma^3 + d_3\frac{(a+3)}{4(c+3)}\gamma^4 +$$
$$d_4\frac{(a+4)}{5(c+4)}\gamma^5 + d_5\frac{(a+5)}{6(c+5)}\gamma^6 - \cdots \quad (3.55)$$

Now, let us substitute the coefficients in terms of the preceding one using the recurrence relation, *e.g.* $d_0 \equiv 1$, $d_1 = a/c$, etc. Then, we have the following series for the first independent solution S_0:

$$S_0(\gamma;a,c) = 1 + \frac{1}{1!}\frac{a}{c}\gamma + \frac{1}{2!}\frac{a(a+1)}{c(c+1)}\gamma^2 +$$

$$\frac{1}{3!}\frac{a(a+1)(a+2)}{c(c+1)(c+2)}\gamma^3 + \frac{1}{4!}\frac{a(a+1)(a+2)(a+3)}{c(c+1)(c+2)(c+3)}\gamma^4 +$$

$$\frac{1}{5!}\frac{a(a+1)(a+2)(a+3)(a+4)}{c(c+1)(c+2)(c+3)(c+4)}\gamma^5 +$$

$$\frac{1}{6!}\frac{a(a+1)(a+2)(a+3)(a+4)(a+5)}{c(c+1)(c+2)(c+3)(c+4)(c+5)}\gamma^6 + \cdots \quad (3.56)$$

The infinite series given by Eq. (3.56) is a type of *Confluent Hypergeometric Series* (CHS). This series form is known as the *Confluent Hypergeometric Function*(*CHF*) (of the first kind).

The CHF is often denoted by ${}_1F_1(x;a,c)$. The first argument x is the independent variable, and the second and third are only constant input parameters over the range of x. The *CHF* is sometimes written using a shorthand notation with symbols known as *Pochhammer symbols* or *rising factorials*, defined as

$$a(a+1)(a+2)...(a+r) \equiv (a)_r \quad (3.57)$$

In terms of *Pochhammer symbols*, Eq. (3.56) becomes

$$S_0(\gamma;a,c) = {}_1F_1(x;a,c) = 1 + \sum_{r=1}^{\infty}\frac{1}{r!}\frac{(a)_{r-1}}{(c)_{r-1}}\gamma^r \quad (3.58)$$

Alternatively, any coefficient d_r, starting from $r = 1$, can be written as

S_0 Coefficients

$$d_r = \frac{1}{r!}\frac{\prod_{k=1}^{r}(a+k-1)}{\prod_{k=1}^{r}(c+k-1)} \tag{3.59}$$

$\prod$ denotes the multiplication product symbol for multiplication of terms. Then, we have the *first* independent solution S_0 given by

$$S_0 = {}_1F_1(\gamma;a,c) = \sum_{r=0}^{\infty} d_r\gamma^{p+r} = \gamma^p \sum_{r=0}^{\infty} d_r\gamma^r = \sum_{r=0}^{\infty} d_r\gamma^r \tag{3.60}$$

or

First Independent Solution

$$S_0 = {}_1F_1\left(\gamma;a=\frac{1-\Omega}{4},c=\frac{1}{2}\right) = \sum_{r=0}^{\infty} d_r\gamma^r \tag{3.61}$$

where d_r is given by Eq. (3.59). The zero order coefficient is taken to be $d_0 = 1$, so all subsequent coefficients can be determined using Eq. (3.53).

To obtain the *second* independent solution S_{1-c}, we can take advantage of the fact that a and c are unspecified parameters at this point. When $p = 0$, Eq. (3.53) gives

$$\frac{d_r}{d_{r-1}} = \frac{(p+r-1+a)}{(p+r)(p+r+c-1)} = \frac{(r-1+a)}{r(r+c-1)} \tag{3.62}$$

But, if we substitute $p = 1-c$, we have

$$\frac{d_r}{d_{r-1}} = \frac{((1-c)+r-1+a)}{((1-c)+r)((1-c)+r+c-1)} = \frac{(r-1+(a-c+1))}{r(r+(2-c)-1)} \tag{3.63}$$

Eq. (3.63) takes the same form as Eq. (3.62) only with modified parameters a' and c' in terms of $a = (1-\Omega)/4$, $c = 1/2$ given by

$$a' = a-c+1 = \frac{3-\Omega}{4} \tag{3.64a}$$

$$c' = 2-c = \frac{3}{2} \tag{3.64b}$$

Similar to our first solution, we now have

$$\frac{d_r}{d_{r-1}} = \frac{(r-1+a')}{r(r+c'-1)} \tag{3.65}$$

Then, the **second independent solution** $S_{1-c} = S_{1/2}$ becomes

Second Independent Solution

$$S_{1/2} = {}_1F_1\left(\gamma; a' = \frac{3-\Omega}{4}, c' = \frac{3}{2}\right) = \gamma^{1/2}\sum_{r=0}^{\infty} d_r\gamma^r \tag{3.66}$$

Now, the coefficient outside of the summation $\gamma^p = \gamma^{1-c} = \gamma^{1/2}$. Given the modified parameters for the second solution, the coefficients maintain consistent form as

$$d_r = \frac{1}{r!}\frac{\prod_{k=1}^{r}(a'+k-1)}{\prod_{k=1}^{r}(c'+k-1)} \tag{3.67}$$

Eq. (3.67) must be used with Eq. (3.66).

This *almost* completes the spatial solution for the harmonic oscillator. There remains one more consideration, which turns out to be a big one. This is, the boundary condition must also be satisfied, which is given by

$$\psi_x(x=\infty) = 0$$

Let us discuss this point of satisfying the boundary condition next.

3.2.3 Conditions For CHF To Satisfy Boundary Condition

The Confluent Hypergeometric function (CHF) is a generalization of the infinite series of the *exponential function* $\exp(x) = e^x$. This can be seen if we consider the specific case when $a = c$. Then, the CHF becomes

$${}_1F_1(\gamma; a, c) = 1 + \frac{1}{1!}\frac{a}{c}\gamma + \frac{1}{2!}\frac{a(a+1)}{c(c+1)}\gamma^2 +$$

$$\frac{1}{3!}\frac{a(a+1)(a+2)}{c(c+1)(c+2)}\gamma^3 + \frac{1}{4!}\frac{a(a+1)(a+2)(a+3)}{c(c+1)(c+2)(c+3)}\gamma^4 +$$

$$\frac{1}{5!}\frac{a(a+1)(a+2)(a+3)(a+4)}{c(c+1)(c+2)(c+3)(c+4)}\gamma^5+$$

$$\frac{1}{6!}\frac{a(a+1)(a+2)(a+3)(a+4)(a+5)}{c(c+1)(c+2)(c+3)(c+4)(c+5)}\gamma^6+...$$

$$=1+\frac{1}{1!}\gamma+\frac{1}{2!}\gamma^2+\frac{1}{3!}\gamma^3+\frac{1}{4!}\gamma^4+\frac{1}{5!}\gamma^5+\frac{1}{6!}\gamma^6+... \quad (3.68)$$

So,

$$_1F_1(\gamma;a,c)=1+\frac{1}{1!}\gamma+\frac{1}{2!}\gamma^2+\frac{1}{3!}\gamma^3+\frac{1}{4!}\gamma^4+\frac{1}{5!}\gamma^5+\frac{1}{6!}\gamma^6+...=e^\gamma$$

This means that the CHF ($_1F_1$) inherently has e^γ behavior while $\gamma=\zeta^2=\alpha^2x^2>0$. Since our solution also involves a factor of $\psi_{x\infty}\propto e^{-\gamma/2}$ (c.Eq. 3.39), if we leave $_1F_1$ in its *infinite* series form, the solution behaves like $e^{-\gamma/2}\cdot e^\gamma=e^{\gamma/2}$. This function blows up at ∞ and, therefore, *cannot* satisfy the boundary condition for a wave function $\lim_{|x|\to\infty}\psi^*\psi=0$. The *only* way that the CHF can satisfy the wave-function's boundary condition is if it is turned into a polynomial p_n of *finite order n* because then $\lim_{|x|\to\infty}\psi^*\psi=0$ *is* satisfied because any finite polynomial in γ when multiplied by $e^{-\gamma/2}$ *is* finite at $\gamma=\infty$ because the exponential terms converges more rapidly, compared to a finite order polynomial p_n. Thus, the infinite series *must* be truncated.

$_1F_1$ can be truncated to a finite order polynomial if the numerators d_r vanish at some finite order n in the series. So, the condition to render a finite polynomial from $_1F_1$ becomes

$$a=\frac{1-\Omega}{4}=-n \quad (n\text{=0,1,2,3,..}) \quad (3.69)$$

Likewise, since $a'=a-c+1$ is a function of a and c, we have

$$a'=a-c+1=a+\frac{1}{2}=-n \quad (3.70)$$

If a is a negative integer, $_1F_1\times\psi_{x\infty}$ satisfies the boundary condition.

Notice that when $a=-n$, it follows that $a'=-n+1/2$ is *not* an integer. Since a and a' cannot both be negative integers simultaneously, it is not possible for both independent solutions S_0 and $S_{1/2}$ to satisfy the boundary condition *simultaneously*. Then, *only one* of the two independent solutions can be nonzero. For the second solution, the condition is

$$a'=a-c+1=a+\frac{1}{2}=\frac{1-\Omega}{4}+\frac{1}{2}=\frac{3-\Omega}{4}=-n \quad (3.71)$$

With both cases combined, when we solve for Ω, we have

$$1-\Omega=-4n \Rightarrow \Omega=4n+1=1,5,9,13,\ldots \tag{3.72a}$$

$$3-\Omega=-4n \Rightarrow \Omega=4n+3=3,7,11,15,\ldots \tag{3.72b}$$

Consequently, the Ω *only* takes *odd integer values, i.e.*

$$\Omega=2n+1 \tag{3.73}$$

Then, using the definition of Ω given by Eq. (3.28), we determine the corresponding *real* eigenenergy values E_R as

$$\Omega=\frac{2E_R}{\hbar\omega_0}=2n+1 \tag{3.74}$$

Solving for the allowed energy states of the harmonic oscillator, we find that E_R has *discrete* energy values given by

1D Harmonic Oscillator Energy

$$E_R=\left(n+\frac{1}{2}\right)\hbar\omega \tag{3.75}$$

The solution ψ_x for the harmonic oscillator is given by

$$\psi_x(\gamma;a,c)=\left[C_1\cdot{}_1F_1\left(\gamma;\frac{1-\Omega}{4},\frac{1}{2}\right)+C_2\cdot\gamma^{\frac{1}{2}}{}_1F_1\left(\gamma;\frac{3-\Omega}{4},\frac{3}{2}\right)\right]e^{-\frac{1}{2}\gamma}$$

Substituting $\gamma=\zeta^2$, where $\zeta=\alpha x$, in terms of the spatial variable x, we have

Harmonic Oscillator Spatial Eigensolution

$$\psi_{x,n}=\left[C_1\cdot{}_1F_1\left(\alpha x;-n,\frac{1}{2}\right)+C_2\alpha\cdot x\cdot{}_1F_1\left(\alpha x;-n',\frac{3}{2}\right)\right]e^{-\frac{1}{2}\alpha^2x^2} \tag{3.76}$$

and

1D Harmonic Oscillator Space+Time Eigensolution

$$\psi_n(x,t) = C_n \psi_{x,n}(x) e^{-\frac{E_I}{\hbar}t} \tag{3.77}$$

where $\alpha = \sqrt{(m\omega_0)/\hbar}$ and we must also remember that either $C_1 = 0$ or $C_2 = 0$ for any single solution.

There is another way to represent the harmonic oscillator solution given by Eq. (3.76) *without* having to keep track of whether $C_1 = 0$ or $C_2 = 0$ when n or n' is chosen. This condition of a and a' as negative integers turns out to generate a well-known set of polynomials known as *Hermite polynomials*. Let's see how.

3.2.4 Truncated CHF Generates Hermite Polynomials H_n

In the previous section, we showed that in order for the wave function ψ to satisfy the boundary condition, it is necessary that a and/or a' be a negative integer $-n$. This results in the truncation of the *Confluent Hypergeometric Function* (CHF) because all terms beyond the nth become zero because they have a $a+n = -n+n = 0$ factor. There is no approximation in satisfying these conditions. The solution is still exact. The truncation of the CHF results in polynomials of order n, where $n = |a|$ or $n = |a'|$. To better see how the truncated CHFs produce Hermite polynomials, in particular, let us compute the first five ($n = 0, 1, 2, 3, 4$).

$n = 0$: In the case where n takes its smallest value $n = 0$, , Eq. (3.73) gives $\Omega = 1$. Since truncating ${}_1F_1$ leads to a polynomial, the second independent solution with C_2 is at least linear in x, due to the αx factor. For a constant polynomial (of order $n = 0$), since we have established that one of the coefficients C_1 and C_2 *must* be zero, it follows, in this case, that $C_2 = 0$ (otherwise, there'd be a linear term). The truncated series then gives

$${}_1F_1\left(\gamma; a=0, \frac{1}{2}\right) = 1 + \frac{1}{1!}\frac{a}{c}\gamma + \frac{1}{2!}\frac{a(a+1)}{c(c+1)}\gamma^2 + ... = 1 + 0\gamma + 0\gamma^2 + ... = 1$$

To express the resulting polynomials in the variable ζ, the convention is to scale the polynomial such that the coefficient of the highest power (the leading coefficient) of γ, which is n,is equal to $+2^n$. Doing this scales the equation by a constant. Scaling the truncated polynomials in this way enables a systematic production of *Hermite Polynomials* H_n. This means that by knowing the CHF ${}_1F_1$, any Hermite polynomial can be generated. For the case of $|a| = n = 0$, the leading coefficient becomes $2^0 = 1$, which is what we already have. Then, the lowest order Hermite polynomial H_0 is given by

$$H_0(\zeta) = 1 \tag{3.78}$$

This turns out to be the only case where ${}_1F_1 = H_n$.

$n = 1$: In this case, using Eq. (3.74) gives $\Omega = 2n + 1 = 3$, so $a' = 0$. This also leads to $a = -1/2$. As expected, when one is integer, the other is not. In this case, C_1 must be zero in order to satisfy the boundary conditions. Using the above result for $a' = 0$, we know

$${}_1F_1\left(\gamma; 0, \frac{3}{2}\right) = 1$$

The second solution, corresponding to $p = 1 - c$ also has a factor $\gamma^{1-c} = \gamma^{\frac{1}{2}} = \zeta$. When we include this, the truncated polynomial becomes

$$S_{1-c} = \gamma^{1-c}{}_1F_1\left(\gamma; a' = 0, \frac{3}{2}\right) = \gamma^{\frac{1}{2}}(1 - 0\gamma + 0...) = \zeta \tag{3.79}$$

Scaling the resulting polynomial to obtain the Hermite polynomial H_1, the highest power coefficient is $2^n = 2^1 = 2$. Then, for $n = 1$, we have

$$H_1(\zeta) = 2^1\gamma^{\frac{1}{2}} = 2\zeta \tag{3.80}$$

$n = 2$: Now, we have $\Omega = 2n+1 = 5$. This yields $a = -1$, thus $a' = a-c+1 = -1 + 1/2 + 1 = 1/2$ is not an integer. Therefore, $C_2 = 0$. This gives

$$S_0\left(\gamma; a = -1, \frac{1}{2}\right) = 1 - \frac{1}{1/2}\gamma + 0... = 1 - 2\gamma \tag{3.81}$$

Scaling Eq. (3.81) so that the highest power has a coefficient $2^n = 2^2 = 4$, the *Hermite polynomial* H_2 becomes

$$H_2(\zeta) = 4\gamma - 2 = 4\zeta^2 - 2 \tag{3.82}$$

Hopefully, the pattern is becoming clear. Let's do two more examples.

$n = 3$: In this case, since $\Omega = 2n + 1 = 7$, $a' = (3 - \Omega)/4 = -1$ while $a = (1 - \Omega)/4 = -3/2$. Therefore, since a is not an integer, $C_1 = 0$. We then have

$$S_{1-c}\left(\gamma; a' = -1, \frac{3}{2}\right) = \gamma^{\frac{1}{2}}\left(1 - \frac{1}{3/2}\gamma + 0...\right) = \gamma^{\frac{1}{2}}\left(1 - \frac{2}{3}\gamma\right) \tag{3.83}$$

We want the leading coefficient to be $+2^3$. For this, we can scale by $-\frac{3}{2} \times +2^3 = -12$. With this, further substituting $\gamma = \zeta^2$, the Hermite polynomial H_3 becomes

$$H_3(\zeta) = 8\zeta^3 - 12\zeta \tag{3.84}$$

$n = 4$: In this case, $\Omega = 2 \cdot (n = 4) + 1 = 9$, leading to $a = -2$, and $a' = 6/4$. Since a' is not an integer, $C_2 = 0$ and we have

$$S_0\left(\gamma; a = -2, \frac{1}{2}\right) = 1 + \frac{-2}{1/2}\gamma + \frac{1}{2!}\frac{-2(-2+1)}{1/2(1/2+1)}\gamma^2 + 0...) = 1 - 4\gamma + \frac{8}{6}\gamma^2$$

Scaling the result so that the leading coefficient is $+2^4 = 16$, we can use $+\frac{6}{8} \times 16 = 12$. With this, substituting $\gamma = \zeta^2$, we obtain a result for H_4 given by

$$H_4(\zeta) = 16\zeta^4 - 48\zeta^2 + 12 \tag{3.85}$$

This process can continue indefinitely to obtain any Hermite polynomial H_n, provided the values for a and c are known. The first ten Hermite polynomials are plotted in Fig. 3.3. The first fourteen ($n = 0, 1, ..., 13$) Hermite polynomials are also given in Table A.2 in the end of the Chapter.

Figure 3.3: First ten Hermite polynomials H_n. The polynomials iterate through symmetric and anti-symmetric states, as n increases. This pattern establishes sequential symmetric and anti-symmetric quantum states for the harmonic oscillator.

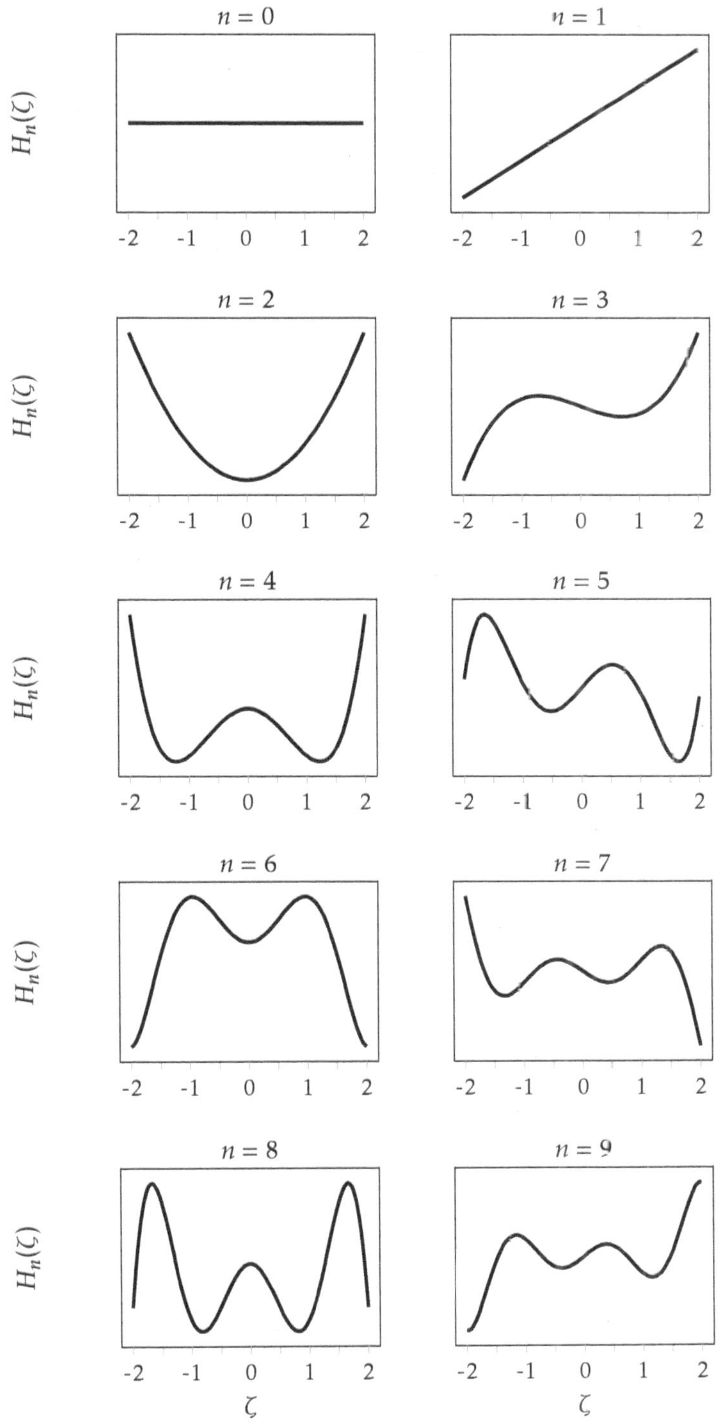
n = 0
n = 1
n = 2
n = 3
n = 4
n = 5
n = 6
n = 7
n = 8
n = 9
$H_n(\zeta)$
-2 -1 0 1 2
ζ

The Hermite polynomials iterate through symmetric and anti-symmetric eigenstates as n increases. With the Hermite polynomials H_n being proportional to the Confluent Hypergeometric Function ${}_1F_1$, we can also express the eigensolution to the Confluent Hypergeometric Equation of Eq. (3.43) as

Oscillator Eigensolution (Hermite)

$$\psi_{x,n}(\zeta) = C_n H_n(\zeta) e^{-\frac{1}{2}\zeta^2} \tag{3.86}$$

The first ten eigensolutions $\psi_{x,n}(\zeta)$ given by Eq. (3.86) are plotted in Fig. 3.4.

Figure 3.4: First ten eigensolutions $\psi_{x,n}(\zeta)$ of the harmonic oscillator problem. Owing to the Hermite polynomials being multiplied by a symmetric Gaussian function, the eigenstates iterate through symmetric and anti-symmetric spatial states as n increases.

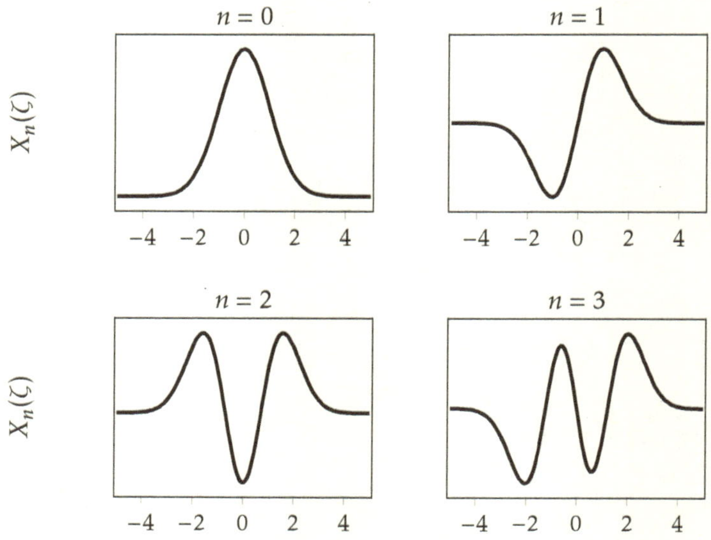

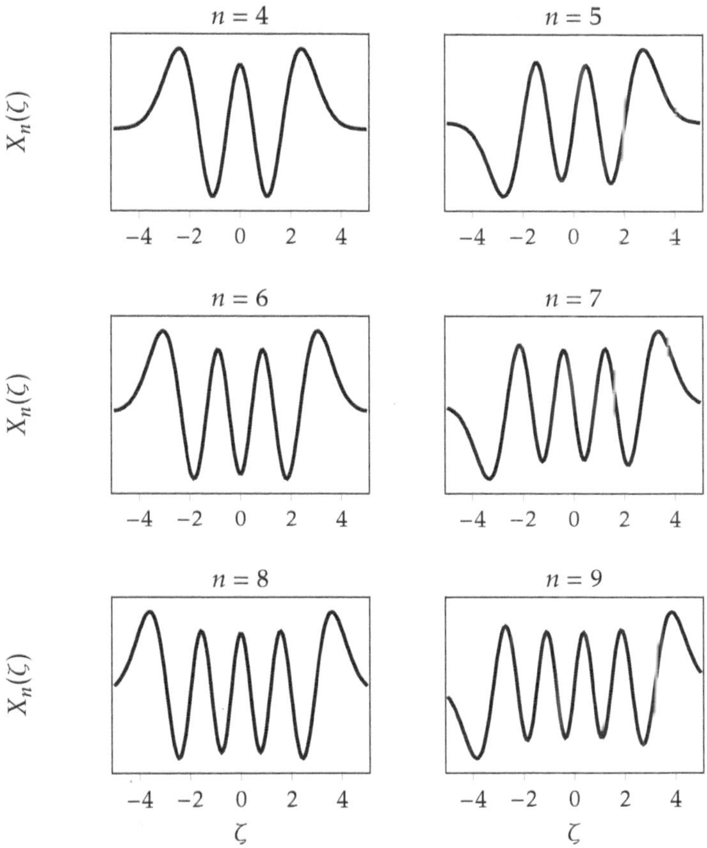

Because the Hermite polynomials are multiplied by a symmetric Gaussian function, we have that the eigenstates iterate through symmetric and anti-symmetric states, as n increases. Being able to generate Hermite polynomials from the CFE is useful because the CHF is easy to remember and we do not have to be concerned with the two independent solutions S_0 and $S_{1/2}$, tracking whether $C_1 = 0$ or $C_2 = 0$.

Having determined the eigenfunctions, we can also determine normalization constant C_n by evaluating the normalization integral using a well-known property of the Hermite polynomials, given by

Orthogonality Property Of Hermite Polynomials

$$\int_{-\infty}^{\infty} H_m(x)H_n(x)e^{-x^2}dx = 2^n n!\sqrt{\pi}\delta_{mn} \tag{3.87}$$

Using the *orthogonality property of the Hermite polynomials,* evaluation of the normalization condition leads to

$$\begin{aligned}
1 &= \int_{-\infty}^{+\infty} \psi_n(\zeta)^\dagger \psi_n(\zeta)dx \\
&= \int_{-\infty}^{+\infty} C_n^\dagger H_n^\dagger(\zeta)e^{-\zeta^2/2}C_nH_n(\zeta)e^{-\zeta^2/2}dx \\
&= \frac{|C_n|^2}{\alpha}\int_{-\infty}^{+\infty} H_n(\zeta)H_n(\zeta)e^{-\zeta^2}d\zeta \\
&= \frac{|C_n|^2}{\alpha}\left(2^n n!\sqrt{\pi}\right)
\end{aligned}$$

We have used the earlier relation $\zeta = \alpha x$. This leads us to our final result for the magnitude of the normalization factor $|C_n|$, given by

1D Oscillator Normalization Factor

$$|C_n| = \sqrt{\frac{\alpha}{2^n n!\sqrt{\pi}}} \tag{3.88}$$

where $\alpha = \sqrt{(m\omega_0)/\hbar}$ having units of m^{-1}(*c.* Eq. 3.26). m is the mass, $\omega_0 = \sqrt{k/m}$ is the natural frequency of the and $\hbar = h/2\pi$ is Planck's reduced constant. While we have the complete eigensolution now, the *most* general solution for the harmonic oscillator is a linear combination of the countably infinite eigenstates $\psi_n(\zeta, t)$. When we use the general solution, we find that it leads to the classical time-dependent solution for the oscillating position $x = x_0 \cos\omega_0 t$. The most general harmonic oscillator solution is, therefore, given by

1D Harmonic Oscillator General Solution

$$\Psi(x,t) = \sum_n d_n \psi_n e^{-E_l t/\hbar} e^{-i\left(n+\frac{1}{2}\right)\omega_0 t} \tag{3.89}$$

Based on Eq. (3.89), the initial state $\Psi(x, t = 0)$ is

$$\Psi(x,0) = \sum_n d_n \psi_n \tag{3.90}$$

This result follows because any pair of eigensolutions are *mutually orthogonal* with respect to spatial variables. Therefore, that for any two distinct integers n and m, it must be that

$$\int_{-\infty}^{\infty} \psi_m^\dagger \psi_n dx = \delta_{mn}$$

$\psi_m^\dagger$ is more general notation for the *transposed complex conjugate* of ψ_m. Since ψ_m is a scalar function in this case, it is identical to the complex conjugate of ψ_m, or $\psi_m^\dagger = \psi_m^*$. δ_{mn} is the Kronecker delta function. The integral is nonzero only when $m = n$, and in this case, the integral becomes unity since we have normalized eigensolutions. Otherwise, the result is zero, which means that the eigenfunctions are *orthogonal*. This property can be used to obtain the coefficients d_n. Multiplying the initial state by $X_m^\dagger$ and integrating over space, we obtain

$$\begin{aligned}
\int_{-\infty}^{\infty} X_m^\dagger \Psi(x,0) dx &= \int_{-\infty}^{+\infty} \sum_n d_n \psi_m^\dagger \psi_n \\
&= \sum_n d_n \int_{-\infty}^{+\infty} \psi_m^\dagger \psi_n dx \\
&= \sum_n d_n \delta_{mn} \\
&= d_m
\end{aligned}$$

Then, having the initial condition of the general wave function, we can obtain any of the coefficients by the result

State Coefficient Formula

$$d_n = \int_{-\infty}^{\infty} \psi_n^{\dagger}\Psi(x,0)dx \tag{3.91}$$

Therefore, the initial condition of the most general solution provides sufficient information to obtain the most general solution given by Eq. (3.89). After introducing an alternative notation, we can use the most general solution to find the expectation value for the position x.

3.2.5 Time-dependent Expectation Value

The conventional classical harmonic oscillator is a well-known problem. Under *non-dissapative conditions*, the time-dependence of the position of the mass $x(t)$ is given by

$$x(t) = \langle x\rangle(t) = x_0\cos(\omega_0 t) \tag{3.92}$$

x_0 is the initial condition and is related to the system energy as

$$x_0 = \sqrt{\frac{2E}{m_0\omega_0^2}} \tag{3.93}$$

E is the conserved energy of the system without dissipation. If the system also has *dissipation* or damping, $\langle x\rangle(t)$ is then given by

$$\langle x\rangle(t) = x_0 e^{-t/\tau}\cos(\omega_0 t) \tag{3.94}$$

τ is the time constant of the amplitude decay. To examine how consistent our harmonic oscillator results are with the classical result, we can evaluate $\langle x\rangle(t)$ to determine if the wave function leads to the one of the forms given by Eqs. (3.92) and (3.94). To find the time dependence of the expectation value of the position $\langle x\rangle(t)$, we need two important relations. They actually follow from the following defined *operators*

$$\widehat{a} = \frac{m\omega_0 x + i\widehat{p}}{\sqrt{2m_0\hbar\omega_0}} \quad \text{and} \quad \widehat{a}^{\dagger} = \frac{m\omega_0 x - i\widehat{p}}{\sqrt{2m_0\hbar\omega_0}} \tag{3.95}$$

$\widehat{a}$ is known as the *annihilation* operator. Using Eq. (3.95), the following useful relations may be obtained, namely

$$\widehat{a}|n\rangle = \sqrt{n}|n-1\rangle \tag{3.96}$$

$$\widehat{a}^{\dagger}|n\rangle = \sqrt{n+1}|n+1\rangle \tag{3.97}$$

$$x = \sqrt{\frac{\hbar}{2m_0\omega_0}}\left(\widehat{a}+\widehat{a}^{\dagger}\right) \tag{3.98}$$

In Eq. (3.96), $|\cdot\rangle$ is an alternative way to write a quantum eigenstate. For example

$$\psi_0 = |0\rangle \quad \psi_0^{\dagger} = \langle 0| \quad (n = 0)$$

$$\psi_1 = |1\rangle \quad \psi_1^{\dagger} = \langle 1| \quad (n = 1)$$

$$\psi_2 = |2\rangle \quad \psi_2^{\dagger} = \langle 2| \quad (n = 2)$$

$$\vdots$$

$|n\rangle$ is an example of a *Ket*, while $\langle n|$ is called a *Bra*. This alternative notation is known as the *Dirac notation*. Since its more compact, we will use it in finding the time-dependent behavior. For a more detailed introduction to the Dirac notation and its use in the harmonic oscillator problem, see the Appendix C. Then, using the Dirac notation, we can write the initial condition as

$$|C\rangle = \sum_{n=0}^{\infty} d_n|n\rangle \tag{3.99}$$

Using the *annihilation* operator $\widehat{a}$ on $|C\rangle$, and equating to its respective eigenvalue β, we have

$$\widehat{a}|C\rangle = \sum_{n=0}^{\infty} d_n\widehat{a}|n\rangle = \sum_{n=0}^{\infty} d_n\sqrt{n}|n-1\rangle = \beta|C\rangle = \beta\sum_{n=0}^{\infty} d_n|n\rangle$$

β is generally complex because the operator $\widehat{a}$ is *not Hermitian*. This is also inferrable from the considering that $\widehat{p}$ *has* real eigenvalues, and $\widehat{a} \propto i\widehat{p}$. Since the oscillator has a ground state $|0\rangle$, the very first term in the series involving $|n-1\rangle$ corresponds to $|0-1\rangle = |-1\rangle$, which is the *null-Ket* since it corresponds to an energy *below* that of the ground state. Therefore, we have

$$\sum_{n=1}^{\infty} d_n\sqrt{n}|n-1\rangle = \beta\sum_{n=0}^{\infty} d_n|n\rangle \tag{3.100}$$

If we write out the terms explicitly from Eq. 3.100, we obtain the following:

$$d_1\sqrt{1}|0\rangle + d_2\sqrt{2}|1\rangle + d_3\sqrt{3}|2\rangle + d_4\sqrt{4}|3\rangle + d_5\sqrt{5}|4\rangle + \cdots =$$

$$\beta d_0|0\rangle + \beta d_1|1\rangle + \beta d_2|2\rangle + \beta d_3|3\rangle + \beta d_4|4\rangle + \cdots$$

Equating coefficients of the eigenkets gives the following relations:

$$d_1\sqrt{1} = \beta d_0, \quad d_2\sqrt{2} = \beta d_1, \quad d_3\sqrt{3} = \beta d_2, \quad d_4\sqrt{4} = \beta d_3, \quad d_5\sqrt{5} = \beta d_4 \cdots$$

We can solve for each coefficient in terms of d_0 by solving for d_1, then successively substituting from the lowest order, *etc.* Doing this, we obtain

$$d_n = \frac{\beta}{\sqrt{n}} d_{n-1} \Rightarrow d_1 = \frac{\beta}{\sqrt{1}} d_0$$

$$d_2 = \frac{\beta}{\sqrt{2}} d_1 = \frac{\beta^2}{\sqrt{1\cdot 2}} d_0$$

$$d_3 = \frac{\beta}{\sqrt{3}} d_2 = \frac{\beta^3}{\sqrt{1\cdot 2\cdot 3}} d_0$$

$$\vdots$$

$$d_n = \frac{\beta^n}{\sqrt{n!}} d_0$$

Since β is complex, we have that d_n is, generally, complex. With the solutions for the coefficients d_n, the state $|C\rangle$ can be written as

$$|C\rangle = d_0 \sum_{n=0}^{\infty} \frac{\beta^n}{\sqrt{n!}} |n\rangle \tag{3.101}$$

We can solve for d_0 in terms of β by imposing the *normalization condition* on $|B\rangle$. Since each of the eigenstates ψ_n are orthogonal, or satisfy the condition

$$\int_{-\infty}^{+\infty} \psi_m^{\dagger} \psi_n dV = \langle m|n\rangle = \delta_{mn} \tag{3.102}$$

where $\delta_{mn} = 1$ if $m = n$, 0 otherwise. This leads to

$$1 = \langle B|B\rangle = d_0^* d_0 \sum_{m=0}^{\infty} \sum_{n=0}^{\infty} \frac{(\beta^*)^m}{\sqrt{m!}} \frac{\beta^n}{\sqrt{n!}} \langle m|n\rangle$$

$$= |d_0|^2 \sum_{m=0}^{\infty} \sum_{n=0}^{\infty} \frac{(\beta^*)^m}{\sqrt{m!}} \frac{\beta^n}{\sqrt{n!}} \delta_{mn}$$

$$= |d_0|^2 \sum_{n=0}^{\infty} \frac{\left(|\beta|^2\right)^n}{n!}$$

The MacLauren (Taylor) series expansion for e^x is given by

$$e^x = 1 + x + \frac{x^2}{2!} + \cdots = \sum_{n=0}^{\infty} \frac{x^n}{n!}$$

Exploiting this form in our normalization condition above, we have for $|d_0|^2$

$$1 = |d_0|^2 \sum_{n=0}^{\infty} \frac{\left(|\beta|^2\right)^n}{n!} = |d_0|^2 e^{|\beta|^2} \Rightarrow \underline{|d_0|^2 = e^{-|\beta|^2}}$$

or

$$|d_0| = \sqrt{e^{-|\beta|^2}} = e^{-\frac{|\beta|^2}{2}} \tag{3.103}$$

Therefore, the initial state $|C\rangle$ becomes

Initial State Of The LHO (Coherent State)

$$|C\rangle = e^{-\frac{|\beta|^2}{2}} \sum_{n=0}^{\infty} \frac{\beta^n}{\sqrt{n!}} |n\rangle \tag{3.104}$$

We can write the complete time-dependent solution $|\Psi(x,t)\rangle = |C\rangle e^{-i\frac{E}{\hbar}t}$ as

Complete Time-Dependent LHO Solution

$$|\Psi(x,t)\rangle = e^{-\frac{|\beta|^2}{2}} \sum_{n=0}^{\infty} \frac{\beta^n}{\sqrt{n!}} |n\rangle e^{-i\left(n+\frac{1}{2}\right)\omega_0 t} e^{-E_I t/\hbar} \tag{3.105}$$

Eq. (3.105) can be used to determine the time-dependent expectation value of x or $\langle x\rangle(t)$. Using the definition, $\langle x\rangle(t)$ is given by

$$\langle x\rangle(t) = \langle\Psi|x|\Psi\rangle$$

$$= \sqrt{\frac{\hbar}{2m_0\omega_0}}\langle\Psi|\widehat{a}+\widehat{a}^{\dagger}|\Psi\rangle$$

$$= \sqrt{\frac{\hbar}{2m_0\omega_0}}e^{-|\beta|^2}e^{-E_I t/2\hbar}\sum_{m=0}^{\infty}\sum_{n=0}^{\infty}\frac{(\beta^*)^m}{\sqrt{m!}}\frac{\beta^n}{\sqrt{n!}}(\langle m|\widehat{a}+\widehat{a}^{\dagger}|n\rangle)e^{-i(m-n)\omega_0 t}$$

Evaluating $\langle m|\widehat{a}+\widehat{a}^{\dagger}|n\rangle$, we have

$$\langle m|\widehat{a}+\widehat{a}^{\dagger}|n\rangle = \langle m|\widehat{a}|n\rangle + \langle m|\widehat{a}^{\dagger}|n\rangle$$

$$= \sqrt{n}\langle m|n-1\rangle + \sqrt{n+1}\langle m|n+1\rangle$$

$$= \sqrt{n}\delta_{m,n-1} + \sqrt{n+1}\delta_{m,n+1}$$

Substitution of this result into our expectation value expression leads to

$$\langle x\rangle(t) = \sqrt{\frac{\hbar}{2m_0\omega_0}}e^{-|\beta|^2}e^{-E_I t/2\hbar}\sum_{m=0}^{\infty}\sum_{n=0}^{\infty}\frac{(\beta^*)^m}{\sqrt{m!}}\frac{\beta^n}{\sqrt{n!}}(\langle m|\widehat{a}+\widehat{a}^{\dagger}|n\rangle)e^{-i(m-n)\omega_0 t}$$

$$= \sqrt{\frac{\hbar}{2m_0\omega_0}}e^{-|\beta|^2}e^{-E_I t/2\hbar}\sum_{m=0}^{\infty}\sum_{n=0}^{\infty}\frac{(\beta^*)^m}{\sqrt{m!}}\frac{\beta^n}{\sqrt{n!}}(\sqrt{n}\delta_{m,n-1} + \sqrt{n+1}\delta_{m,n+1})e^{-i(m-n)\omega_0 t}$$

The Kronecker delta terms dictate the values m must take in our summations. In the first term, we set $m = n-1$ and in the second term $m = n+1$, leading to

$$\langle x\rangle(t) = \sqrt{\frac{\hbar}{2m_0\omega_0}}e^{-|\beta|^2}e^{-E_I t/2\hbar}e^{+i\omega_0 t}\sum_{n-1=0}^{\infty}\frac{(|\beta|)^{2n-1}}{\sqrt{(n-1)!n!}}\sqrt{n} \quad +$$

$$\sqrt{\frac{\hbar}{2m_0\omega_0}}e^{-|\beta|^2}e^{-E_I t/2\hbar}e^{-i\omega_0 t}\sum_{n+1=0}^{\infty}\frac{|\beta|^{2n+1}}{\sqrt{n!(n+1)!}}\sqrt{n+1}$$

To simplify further, we can use the following relations:

$$\sqrt{\frac{n}{n!}} = \sqrt{\frac{1}{(n-1)!}} \quad \text{and} \quad \sqrt{\frac{n+1}{(n+1)!}} = \sqrt{\frac{1}{n!}}$$

Then, the expectation value $\langle x\rangle$ becomes

$$\langle x\rangle(t) = \sqrt{\frac{\hbar}{2m_0\omega_0}}e^{-|\beta|^2}e^{-E_I t/\hbar}\left[e^{+i\omega_0 t}\sum_{n-1=0}^{\infty}\frac{(|\beta|)^{2n-1}}{(n-1)!} + e^{-i\omega_0 t}\sum_{n+1=0}^{\infty}\frac{|\beta|^{2n+1}}{n!}\right]$$

$$= \sqrt{\frac{\hbar}{2m_0\omega_0}}|\beta|e^{-|\beta|^2}e^{-E_I t/\hbar}\left[e^{+i\omega_0 t}\sum_{n=1}^{\infty}\frac{(|\beta|)^{2(n-1)}}{(n-1)!} + e^{-i\omega_0 t}\sum_{n=0}^{\infty}\frac{|\beta|^{2n}}{n!}\right]$$

In the form that we now have, the two series in the last line are identical to one another, and as we discussed above, is in the form of an exponential. This leads to

$$\begin{aligned}
\langle x\rangle(t) &= \sqrt{\frac{\hbar}{2m_0\omega_0}}|\beta|e^{-|\beta|^2}e^{-E_I t/2\hbar}\left[e^{+i\omega_0 t}e^{|\beta|^2}+e^{-i\omega_0 t}e^{|\beta|^2}\right]\\
&= \sqrt{\frac{\hbar}{2m_0\omega_0}}|\beta|e^{-E_I t/2\hbar}\left[e^{+i\omega t}+e^{-i\omega_0 t}\right]\\
&= \sqrt{\frac{\hbar}{2m_0\omega_0}}|\beta|e^{-E_I t/2\hbar}\left[2\cos(\omega_0 t)\right] = x_0 e^{-E_I t/2\hbar}\cos(\omega_0 t)
\end{aligned}$$

Using Eq. (3.93), if we define x_0 as

$$x_0 = \sqrt{\frac{2\hbar}{m_0\omega_0}}|\beta| = \sqrt{\frac{2E}{m_0\omega_0^2}} \Rightarrow |\beta| = \sqrt{\frac{E}{\hbar\omega_0}} = 1 \tag{3.106}$$

We have used the relation $E = \hbar\omega$. Using the above definitions, the expectation value of the position for a harmonic oscillator in 4D, whose Hamiltonian is given by Eq. (3.14), is

Damped Linear Harmonic Oscillator

$$\langle x\rangle(t) = x_0 e^{-E_I t/2\hbar}\cos(\omega_0 t) \tag{3.107}$$

Eq. (3.107) describes the oscillatory motion of a *damped harmonic oscillator*, having a monotonically decaying peak amplitude. If $E_I = 0$, Eq. (3.107) reduces to the non-dissipative steady-state motion described by $x_0\cos\omega_0 t$. Fig. 3.5 illustrates both cosine and damped cosine oscillator motions.

The previous two examples demonstrate problems involving a single *spatial* degree of freedom, *i.e.* x. When more degrees of freedom are added, we begin to observe richer dynamical phenomena, such as dynamical transitions between the quantum states. Let us analyze an ideal example of quantum state transitions in the next example.

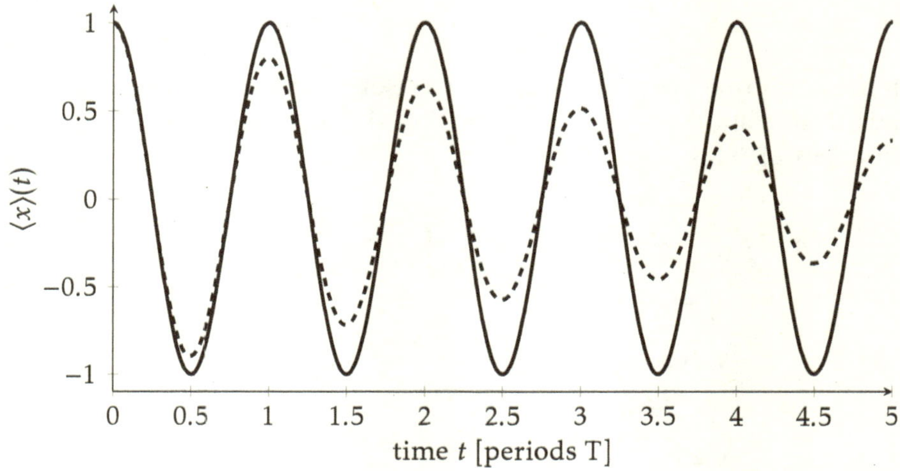

Figure 3.5: Expectation value of the harmonic oscillator as a function time t. The solid-line corresponds to no-damping or $\tau \to \infty$, while the dashed line corresponds to finite $\tau = 4.5$.

3.3 Electron Spin Dynamics Using Time-Symmetry

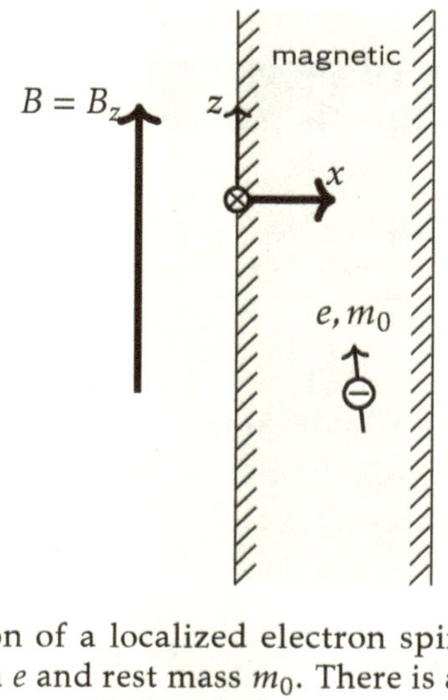

Figure 3.6: Illustration of a localized electron spin in a magnetic material having charge strength e and rest mass m_0. There is an external magnetic field B along z, which interacts with the electron spin.

Let us apply a 4D Hamiltonian to a localized system of *angular momentum*(*e.g.* spin) like that shown in Fig. 3.6. The contributions to the *real* part of the Hamiltonian $\widehat{H}_R$ for an electron in a magnetic field includes the following contributions:

$$\widehat{H}_L = -\widehat{\mu}_L \cdot \mathbf{B} \quad \text{(orbital)} \quad \widehat{H}_S = -\widehat{\mu}_S \cdot \mathbf{B} \quad \text{(spin)} \tag{3.108}$$

The first is energy due to *orbital* angular momentum (*c*.Eq. 2.141) which we derived directly from the relativistic kinetic energy. The second is due to electron *spin* angular momentum (*c*.Eq. (2.128)), also obtained directly from the relativistic kinetic energy. $\widehat{\mu}_L = (e\hbar/2m_0)\widehat{\mathbf{L}}$ and $\widehat{\mu}_S = (e\hbar/2m_0)\widehat{\mathbf{S}}$ are the respective magnetic moments, being proportional to their corresponding angular momentum $\widehat{\mathbf{L}}$ and $\widehat{\mathbf{S}}$; B is the magnetic field strength.

Including the relevant terms which contribute to time-dependent energy, the 4D Hamiltonian from Eq. (2.120) becomes

$$\widehat{H} = \left[\widehat{H}_S - \mathbf{I}(2)\left(i\hbar\frac{\partial}{\partial t}\right)\right]\psi \tag{3.109}$$

Eq. (3.110) is applied to a spinor state ψ which we write as

$$\psi = \begin{bmatrix} \psi_\uparrow \\ \psi_\downarrow \end{bmatrix} \tag{3.110}$$

Thus, in describing the spinor state of *two distinct eigenstates,* we have

$$\left[-\widehat{\mu}_S \cdot \mathbf{B} - i\hbar\frac{\partial}{\partial t}\right]\begin{bmatrix} \psi_\uparrow \\ \psi_\downarrow \end{bmatrix} = i\begin{bmatrix} E_I & 0 \\ 0 & E_I \end{bmatrix}\begin{bmatrix} \psi_\uparrow \\ \psi_\downarrow \end{bmatrix} \tag{3.111}$$

Let us write the imaginary component as

$$E_I = \alpha E_R \tag{3.112}$$

Then, we have

$$-\widehat{\mu}_S \cdot \mathbf{B}\begin{bmatrix} \psi_\uparrow \\ \psi_\downarrow \end{bmatrix} + \begin{bmatrix} -i\alpha E_R & 0 \\ 0 & -i\alpha E_R \end{bmatrix}\begin{bmatrix} \psi_\uparrow \\ \psi_\downarrow \end{bmatrix} = i\hbar\frac{\partial}{\partial t}\begin{bmatrix} \psi_\uparrow \\ \psi_\downarrow \end{bmatrix} \tag{3.113}$$

Using the relation for the Hermitian Hamiltonian

$$\widehat{H}_R = -\widehat{\mu}_S \cdot \mathbf{B}\psi = E_R\psi$$

we have an equivalent form given by

$$(1 - i\alpha)\widehat{H}_R\begin{bmatrix} \psi_\uparrow \\ \psi_\downarrow \end{bmatrix} = i\hbar\frac{\partial}{\partial t}\begin{bmatrix} \psi_\uparrow \\ \psi_\downarrow \end{bmatrix} \tag{3.114}$$

Using the spin vector $\sigma = [\sigma_x, \sigma_y, \sigma_z]^T$ and taking the magnetic field to be along the z direction, Eq. (3.114) then leads to

$$i\hbar\frac{\partial\psi}{\partial t} = (1 - i\alpha)(-\widehat{\mu}_S \cdot \mathbf{B})\psi \tag{3.115a}$$

$$= \left[-\frac{e\hbar B}{2m_0}\widehat{\sigma}_z + \frac{i\alpha e\hbar B}{2m_0}\widehat{\sigma}_z\right]\psi \tag{3.115b}$$

$$= [-\gamma B\widehat{\sigma}_z + i\alpha\gamma B\widehat{\sigma}_z]\psi \tag{3.115c}$$

The parameter γ is known as the electron's *gyromagnetic ratio* defined as

$$\gamma = \frac{e\hbar}{2m_0} \quad \text{(electron gyromagnetic ratio)} \tag{3.116}$$

Resolving Eq. (3.115) into two equations describing $\psi_\uparrow$ and $\psi_\downarrow$ gives

$$\frac{d\psi_\uparrow}{dt} = -(i\gamma B + \alpha\gamma B)\psi_\uparrow \tag{3.117a}$$

$$\frac{d\psi_\downarrow}{dt} = (\alpha\gamma B + i\gamma B)\psi_\downarrow \tag{3.117b}$$

The solutions to the above equations for $\psi_\uparrow$ and $\psi_\downarrow$ are given by

$$\psi_\uparrow = \psi_\uparrow(0)e^{-(\alpha\gamma B + i\gamma B)\cdot t} = \psi_\uparrow(0)e^{-\alpha\gamma Bt}e^{-i\gamma B\cdot t} \tag{3.118a}$$

$$\psi_\downarrow = \psi_\downarrow(0)e^{+(\alpha\gamma B + i\gamma B)\cdot t} = \psi_\downarrow(0)e^{\alpha\gamma Bt}e^{i\gamma Bt} \tag{3.118b}$$

Before we can make use of the solutions given by Eqs. (3.118), we must also identify the initial conditions $\psi_\uparrow(0)$ and $\psi_\downarrow(0)$. Because we have taken the magnetic field along $+z$, the appropriate spinor *initial condition* is given by

$$\psi(0) = \left[\psi_\uparrow(0) \quad \psi_\downarrow(0)\right]^T = \left[\cos\tfrac{\theta}{2}e^{-i\frac{\varphi}{2}} \quad \sin\tfrac{\theta}{2}e^{i\frac{\varphi}{2}}\right]^T \tag{3.119}$$

These arise from the eigenvector of a generalized spin matrix $\widehat{\sigma}$ for an arbitrary field direction, and can be used to specify initial conditions for a *spinor*. θ is the polar angle taken relative to the $+z$ axis, while φ is azimuthal taken from the $+x$ axis. The spin eigensolutions $\psi_\uparrow$ and $\psi_\downarrow$ are products of a real exponential and complex exponential term.

Now, let's use the above results to calculate the expectation value of the *spin angular momentum vector* **S**.

3.3.1 Time-dependent Expectation Values of Spin

Since we have found wave functions that explicitly depend on time, they can be used to determine the time-dependent expectation values of the components of the intrinsic angular momentum $\widehat{S}_x$, $\widehat{S}_y$, and $\widehat{S}_z$. The expectation value of the first component $\langle\widehat{S}_x\rangle$ is defined as

$$\langle\widehat{S}_x\rangle = \frac{\hbar}{2}\frac{\begin{bmatrix}\psi_\uparrow^* & \psi_\downarrow^*\end{bmatrix}\widehat{\sigma}_x\begin{bmatrix}\psi_\uparrow \\ \psi_\downarrow\end{bmatrix}}{\begin{bmatrix}\psi_\uparrow^* & \psi_\downarrow^*\end{bmatrix}\begin{bmatrix}\psi_\uparrow \\ \psi_\downarrow\end{bmatrix}} = \frac{\hbar}{2}\frac{\psi_\uparrow^*\psi_\downarrow + \psi_\downarrow^*\psi_\uparrow}{\psi_\uparrow^*\psi_\uparrow + \psi_\downarrow^*\psi_\downarrow} \tag{3.120}$$

Substitution of the initial condition given by Eq. (3.119), with the solutions given by Eqs. (3.118a)-(3.118b), $\langle\widehat{S}_x\rangle$ becomes

$$\langle\widehat{S}_x\rangle = \frac{\hbar}{2}\frac{\cos\frac{\theta}{2}\sin\frac{\theta}{2}\left(e^{i\omega_p t}e^{i\varphi} + e^{-i\omega_p t}e^{-i\varphi}\right)}{\left(\cos^2\frac{\theta}{2}e^{-\alpha\omega_p t} + \sin^2\frac{\theta}{2}e^{\alpha\omega_p t}\right)}$$

$$= \frac{\hbar}{2}\frac{\cos\frac{\theta}{2}\sin\frac{\theta}{2}\left(e^{i\omega_p t}e^{i\varphi} + e^{-i\omega_p t}e^{-i\varphi}\right)}{\left(\left[\frac{1+\cos\theta}{2}\right]e^{-\alpha\omega_p t} + \left[\frac{1-\cos\theta}{2}\right]e^{\alpha\omega_p t}\right)}$$

$$= \frac{\hbar}{2}\frac{\cos\frac{\theta}{2}\sin\frac{\theta}{2}(e^{i\omega_p t}e^{i\varphi} + e^{-i\omega_p t}e^{-i\varphi})}{\frac{\left(e^{-\alpha\omega_p t}+e^{\alpha\omega_p t}\right)}{2} + \cos\theta\frac{\left(e^{-\alpha\omega_p t}-e^{\alpha\omega_p t}\right)}{2}}$$

$$= \frac{\hbar}{2}\frac{(\sin\theta/2)(e^{i\omega_p t}e^{i\varphi} + e^{-i\omega_p t}e^{-i\varphi})}{\frac{\left(e^{-\alpha\omega_p t}+e^{\alpha\omega_p t}\right)}{2} + \cos\theta\frac{\left(e^{-\alpha\omega_p t}-e^{\alpha\omega_p t}\right)}{2}}$$

$$= \frac{\hbar}{2}\frac{\sin\theta\cos(\omega_p t + \varphi)}{\cosh\left(\alpha\omega_p t\right) - \cos\theta\sinh\left(\alpha\omega_p t\right)}$$

In the above, the parameter ω_p is known as the electron's *precessional frequency* defined as

$$\omega_p = 2\gamma B \tag{3.121}$$

We have also made use of some trigonometric identities given by

$$\cos x = \frac{e^{ix}+e^{-ix}}{2} \quad \cos^2\frac{\theta}{2} = \frac{1+\cos\theta}{2} \quad \sin^2\frac{\theta}{2} = \frac{1-\cos\theta}{2}$$

$$\cosh x = \frac{e^{x}+e^{-x}}{2} \quad \sinh x = \frac{e^{x}-e^{-x}}{2} \quad \sin x\cos x = \frac{\sin(2x)}{2}$$

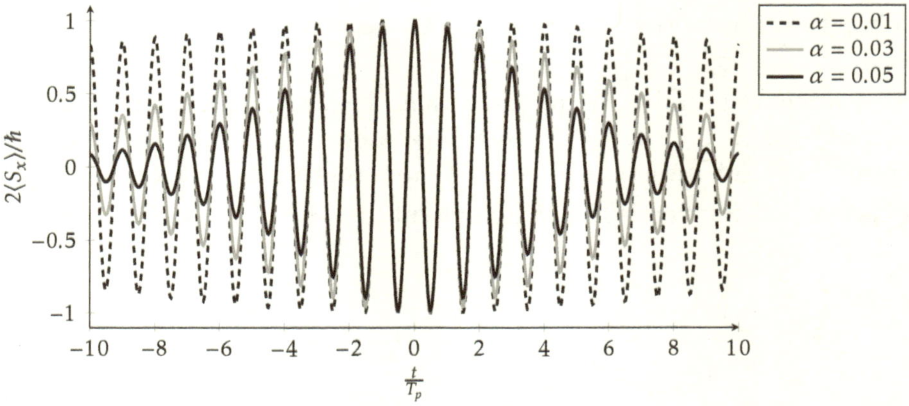

Figure 3.7: Plot of the x-component of the electron spin momentum $\langle S_x\rangle(t)$. $\langle S_x\rangle$ oscillates with frequency $\omega_p = g|\gamma|B = 2\pi/T_p$, with damping leading to decaying away from $t = 0$. The larger the damping parameter α, the faster the damping in the damped oscillations.

Since the denominator of $\langle\widehat{S_x}\rangle$ is identical for all three expectation values $\langle S_x\rangle$, $\langle S_y\rangle$, and $\langle S_z\rangle$, we won't repeat calculation of the denominator for the remaining two components. We will just use the result above.

The initial spinor condition for (θ, φ) *is not completely arbitrary*. It must be consistent with the chosen Hamiltonian, and these conditions define the canonical problem of the magnetic moment, [3] which gives $\theta = \pi/2$ while $\varphi = \varphi_i$ is arbitrary. Using these spinor parameters, we have the following result for $\langle S_x\rangle$.

$$\langle\widehat{S_x}\rangle(t) = \frac{\hbar}{2}\frac{\cos(\omega_p t + \varphi)}{\cosh\left(\alpha\omega_p t\right)} \tag{3.122}$$

Fig. 6.14 illustrates $\langle\widehat{S_x}\rangle(t)$ for three different parameter values $\alpha = 0.01$, $\alpha = 0.03$, and $\alpha = 0.05$. A similar analysis can be done for $\langle S_y\rangle$ using the corresponding definition

$$\langle\widehat{S_y}\rangle(t) = \frac{\hbar}{2}\frac{\begin{bmatrix}\psi_\uparrow^* & \psi_\downarrow^*\end{bmatrix}\begin{bmatrix}0 & -i\\ i & 0\end{bmatrix}\begin{bmatrix}\psi_\uparrow\\ \psi_\downarrow\end{bmatrix}}{\begin{bmatrix}\psi_\uparrow^* & \psi_\downarrow^*\end{bmatrix}\begin{bmatrix}\psi_\uparrow\\ \psi_\downarrow\end{bmatrix}} = \frac{i\hbar}{2}\frac{\psi_\downarrow^*\psi_\uparrow - \psi_\uparrow^*\psi_\downarrow}{\psi_\uparrow^*\psi_\uparrow + \psi_\downarrow^*\psi_\downarrow} \tag{3.123}$$

[3] for a detailed introduction of the canonical moment problem, see Chap. 7, Sec. 7.2, Eason, Quantum Mechanics With A SPIN On It

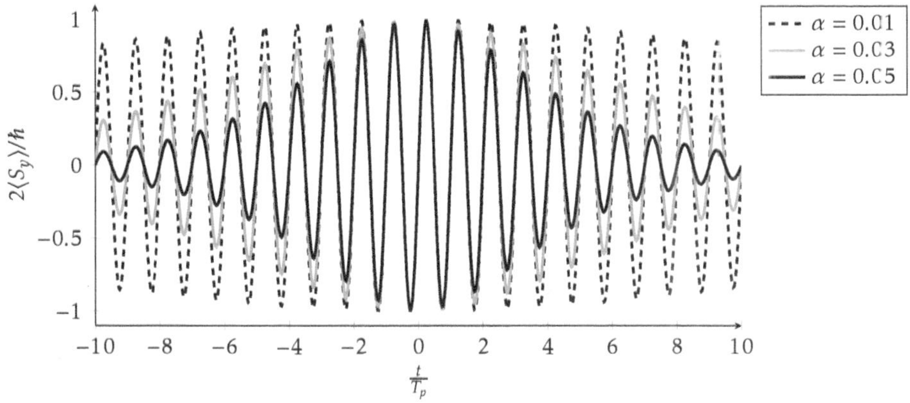

Figure 3.8: Plot of the y-component of the electron spin momentum. Like $\langle \widehat{S}_x \rangle$, $\langle \widehat{S}_y \rangle$ oscillates with frequency $\omega_p = 2\pi/T_p$, with larger dissipation parameter α leading to more rapidly decaying oscillations.

Substitution of the wave functions from Eq. (3.118) into Eq. (3.123) leads to

$$\langle \widehat{S}_y \rangle = \frac{i\hbar}{2} \frac{\cos\frac{\theta}{2}\sin\frac{\theta}{2}\left(e^{-i\omega_p t}e^{-i\varphi} - e^{i2\omega_p t}e^{i\varphi}\right)}{\frac{\left(e^{-\alpha\omega_p t}+e^{\alpha\omega_p t}\right)}{2} + \cos\theta\frac{\left(e^{-\alpha\omega_p t}-e^{\alpha\omega_p t}\right)}{2}}$$

$$= \frac{\hbar}{2}\frac{\sin\theta\sin(\omega_p t + \varphi)}{\cosh\left(\alpha\omega_p t\right) - \cos\theta\sinh\left(\alpha\omega_p t\right)}$$

We have used the relation $-i = 1/i$ along with some of the trigonometric identities given above. Substitution of the initial boundary condition $(\theta, \varphi) = (\pi/2, \varphi_i)$, we obtain the following result for $\langle \widehat{S}_y \rangle$:

$$\langle \widehat{S}_y \rangle = \frac{\hbar}{2}\frac{\sin(\omega_p t + \varphi)}{\cosh\left(\alpha\omega_p t\right)} \tag{3.124}$$

Fig. 6.15 illustrates $\langle \widehat{S}_y \rangle$ for the same parameter values of $\alpha = 0.01, 0.03,$ and 0.05 : Lastly, for $\langle \widehat{S}_z \rangle$, we have

$$\langle \widehat{S}_z \rangle(t) = \frac{\hbar}{2}\frac{\begin{bmatrix}\psi_\uparrow^* & \psi_\downarrow^*\end{bmatrix}\begin{bmatrix}1 & 0\\0 & -1\end{bmatrix}\begin{bmatrix}\psi_\uparrow\\\psi_\downarrow\end{bmatrix}}{\begin{bmatrix}\psi_\uparrow^* & \psi_\downarrow^*\end{bmatrix}\begin{bmatrix}\psi_\uparrow\\\psi_\downarrow\end{bmatrix}} = \frac{\hbar}{2}\frac{\psi_\uparrow^*\psi_\uparrow - \psi_\downarrow^*\psi_\downarrow}{\psi_\uparrow^*\psi_\uparrow + \psi_\downarrow^*\psi_\downarrow}$$

Substitution of the wave functions from Eqs. (3.118) gives

$$\langle\widehat{S_z}\rangle = \frac{\hbar}{2}\frac{\left(e^{-\alpha\omega_p t} - e^{\alpha\omega_p t}\right) + \cos\theta\left(e^{-\alpha\omega_p t} + e^{\alpha\omega_p t}\right)}{\left(e^{-\alpha\omega_p t} + e^{\alpha\omega_p t}\right) + \cos\theta\left(e^{-\alpha\omega_p t} - e^{\alpha\omega_p t}\right)}$$

$$= \frac{\hbar}{2}\frac{-\sinh\left(\alpha\omega_p t\right) + \cos\theta\cosh\left(\alpha\omega_p t\right)}{\cosh\left(\alpha\omega_p t\right) - \cos\theta\sinh\left(\alpha\omega_p t\right)}$$

Substituting the initial spinor condition into $\langle S_z\rangle$, we have

$$\langle\widehat{S_z}\rangle = -\frac{\hbar}{2}\frac{\sinh\left(\alpha\omega_p t\right)}{\cosh\left(\alpha\omega_p t\right)} = -\frac{\hbar}{2}\tanh\left(\alpha\omega_p t\right) \tag{3.125}$$

Because **B** is along this axis, it is from this component specifically, we may obtain a time-scale for *spin reversal* denoted by τ_S for $\widehat{\mathbf{S}}$ given by

$$\tau_S = \frac{1}{\alpha\omega_p} \tag{3.126}$$

or

Spin Reversal Time-Constant

$$\tau_S = \frac{1}{\alpha g\gamma B} \tag{3.127}$$

Fig. 6.16 shows $\langle\widehat{S_z}\rangle(t)$ for $\alpha = 0.01, 0.03, 0.05$. The component of the spin angular momentum along the z–axis (or field axis) *aligns* itself *anti-parallel* to the magnetic field. This is equivalent the magnetic moment (anti-parallel with spin) being aligned with **B**.

So we find that applying the 4D non-Hermitian Hamiltonian to the spin problem, we are able to determine a dynamical response given by

Spin Angular Momentum Expectation Values

$$\langle\widehat{S_x}\rangle = \frac{\hbar}{2}\frac{\cos\left(\omega_p t + \varphi\right)}{\cosh\left(\alpha\omega_p t\right)}, \langle\widehat{S_y}\rangle = \frac{\hbar}{2}\frac{\sin\left(\omega_p t + \varphi\right)}{\cosh\left(\alpha\omega_p t\right)}, \langle\widehat{S_z}\rangle = -\frac{\hbar}{2}\tanh\left(\alpha\omega_p t\right)$$

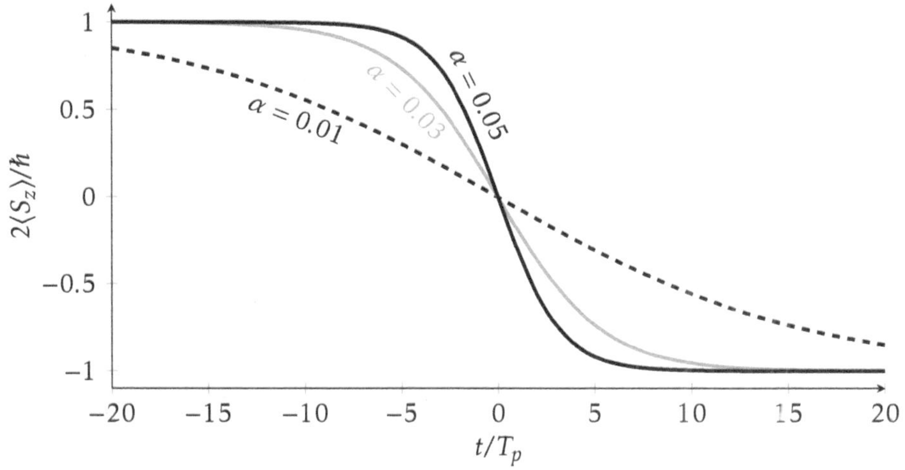

Figure 3.9: Plot of the z-component of the electron spin momentum, which is the component along the field axis. This component reverses from alignment with B to being anti-aligned (the magnetic moment is what aligns). We also see that the larger the parameter α, the faster the transition.

(3.128)

The two components $\langle \widehat{S}_x\rangle$ and $\langle \widehat{S}_y\rangle$, perpendicular to the field, oscillate about **B** at a frequency $\omega_p = g\gamma B$, which is known as the *electron spin precessional frequency*.

We may use the above results to examine the conservation of the *spin angular momentum* in time t. Let us consider the length of $\langle S\rangle(t)$ while undergoing a dynamic transition in a magnetic field. The quantity $\langle \widehat{S}\rangle^2$ is given by

$$\langle \widehat{S}\rangle^2 = \langle \widehat{S}_x\rangle^2 + \langle \widehat{S}_y\rangle^2 + \langle \widehat{S}_z\rangle^2 =$$

$$\left[\frac{\hbar}{2}\frac{\cos\left(\omega_p t + \varphi\right)}{\cosh\left(\alpha\omega_p t\right)}\right]^2 + \left[\frac{\hbar}{2}\frac{\sin\left(\omega_p t + \varphi\right)}{\cosh\left(\alpha\omega_p t\right)}\right]^2 + \left[-\frac{\hbar}{2}\tanh\left(\alpha\omega_p t\right)\right]^2 =$$

$$\frac{\hbar^2}{4}\left[\frac{\cos^2\left(\omega_p t + \varphi\right) + \sin^2\left(\omega_p t + \varphi\right) + \sinh^2\left(\alpha\omega_p t\right)}{\cosh^2\left(\alpha\omega_p t\right)}\right] =$$

$$\frac{\hbar^2}{4}\left[\frac{1+\sinh^2\left(\alpha\omega_p t\right)}{\cosh^2\left(\alpha\omega_p t\right)}\right] = \frac{\hbar^2}{4}\left[\frac{\cosh^2\left(\alpha\omega_p t\right)}{\cosh^2\left(\alpha\omega_p t\right)}\right] = \frac{\hbar^2}{4} \tag{3.129}$$

Eq. (6.103) reveals that the length of the expectation value of **S** *is conserved* as it relaxes over time, or

$$|\langle\widehat{\mathbf{S}}\rangle(t)| = \frac{\hbar}{2} \tag{3.130}$$

Since the electron spin magnetic moment μ_S is directly proportional to the spin angular momentum as **S**

$$\widehat{\mu}_S = \frac{e}{2m_0}\widehat{\mathbf{S}} \tag{3.131}$$

then $\langle\mu_x\rangle(t)$, are then given by

$\mu_x(t)$ Expectation Value

$$\langle\mu_x\rangle(t) = -\gamma\hbar\frac{\cos\left(\omega_p t+\varphi\right)}{\cosh\left(\alpha\omega_p t\right)} \tag{3.132}$$

$\mu_y(t)$ Expectation Value

$$\langle\mu_y\rangle(t) = -\gamma\hbar\frac{\sin\left(\omega_p t+\varphi\right)}{\cosh\left(\alpha\omega_p t\right)} \tag{3.133}$$

$\mu_z(t)$ Expectation Value

$$\langle\mu_z\rangle(t) = \gamma\hbar\tanh\left(\alpha\omega_p t\right) \tag{3.134}$$

Using the relation $\mu_B = \gamma\hbar$, it also follows that

$$|\boldsymbol{\mu}(t)| = \mu(t) = \mu_B \tag{3.135}$$

μ_B is known as the *Bohr magneton*. Thus, we find that by inclusion of the fourth dimension, we incorporate a time-dependence in the wave functions which allow time dependent expectation values which can describe transitions between quantum states $+\hbar/2$ to $-\hbar/2$.

It is also convenient to have a time-dependent differential equation describing the vector components of $\widehat{\mu}_S$ or $\widehat{\mathbf{S}}$. In quantum mechanics, the time-derivative of the expectation of an operator $\langle \mu_S \rangle$ can be obtained by using a special formula known as the Generalized Ehrenfest Theorem. *However,* the conventional form of the theorem will *not* suffice because *it assumes a system described by a Hermitian Hamiltonian,* or that the Hamiltonian has strictly real eigenvalues. Starting in the next Chapter, we will generalize this result to accommodate a non-Hermitian Hamiltonian, as we have in 4D. We can then use the result to find the corresponding classical differential equation describing the above system.

3.4 Chapter Summary

In this chapter, we have applied the 4D Hamiltonian obtained in Chap 2 to several problems to illustrate both how to apply it to problems, as well as the kind of results that can be derived from solving problems using this Hamiltonian.

We studied the 1D propagating wave problem where we found the wave picks up a decaying amplitude in its propagation. We also examined the harmonic oscillator and found that it generalizes to a damped harmonic oscillator, having up a generally decaying amplitude. We then studied a problem allowing superposition and obtained time dependent expectation values, with the electron spin in an external magnetic field. In this case, we found the description of a transition between quantum states in the expectation value of the spin angular momentum. Hopefully, the reader gained some insight into application of a 4D Hamiltonian. We'll see more applications in later chapters, as well.

Table 3.3 lists some of the key formulas from this chapter.

Table 3.3: Chapter 3 Summary Equations.

Name	Equation

Table 3.3

Recurrence Relation	$\frac{d_r}{d_{r-1}} = \frac{(p+r-1+a)}{(p+r)(p+r+c-1)}$
First Independent Solution	$S_0 = {}_1F_1\left(\gamma; a = \frac{1-\Omega}{4}, c = \frac{1}{2}\right)$
Second Independent Solution	$S_{1/2} = {}_1F_1\left(\gamma; a' = \frac{3-\Omega}{4}, c' = \frac{3}{2}\right)$
Damped Harmonic Oscillator	$\langle x\rangle(t) = x_0 e^{-E_I t/2\hbar}\cos(\omega_0 t)$
Spin in Magnetic Field (x)	$\langle\widehat{S_x}\rangle = \frac{\hbar}{2}\frac{\cos(\omega_p t+\varphi)}{\cosh(\alpha\omega_p t)}$
Spin in Magnetic Field (y)	$\langle\widehat{S_y}\rangle = \frac{\hbar}{2}\frac{\sin(\omega_p t+\varphi)}{\cosh(\alpha\omega_p t)}$
Spin in Magnetic Field (z)	$\langle\widehat{S_z}\rangle = -\frac{\hbar}{2}\tanh\left(\alpha\omega_p t\right)$
Precessional Frequency	$\omega_p = 2\gamma B$
Spin Reversal Time-Constant	$\tau_S = \frac{1}{\alpha g\gamma B}$

3.5 Table of Hermite Polynomials

n	Hermite Polynomial H_n
0	$H_0(\zeta) = 1$
1	$H_1(\zeta) = 2\zeta$
2	$H_2(\zeta) = 4\zeta^2 - 2$
3	$H_3(\zeta) = 8\zeta^3 - 12\zeta$
4	$H_4(\zeta) = 16\zeta^4 - 48\zeta^2 + 12$
5	$H_5(\zeta) = 32\zeta^5 - 160\zeta^3 + 120\zeta$
6	$H_6(\zeta) = 64\zeta^6 - 480\zeta^4 + 720\zeta^2 - 120$
7	$H_7(\zeta) = 128\zeta^7 - 1344\zeta^5 + 3360\zeta^3 - 1680\zeta$
8	$H_8(\zeta) = 256\zeta^8 - 3584\zeta^6 + 13440\zeta^4 - 13440\zeta^2 + 1680$
9	$H_9(\zeta) = 512\zeta^9 - 9216\zeta^7 + 48384\zeta^5 - 80640\zeta^3 + 30240\zeta$
10	$H_{10}(\zeta) = 1024\zeta^{10} - 23040\zeta^8 + 161280\zeta^6 - 403200\zeta^4 + 302400\zeta^2 - 30240$
11	$H_{11}(\zeta) = 2048\zeta^{11} - 56320\zeta^9 + 506880\zeta^7 - 1774080\zeta^5 + 2217600\zeta^3 - 665280\zeta$
12	$H_{12}(\zeta) = 4096\zeta^{12} - 135168\zeta^{1}0 + 1520640\zeta^8 - 7096320\zeta^6 + 13305600\zeta^4 - 7983360\zeta^2 + 665280$
13	$H_{13}(\zeta) = 8192\zeta^{13} - 319488\zeta^{1}1 + 4392960\zeta^9 - 26357760\zeta^7 + 69189120\zeta^5 - 7983360\zeta^2 + 17297280\zeta$

Table 3.4: List of First 14 Hermite Polynomials.

3.6 Chapter Problems

Problem 3.1 Consider the 3D and 4D solutions given by

$$\psi(x,t) = Ae^{-i\frac{E_R}{\hbar}t}\left(e^{+i\frac{\sqrt{2m_0E_R}}{\hbar}x} + Be^{-i\frac{\sqrt{2m_0E_R}}{\hbar}x}\right)$$

and

$$\psi(x,t) = Ce^{-i\frac{E_R}{\hbar}t}e^{-\frac{E_I}{\hbar}t}\left(Ae^{+i\frac{\sqrt{2m_0E_R}}{\hbar}x} + Be^{-i\frac{\sqrt{2m_0E_R}}{\hbar}x}\right)$$

For both eigenfunctions, calculate $\psi^*\psi$.

Problem 3.2 The recurrence relation for the spatial part of the harmonic oscillator is given by

$$\frac{d_r}{d_{r-1}} = \frac{(p+r-1+a)}{(p+r)(p+r+c-1)}$$

For $a = -1/2$ and $a = -5/2$, letting $c = 1/2$ and $d_0 = 1$, compute the next five elements for $r = 1, 2, 3, 4$ and 5.

Problem 3.3 Using the Hermite polynomials given by Table 3.4, first calculate the derivatives, then evaluate them at $x = 1$ for $n = 0, 1, 2, 3,$ and 4.

Is there a trend in the values as n increases? Since the quantum mechanics momentum operator is proportional to the spatial derivatives, how would the observed trend translate to the momenta trend for increasing n?

Problem 3.4 For the harmonic oscillator problem, we found a normalization constant to be

$$|C_n| = \sqrt{\frac{\alpha}{2^n n! \sqrt{\pi}}}$$

where $\alpha = \sqrt{(m\omega_0)/\hbar}$ and $\omega_0 = \sqrt{k/m}$ is the natural frequency. Use the harmonic oscillator relations to solve for the *quantized* stiffness $k(n)$.

Based on your result, discuss whether larger stiffness accommodates lower or higher energy quantum states? How will the quantum number affect the oscillation frequency ω_0?

Problem 3.5 The first two eigenstates for the harmonic oscillator can be written as

$$\psi_1 = \left(\frac{\alpha}{\pi}\right)^{1/4} e^{-\alpha x^2/2} \quad \text{and} \quad \psi_2 = \left(\frac{4\alpha^3}{\pi}\right)^{1/4} x e^{-\alpha x^2/2}$$

Show that these two states are orthogonal, or that

$$\int_{-\infty}^{\infty} \psi_1^* \psi_2 dx = 0$$

Hint: Use the integral result

$$\int x^2 e^{\alpha x} = \left(\frac{x^2}{\alpha} - \frac{2x}{\alpha^2} + \frac{2}{\alpha^3}\right) e^{\alpha x}$$

Problem 3.6 Show that for the harmonic oscillator quantum state $n = 2$, the expectation $\langle x^2 \rangle$ is given by

$$\langle x^2 \rangle = \frac{5}{2} \frac{\hbar}{(mk)^{1/2}}$$

Problem 3.7 Show that the expectation of p is zero, and the squared momentum $\langle p^2 \rangle$ for $n = 2$ is given by

$$\langle p^2 \rangle = \frac{5}{2} \hbar (mk)^{1/2}$$

Problem 3.8 In this chapter, we found the expectation values for the electron spin in a magnetic field B are given by

$$\langle \widehat{S}_x \rangle = \frac{\hbar}{2} \frac{\cos(g\gamma Bt + \varphi)}{\cosh(\alpha g\gamma Bt)}, \langle \widehat{S}_y \rangle = \frac{\hbar}{2} \frac{\sin(g\gamma Bt + \varphi)}{\cosh(\alpha g\gamma Bt)}, \langle \widehat{S}_z \rangle = -\frac{\hbar}{2} \tanh(\alpha g\gamma Bt)$$

First, calculate the time-derivatives of each component of spin angular momentum. Then, evaluate the following:

$$-g\gamma \mathbf{S} \times \mathbf{B}$$

and

$$-\frac{g\gamma\alpha}{S} \mathbf{S} \times \mathbf{S} \times \mathbf{B}$$

where $S = \hbar/2$.

Add the two results together from the above cross-products, then compare the resulting vector components to the time-derivatives you calculated, respectively. How do they compare?

Problem 3.9 Using the result for the spin reversal time

$$\tau_S = \frac{1}{\alpha g \gamma B}$$

calculate τ_S for the following parameters:

$$\alpha = 0.001, B = 0.1$$

$$\alpha = 0.002, B = 0.4$$

$$\alpha = 0.01, B = 0.5$$

$$\alpha = 0.05, B = 0.5$$

3.7 Suggested Readings & References

[1] K. Eason, *Quantum Mechanics With A SPIN On It*, Sym. Pub. (2019).

CHAPTER 4

Non-Hermitian Operators

In this chapter, we examine more *consequences* of including the fourth dimension into quantum mechanics. As we have seen, this leads to a generalization of the operators working on ψ. In particular, the energy operators turn out to have *complex* eigenvalues (*i.e.* Non-Hermitian operator). The non-Hermitian property contains nonequilibrium time dependence such as transitions *between* quantum states, and more. However, in order to make full use of a non-Hermitian Hamiltonian, we must also generalize some well-known results developed for Hermitian operators. Generalization for non-Hermitian operators will allow us to calculate quantities like current density for systems described by non-Hermitian operators, as well as construction of time-dependent classical differential equations consistent. We will also demonstrate the construction of a differential equation by applying our results for non-Hermitian operators to the electron spin problem from Sec. 3.3.

4.1 Hermitian & Non-hermitian Operators

A Hermitian operator $\widehat{O}$ is an operator that has real-eigenvalues, or eigenvalues that are real numbers. In quantum mechanics, this means that the *eigenoperator equations* are of the general form

$$\widehat{O}\psi = \lambda\psi \tag{4.1}$$

Eq. (4.1) expresses that one first applies the operator $\widehat{O}$ to the wave function ψ. The operator is a well-defined composition of mathematical operations *e.g.* vector products, derivatives, *etc.* When the operation on ψ is complete, the result equals to the RHS, being a constant λ times the wave function psi. When $\lambda = \lambda_R$ is real, then $\widehat{O}$ is said to be *Hermitian.* To the contrary, if λ is a complex number or $\lambda_R + i\lambda_I$, then $\widehat{O}$ is *non-Hermitian.* It is the non-Hermitian form that emerges when the fourth dimension $x_4 = ict$ is included with the other three $x_1 = x, x_2 = y$,and $x_3 = z$. The equilibrium eigevalues only manifest strictly in the limit as $t \to \infty$, *i.e.* equilibrium and/or steady-state. The steady-state condition can be regarded as a generalization of an equilibrium condition. In both, there exists a finite period of time T_p over which the system behavior is completely described. In equilibrium, for example, *any* finite period of time will suffice as the system is *constant* over all time in any interval $-T_p/2 < t \le T_p/2$. However, in steady-state, there only need exist a unique finite period of time T_p over which the time-dependent behavior is completely described. An example would be a stable periodic system such as the *harmonic oscillator* or cyclotron motion, *etc.* With the fourth dimension, one finds that the system description goes beyond equilibrium and steady-state conditions to include *nonequilibrium* conditions, where the system is *only* described over an *infinite* interval of time. So we have the following three dynamical conditions contained in a 4D description:

$$T_p = 0 \qquad \text{(equilibrium)} \tag{4.2a}$$

$$0 < T_p < \infty \qquad \text{(steady state)} \tag{4.2b}$$

$$T_p = \infty \qquad \text{(nonequilibrium)} \tag{4.2c}$$

We found from including the 4th dimension, the relativistic kinetic energy E becomes

$$E = \frac{\gamma_r}{2}\sqrt{p^2c^2 + m_0^2c^4} + ic\gamma_r p_4 + V = E_R + iE_I \tag{4.3}$$

When we applied the corresponding 4D Hamiltonian to the electron spin in an external magnetic field, for example, we found a Hamiltonian that could be expressed in the form

$$\widehat{H} = (1 - i\alpha)\widehat{H}_R \tag{4.4}$$

$\alpha > 0$ is a real number. This is an example of a non-Hermitian Hamiltonian. Recognizing 4D generally leads to complex eigenvalues, we now revisit the continuity equation to obtain a more general form that also accommodates wave functions obtained from *non-Hermitian* operator equations.

4.2 Continuity Equation Using Non-Hermitian Operators

To derive the continuity equation for a quantum mechanical system consistent with four dimensions, let us start with the 4D energy operator given by

$$iE_I\psi = -\frac{\hbar^2}{2m}\nabla^2\psi + V\psi - i\hbar\frac{\partial\psi}{\partial t} \tag{4.5}$$

Since E_I is real, we can use the relation $E_I = \alpha V$, so Eq. (4.5) can be rearranged to

$$i\hbar\frac{\partial\psi}{\partial t} = -\frac{\hbar^2}{2m}\nabla^2\psi + (1 - i\alpha)V\psi \tag{4.6}$$

The complex conjugate of Eq. (4.6) is

$$-i\hbar\frac{\partial\psi^*}{\partial t} = -\frac{\hbar^2}{2m}\nabla^2\psi^* + (1 + i\alpha)V\psi^* \tag{4.7a}$$

If we multiply Eq. (4.6) by ψ^* and Eq. (4.7a) by ψ, we obtain[1]

$$i\hbar\psi^*\frac{\partial\psi}{\partial t} = -\frac{\hbar^2}{2m}\psi^*\nabla^2\psi + (1 - i\alpha)V\psi^*\psi \tag{4.8a}$$

$$-i\hbar\psi\frac{\partial\psi^*}{\partial t} = -\frac{\hbar^2}{2m}\psi\nabla^2\psi^* + (1 + i\alpha)V\psi\psi^* \tag{4.8b}$$

Taking the difference between the two equations leads to the following:

$$i\hbar\left(\psi^*\frac{\partial\psi}{\partial t} + \psi\frac{\partial\psi^*}{\partial t}\right) = \frac{\hbar^2}{2m}\left(\psi^*\nabla^2\psi - \psi\nabla^2\psi^*\right) - i2\alpha V\psi^*\psi \tag{4.9}$$

We define the *probability density* $\rho(x,t)$ as

$$\rho \equiv \psi\cdot\psi^* = \psi^*\cdot\psi = |\psi|^2 \tag{4.10}$$

[1] As a scalar equation left and right multiplication are identical, however, for tensors, the conjugate must be right-multiplied.

Then, dividing both sides by $i\hbar$, Eq. (4.9) becomes

$$\frac{\partial\rho}{\partial t}+\frac{i\hbar}{2m}\left(\psi\nabla^2\psi^*-\psi^*\nabla^2\psi\right)=-\frac{2\alpha}{\hbar}V\rho \tag{4.11}$$

The term in parentheses on the LHS can be expressed as a divergence by using the relation

$$\nabla\cdot(\psi\nabla\psi^*)=\psi(\nabla^2\psi^*)+\nabla\psi\nabla\psi^*$$

With this result, we have

$$\frac{\partial\rho}{\partial t}+\frac{i\hbar}{2m}\nabla\cdot(\psi\nabla\psi^*-\psi^*\nabla\psi)=-\frac{2\alpha}{\hbar}V\rho \tag{4.12}$$

From this result, a *current density vector operator* can be defined as

Current Density

$$\widehat{\mathbf{J}}=\frac{i\hbar}{2m}(\psi\nabla\psi^*-\psi^*\nabla\psi) \tag{4.13}$$

Finally, we then have a *generalized continuity equation* given by

Continuity Equation

$$\frac{\partial\rho}{\partial t}+\nabla\cdot\mathbf{J}=-\frac{2\alpha}{\hbar}V\rho \tag{4.14}$$

When $\alpha=0$, Eq. (4.14) reduces to the *conserved* form given by

$$\frac{\partial\rho}{\partial t}+\nabla\cdot\mathbf{J}=0 \tag{4.15}$$

As mentioned, another equation that is useful in quantum mechanics, but assumes a Hermitian Hamiltonian, is the Generalized Ehrenfest Theorem. It allows us to find the time-derivative of an expectation value of an operator. Let us generalize this further so that we can later explore consequences using the result.

4.3 Most General Ehrenfest Theorem

It is useful to find the time derivative of a quantum mechanical operator. In next section, this will be demonstrated, however, we must first generalize the theorem that allows us to find this quantity. Using the definition of the expectation value of an operator, the time-derivative of the expectation value is given by

$$\frac{\partial\langle\widehat{O}\rangle}{\partial t}=\frac{\partial}{\partial t}\left[\frac{\int_{-\infty}^{+\infty}\psi^{\dagger}\widehat{O}\psi dx}{\int_{-\infty}^{+\infty}\psi^{\dagger}\psi dx}\right] \tag{4.16}$$

Because $\langle\widehat{O}\rangle$ is a ratio of two integral functions, we can use the quotient rule to evaluate the derivative. The quotient rule is given by

$$\frac{\partial}{\partial t}\left(\frac{N(t)}{D(t)}\right)=\frac{\frac{\partial N}{\partial t}D-N\frac{\partial D}{\partial t}}{D^2}=\frac{1}{D}\frac{\partial N}{\partial t}-\frac{N}{D}\frac{\frac{\partial D}{\partial t}}{D} \tag{4.17}$$

Let's start with the numerator term $N(t)$. Bringing the time derivative into the integral (which is not integrated over time, but space), we have

$$\begin{aligned}\frac{\partial N}{\partial t}&=\frac{\partial}{\partial t}\int_{-\infty}^{+\infty}\psi^{\dagger}\widehat{O}\psi dx\\&=\int_{-\infty}^{+\infty}\frac{\partial\psi^{\dagger}}{\partial t}\widehat{O}\psi dx+\int_{-\infty}^{+\infty}\psi^{\dagger}\frac{\partial\widehat{O}}{\partial t}\psi dx+\int_{-\infty}^{+\infty}\psi^{\dagger}\widehat{O}\frac{\partial\psi}{\partial t}dx\end{aligned}$$

The second term in the last line is the expectation value of the derivative of the operator. This applies to any operator that has an explicit time dependence. Fortunately, many operators are independent of the time parameter, and therefore, this term often vanishes in practice. Nonetheless, $\partial N/\partial t$ becomes

$$\frac{\partial N}{\partial t}=\int_{-\infty}^{+\infty}\frac{\partial\psi^{\dagger}}{\partial t}\widehat{O}\psi dx+\int_{-\infty}^{+\infty}\psi^{\dagger}\widehat{O}\frac{\partial\psi}{\partial t}dx+\left\langle\frac{\partial\widehat{O}}{\partial t}\right\rangle \tag{4.18}$$

In the next step, the time derivatives of the wave function are replaced by using the *energy operator equation* which relates the time derivatives to the Hamiltonian $\widehat{H}$. The partial derivatives of ψ and $\psi^{\dagger}$ can be replaced by using the following relations:

$$i\hbar\frac{\partial\psi}{\partial t}=\widehat{H}\psi\Rightarrow\frac{\partial\psi}{\partial t}=-\frac{i}{\hbar}\widehat{H}\psi$$

For $\psi^\dagger$, we also have

$$\frac{\partial \psi^\dagger}{\partial t} = \frac{i}{\hbar}\psi^\dagger \widehat{H}^\dagger$$

Multiplying by $1/D$, and substituting these relations into Eq. (4.18), we obtain

$$\frac{1}{D}\frac{\partial N}{\partial t} = \frac{i}{\hbar}\left[\frac{\int_{-\infty}^{+\infty}\psi^\dagger \widehat{H}^\dagger \widehat{O}\psi dx}{\int_{-\infty}^{+\infty}\psi^\dagger \psi dx} - \frac{\int_{-\infty}^{+\infty}\psi^\dagger \widehat{O}\widehat{H}\psi dx}{\int_{-\infty}^{+\infty}\psi^\dagger \psi dx}\right] + \left\langle \frac{\partial \widehat{O}}{\partial t}\right\rangle \tag{4.19}$$

We can carry out a similar process for the denominator integral $D(t)$, using the relations for ψ and $\psi^\dagger$. The partial derivative of the denominator becomes

$$\frac{\partial D}{\partial t} = \frac{\partial}{\partial t}\int_{-\infty}^{+\infty}\psi^\dagger \psi dx$$

$$= \int_{-\infty}^{+\infty}\frac{\partial \psi^\dagger}{\partial t}\psi dx + \int_{-\infty}^{+\infty}\psi^\dagger \frac{\partial \psi}{\partial t}dx$$

$$= \frac{i}{\hbar}\left[\int_{-\infty}^{+\infty}\psi^\dagger \widehat{H}^\dagger \psi dx - \int_{-\infty}^{+\infty}\psi^\dagger \widehat{H}^\dagger \psi dx\right]$$

Thus, expressing the second term on the RHS of Eq. (4.17), we have

$$\frac{N}{D}\frac{\frac{\partial D}{\partial t}}{D} = \frac{i}{\hbar}\left[\frac{\int_{-\infty}^{+\infty}\psi^\dagger \widehat{O}\psi dx}{\int_{-\infty}^{+\infty}\psi^\dagger \psi dx}\right]\left[\frac{\int_{-\infty}^{+\infty}\psi^\dagger \widehat{H}^\dagger \psi dx}{\int_{-\infty}^{+\infty}\psi^\dagger \widehat{O}\psi dx} - \frac{\int_{-\infty}^{+\infty}\psi^\dagger \widehat{H}^\dagger \psi dx}{\int_{-\infty}^{+\infty}\psi^\dagger \widehat{O}\psi dx}\right] \tag{4.20}$$

Putting the results for both terms together using the definition of an expectation value, we can finally express the time derivative of the expectation value $\langle O\rangle$ by the following:

Most General Ehrenfest Equation

$$\frac{\partial\langle\widehat{O}\rangle}{\partial t} = \frac{i}{\hbar}\left[\langle\widehat{H}^\dagger\widehat{O}\rangle - \langle\widehat{O}\widehat{H}\rangle\right] - \frac{i}{\hbar}\langle O\rangle\left[\langle\widehat{H}^\dagger\rangle - \langle\widehat{H}\rangle\right] + \left\langle\frac{\partial\widehat{O}}{\partial t}\right\rangle \tag{4.21}$$

Eq. (4.21) can be used with energy operators either Hermitian or *non-Hermitian*. By inspection, we can see that in the case that $\widehat{H}^\dagger = \widehat{H}$ (i.e.

$\widehat{H}$ is Hermitian, adjoint, or real), the term in brackets then becomes the commutator, and Eq. (4.21) reduces to

$$\frac{\partial\langle\widehat{O}\rangle}{\partial t}=\frac{i}{\hbar}\left\langle\left[\widehat{H},\widehat{O}\right]\right\rangle+\left\langle\frac{\partial\widehat{O}}{\partial t}\right\rangle \tag{4.22}$$

Eq. (4.22) is known as the *Generalized Ehrenfest Theorem.*

4.4 Dynamical Equation For Single Spin Dynamics

We can use Eq. (4.21) to find the time-dependence of the spin angular momentum operator $\langle\widehat{\mathbf{S}}\rangle$. The result can be used as a *classical* equation of motion in the form of an ordinary differential equation (ODE). In Sec. 3.3, we wrote out the Hamiltonian of the system using the 4D kinetic energy operator (*c*.Eq 3.114), where we found

$$i\hbar\frac{\partial\psi}{\partial t}=(1-i\alpha)\widehat{H}_{\sigma}\psi=[\gamma\hbar\widehat{\boldsymbol{\sigma}}\cdot\mathbf{B}-i\alpha\gamma\hbar\widehat{\boldsymbol{\sigma}}\cdot\mathbf{B}]\psi \tag{4.23}$$

With Eq. (4.23), we can apply the *Most General Ehrenfest equation* given by

$$\frac{d\langle\widehat{\mu}_S\rangle}{dt}=\frac{i}{\hbar}\left[\langle\widehat{H}^{\dagger}\widehat{\mu}_S\rangle-\langle\widehat{\mu}_S\widehat{H}\rangle\right]-\frac{i}{\hbar}\langle\widehat{\mu}_S\rangle\left[\langle\widehat{H}^{\dagger}\rangle-\langle\widehat{H}\rangle\right]+\left\langle\frac{\partial\widehat{\mu}_S}{\partial t}\right\rangle \tag{4.24}$$

The time-derivative term on the RHS of Eq. (4.24) is taken to be zero since the operator does not contain an explicit time dependence. Because the Hamiltonian is the sum of two terms, let's consider each of the terms separately, then combine both results in the end. The first term of the Hamiltonian $\widehat{H}=(1-i\alpha)$ is just $1\cdot H_{\sigma}$, which is *Hermitian*. Therefore, Eq. (4.24) simplifies to

$$\frac{d\widehat{\mu}_S}{dt}=\frac{i}{\hbar}\left[\widehat{H}_{\sigma}\widehat{\mu}_S-\widehat{\mu}_S\widehat{H}_{\sigma}\right]=\frac{i}{\hbar}\left[\widehat{H}_{\sigma},\widehat{\mu}_S\right] \tag{4.25}$$

Substitution of $\widehat{\mu}_S=(-\gamma\hbar)\widehat{\boldsymbol{\sigma}}$ into the above along with the spin matrices $\widehat{\sigma}_x,\widehat{\sigma}_y,\widehat{\sigma}_z$, we have

$$\frac{d\widehat{\mu}_S}{dt}=-i\gamma^2\hbar[(\widehat{\boldsymbol{\sigma}}\cdot\mathbf{B})\widehat{\boldsymbol{\sigma}}-\widehat{\boldsymbol{\sigma}}(\widehat{\boldsymbol{\sigma}}\cdot\mathbf{B})]$$

Expanding the dot products lead to

$$\frac{d\widehat{\mu}_S}{dt} = -i\gamma^2\hbar\left[(\widehat{\sigma}_x B_x + \widehat{\sigma}_y B_y + \widehat{\sigma}_z B_z)\widehat{\boldsymbol{\sigma}} - \widehat{\boldsymbol{\sigma}}(\widehat{\sigma}_x B_x + \widehat{\sigma}_y B_y + \widehat{\sigma}_z B_z)\right]$$

$$= -i\gamma^2\hbar\begin{bmatrix}(\widehat{\sigma}_y\widehat{\sigma}_x - \widehat{\sigma}_x\widehat{\sigma}_y)B_y + (\widehat{\sigma}_z\widehat{\sigma}_x - \widehat{\sigma}_x\widehat{\sigma}_z)B_z \\ (\widehat{\sigma}_x\widehat{\sigma}_y - \widehat{\sigma}_y\widehat{\sigma}_x)B_x + (\widehat{\sigma}_z\widehat{\sigma}_y - \widehat{\sigma}_y\widehat{\sigma}_z)B_z \\ (\widehat{\sigma}_x\widehat{\sigma}_z - \widehat{\sigma}_z\widehat{\sigma}_x)B_x + (\widehat{\sigma}_y\widehat{\sigma}_z - \widehat{\sigma}_z\widehat{\sigma}_y)B_z\end{bmatrix}$$

Making use of the spin matrix relation $\widehat{\sigma}_i\widehat{\sigma}_j - \widehat{\sigma}_j\widehat{\sigma}_i = 2i\widehat{\sigma}_k$, the above leads to the familiar *precessional* motion given by

$$\frac{d\widehat{\mu}_S}{dt} = -i\gamma^2\hbar\begin{bmatrix}-2i\widehat{\sigma}_z B_y + 2i\widehat{\sigma}_y B_z \\ 2i\widehat{\sigma}_z B_x - 2i\widehat{\sigma}_x B_z \\ -2i\widehat{\sigma}_y B_x + 2i\widehat{\sigma}_x B_z\end{bmatrix} = 2\gamma^2\hbar\widehat{\boldsymbol{\sigma}} \times \mathbf{B} = -2\gamma\widehat{\mu}_S \times \mathbf{B}$$

This accounts for *one of the two contributions* to the dynamic equation for $\widehat{\mu}_S$.

Next, let's determine the contribution from the imaginary part of the Hamiltonian $-i\alpha\widehat{H}_\sigma$ using both terms of Eq. (4.21). There is no simplification here because *this contribution to the Hamiltonian is non-Hermitian.* Therefore, the first part of the time-derivative due to the imaginary contribution becomes

$$\frac{d\widehat{\mu}_S}{dt} = \frac{i}{\hbar}\left[\widehat{H}_\sigma^\dagger\widehat{\mu}_S - \widehat{\mu}_S\widehat{H}_\sigma\right] = \frac{i}{\hbar}\left[i\alpha\widehat{H}_\sigma\widehat{\mu}_S + i\alpha\widehat{\mu}_S\widehat{H}_\sigma\right]$$

$$= \frac{-\alpha}{\hbar}\left(\widehat{H}_\sigma\widehat{\mu}_S + \widehat{\mu}_S\widehat{H}_\sigma\right) = \alpha\gamma^2\hbar(\widehat{\boldsymbol{\sigma}} \cdot \mathbf{B}\widehat{\boldsymbol{\sigma}} + \widehat{\boldsymbol{\sigma}}\widehat{\boldsymbol{\sigma}} \cdot \mathbf{B})$$

$$= \alpha\gamma^2\hbar\begin{bmatrix}2\widehat{\sigma}_x^2 B_x + (\widehat{\sigma}_y\widehat{\sigma}_x + \widehat{\sigma}_x\widehat{\sigma}_y)B_y + (\widehat{\sigma}_z\widehat{\sigma}_x + \widehat{\sigma}_x\widehat{\sigma}_z)B_z \\ 2\widehat{\sigma}_y^2 B_y + (\widehat{\sigma}_x\widehat{\sigma}_y + \widehat{\sigma}_y\widehat{\sigma}_x)B_x + (\widehat{\sigma}_z\widehat{\sigma}_y + \widehat{\sigma}_y\widehat{\sigma}_z)B_z \\ 2\widehat{\sigma}_z^2 B_z + (\widehat{\sigma}_x\widehat{\sigma}_z + \widehat{\sigma}_z\widehat{\sigma}_x)B_x + (\widehat{\sigma}_y\widehat{\sigma}_z + \widehat{\sigma}_z\widehat{\sigma}_y)B_z\end{bmatrix}$$

Using the relations $\widehat{\sigma}_i^2 = \mathbf{I}(2)$ and the spin matrix anti-commutation relation $\widehat{\sigma}_i\widehat{\sigma}_j + \widehat{\sigma}_j\widehat{\sigma}_i = 0$, the above becomes

$$\frac{d\widehat{\mu}_S}{dt} = \alpha\gamma^2\hbar\begin{bmatrix}2B_x \\ 2B_y \\ 2B_z\end{bmatrix} = 2\alpha\gamma^2\hbar\mathbf{B} \tag{4.26}$$

For the second part, we have

$$\frac{-i}{\hbar}\langle\boldsymbol{\mu}_S\rangle\left[\widehat{H}_\sigma^\dagger - \widehat{H}_\sigma\right] = \frac{-i}{\hbar}\langle\boldsymbol{\mu}_S\rangle[i\alpha\gamma\hbar + i\alpha\gamma\hbar](\boldsymbol{\sigma}\cdot\mathbf{B}) \tag{4.27}$$

$$= \frac{2\alpha}{\hbar}\langle\boldsymbol{\mu}_S\rangle[\gamma\hbar](\boldsymbol{\sigma}\cdot\mathbf{B}) \tag{4.28}$$

$$= -\frac{2\alpha}{\hbar}(\boldsymbol{\mu}_S\cdot\mathbf{B})\,\boldsymbol{\mu}_S \tag{4.29}$$

To tie the results together, we can write the magnetic field **B** in another way using of a vector identity. This identity states that with a 3D unit vector **u**$= \boldsymbol{\mu}_S/\mu_S$ so $|\mathbf{u}| = 1$, a vector **B** can be decomposed with **u** as

$$\mathbf{B} = \frac{1}{\mu_S^2}[(\boldsymbol{\mu}_S\cdot\mathbf{B})\boldsymbol{\mu}_S - \boldsymbol{\mu}_S\times\boldsymbol{\mu}_S\times\mathbf{B}] \tag{4.30}$$

The idea is illustrated in Fig. 4.1. For a single electron spin, the satura-

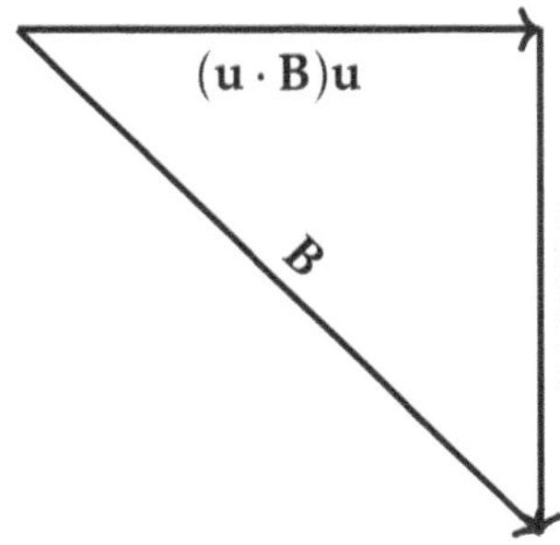

Figure 4.1: Illustration of the vector relations between **B**, **u** × **u** × **B**, and (**u** · **B**)**u**.

tion moment μ_S along the axis of the field is given by

$$|\boldsymbol{\mu}_S| = \mu_S = \gamma\hbar \tag{4.31}$$

Substitution into Eqs. (4.26) gives

$$\mathbf{B} = \frac{2\alpha}{\hbar}[(\boldsymbol{\mu}_S\cdot\mathbf{B})\boldsymbol{\mu}_S - \boldsymbol{\mu}_S\times\boldsymbol{\mu}_S\times\mathbf{B}] \tag{4.32}$$

The first term in Eq. (4.32) *cancels* the result from Eq. (6.114). Finally, we are left with the following for the time-derivative of the operator $\boldsymbol{\mu}_S$:

$$\frac{d\langle\boldsymbol{\mu}_S\rangle}{\partial t} = -2\gamma\boldsymbol{\mu}_S\times\mathbf{B} - \frac{2\alpha}{\hbar}\boldsymbol{\mu}_S\times\boldsymbol{\mu}_S\times\mathbf{B}$$

$$= -2\gamma\boldsymbol{\mu}_S\times\mathbf{B} - \frac{2\alpha\gamma}{\mu_S}\boldsymbol{\mu}_S\times\boldsymbol{\mu}_S\times\mathbf{B}$$

Using the definition for the electron g-factor $g = 2$, we find that the components of the electron magnetic moment μ_S given by Eqs. (3.133),(3.134), and (3.135) also satisfy an equation of motion given by

Electron Magnetic Moment Dynamical Equation

$$\frac{d\boldsymbol{\mu}_S}{dt} = \underbrace{-g\gamma\boldsymbol{\mu}_S \times \mathbf{B}}_{\text{precession}} - \underbrace{\frac{\alpha g\gamma}{\mu_S}\boldsymbol{\mu}_S \times \boldsymbol{\mu}_S \times \mathbf{B}}_{\text{transverse damping}} \tag{4.33}$$

The first term in Eq. (4.33) describes a type of cyclical motion called *precessional* motion, contained by the components of μ_S *perpendicular* to B. The second term is known as *transverse damping* and corresponds to the rotation of the electron magnetic moment vector μ *towards* the magnetic field $\mathbf{B}$. Eq. (4.33) is known as a *Landau-Lifshitz*-type equation.

Transverse damping was originally proposed as a phenomenological term to add to any precessional motion to more correctly describe magnetic materials, for example. However, we see that 4D relativistic quantum mechanics produces this term. In general, the the LLG-like equation is still approximate because the longitudinal relaxation term discussed in Sec. 1.7 is not included.

The electron spin angular momentum is generally accompanied by orbital angular momentum $\mathbf{L}$ where the electron's total angular momentum is $\mathbf{S} + \mathbf{L}$. Let us look for an example of how non-Hermitian operator emerges in a practical problem involving angular momentum. In order to connect the orbital angular momentum to the non-Hermitian 4D Hamiltonian, another set of *relativistic* equations can be used, namely the laws of electromagnetics. Let's see how.

4.5 Adding Lenz' Law Yields Non-hermitian Operator

Consider an electron's classical orbital magnetic moment $\mu_L = \gamma\mathbf{L}$ coupled to a magnetic field. Recall *Lenz' law* which states:

The direction of the current induced in a conductor by a changing magnetic

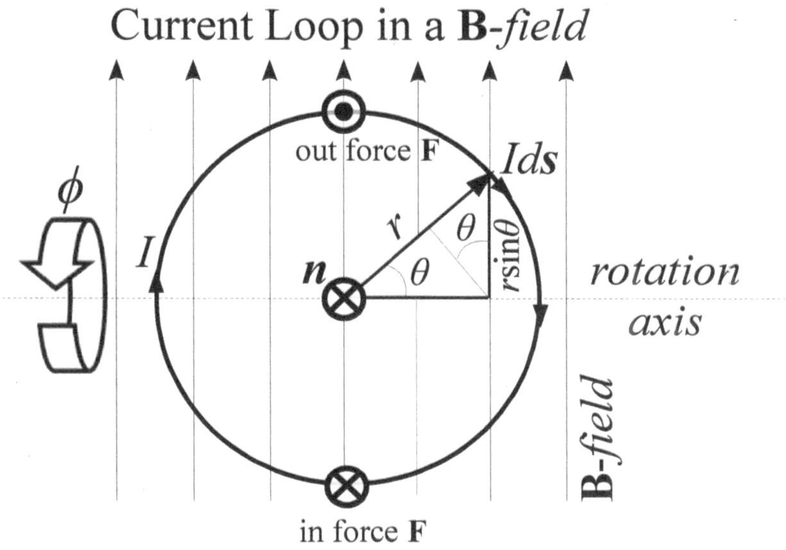

Figure 4.2: Illustration of magnetic moment's current I_0 circulation in a closed circular loop in a uniform magnetic field **B**. A Lorentz force will begin to act on the loop, imparting a torque τ or moment on the loop.

field is such that the magnetic field created by the induced current opposes the initial changing magnetic field.
H. Lenz

In mathematical notation, Lenz' law is expressed as

Lenz Law

$$V_{\text{emf}} = -\frac{d\Phi}{dt} \tag{4.34}$$

$\Phi = \mathbf{B} \cdot \mathbf{A}_o$ is the *magnetic flux* through the orbital area $\mathbf{A}_o$ of the magnetic moment. Even though **B** is static and uniform, the effect described by Lenz' law *still* manifests because the electromotive potential force V is the derivative of the product between the flux *area* and the magnetic field. When the loop rotates, because the *flux area changes,* there is *still* an effect on the motion of the magnetic moment μ_L and angular momentum **L**. This property gives rise to a *non-conservative magnetic field* which creates a nonzero electromotive force V_{emf} around the closed current loop of the magnetic moment μ_L. The electric potential around the loop is *not* conserved in the case that a magnetic field is driving charge motion.

Eq. (4.34) can be used to describe an additional dynamical effect on the motion of μ_L in a magnetic field **B**. Since **B** is uniform and constant, what matters here is the amount of area that is *normal* to **B**. This, of course, changes as the loop undergoes rotation from any angle ϕ away from the equilibrium energy position. The interacting orbital flux area is given by $A_o\cos\phi$. Thus, when **B** and μ_L are parallel (or **L** is antiparallel), the flux is maximum and the induced emf is *most* negative, or minimum. However, when they are perpendicular (as in the initial condition), the flux Φ vanishes because there is no component of the area normal vector **n** parallel to **B**. This is the condition illustrated in Fig. 4.2.

The condition shown is also analogous to the initial condition used for the spin angular momentum in Sec. 3.3. As the loop rotates, the induced current I is dependent on the rotation angle ϕ.[2] Lenz' law is used to quantify this variation, where we have

$$V_{\mathrm{emf}} = -\frac{d}{dt}(BA_o\cos\phi) = BA_o\sin\phi\cdot\dot{\phi} = -BA_o\sin\phi\cdot|\dot{\phi}| \tag{4.35}$$

$\dot{\phi} = d\phi/dt$ is the *rotational rate* of μ_L towards the magnetic field orientation. Since the rotational rate towards B goes to zero approaching equilibrium, $\Delta\phi < 0$. Then, for any angle between $\phi = \pi/2$ and $\phi = 0$, it follows that $V_{\mathrm{emf}} < 0$ or $\Delta V_{\mathrm{emf}} = V_{\mathrm{emf}}(\phi) - V_{\mathrm{emf}}(\phi_i = \pi/2) = -(BA_o\sin\phi\dot{\phi}) - 0 < 0$. This indicates the system is *loosing* emf, acting as a *non-conservative* system.

In any *nonzero* magnetic field with misalignment, or out of equilibrium $\phi = 0$, then $|V_{\mathrm{emf}}| > 0$ for all angles $\phi > 0$. Therefore, equivalently, we can define a positive-definite or *nonzero* effective **orbital resistance** R_o using the relation

$$|V_{\mathrm{emf}}| \equiv \Delta I R_o \tag{4.36}$$

---

[2] ϕ is also the polar angle when **B** is along the z direction.

ΔI is the *induced* current generated within the current loop arising from the changing flux during rotation. The *orbital resistance* is associated with the motion along the circumference of the orbital area A_o. Since $V_{\text{emf}} < 0$, it follows that $\Delta I < 0$, which constitutes a *reduction in the loop current*. The induced current ΔI variation due to rotation of the moment in a **B**-field (taken along the positive axis direction) then becomes

$$\Delta I = -\frac{BA_o}{R_o}|\dot{\phi}|\text{sin}\phi = \Delta I_0 \text{sin}\phi \tag{4.37}$$

And the *amplitude* of the *induced current* ΔI_0 and corresponding magnetic moment $\Delta\mu_L$ become

$$\Delta I_0 = -\frac{BA_o|\dot{\phi}|}{R_o} \quad \text{(units of Amps)} \tag{4.38a}$$

$$\Delta\mu_L = \Delta I_0 A_o = -\frac{BA_o^2|\dot{\phi}|}{R_o} \quad (\text{units of Amps} \cdot \text{m}^2) \tag{4.38b}$$

We are interested in the *total* instantaneous magnetic moment and resulting torque τ. In equilibrium, a magnetic moment already has an associated current I_0 constituting the magnetic moment. So, the *total* current is now given by

$$I = I_0 + \Delta I \tag{4.39a}$$

$$\mu_T = \mu I = I_0 A_0 + \Delta I A_0 \tag{4.39b}$$

Then, the net torque $\boldsymbol{\tau} = \boldsymbol{\mu}_L \times \mathbf{B}$ based on the *total* spontaneous current I, when we include dynamical effects described by Lenz' Law, becomes

$$\tau = A_o I \text{Bsin}\phi = A_o \left(I_0 - \frac{BA_o|\dot{\phi}|}{R_o} \text{sin}\phi \right) \text{Bsin}\phi \tag{4.40a}$$

$$= A_o I_0 \text{Bsin}\phi - A_o \left[\frac{BA_o|\dot{\phi}|}{R_o} \right] \text{Bsin}\phi\text{sin}\phi \tag{4.40b}$$

The units of the term in brackets on the RHS are the same as the current I_0. Then, we can define the *dimensionless dynamical coefficient* $\alpha(t)$ by

$$\alpha(\phi(t)) = \frac{|\Delta\mu_L|}{\mu_L} = \left[\frac{BA_o^2|\dot{\phi}|}{R_o \mu_L} \right] \tag{4.41}$$

Substituting α from Eq. (4.41), the torque equation in Eq. (4.40) becomes

$$\tau = \mu_L \text{Bsin}\phi - \alpha(\phi)\mu_L \text{Bsin}^2\phi \tag{4.42}$$

$$= \mu_L B\sin\phi - \frac{\alpha(\phi)}{\mu_L}\mu_L^2 B\sin^2\phi \tag{4.43}$$

In this form, α still depends on the angular rate $\dot{\phi}$, the magnetic field B, and magnetic moment μ_L. α is also proportional to the maximum flux amplitude of the loop $\Phi_0 = BA_o$. From this, we have a corresponding total torque $\boldsymbol{\tau}$ in vector form in terms of $\boldsymbol{\mu}_L$, given by

Torque Equation With Lenz' Law Correction

$$\boldsymbol{\tau} = \boldsymbol{\mu}_L \times \mathbf{B} - \frac{\alpha(\phi)}{\mu_L}\boldsymbol{\mu}_L \times \mathbf{B} \times \boldsymbol{\mu}_L \tag{4.44}$$

Using Eq. (4.44), the corresponding magnetic potential energy V_M can be found via the relation $V_M = \int \boldsymbol{\tau} d\phi$. Thus, evaluating this integral using Eq. (4.44), we have

$$V_M = \int_{\pi/2}^{\phi} \tau(\phi)d\phi = \int_{\pi/2}^{\phi} \mu_L B\sin\phi - \mu_L B \int_{\pi/2}^{\phi} \alpha(\phi)\sin^2\phi d\phi \tag{4.45}$$

In order to evaluate Eq. (4.45), we can use a mathematical theorem known as the *Generalized Mean-Value Theorem* which states that for a definite *integral of the product of any two integrable functions* f_1 and f_2, we have

$$\int_a^b f_1(x)f_2(x)dx = \overline{f_1(x^*)}\int_a^b f_2(x)dx \quad \text{where} \quad a \le x^* \le b$$

$\overline{f_1(x^*)}$ is the *average* or *mean* value over the limits of the definite integral. Our limits are from $\phi = \pi/2$ to $\phi = 0$. In this way, we have not introduced any *approximation errors* in the evaluation of the integral. The function in the integrand to be averaged is the rotational rate $\dot{\phi}$. Using this theorem, the magnetic potential energy V_M then becomes

$$V_M = \int_{\pi/2}^{\phi} \tau(\phi)d\phi = \int_{\pi/2}^{\phi} \mu_L B\sin\phi - \overline{\alpha}\mu_L B \int_{\pi/2}^{\phi} \sin^2\phi d\phi \tag{4.46}$$

The second integral on the RHS can be evaluated using the relation $\int \sin^2 x dx = \sin(2x)/4 - x/2$ and $\sin(2x) = 2\sin x\cos x$. Eq. (4.46) then becomes

$$V_M = -\mu_L B\cos\phi - \frac{1}{2}\overline{\alpha}\mu_L B\sin\phi\cos\phi + \frac{1}{2}\overline{\alpha}\mu_L B\left(\phi - \frac{\pi}{2}\right) \tag{4.47}$$

Alternatively, a vector representation of the magnetic potential energy V_M is given by

Orbital Magnetic Potential Energy

$$V_M(\mathbf{B}) = -\boldsymbol{\mu}_L \cdot \mathbf{B} - \frac{1}{2}\frac{\overline{\alpha}}{\mu_L}\boldsymbol{\mu}_L \times \mathbf{B} \cdot \boldsymbol{\mu}_L + \frac{1}{2}\overline{\alpha}\mu_L B\left(\phi - \frac{\pi}{2}\right) \tag{4.48}$$

Fig. 4.3 illustrates examples of the potential energy function $V(\phi)$ given by Eq. (4.48) as a function of the inclusive angle ϕ and the external magnetic field amplitude B, using a parameter value $\overline{\alpha} = 0.01$.

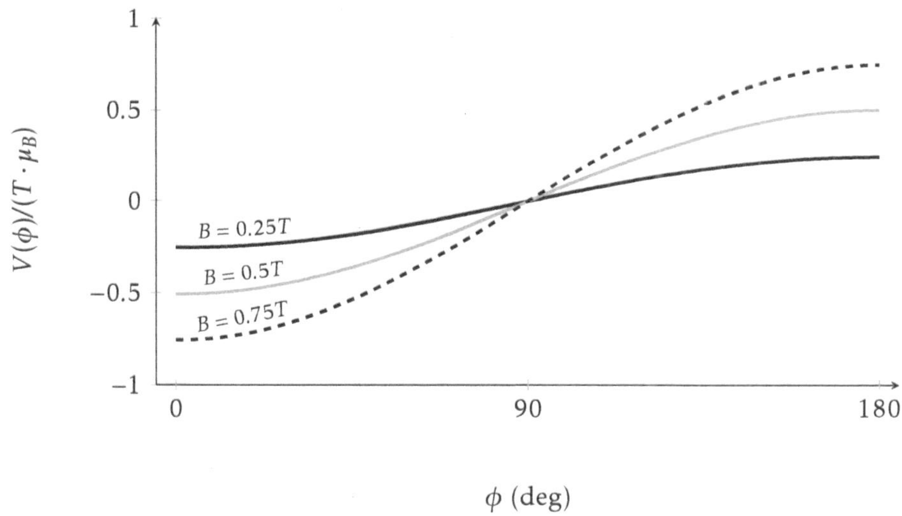

Figure 4.3: Potential energy V per Tesla-Bohr magneton as a function of external magnetic field amplitude B and inclusive angle ϕ. The magnetic field values used are $B = 0.25, 0.5$, and 0.75 Tesla. Notice that the energy minimum for all field amplitudes is at $\phi = 0$, corresponding to alignment.

The dominating term in Eq. (4.48) is the first term, proportional to $-\cos\phi$. This is because the other terms are higher order and multiply by $\overline{\alpha} << 1$, so the curve is nearly a $-\cos\phi$ plot. Based on the above illustration, it could be tempting to ignore the second and third terms altogether. However, we shall find that the second term, in particular, relates to dynamical transitions for the angular momentum. To see this, let us now determine the corresponding quantum mechanical energy operator for the potential energy term V_I, given by

$$V_I = -\frac{\overline{\alpha}}{2\mu_L}\boldsymbol{\mu}_L \times \mathbf{B} \cdot \boldsymbol{\mu}_L \tag{4.49}$$

Replacing the magnetic moment with $\mu_L = -\gamma\mathbf{L}$, the first of the two terms becomes

$$V_I = -\frac{\overline{\alpha}\gamma^2}{2\gamma L}\mathbf{L}\times\mathbf{B}\cdot\mathbf{L} = \frac{\overline{\alpha}\gamma}{2L}\mathbf{L}\times\mathbf{B}\cdot\mathbf{L}$$

Using the quantum mechanical operator for $\widehat{\mathbf{L}} = \mathbf{r}\times\widehat{\mathbf{p}}$, the quantum mechanical eigenvalue equation becomes

$$\widehat{H}_I\psi = \left[\frac{\overline{\alpha}|\gamma|}{2L}\widehat{\mathbf{L}}\times\mathbf{B}\cdot\widehat{\mathbf{L}}\right]\psi = E_I\psi \tag{4.50}$$

The Hamiltonian in Eq. (4.50) is well-defined since we know both operators involved between the magnetic field $\mathbf{B}$ and $\widehat{\mathbf{L}} = \mathbf{r}\times\widehat{\mathbf{p}}$. Taking $\mathbf{B}$ along the $+z$ direction, evaluatiion of only the cross product in Eq. (4.50) gives

$$\widehat{\mathbf{L}}\times\mathbf{B} = \begin{bmatrix} i & j & k \\ \widehat{L}_x & \widehat{L}_y & \widehat{L}_z \\ 0 & 0 & B_z \end{bmatrix} = \begin{bmatrix} \widehat{L}_y B_z - \widehat{L}_z B_y \\ -(\widehat{L}_x B_z - \widehat{L}_z B_x) \\ \widehat{L}_x B_y - \widehat{L}_y B_x \end{bmatrix} = \begin{bmatrix} \widehat{L}_y B_z \\ -\widehat{L}_x B_z \\ 0 \end{bmatrix}$$

Then, dotting this result with $\widehat{\mathbf{L}}$ gives

$$\widehat{\mathbf{L}}\times\mathbf{B}\cdot\widehat{\mathbf{L}} = \widehat{L}_y\widehat{L}_x B_z - \widehat{L}_x\widehat{L}_y B_z \tag{4.51}$$

$$= -B(\widehat{L}_x\widehat{L}_y - \widehat{L}_y L_x) \tag{4.52}$$

$\mathbf{B}$ is taken along the $+z$ axis, or $\mathbf{B} = B_z\mathbf{e}_z$. The term in parentheses can be replaced by the quantum mechanical *commutation relations* for the angular momentum, namely[3]

$$\widehat{L}_x\widehat{L}_y - \widehat{L}_y\widehat{L}_x = i\hbar\widehat{L}_z$$

Using this result in Eq. (4.51), we have

$$\widehat{H}_I\psi = -i\left[\frac{\overline{\alpha}|\gamma|\hbar B}{2L}\widehat{L}_z\right]\psi$$

$$= i\left(\frac{\overline{\alpha}}{2}\right)\widehat{\mu}_L\cdot\mathbf{B}\psi$$

$$= i\alpha^* V_R$$

[3]For a complete derivation of these relations, see, for example, Sec. 7.8 of Quantum Mechanics With A Spin On It (Eason)

$V_R = -\boldsymbol{\mu}_L \cdot \mathbf{B}$ is the *ordinary* magnetic energy in equilibrium, ignoring the additional contribution from V_I. We have also replaced the mean classical dimensionless dynamical coefficient $\overline{\alpha}$ with a *microscopic dimensionless coefficient* α^* related by

$$\alpha^* = \frac{\overline{\alpha}}{2} \tag{4.53}$$

Therefore, the potential energy given by Eq. (4.49) leads to a purely imaginary scalar $-i\alpha^*$ times (×) the *Hermitian Hamiltonian operator*. Adding both to the Hamiltonian $\widehat{H}_T$ gives the result

$$\widehat{H}_T\psi = [-\boldsymbol{\mu}_L \cdot \mathbf{B} + i\alpha^* \boldsymbol{\mu}_L \cdot \mathbf{B}]\psi = (1 - i\alpha^*)(-\boldsymbol{\mu}_L \cdot \mathbf{B})\psi \tag{4.54}$$

Equivalently, the complex eigenenergy E_λ for the orbital angular momentum becomes

Non-Hermitian Eigenvalues Of Total Hamiltonian

$$E_\lambda = (1 - i\alpha^*)|\gamma|Bm_\ell\hbar \tag{4.55}$$

Thus, we find that making a correction to the current of a magnetic moment, based on Lenz' Law, we obtain a fully consistent with the non-Hermitian operators of 4D quantum mechanics.

4.6 Deriving Faraday's Law From Quantum Mechanics

In the previous section, we went from an electromagnetic law to its implications in quantum mechanics. Here, we'll go in the other direction by deriving the dynamical Faraday's Law from quantum mechanics. This becomes possible now that we have a fourth dimension. Because of this, we can add a new potential function that equivalently depends on the fourth dimension. From here, we can construct higher order operators that lead to the dynamical electromagnetic laws.

4.6.1 Defining the Quantum Electric Field

To obtain any of the dynamical electromagnetic laws, the *commutation operator* must be used. It is defined as

$$[\widehat{A}, \widehat{B}] = \widehat{A}\widehat{B} - \widehat{B}\widehat{A} \tag{4.56}$$

When $[\widehat{A},\widehat{B}] = 0$, it is said that the two operators commute. We can use a commutation operator to define an electric field. Let V be a potential energy, and the $\widehat{\mathbf{p}}$ be the momentum operator. Then, evaluating the commutation operator with these two, we have

$$[V,\widehat{\mathbf{p}}]\psi = [V(-i\hbar\nabla) - (-i\hbar\nabla V)]\psi \tag{4.57a}$$

$$= V(-i\hbar\nabla\psi) - (-i\hbar\nabla(V\cdot\psi) \tag{4.57b}$$

$$= -i\hbar V\nabla\psi - (-i\hbar\nabla V\cdot\psi + -i\hbar V\nabla\psi) \tag{4.57c}$$

$$= -i\hbar V\nabla\psi + i\hbar\nabla V\cdot\psi + i\hbar V\nabla\psi \tag{4.57d}$$

$$= i\hbar\nabla V\cdot\psi \tag{4.57e}$$

The shorthand for this result is that

$$[V,\widehat{\mathbf{p}}] = i\hbar\nabla V \tag{4.58}$$

For any potential that affects a particle of charge q, an *electric field* vector $\widehat{\mathbf{E}}$ can be defined as the (negative) gradient of that potential *per unit charge*. This has units of force per unit charge. Since, from

$$\nabla V = \frac{1}{i\hbar}[V,\widehat{\mathbf{p}}] = -\frac{i}{\hbar}[V,\widehat{\mathbf{p}}]$$

The negative of the gradient per unit charge then becomes

Electric Field Operator

$$\widehat{\mathbf{E}} = \frac{i}{q\hbar}[V,\widehat{\mathbf{p}}] \tag{4.59}$$

where q is a quanta of charge for the associated quantum particle.

4.6.2 The 4th Dimensional Potential Condition

In the definition of the electric field, we used the 3D momentum so that changes in the potential are only with respect to these three. Having the fourth dimension, we can add a contribution to the potential that arises from changes in the fourth dimension.

Fourth Dimensional Real Potential Condition

$$d\phi = \mathbf{v} \cdot d\mathbf{p}_B \tag{4.60}$$

$\mathbf{p}_B$ is an *additional* momentum existing *only* in the presence of a magnetic field $\mathbf{B}$. Then, we can write

$$d\phi = v_x dp_{B,x} + v_y dp_{B,y} + v_z dp_{B,z} \tag{4.61}$$

As independent coordinates, there are three contributions to a change in the potential ϕ where each contribution can be written as

$$d\phi|_x = v_x dp_{B,x} = \frac{dx}{dt} dp_{B,x} \tag{4.62a}$$

$$d\phi|_y = v_y dp_{B,y} = \frac{dy}{dt} dp_{B,y} \tag{4.62b}$$

$$d\phi|_z = v_z dp_{B,z} = \frac{dz}{dt} dp_{B,z} \tag{4.62c}$$

From the above relations, we have the following:

$$\frac{\partial \phi}{\partial x} = \frac{\partial p_{B,x}}{\partial t} \tag{4.63a}$$

$$\frac{\partial \phi}{\partial y} = \frac{\partial p_{B,y}}{\partial t} \tag{4.63b}$$

$$\frac{\partial \phi}{\partial z} = \frac{\partial p_{B,z}}{\partial t} \tag{4.63c}$$

so that Eq. (4.60) holds if and only if

Fourth Dimensional Potential Gradient

$$\nabla \phi = \frac{\partial \mathbf{p}_B}{\partial t} \tag{4.64}$$

Then, we can define a new potential V' by

$$V' = V + \phi \Rightarrow \nabla V' + \nabla \phi \tag{4.65}$$

Then, from Eq. (4.64), the new electric field $\mathbf{E}'$, which is the potential per unit charge q, becomes

Generalized Electric Field Operator

$$\widehat{\mathbf{E}'} = \frac{1}{q}\left[-\nabla V - \frac{\partial \mathbf{p}_B}{\partial t}\right] \tag{4.66}$$

We may recognize that $-\nabla V/q = \nabla V_e$ is equivalent to an electrical potential gradient. If we define a magnetic momentum per unit charge as

$$\mathbf{A} = \frac{1}{q}\mathbf{p}_B \tag{4.67}$$

Then, an equivalent form of the general electric field is given by

$$\widehat{\mathbf{E}'} = -\nabla V_e - \frac{\partial \mathbf{A}}{\partial t} \tag{4.68}$$

One of the conveniences of this approach is that $\nabla V \rightarrow \nabla V'$, so we can use the same analysis above involving the gradient in a straightforward fashion, while keeping in mind that there is an additional contribution in V'.

4.6.3 CURL Of Electric Field Operator

Using V' from Eq. (4.65) instead of V, let us evaluate the following operator

$$\widehat{O} = \frac{1}{q\hbar^2}\left[\widehat{p_y}[V',\widehat{p_z}] - \widehat{p_z}[V',\widehat{p_y}]\right] \tag{4.69}$$

Although this analysis gets a little congested, we will not omit many details because it is important to follow each step. We will, however, use the shorthand instead of explicitly writing the wave function ψ. So, we have

$$\frac{1}{q\hbar^2}\left[\cdot\widehat{p_y}[V',\widehat{p_z}] - \widehat{p_z}[V',\widehat{p_y}]\right] =$$

$$\frac{1}{q\hbar^2}\left[\widehat{p_y}\left(V'\widehat{p_z} - \widehat{p_z}V'\right) - \widehat{p_y}\left(V'\widehat{p_y} - \widehat{p_y}V'\right)\right] \tag{4.70}$$

Using the definition for only the outermost momentum operator, we pickup a factor of $i\hbar$. Multiplying this by the constant in front, our factor becomes

$$\frac{1}{q\hbar^2} \times i\hbar = +\frac{i}{q\hbar}$$

Then, Expanding Eq. (4.70) a bit (omitting the factor), we have

$$\cdots = \left(\frac{\partial V'}{\partial y}\widehat{p}_z + V'\frac{\partial \widehat{p}_z}{\partial y}\right) - \left(\frac{\partial \widehat{p}_z}{\partial y}V' + \widehat{p}_z\frac{\partial V'}{\partial y}\right) -$$

$$\left(\left[\frac{\partial V'}{\partial z}\widehat{p}_y + V'\frac{\partial \widehat{p}_y}{\partial z}\right] - \left[\frac{\partial \widehat{p}_y}{\partial z}V' + \widehat{p}_y\frac{\partial V'}{\partial z}\right]\right) \tag{4.71}$$

From this expanded form, the 2nd and 6th terms cancel each other because of the equality of mixed partials. Likewise, the 3rd and 7th terms cancel. Left only with the 1st, 3rd, 4th, and 5th terms, Eq. (4.71) reduces to

$$\cdots = \frac{i}{\hbar q}\left[-\left(\frac{\partial V'}{\partial y}\widehat{p}_z - \widehat{p}_z \cdot -\frac{\partial V'}{\partial y}\right) - \left(-\frac{\partial V'}{\partial z}\widehat{p}_y - \widehat{p}_y \cdot -\frac{\partial V'}{\partial z}\right)\right] \tag{4.72a}$$

$$= \frac{i}{\hbar}\left[\left(\widehat{E}'_y\widehat{p}_z - \widehat{p}_z\widehat{E}'_y\right) - \left(\widehat{E}'_z\widehat{p}_y - \widehat{p}_y\widehat{E}'_z\right)\right] \tag{4.72b}$$

$$= \frac{i}{\hbar}\left[\left[\widehat{E}'_y, \widehat{p}_z\right] - \left[\widehat{E}'_z, \widehat{p}_y\right]\right] \tag{4.72c}$$

Expanding the term in square brackets from the last equation (omitting the factor), we have

$$\cdots = \widehat{E}'_y\widehat{p}_z - \left[-i\hbar\frac{\partial \widehat{E}'_y}{\partial z} + \widehat{E}'_y(-i\hbar)\frac{\partial}{\partial z}\right] -$$

$$\widehat{E}'_z\widehat{p}_y - \left[-i\hbar\frac{\partial \widehat{E}'_z}{\partial y} + \widehat{E}'_z(-i\hbar)\frac{\partial}{\partial y}\right] \tag{4.73}$$

The 1st and 3rd terms cancel each other. Likewise, so do the 4th and 6th terms, reducing Eq. (4.72) to

$$\cdots = \frac{i}{\hbar}\left[i\hbar\frac{\partial \widehat{E}'_y}{\partial z} - i\hbar\frac{\partial \widehat{E}'_z}{\partial y}\right] \tag{4.74}$$

$$= \frac{\partial \widehat{E}'_z}{\partial y} - \frac{\partial \widehat{E}'_y}{\partial z} \tag{4.75}$$

The last term takes the form of a familiar operator known as the *vector curl operator*. The vector curl operator $\nabla\times$ of a vector $\mathbf{F}$ is defined as

$$\nabla \times \mathbf{F} = \begin{bmatrix} \frac{\partial F_z}{\partial y} - \frac{\partial F_y}{\partial z} \\ \frac{\partial F_x}{\partial z} - \frac{\partial F_z}{\partial x} \\ \frac{\partial F_y}{\partial x} - \frac{\partial F_x}{\partial y} \end{bmatrix} \tag{4.76}$$

We have obtained the *first* component of the curl operator by using the y and z components. The procedure is the same for the other two coordinates rotating the them appropriately. The general result can be written as

Generalized Electric Field CURL Operator

$$\left(\nabla \times \widehat{\mathbf{E}}'\right)_k = \frac{i}{\hbar}\left[\left[\widehat{E}'_i, \widehat{p}_j\right] - \left[\widehat{E}'_j, \widehat{p}_i\right]\right] \tag{4.77}$$

Since the potential has the form

$$\nabla V' = \nabla V + \nabla\phi = \nabla V + \frac{\partial \mathbf{p}_B}{\partial t} \tag{4.78}$$

let us, now, evaluate the equivalent operator using the latter form having a *time*-dependence.

4.6.4 Relation To Magnetic Field

Using Eq. (4.78), let us write the operator of Eq. (4.77) as

$$\frac{i}{\hbar}\left[\left[\widehat{E}'_i, \widehat{p}_j\right] - \left[\widehat{E}'_j, \widehat{p}_i\right]\right] = \frac{i}{\hbar}\left[\left[\widehat{E}_i - \frac{\partial A_j}{\partial t}, \widehat{p}_j\right] - \left[\widehat{E}_j - \frac{\partial A_j}{\partial t}, \widehat{p}_i\right]\right] \tag{4.79}$$

The LHS of Eq. (4.80) can also be written as

$$\frac{i}{\hbar}\left[\left[\widehat{E}'_i, \widehat{p}_j\right] - \left[\widehat{E}'_j, \widehat{p}_i\right]\right] = \cdots$$

$$\frac{i}{\hbar}\left[\left[\widehat{E}_i, \widehat{p}_j\right] - \left[\widehat{E}_j, \widehat{p}_i\right]\right] + \frac{i}{\hbar}\left[\left[-\frac{\partial A_j}{\partial t}, \widehat{p}_j\right] - \left[-\frac{\partial A_j}{\partial t}, \widehat{p}_i\right]\right]$$

$$\cdots = (\nabla \times \mathbf{E})_k + \frac{i}{\hbar}\left[\left[-\frac{\partial A_j}{\partial t}, \widehat{p}_j\right] - \left[-\frac{\partial A_j}{\partial t}, \widehat{p}_i\right]\right] \tag{4.80}$$

We can show that the first operator on the RHS is null, or that

$$(\nabla \times \mathbf{E})_k = 0 \tag{4.81}$$

so that we need only evaluate the second operator on the RHS. It is well-known in vector calculus that

$$\nabla \times (\nabla \varphi) = 0 \tag{4.82}$$

or that the curl of the gradient of a scalar field is null. Since the electric field $qE = \nabla V$, it *is* a gradient of a scalar field so that Eq. (4.81) is valid.

Note, however, it is *not* valid for $\mathbf{E}'$ because it is made up of *two* independent potential energies V and ϕ, where ϕ depends on a coordinate/variable that V does not, or that are beyond just x, y, and z. This quirky point is equivalent to the addition of the fourth dimension, which depends on time, but is independent of *real*-valued coordinates x, y, and z. The reason ϕ is characterized this way depending on *four* parameters is only to assist in deriving a form consistent with the electromagnetic laws which essentially only describes the *four* parameters $A_x(t), A_y(t), A_z(t)$, and V'. Note that it would be equivalent to introduce a potential energy that depends *only* on the fourth dimension.

Thus, we only need to evaluate the operator given by

$$\widehat{O} = \frac{i}{\hbar}\left[\left[-\frac{\partial A_i}{\partial t}, \widehat{p}_j\right] - \left[-\frac{\partial A_j}{\partial t}, \widehat{p}_i\right]\right] \tag{4.83}$$

Evaluating Eq. (4.82) without the factor leads to

$$\left[-\frac{\partial A_i}{\partial t}, \widehat{p}_j\right] - \left[-\frac{\partial A_j}{\partial t}, \widehat{p}_i\right] = \cdots$$

$$\cdots - \frac{\partial A_i}{\partial t} \cdot (-i\hbar)\frac{\partial}{\partial x_j} - \left[-i\hbar\frac{\partial}{\partial x_j}\left(-\frac{\partial A_i}{\partial t}\right) + -\frac{\partial A_i}{\partial t}\left(-i\hbar\frac{\partial}{\partial x_j}\right)\right] - \cdots$$

$$\cdots - \frac{\partial A_j}{\partial t} \cdot (-i\hbar)\frac{\partial}{\partial x_i} - \left[-i\hbar\frac{\partial}{\partial x_i}\left(-\frac{\partial A_j}{\partial t}\right) + -\frac{\partial A_j}{\partial t}\left(-i\hbar\frac{\partial}{\partial x_i}\right)\right] \tag{4.84}$$

In Eq. (4.84), the 1st and 3rd elements cancel. And so do the 4th and 6th elements. Left with the 2nd and 5th terms. Returning the factor to the operator, we then have

$$\frac{i}{\hbar}\cdot(-i\hbar)\cdot\left[\frac{\partial}{\partial x_j}\left(-\frac{\partial A_i}{\partial t}\right)-\frac{\partial}{\partial x_i}\left(-\frac{\partial A_j}{\partial t}\right)\right]=-\left[\frac{\partial}{\partial x_j}\left(\frac{\partial A_i}{\partial t}\right)-\frac{\partial}{\partial x_i}\left(\frac{\partial A_j}{\partial t}\right)\right] \quad (4.85)$$

Using the equality of mixed partials stating that

$$\frac{\partial^2}{\partial x_1 \partial x_2}=\frac{\partial^2}{\partial x_2 \partial x_1} \quad (4.86)$$

we have

$$-\left[\frac{\partial}{\partial x_j}\left(\frac{\partial A_i}{\partial t}\right)-\frac{\partial}{\partial x_i}\left(\frac{\partial A_j}{\partial t}\right)\right]=\left[\frac{\partial}{\partial t}\left(\frac{\partial A_j}{\partial x_i}\right)-\frac{\partial}{\partial t}\left(\frac{\partial A_i}{\partial x_j}\right)\right] \quad (4.87)$$

$$=\frac{\partial}{\partial t}\left[\frac{\partial A_j}{\partial x_i}-\frac{\partial A_i}{\partial x_j}\right] \quad (4.88)$$

For a magnetic field in the coordinate system $[x_i, x_j, x_k]$, defined as $\mathbf{B} = \nabla \times \mathbf{A}$, referring to the curl operator from Eq. (4.76), it follows that

$$\frac{\partial}{\partial t}\left[\frac{\partial A_j}{\partial x_i}-\frac{\partial A_i}{\partial x_j}\right]=-\frac{\partial B_k}{\partial t} \quad (4.89)$$

Since our operator has also been shown to equal $\nabla \times \widehat{\mathbf{E}'}$, by transitivity, we have

$$\left(\nabla \times \widehat{\mathbf{E}'}\right)_k=\frac{i}{\hbar}\left[\left[\widehat{E}'_i,\widehat{p}_j\right]-\left[\widehat{E}'_j,\widehat{p}_i\right]\right]=-\frac{\partial B_k}{\partial t} \quad (4.90)$$

or in vector form

Faraday's Electromagnetic Law

$$\nabla \times \widehat{\mathbf{E}'}=-\frac{\partial \mathbf{B}}{\partial t} \quad (4.91)$$

4.7 Deriving The Maxwell-Ampere Law

We have already seen that Faraday's law can be derived from quantum mechanical operators and/or wave functions. Faraday's law is only one of two essential electromagnetic laws. This section establishes a path towards the other electromagnetic law known as the Maxwell-Ampere's law. To get started, let us defined a *electric displace vector field* **D** as

$$\nabla \cdot \mathbf{D} = \rho \tag{4.92}$$

In this context, ρ is the volume density which is a function of the wave function ψ, defined as (*c*.Sec. 4.2)

$$\rho = \psi^* \psi \tag{4.93}$$

There, the scalar version of the continuity equation was also derived, where we found

$$\frac{\partial \rho}{\partial t} + \nabla \cdot \mathbf{J} = -\frac{2\alpha V}{\hbar}\rho \tag{4.94}$$

It is the continuity equation that leads us to the Maxwell-Ampere law. Because the 4th dimension has been included, notice the RHS is *nonzero*. We'll see that this is *necessary* to obtain the law for *real* materials or non-vacuum environments. Substituting Eq. (4.92) into Eq. (4.94), we have

$$\frac{\partial}{\partial t}(\nabla \cdot \mathbf{D}) + \nabla \cdot \mathbf{J} = -\frac{2\alpha V}{\hbar}\rho \tag{4.95}$$

Before continuing, it should be mentioned that Eq. (4.95) can be combined with Faraday's law and/or its equivalent operators for a complete quantum mechanical time+space dependent electromagnetic calculation. It could end here. However, the objective here is to demonstrate that by including the 4th dimension as we have, it is wholly consistent with the established relativistic electromagnetic laws, given by *both* Faraday's law as well as the Ampere-Maxwell law. So, towards this aim, let's continue exploring how quantum mechanics leads to latter of these. We can use the mathematical identity given by

$$\nabla \cdot (\nabla \times \mathbf{F}) = 0 \tag{4.96}$$

which says that the divergence of the curl of a vector is always null. Using Eq. (4.96), we may add the following to the RHS of Eq. (4.95) using $\mathbf{F} = \mathbf{H}$ so we have

$$\frac{\partial}{\partial t}(\nabla \cdot \mathbf{D}) + \nabla \cdot \mathbf{J} = \nu \nabla \cdot (\nabla \times \mathbf{H}) - \frac{2\alpha V}{\hbar}\rho \tag{4.97}$$

or

$$\nabla \cdot \left(\frac{\partial \mathbf{D}}{\partial t}\right) + \nabla \cdot \mathbf{J} = \nu \nabla \cdot (\nabla \times \mathbf{H}) - \frac{2\alpha V}{\hbar}\rho \tag{4.98}$$

The constant parameter ν is required *at least* to ensure consistent units for the vector $\nu\mathbf{H}$ in the equation.

Let us first consider the limiting case of $\alpha = 0$, so we can ignore the material dependent term on the RHS. This corresponds to the ideal concept of vacuum, where no atoms to absorb/emit light are present. In this case, we have

$$\nabla \cdot \left(\frac{\partial \mathbf{D}}{\partial t} \right) + \nabla \cdot \mathbf{J} = \nu \nabla \cdot (\nabla \times \mathbf{H}) \tag{4.99}$$

Since all terms are a divergence of a vector, the vectors upon which the divergences act are all to within a constant of each other. So, we can write

$$\frac{\partial \mathbf{D}}{\partial t} + \mathbf{J} = \nu \nabla \times \mathbf{B} + \mathbf{C}_0 \tag{4.100}$$

$\mathbf{C}_0$ is a constant *vector*. Moreover, we can express this constant vector as a *uniform* curl, where

$$\mathbf{C}_0 = -\chi \nabla \times \mathbf{H} \tag{4.101}$$

This quantity represents the contribution to $\mathbf{H}$ arising from the magnetization (magnetic moment or moment per unit volume) vector associated with the propagation medium. χ is the *magnetic susceptibility* defined as

$$\chi = \frac{dM}{dH} \quad \text{or} \quad [\chi] = \frac{d\mathbf{M}}{d\mathbf{H}} \tag{4.102}$$

We'll assume the scalar form first since its simpler. Then, for the ideal case of vacuum, since there are *no* atoms interacting with the wave, then $\chi = 0$ since either $\mathbf{M} = 0$ or $d\mathbf{M} = 0$. So, was this choice for nothing? Actually, it becomes useful when we consider the case of $\alpha \neq 0$, which is more practical. Still, substitution into Eq. (4.100) then gives

$$\frac{\partial \mathbf{D}}{\partial t} + \mathbf{J} = \nu \nabla \times \mathbf{H} \tag{4.103}$$

Then, taking $\mathbf{H}$ to have units of current per unit length *e.g.* A/m, we can set the constant parameter ν to

$$\nu = \frac{1}{\mu_0} \tag{4.104}$$

where μ_0 the *magnetic permeability in vacuum* whose experimental value is given by

$$\mu_0 = 4 \times \pi \times 10^{-7} \mathbf{N}\text{-}\mathbf{A}^{-7} \tag{4.105}$$

Let's define the magnetic field to have consistent units in the given form, as $\mathbf{H} = \nu\mathbf{B} - \chi\mathbf{M}$). Then, substituting the value for ν, we obtain the Ampere-Maxwell Law *in vacuum*, given by

Ampere-Maxwell Electromagnetic Law (Vacuum)

$$\frac{\partial \mathbf{D}}{\partial t} + \mathbf{J} = \frac{1}{\mu_0} \nabla \times \mathbf{B} \tag{4.106}$$

For the case of $\alpha \neq 0$, Eq. (4.98) can be generalized using the concept of susceptibility. In this case, the potential energy V is of the form

$$V = -\boldsymbol{\mu} \cdot \mathbf{B} \tag{4.107}$$

$\boldsymbol{\mu}$ is the magnetic moment for an electron. Therefore, the product $\rho\boldsymbol{\mu} = \mathbf{M}$ is the magnetization, or moment per unit volume. So, we have

$$\nabla \cdot \left(\frac{\partial \mathbf{D}}{\partial t} \right) + \nabla \cdot \mathbf{J} = \frac{1}{\mu_0} \nabla \cdot (\nabla \times \mathbf{B}) + \frac{2\alpha \mathbf{M} \cdot \mathbf{B}'}{\hbar} \tag{4.108}$$

$$= \frac{1}{\mu_0} \nabla \cdot (\nabla \times \mathbf{B}) - \frac{2\alpha BM \cos\phi}{\hbar} \tag{4.109}$$

Note that $\mathbf{B}' \neq \mathbf{B}$, however, they do overlap because $\mathbf{B}'$ includes any external magnetic field $\mu_0\mathbf{H}$ as well as the *internal* magnetic field contributions coupling to the electrons of the atom. So

$$\mathbf{B}' = \mu_0 \mathbf{H} + \mathbf{B}_{\text{int}} \tag{4.110}$$

$\mathbf{B}_{\text{int}}$ denotes the internal magnetic field of the atoms of the propagating medium. We can see that the general case starts to become more complex, so the utility of the magnetic susceptibility allows a simpler approach by expressing the above relations in an alternative form, *etc.*

Since the additional term is proportional to $\mathbf{M}$, we can express this relationship in another way. Also depending on the vector $\mu_0\mathbf{H}$, it is equivalent to defining an expansion of the magnetization **M** in terms of **H**. For example, let us express the magnetization using the expansion

$$\mathbf{M} = \mathbf{M}_i + \frac{d\mathbf{M}}{d\mathbf{H}}\mathbf{H} = \mathbf{M}_i + [\chi]\mathbf{H} \tag{4.111}$$

or

$$\Delta\mathbf{M} = \mathbf{M} - \mathbf{M}_i = [\chi]\mathbf{H} \tag{4.112}$$

$\mathbf{M}_i$ is the initial magnetization vector before any excitation, *e.g.* along the magnetization easy axis. For a magnetization vector of *conserved length*, a rotation $\Delta\mathbf{M}$ is *perpendicular* to **M**, but so is $\nabla\times\mathbf{M}$. They both lie the plane perpendicular to **M**. With the added susceptibility tensor $[chi]$, additional rotation is permitted for the solution. Then, let us define a *nonzero magnetic* divergence of a curl equation as

$$\frac{2\alpha BM(1-\cos\phi)}{\hbar} = -\frac{1}{\mu_0}\nabla\cdot\nabla\times\mathbf{M} = -\frac{1}{\mu_0}\nabla\cdot\nabla\times([\chi]\mathbf{H}) \tag{4.113}$$

ϕ is the angle between **M** and $\mathbf{B}'$. The use of $1-\cos\phi$ arises from setting the initial condition to zero rotation when $\phi(t=0)=0$. Here, $[\chi]$ is the 3×3 *susceptibility tensor*. Because the LHS is *non-zero* except for the initial condition, the resulting field is *non-conservative*. This is a well-known property of a general magnetic field. $[\chi]$ only depends on time t, but *not* space. Then, a more general *magnetic induction* **B** , containing the additional magnetization contribution, can be defined as

$$\mathbf{B} = \mu_0\mathbf{H} + \mu_0\Delta\mathbf{M} = \mu_0\mathbf{H} + \mu_0[\chi]\mathbf{H} \rightarrow \mathbf{H} = [\mu]\mathbf{H} \tag{4.114}$$

Then, substituting the RHS of Eq. (4.113) into Eq. (4.108), we have

$$\nabla\cdot\left(\frac{\partial\mathbf{D}}{\partial t}\right) + \nabla\cdot\mathbf{J} = [\mu]^{-1}\nabla\cdot(\nabla\times\mathbf{B}) \tag{4.115}$$

$[\mu]^{-1}$ is the *inverse* of the magnetic permeability tensor $[\mu] = \mu_0(I(3)+[\chi])$, and it only depends on time explicitly. From this, we then have

$$\frac{\partial\mathbf{D}}{\partial t} + \mathbf{J} = [\mu]^{-1}\,(\nabla\times\mathbf{B}) \tag{4.116}$$

This is the generalization of Eq. (4.106). It is worth pointing out again, that when removing the divergence, there is an integration constant. This has been handled by using the angular form $1-\cos\phi$ to satisfy the initial condition set to zero rotation from the internal field B_{int}.[4]

[4]The internal field is not necessarily parallel to any external field.

Summing it all up, we have obtained the second electromagnetic law, being the Ampere-Maxwell electromagnetic law. It is a subtle but crucial detail that it has been done for both cases of vacuum ($\alpha = 0$) as well as non-vacuum ($\alpha \neq 0$). For the latter case, it is enabled by the extended continuity equation based on the inclusion of the 4th dimension.

4.8 Chapter Summary

In this chapter, we have given further consideration of the implications of non-Hermitian operators in quantum mechanics. This is relevant to 4th dimension because it leads to non-Hermitian energy operators. We generalized some important relations that would be necessary to demonstrate more of the properties of non-Hermitian systems.

First, we obtained the continuity equation based on a quantum mechanical wave function of a non-Hermitian operator. We then generalized the Ehrenfest Theorem to allow for determination of the time-derivative of a quantum mechanical operator used in a non-Hermitian Hamiltonian. The result was then used to demonstrate that the obtained dynamical equation includes two torque terms that together describe the spin transitioning between two separate states, e.g. misaligned to aligned with **B**. We applied Lenz Law to the orbital magnetic moment in an external magnetic field making a correction to the orbital current. This lead to an additional torque term. The additional term enables the full transition between two states. The resulting energy was also shown to have complex eigenvalues, consistent with a 4D Hamiltonian. Finally, we used the results specific to the 4th dimension to derive the full dynamical electromagnetic laws. We are working our way towards another application in the field of spintronics.

However, to prepare for the discussion of spintronics, the topic of *transport* must be reviewed. This is the focus of the next chapter.

Table 4.1 summarizes some of the key formulas from this chapter.

Table 4.1: Chapter 4 Summary Equations.

Name	Equation

Table 4.1

Prob. Density	$\rho = \psi \cdot \psi^*$
Current Density	$\widehat{\mathbf{J}} = \frac{i\hbar}{2m}(\psi \nabla \psi^* - \psi^* \nabla \psi)$
Continuity Eq.	$\frac{\partial \rho}{\partial t} + \nabla \cdot \mathbf{J} = -\frac{2\alpha}{\hbar} V \rho$
M. G. Ehrenfest	$\frac{\partial \langle \widehat{O} \rangle}{\partial t} = \frac{i}{\hbar}\left[\langle \widehat{H}^{\dagger} \widehat{O} \rangle - \langle \widehat{O} \widehat{H} \rangle\right] - \frac{i}{\hbar}\langle O \rangle \left[\langle \widehat{H}^{\dagger} \rangle - \langle \widehat{H} \rangle\right] + \left\langle \frac{\partial \widehat{O}}{\partial t} \right\rangle$
Spin Dynamics	$\frac{\partial \boldsymbol{\mu}_S}{\partial t} = -g\gamma \boldsymbol{\mu}_S \times \mathbf{B} \quad - \frac{\alpha g \gamma}{\mu_S} \boldsymbol{\mu}_S \times \boldsymbol{\mu}_S \times \mathbf{B}$
E-Field	$\widehat{\mathbf{E}} = \frac{i}{q\hbar}[V, \widehat{\mathbf{p}}]$
E-CURL	$\left(\nabla \times \widehat{\mathbf{E}}\right)_k = \frac{i}{\hbar}\left[\left[\widehat{E}_i, \widehat{p}_j\right] - \left[\widehat{E}_j, \widehat{p}_i\right]\right]$

4.9 Chapter Problems

Problem 4.1 In this chapter, we found the operator for the *current density* to be

$$\widehat{\mathbf{J}} = \frac{i\hbar}{2m}(\psi \nabla \psi^* - \psi^* \nabla \psi)$$

Let $\psi = f(x) A e^{ikx}$. Calculate the current density for ψ and compare it to the current density for the case $f(x) = 1$.

Problem 4.2 Let $\psi_1 = Ae^{ikx}$, $\psi_2 = Ae^{-\alpha kx}$ and $\psi_3 = Ae^{ikx}e^{-\alpha kx}$. Calculate the current densities for ψ_1, ψ_2 and ψ_3.

Problem 4.3 Let $\psi = if(x)$ where $f(x)$ is a real-function and $i = \sqrt{-1}$. Calculate the current density for ψ. Is your result $>, =,$ or < 0?

Problem 4.4 Calculate the prececessional frequency of an electron in magnetic fields of $.5, 1,$ and 1.5 Tesla, and the corr

Problem 4.5 Earlier in the chapter, we found the time-derivative of an operator $\widehat{O}$ is given by

$$\frac{\partial\langle\widehat{O}\rangle}{\partial t}=\frac{i}{\hbar}\left[\langle\widehat{H}^{\dagger}\widehat{O}\rangle-\langle\widehat{O}\widehat{H}\rangle\right]-\frac{i}{\hbar}\langle O\rangle\left[\langle\widehat{H}^{\dagger}\rangle-\langle\widehat{H}\rangle\right]+\left\langle\frac{\partial\widehat{O}}{\partial t}\right\rangle$$

Under which conditions will

$$\frac{\partial\langle\widehat{O}\rangle}{\partial t}=\left\langle\frac{\partial\widehat{O}}{\partial t}\right\rangle$$

Discuss the properties of the energy operator $\widehat{H}$ and the relation $\widehat{O}$ must have to the operator $\widehat{H}$.

Problem 4.6 The *Most General Ehrenfest* theorem was used to obtain the equation

$$\frac{d\boldsymbol{\mu}_S}{dt}=-g\gamma\boldsymbol{\mu}_S\times\mathbf{B}-\frac{\alpha g\gamma}{\mu_S}\boldsymbol{\mu}_S\times\boldsymbol{\mu}_S\times\mathbf{B}$$

Use the above equation to prove that

$$\frac{d}{dt}(\boldsymbol{\mu}_S\cdot\boldsymbol{\mu}_S)=0$$

Hint: Use the relation

$$\boldsymbol{\mu}_S\cdot\frac{d\boldsymbol{\mu}_S}{dt}=\frac{1}{2}\frac{d}{dt}(\boldsymbol{\mu}_S\cdot\boldsymbol{\mu}_S)$$

What is the meaning of this result for a moment $\boldsymbol{\mu}_S$ that satisfies this equation of motion?

Problem 4.7 Earlier, we found the relation

$$\tau_S=\frac{1}{\alpha g\gamma B}$$

Based on the equation

$$\frac{d\boldsymbol{\mu}_S}{dt}=-g\gamma\boldsymbol{\mu}_S\times\mathbf{B}-\frac{\alpha g\gamma}{\mu_S}\boldsymbol{\mu}_S\times\boldsymbol{\mu}_S\times\mathbf{B}$$

If we let $\Delta\mu_S=\hbar$, and $\Delta t\approx\tau_S$, compute

$$\frac{\Delta\mu_S}{\Delta t}$$

and discuss any noticeable consistencies between the two forms for the rates of change.

Problem 4.8 In this chapter, we also found the operator for the electric field to be

$$\widehat{\mathbf{E}} = \frac{i}{q\hbar}[V, \widehat{\mathbf{p}}]$$

Using this operator, calculate the electric field for wave functions given by

$$\psi = \psi_0 e^{i(px - E_R t)/\hbar}$$

$$\psi = \psi_0 e^{i(px - E_R t)/\hbar} e^{-E_I t/\hbar} \quad \text{where } E = E_R + iE_I$$

4.10 Suggested Readings & References

[1] K. Eason, *Quantum Mechanics With A SPIN On It*, Sym. Pub. (2019)

[2] L. Landau and E. Lifshitz, *On the theory of the dispersion of magnetic permeability in ferromagnetic bodies*, Physik. Zeits. Sowjetunion **8**, pp. 153-169 (1935)

CHAPTER 5

Fundamentals of Quantum Transport: Evanescence, Scattering, Tunneling & Conduction

It is known that when light travels along an optical wave-guide, the light continually bounces off of the interface of the wave-guide. The interface is formed between a central core and a surrounding material known as cladding. This confinement to the core where the light continually reflects, is concomitant with a *special condition at the core-clad interface*. The electric field of the electromagnetic wave does not vanish instantaneously at the interface. Instead, there is a *finite* distance from the interface extending out into the cladding, over which the electric field decays toward zero amplitude. This decaying behavior occurs whenever waves experience conditions for reflection at an interface.

In optical fiber, the cladding thickness is typically much larger than the length scale of the amplitude decay. However, when layers transporting electrical currents are made that are sufficiently small or thin in comparison to the decay length, additional phenomena can be observed, such as tunneling. These unique conditions of thin layer composites are readily manufactured in the world of electronics. Because of this, additional phenomena relating to transport may be utilized.

Like electromagnetic waves, the eigenfunctions of quantum mechanical systems also can form decaying amplitudes at interfaces. This is mostly observed in electronic transport. Therefore, in this chapter, key problems are analyzed that demonstrate unique transport behavior in quantum mechanics, including wave function amplitude decay at an interface. First, we discuss the problem of a single potential energy step,[1] followed by more complex examples combining potential energy steps. They reveal fundamentally unique and useful properties of quantum transport.

5.1 Creation Of Evanescent wave functions

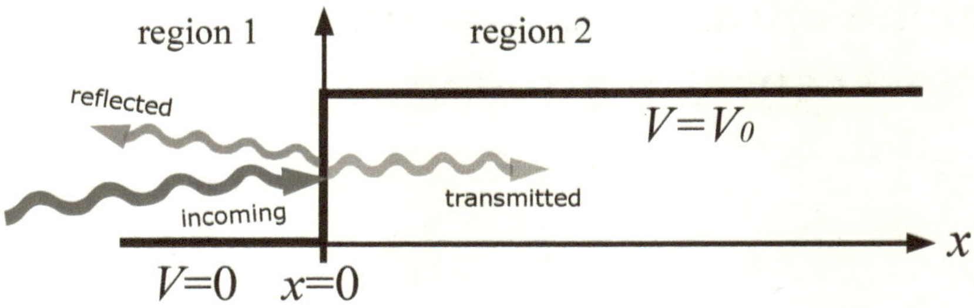

Figure 5.1: Illustration of a elementary particle approaching a step change in the potential energy V. At $x = 0$, there is a discontinuous jump from $V = 0$ to $V = V_0$. For $x < 0$, $V = 0$ and for $x > 0$, $V = V_0$.

Consider a potential energy like that shown in Fig. 5.1, with a fundamental particle in motion to the right towards the potential change. The situation represents an electron approaching an interface between two different materials. The particle encounters a potential energy step function $V(x)$. Because the potential only depends on x, the problem is 1D in space, so we have an energy operator given by

$$\widehat{H}\psi_x = \left[-\frac{\hbar^2}{2m_0}\frac{\partial^2}{\partial x^2} + V(x)\right]\psi_x = E_R\psi_x \tag{5.1}$$

Eq. (5.1) can be written as

$$\frac{d^2\psi_x}{dx^2} + \frac{2m_0(E_R - V)}{\hbar^2}\psi_x = 0 \tag{5.2}$$

[1] this is an example of a potential energy having *spatial* variations

Let us define the parameter k as

$$k^2 = \frac{2m_0(E_R - V)}{\hbar^2} \tag{5.3}$$

k is the so-called wavenumber of the particle (*c*.Eq. 2.5) also given by $k = 2\pi/\lambda$. λ is the associated *wavelength*. Then, we can write Eq. (5.2) as

$$\frac{d^2\psi_x}{dx^2} + k^2\psi_x = 0 \tag{5.4}$$

We know Eq. (5.4) has the general solution given by

$$\psi_x(x) = Ae^{ikx} + Be^{-ikx} \tag{5.5}$$

Eq. (5.5) is a sum of the *two* independent solutions of the *second* order differential equation given by Eq. (5.4). This solution is also consistent with the *type II* general solution of the wave equation (*c*.Eq.(2.56)). The solution is comprised of two propagating waves with wavenumbers $+k$ and $-k$. $+k$ denotes wave motion to the right($\rightarrow$) while $-k$ is to the left($\leftarrow$). However, a wave traveling to the left in the region where $x > 0$ is only possible if there exists an opportunity for the wave to be reflected backwards. This possibility exists at the position $x = 0$ and to the left of the potential jump. Thus, the general solution will contain both waves in the region $x < 0$. However, whatever part of the solution propagates past $x = 0$ will not be reflected as it is assumed that the right side extends indefinitely. Therefore, only a right propagating wave exists in the region $x > 0$. Thus, the boundary conditions, and thus, solutions are different in the two respective regions. Specifically, we have

$$\psi_{x,1}(x) = Ae^{ik_1x} + Be^{-ik_1x} \quad (x < 0) \tag{5.6a}$$

$$\psi_{x,2}(x) = Ce^{ik_2x} \quad (x > 0) \tag{5.6b}$$

Since we have two distinct materials, we also two respective wavenumbers k_i, denoted by k_1 and k_2. k_1 corresponds to region 1 ($x < 0$) and k_2 corresponds to region 2 ($x > 0$). Using Eq. (5.3), k_1^2 and k_2^2 become

$$k_1^2 = \frac{2m_0E_R}{\hbar^2} \quad (x < 0) \tag{5.7a}$$

$$k_2^2 = \frac{2m_0(E_R - V_2)}{\hbar^2} \quad (x > 0) \tag{5.7b}$$

There are *three* unknowns A, B, and C, all being independent of x. However, we can only write two equations using the continuity of both the wave function X_i and its first derivative at $x = 0$. This means we can *only* solve for two of them, simultaneously.

Alternatively, we can, instead, solve for the ratios $A' = A/C$ and $B' = B/C$. This turns out to be sufficient because A/C is inversely proportional to a desired quantity known as the transmission probability T. And $B/A = B'/A'$ turns out to be proportional to the reflection probability R. We will discuss these quantities in more detail shortly. At $x = 0$, equating $\psi_{x,1}(x)$ and $\psi_{x,2}(x)$ and their derivatives $\psi'_{x,1}(x)$ and $\psi'_{x,2}(x)$ leads to

$$A + B = C \qquad (x = 0) \tag{5.8a}$$

$$ik_1 A - ik_1 B = ik_2 C \quad (x = 0) \tag{5.8b}$$

Dividing both equations by C to give the desired form in terms of A' and B', we have

$$\frac{A}{C} + \frac{B}{C} = 1 \tag{5.9a}$$

$$ik_1 \left(\frac{A}{C} - \frac{B}{C} \right) = ik_2 \tag{5.9b}$$

This gives two equations in two unknowns, which can be solved. The solutions are

$$\frac{A}{C} = \frac{k_1 + k_2}{2k_1} \tag{5.10a}$$

$$\frac{B}{C} = 1 - \frac{A}{C} = \frac{k_1 - k_2}{2k_1} \tag{5.10b}$$

These solutions seem sensible. They suggest that if $k_1 = k_2$, then $A = C$ and $B = 0$, meaning no back-scattering or reflection, *etc.* It is as though there is no interface present. Using Eqs. (5.10a) and (5.10b), B/A becomes

$$\frac{B}{A} = \frac{B}{C} \cdot \frac{C}{A} = \frac{k_1 - k_2}{2k_1} \cdot \frac{2k_1}{k_1 + k_2} = \frac{k_1 - k_2}{k_1 + k_2} \tag{5.10c}$$

Using the definitions for k_1 and k_2, given by Eqs. (5.7a) and (5.7b), the three ratios in Eqs. (5.10a), (5.10b), and (5.10c) can be expressed in terms of the parameters in the energy operator equation, which gives

$$\frac{A}{C} = \frac{1}{2} + \frac{1}{2}\sqrt{1 - \frac{V_2}{E_R}} \tag{5.11a}$$

$$\frac{B}{C} = \frac{1}{2} - \frac{1}{2}\sqrt{1 - \frac{V_2}{E_R}} \tag{5.11b}$$

$$\frac{B}{A} = \frac{1 - \sqrt{1 - \frac{V_2}{E_R}}}{1 + \sqrt{1 - \frac{V_2}{E_R}}} \tag{5.11c}$$

B/A is the ratio of the *amplitude* of the reflected wave to that of the incoming wave. By inspection, we see that when $V_0 = 0$, we again have $B/A = 0$, and when $V_0 = E_R$, then $B/A = 1$, meaning the entire wave is back-scattered or reflected. This also indicates the range of B/A where $0 \leq B/A \leq 1$. Fig. 5.2 illustrates the dependence of these three parameters on the potential energy V_2 in the region $0 < V_2 \leq E_R$.

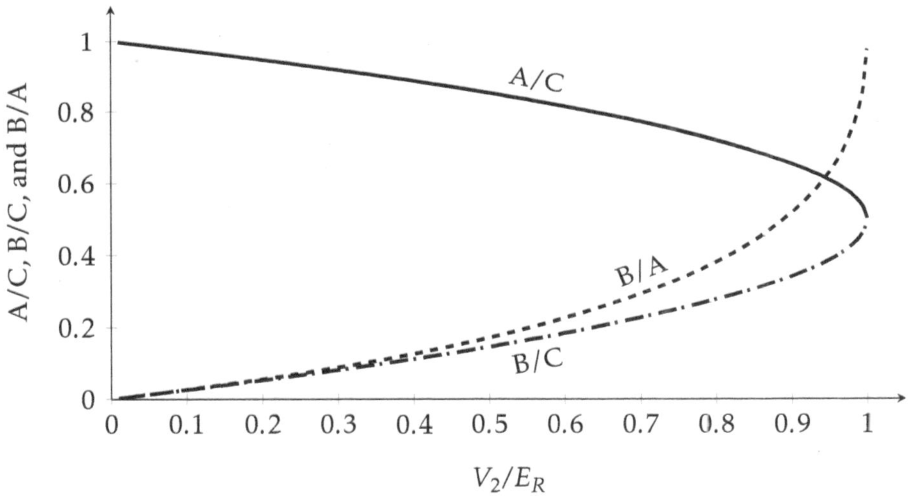

Figure 5.2: Ratios A/C, B/C, and A/B as functions of the dimensionless ratio V_2/E_R. The case of $B/A = 1$ means that the amplitude of the reflecting wave matches that of the incoming wave, or that the entire wave is reflected.

In Fig. 5.2, the potential energy V_2 is assumed positive, but less than the incoming energy E_R. For a free particle encountering a *spatially rising potential energy*, the kinetic energy of the particle is reduced. To the charge carrier, this is an example of a *potential hill.*

R *Positive Potentials Are Hills To Free Particles*: At it's most basic level, for a free quantum particle, a momentum reduction becomes probable whenever there is a spatially rising or uphill potential encountered. It is generally accompanied by a scattering event.

The chosen range of $E_R > V_2 > 0$ in Fig. 5.2 ensures that the terms underneath the radical in A/C, B/C, and B/A are non-negative.

There are two more important cases that can be distinguished that extend beyond this range. For example, another case is when $V_2 < 0$ and thus $k_2^2 = 2m_0(E_R - V_2)/\hbar^2 > 0$, and yields a *larger* wavenumber than in the first case of $E_R > V_2 > 0$. In this case, the term underneath the radical is *real and positive definite*. Therefore, the particle encounters an effectively *downhill* potential variation. When materials satisfy this condition *locally*, the downhill potentials are often referred to as *potential holes*. According to Eq. (5.7b), this leads to an increase in the wavenumber k_2 in region 2, hence, the particle *gains* momentum.

Examples of applications of *holes* can be found in doped semiconductors as well as special materials known as superconductors, which, in proper conditions, conduct charge carriers practically without loss. We'll discuss this topic in more detail in Chap. ??. Unique conditions like this *apply strictly to propagating charge carriers*, because for a *bound* electron, a negative potential tends to stabilize the bound state further, where as for free electrons, their propagation as free particles is made more probable.

Ⓡ *Negative Potentials Are Holes For Free Particles*: When a propagating wave encounters a negative potential step, downward potential or hole, it gains momentum. This can also be interpreted as no scattering.

The other important case to consider is when $V_2 \geq E_R$ and $V_2 > 0$ so that $k_2^2 < 0$ or $k_2 = ik_2'$. Because k_2 becomes imaginary, the amplitude coefficient $B/A = k_1 - k_2/k_1 + k_2$ cannot be zero, i.e. $k_1 \neq k_2$. This condition gives rise to the incoming wave with amplitude A *always* having *some* reflected wave component so that $|B| > 0$. In a 3D system where there is an angular distribution to the incoming waves of particles, this finite reflection corresponds to general *scattering*, or *scattering events* at the interface. Additionally, in a lattice of periodic potential hills, it gives rise to the phenomenon of *band structures* because there is constructive interference between these scattered waves. It becomes possible for the scattering to *cancel* the incoming wave. When this occurs, it corresponds to an inability to conduct consequently creating *gaps* in the values of conductable wavenumbers. The extent of scattering is determined by the extent of the positive potential variation V_2.

So, when $V_2 > E_R$, the wavenumber k_2 is then given by

$$k_2^2 = -\frac{2m_0(V_2 - E_R)}{\hbar^2} \quad (x > 0)$$

Since k_2^2 is negative, the wavenumber k_2 becomes

$$\begin{aligned} k_2 &= \sqrt{-\frac{2m_0(V_2 - E_R)}{\hbar^2}} \\ &= i\sqrt{\frac{2m_0(V_2 - E_R)}{\hbar^2}} \\ &= ik_2' \end{aligned}$$

k_2' is a strictly positive real quantity. The solution in region 2 then becomes

$$\psi_{x,2}(x) = Ce^{ik_2x} = Ce^{i(ik_2')x} = Ce^{-k_2'x} \quad (x > 0) \tag{5.12}$$

Therefore, when $V_2 > E_R$, we obtain a solution with a decreasing amplitude as a function of $+x$, relative to the interface at $x = 0$. The wave function decays more rapidly for larger $V_2 - E_R$. Examples of a decaying $\psi_{x,2}(x)$ are illustrated in Fig. 5.3. This type of decaying wave function is known as an *evanescent wave function*, analogous to the *evanescent electromagnetic wave*, and it is a result of a potential rise, where the height of the potential energy change is *larger* than the incoming energy E_R possessed by the quantum particle. The resulting *evanescent* wave function is purely real-valued, and exponentially decays over a length scale on the order of $\delta \sim 1/k_2'$. As k_2' increases, the decay length decreases. This means that the probability of finding the particle at a point beyond $x = 0$, in region two, diminishes as k_2' increases.

In region one, the real and imaginary components of B/A are also functions of the dimensionless ratio V_2/E_R. The real component starts at $Re(B/A) = 1$, decreasing monotonically while approaching $Re(B/A) = -1, Im(B/A) = 0$ asymptotically. The imaginary component starts at 0,

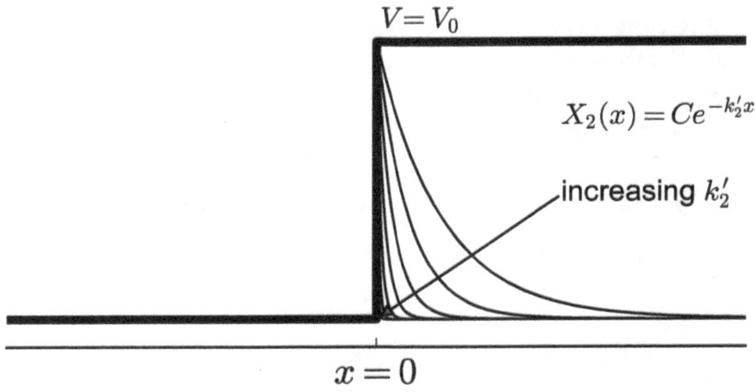

Figure 5.3: Examples of *evanescent* wave functions which have decaying amplitudes with x, resulting from the condition $V_2 > E_R$. The wave function is real-valued with a decay length $\delta \sim 1/k_2'$. As k_2' increases, the decay length decreases, indicating that the probability of finding the particle at a point beyond $x = 0$ diminishes with x.

when $V_2/E_R = 1$ then rises (negatively), proceeding to decay back towards 0 asymptotically. Meanwhile, the amplitude of B/A is fixed at 1, corresponding to full reflection of the wave. As summarized in Fig. 5.4, we find there is *texture* in the process of total reflection. As shown in Fig. 5.4, the amplitude of B/A is constant at unity, however the complex quantity B/A changes phase based on the value of V_R. The real part transitions from +1 to −1, while the imaginary part starts at zero, peaks at −1 when $Re(B/A) = 0$, then transitions back to zero.

While we can observe the behavior of the amplitude ratios, it would be better to connect this information to observable physical quantities. This is done by using the *current density* discussed in Sec. 4.2. Let's do this in the next section.

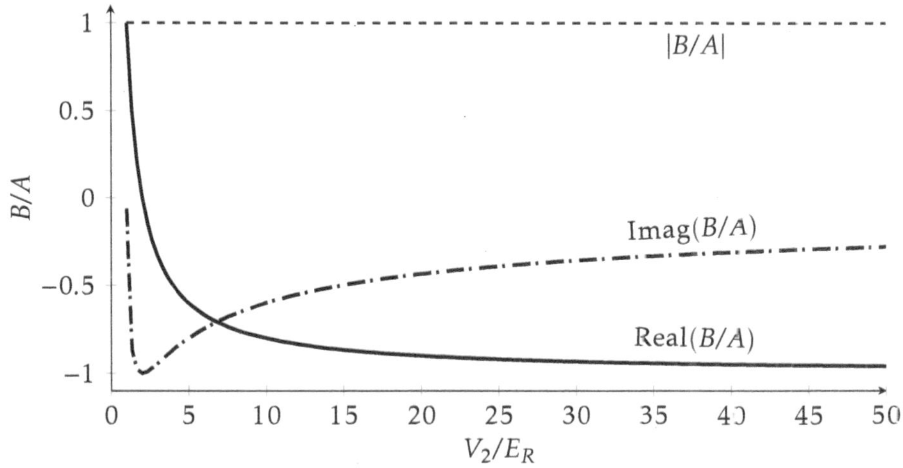

Figure 5.4: Real and imaginary components of B/A as functions of the dimensionless ratio $V_2/E_R \geq 1$. The real component starts at 1, then decreases while approaching −1 asymptotically. The imaginary component starts at 0, when $V_2/E_R = 1$, then rises as it proceeds to decay back towards 0 asymptotically. Across the entire range, the amplitude is always 1, indicating complete reflection.

5.2 Transmission and Reflection Probabilities

Although much of our analyses have included the fourth dimension, note that it is not required to introduce the concepts of quantum transport. This is because with the application of some steady well-controlled external potential (e.g. electrical voltage), it is possible to maintain a steady state current which *does not have nonequilibrium behavior*. This is also why we used the given operator in Eq. (5.1). Here, we extend this assumption to the discussion of current density. With some of the incoming wave function traveling to the right (while some gets reflected), we may view the problem in another way. Instead, we may ask whether a current density $\widehat{J}$ is flowing in region two, particularly, in the conditions where an evanescent wave ($x > 0$) forms. Using the definition of the current density **J** (*c*.Eq. 4.13),[2] we have

$$\widehat{\mathbf{J}} \equiv \frac{i\hbar}{2m}(\psi\nabla\psi^* - \psi^*\nabla\psi) \tag{5.13}$$

[2]because the fourth dimension is excluded $\alpha = 0$.

Because an *evanescent wave* function is real-valued, it equals to its complex conjugate or $\psi^* = \psi$. Therefore, $\nabla\psi = \nabla\psi^*$ which leads to the result

$$J = \frac{i\hbar}{2m_0}(\psi\nabla\psi^* - \psi^*\nabla\psi) = \frac{i\hbar}{2m_0}(\psi\nabla\psi - \psi\nabla\psi) = 0$$

In the context of our discussion, there is *no* current flow in region 2 when $V_2 \geq E_R$ *i.e.* region 2 becomes an *insulator*. *Any* strictly real-valued wave function *always* leads to the term in parentheses vanishing, meaning no current flows.

Ⓡ *Real-Value Wave Function Current Density*: Real-valued wave functions *always* have zero current density.

The evanescent wave function is, therefore, comcommitant with *total reflection* of the incoming wave Ae^{ik_1x}. However, in the reflection process, the reflecting wave still has a finite *penetration depth* $\delta \sim 1/|k_2| = 1/k_2'$ beyond the interface of reflection ($x = 0$). This is where the evanescent wave lives. As we have mentioned, *evanescence* is not exclusive to quantum mechanics. It is a well-known behavior of electromagnetic waves when undergoing what is called *total internal reflection* while propagating through an optical wave-guide. The reflections keep the wave confined to the interior of the wave-guide known as the *core*, while the evanescent wave forms just beyond the surface, where the exterior overcladding material meets the core of the wave-guide. Another example of an evanescent wave forming by electromagnetic waves is in the generation of what is known as a *surface-plasmon-polariton*(SPP). An SPP is a collective interfacial oscillation typically involving certain metallic-dielectric interfaces. Though this process is more complex because the surface-plasmon-polariton exists on both sides of the interface characterized by asymmetric evanescent waves on both sides of the metallic-dielectric interface.[3]

The current density $\widehat{J}$ can be used to define useful parameters known as reflection R and transmission T probabilities. The *reflection probability* is a dimensionless ratio defined as

[3] Generally, the decay length is shorter on the metallic side.

Reflection Probability

$$R = \frac{|\mathbf{J}_R|}{|\mathbf{J}_I|} \tag{5.14}$$

Similar to the reflection probability, we can define the *transmission probability* T as

Transmission Probability

$$T = \frac{|\mathbf{J}_T|}{|\mathbf{J}_I|} \tag{5.15}$$

The current density given by Eq. (5.13) is an appropriate choice because they obey conservation. This follows from the definition to satisfy the continuity equation. Thus, whenever there are no sources or sinks interacting with the propagating wave function within the domain, it follows that along the direction of flow taken as normal to the interface, we have the following relation between the incoming flow J_I, the transmitted flow J_T, and the reflected flow J_R:

$$J_I = J_R + J_T \tag{5.16}$$

From the incoming wave function, some is reflected while the rest is transmitted. From the balance of the current, it follows that

$$R + T = \frac{|J_R|}{|J_I|} + \frac{|J_T|}{|J_I|} = 1 \tag{5.17}$$

Evaluating Eq. (5.13) for the current density J_I in region one, we have

$$\begin{aligned}
\mathbf{J}_I &= \frac{i\hbar}{2m_0}\left[(-ik_1)Ae^{ik_1x}A^*e^{-ik_1x} - ik_1A^*e^{-ik_1x}Ae^{ik_1x}\right] \\
&= \frac{\hbar k_1}{2m_0}\left(2|A|^2\right) \\
&= \frac{\hbar k_1}{m_0}|A|^2
\end{aligned}$$

A similar calculation for the reflected and transmitted wave's component current density J_R and J_T leads to

$$J_R = \frac{\hbar k_1}{m_0}|B|^2$$

$$J_T = \frac{\hbar k_2}{m_0}|C|^2$$

Using the current densities J_I, J_R, and J_T, the reflection and transmission probabilities R and T become

$$R = \frac{J_R}{J_I} = \frac{|B|^2}{|A|^2} \tag{5.18a}$$

$$T = \frac{J_T}{J_I} = \frac{k_2}{k_1}\frac{|C|^2}{|A|^2} \tag{5.18b}$$

From Eq. (5.18b), it follows that *non-penetrating* reflection takes place precisely when $k_2 = 0$, or $V_2 = E_R$. This condition represents the transition into evanescent wave function behavior. Beyond the transition point, the transmission T is not given by Eq. (5.18b) since the wave function becomes real-valued. In this case, $T = 0$. For a better view of the dependence of R and T, they are plotted in Fig. 5.5 over the range $0 < V_2 \leq E_R$. Beyond this range, $R = 1$ and $T = 0$. Using Eqs. (5.10a) and

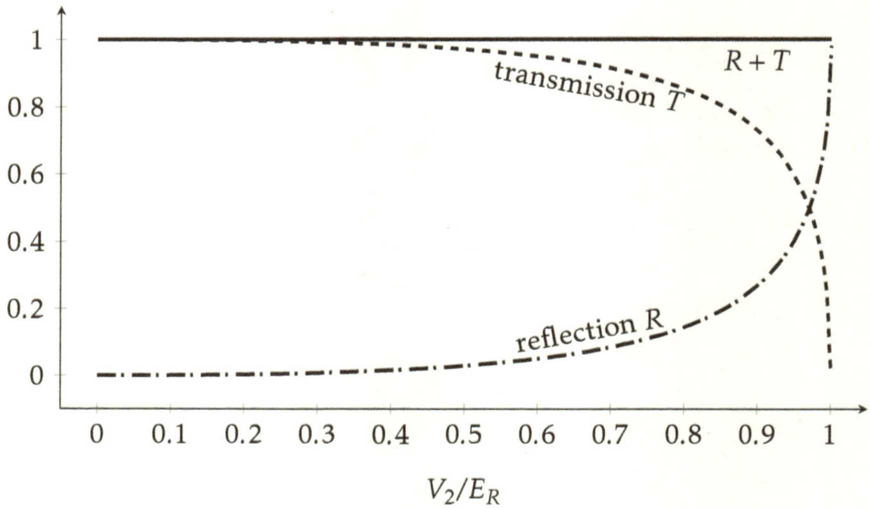

Figure 5.5: Transmission and reflection probabilities T and R. Because $R+T = 1$, they have correlated behavior, where as R increases, T necessarily decreases.

(5.10c), it is straightforward to show that given the solutions for A/C and B/A, that $R+T=1$, also illustrated in Fig. 5.5.

In this section, we have found that when a free particle encounters a spatial variation in potential energy $V(x)$, a multitude of non-intuitive phenomena can be observed. We found that transmission, reflection, or more generally scattering, and evanescent wave functions are possible depending on the conditions of the materials of the interface. Let's extend these ideas from a single step to *two* sequential potential steps, where one is upwards followed by a step down to the original starting potential. This type of potential variation is known as a *rectangular potential barrier* and is the subject of the next section.

5.3 Electrons Across A Rectangular Potential Barrier

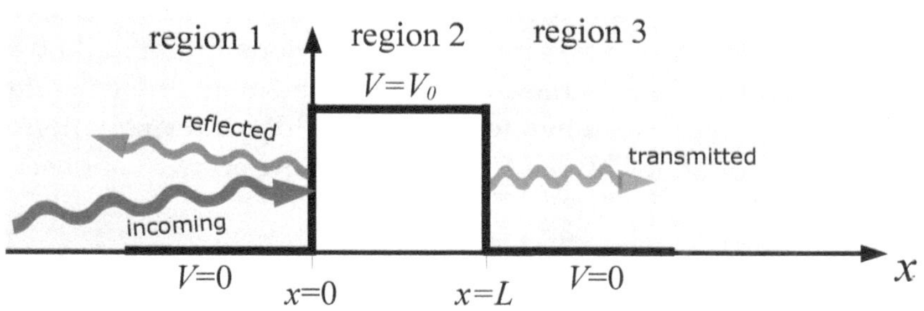

Figure 5.6: Illustration of a rectangular potential barrier. It is two consecutive potential steps (first up, then second one down) separated by a distance L. The incoming wave is traveling to the right in region 1, while the first transmitted wave function exists in region 2, and a second transmitted wave function exists in region 3.

The *rectangular potential barrier* is illustrated in Fig. 5.6. The analysis for the rectangular potential barrier is just more of what we did with the single potential step. However, we now have an *additional* potential step down, located a distance L away from the first potential step at $x=0$.

This simple change to the problem leads to *significantly* more mathematics, though it is straightforward. The second potential step, which behaves as a *potential hole*, goes from V_2 down to 0. The energy operator for any layer is unchanged from that in the single step, where we have

$$\left[-\frac{\hbar^2}{2m_0}\frac{\partial^2}{\partial x^2}+V(x)\right)\psi_x = E_R\psi_x \Rightarrow \frac{d^2\psi_x}{dx}+k^2\psi_x = 0$$

where

$$k^2 = \frac{2m_0(E_R - V)}{\hbar^2}$$

Since we have *three* regions, we have the following solutions corresponding to regions 1, 2, and 3:

$$\psi_{x,1} = Ae^{ik_1x} + Be^{-ik_1x} \quad \text{(region 1)} \tag{5.19a}$$

$$\psi_{x,2} = Ce^{ik_2x} + De^{-ik_2x} \quad \text{(region 2)} \tag{5.19b}$$

$$\psi_{x,3} = Ee^{ik_3x} \quad \text{(region 3)} \tag{5.19c}$$

A, B, C, D, and E are determined from boundary conditions. We can set the wave functions and their derivatives equal at interfaces $x = 0$ and $x = L$. However, this results in four equations in five unknowns. Similar to what was done in the single step problem, in order to reduce the number of unknowns down to four, we divide all the equations by incoming amplitude A. This leads to the following four equations:

$$1+\frac{B}{A} = \frac{C}{A}+\frac{D}{A} \quad \text{(x = 0)} \tag{5.20a}$$

$$ik_1\left(1-\frac{B}{A}\right) = ik_2\left(\frac{C}{A}-\frac{D}{A}\right) \quad \text{(x = 0)} \tag{5.20b}$$

$$\frac{C}{A}e^{ik_2L}+\frac{D}{A}e^{-ik_2L} = \frac{E}{A}e^{ik_3L} \quad \text{(x = L)} \tag{5.20c}$$

$$ik_2\left(\frac{C}{A}e^{ik_2L}-\frac{D}{A}e^{-ik_2L}\right) = ik_3\frac{E}{A}e^{ik_3L} \quad \text{(x = L)} \tag{5.20d}$$

Let us define the four unknowns $B' = B/A$, $C' = C/A$, $D' = D/A$ and $E' = E/A$ so we have

$$1 + B' = C' + D' \quad \text{(x = 0)} \tag{5.21a}$$

$$1 - B' = \frac{k_2}{k_1}(C' - D') \quad \text{(x = 0)} \tag{5.21b}$$

$$C'e^{ik_2L} + D'e^{-ik_2L} = E'e^{ik_3L} \quad \text{(x = L)} \tag{5.21c}$$

$$C'e^{ik_2L} - D'e^{-ik_2L} = \frac{k_3}{k_2}E'e^{ik_3L} \quad (x = L) \tag{5.21d}$$

The above equations can be combined to give the following three equations, where the combinations are shown on the left and resulting equation on the right:

$$(5.21a) + (5.21b) \Rightarrow \quad 2 = (1 + k_2/k_1)C' + (1 - k_2/k_1)D'$$

$$(5.21c) + (5.21d) \Rightarrow \quad 2C'e^{ik_2L} = (1 + k_3/k_2)E'e^{ik_3L}$$

$$(5.21c) - (5.21d) \Rightarrow \quad 2D'e^{-ik_2L} = (1 - k_3/k_2)E'e^{ik_3L}$$

From the last two equations, the amplitude ratios C' and D' become

$$C' = \frac{E'}{2}\left(1 + \frac{k_3}{k_2}\right)e^{ik_3L}e^{-ik_2L} \tag{5.22a}$$

$$D' = \frac{E'}{2}\left(1 - \frac{k_3}{k_2}\right)e^{ik_3L}e^{ik_2L} \tag{5.22b}$$

C' and D' can be substituted into the first combination of equations above to yield the solution for E', given by

$$E' = \frac{4k_1k_2e^{-ik_3L}}{(k_1 + k_2)(k_2 + k_3)e^{-ik_2L} + (k_1 - k_2)(k_2 - k_3)e^{ik_2L}} \tag{5.22c}$$

The probability amplitude is generally a complex number, and it depends on the properties of the material region in question, *as well as* adjacent regions. The quantity $E' = E/A$ is a very general result and provides the *transmission probability* T, being proportional to $(E')^*E' = |E'|^2$. With this solution available, we can also obtain the last amplitude ratio B' from Eq. (5.21a), which gives

$$B' = C' + D' - 1 = \frac{(k_2 + k_3)e^{-ik_2L} + (k_2 - k_3)e^{ik_2L} - 2k_2e^{-ik_3L}}{2k_2e^{-ik_3L}} \tag{5.22d}$$

B' provides the reflection probability R. Though we now have all four amplitude ratios in the most general forms, it is common to treat the two possible cases for $k_2^2 = 2m_0(E_R - V_2)/\hbar^2$. By doing so, we can obtain two distinct expressions for transmission T and reflection R probabilities. Let's do this next.

5.3.1 Evanescent Barrier Solutions $V_2 > E_R$

The first case is when $V_2 > E_R$ in material two. Recall the transmission and reflection probabilities T and R are defined as

$$T = \frac{J_T}{J_I} \quad \text{and} \quad R = \frac{J_R}{J_I}$$

J_I is the current density of the *incoming* current $\psi_I = Ae^{ik_1x}$, while the *transmitted* current is determined from $\psi_T = Ee^{ik_3x}$. The reflected current is determined from $\psi_R = Be^{ik_1x}$. Using the definition for the current density given by Eq. (5.13), we have

$$J_I = \frac{\hbar|A|^2k_1}{m_0}, \; J_R = \frac{\hbar|B|^2k_1}{m_0} \quad \text{and} \quad J_T = \frac{\hbar|E|^2k_3}{m_0}$$

It follows that the probabilities are then given by

$$R = \frac{J_R}{J_I} = \frac{|B|^2}{|A|^2} = |B'|^2 \tag{5.23}$$

$$T = \frac{J_T}{J_I} = \frac{k_3|E|^2}{k_1|A|^2} = \frac{k_3}{k_1}|E'|^2 \tag{5.24}$$

To calculate them analytically, we must know whether k_2 is real or imaginary ($k_2^2 < 0$). If $V_2 > E_R$, then $k_2^2 = 2m_0(E_R - V_2)/\hbar^2 < 0$, and therefore $\sqrt{2m_0(E_R - V_2)}/\hbar = k_2 = ik_2'$. While k_2 is purely imaginary, k_2' is real. This can be substituted into Eqs. (5.22a) - (5.22d) along with Eq. (5.24) to determine an expression for T and R. E' and its complex conjugate become

$$E' = \frac{4ik_1k_2'e^{-ik_3L}}{\left(k_1 + ik_2'\right)\left(ik_2' + k_3\right)e^{k_2L} + \left(k_1 - ik_2'\right)\left(ik_2' - k_3\right)e^{-k_2L}} \tag{5.25a}$$

and

$$(E')^* = \frac{-4e^{ik_3L}ik_1k_2'}{e^{k_2L}\left(-ik_2' + k_3\right)\left(k_1 - ik_2'\right) + e^{-k_2L}\left(-ik_2' - k_3\right)\left(k_1 + ik_2'\right)} \tag{5.25b}$$

The calculation of $|E'|^2$ is straightfoward, but a little tedious. Along the way, one may also make use of the hyperbolic cosine function relations $\cosh x = (e^x + e^{-x})/2$ and $\cosh 2x = 2\sinh^2 x + 1$. T then works out to be

Rectangular Barrier Transmission Probability ($k_2^2 < 0$)

$$T = \frac{4k_1 k_2'^2 k_3}{(k_1^2 + k_2'^2)(k_2'^2 + k_3^2)\sinh^2(k_2' L) + (k_1 k_2' + k_2' k_3)^2} \tag{5.26}$$

Similarly, for the reflection probability R, we obtain

Rectangular Barrier Reflection Probability ($k_2^2 < 0$)

$$R = \frac{(k_1^2 + k_2'^2)(k_2'^2 + k_3^2)\sinh^2(k_2' L) + (k_1 k_2' - k_2 k_3)^2}{(k_1^2 + k_2'^2)(k_2'^2 + k_3^2)\sinh^2(k_2' L) + (k_1 k_2' + k_2' k_3)^2} \tag{5.27}$$

We can check by adding Eqs. (5.26) and (5.27) to confirm that $R + T = 1$. k_1, k_2, and k_3 can be different, but k_1 and k_3 are *real*, while $k_2 = ik_2'$ is purely imaginary.

If we let $k_1 = k_3$ (*i.e.* same material on both sides of the barrier), Eq. (5.26) reduces to

$$T = \frac{4k_1^2 k_2'^2}{(k_1^2 + k_2'^2)^2 \sinh^2(k_2' L) + 4k_1^2 k_2'^2} \tag{5.28a}$$

Likewise, the reflection probability R for the case $k_1 = k_3$ reduces to

$$R = \frac{(k_1^2 + k_2'^2)^2 \sinh^2(k_2' L)}{(k_1^2 + k_2'^2)^2 \sinh^2(k_2' L) + 4k_1^2 k_2'^2} \tag{5.28b}$$

Figs. 5.7, 5.8, and 5.9 illustrate the transmission and reflection probabilities, T and R, varying the strength of the potential barrier V_2. In Figs. 5.8 and 5.9, the vertical dashed line at $V_2/E_R = 1$ represents the classical limit. However, all examples here illustrate nonzero transmission at the classical limit. It is this property leads to the potential for *tunneling* into any subsequent material adjacent to the barrier. As the barrier thickness increases, so does the sharpness of the transitions, which means transmission drops more efficiently and/or reflection increases more efficiently for thicker barriers. This is illustrated more plainly in Fig. 5.10, plotting the dependence on the dimensionless parameter $k_2' L$. The

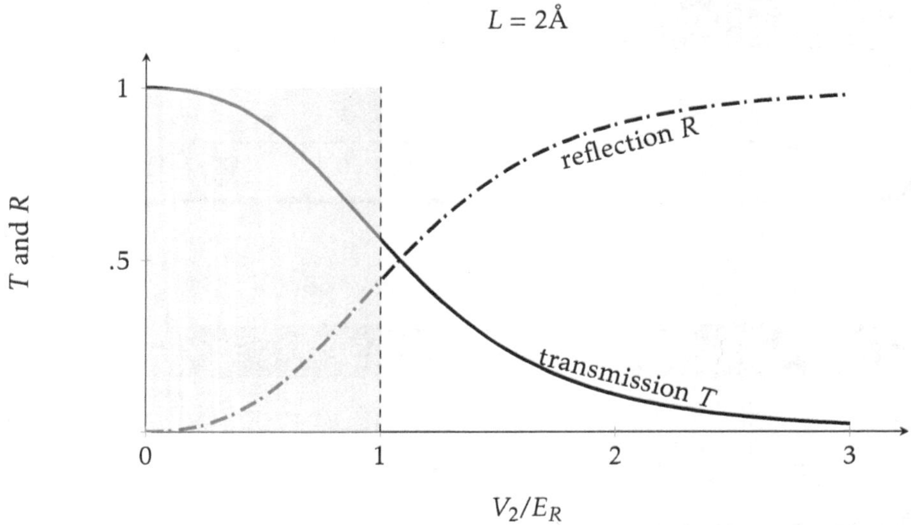

Figure 5.7: Illustration of the transmission and reflection probabilities R and T as functions of V_2/E_R. In the case shown, $L = 2Å$ and $E_R = 3eV$. Note that the region to the right of the shaded rectangle is described by Eqs. (5.28a) and (5.28b).

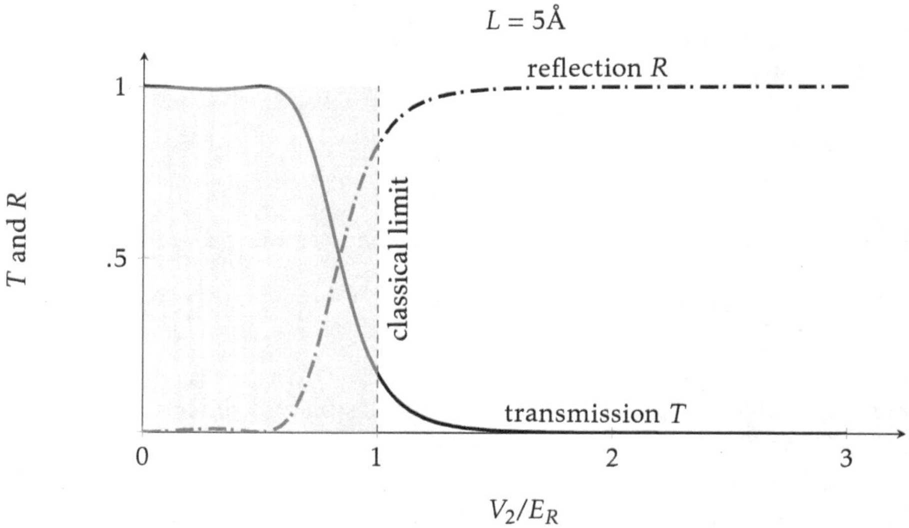

Figure 5.8: Transmission probabilities T versus V_2/E_R for the case $V_2 > E_R$. $L = 5Å$, and $E_R = 3eV$.

abscissa $k_2'L$ indicates that the transmission decays over a length scale on the order of $1/k_2' \sim \lambda_2$, where λ_2 is the wavelength in the rectangu-

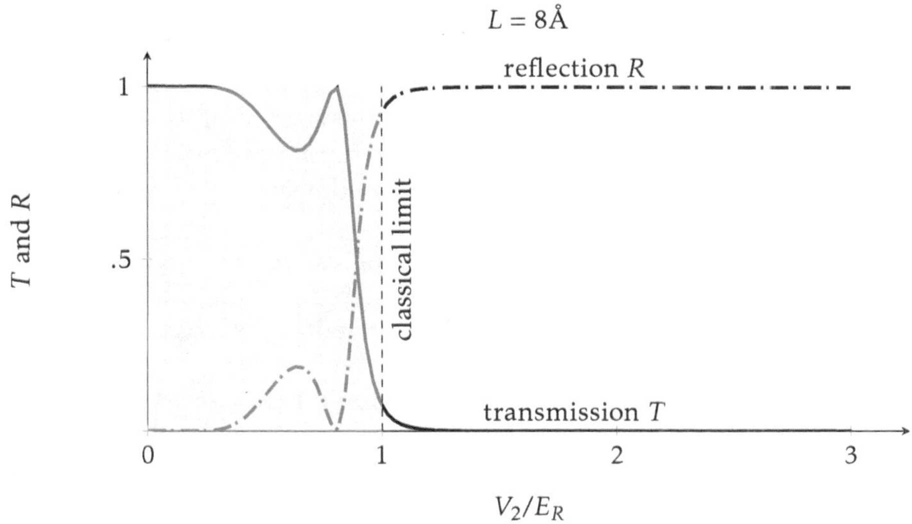

Figure 5.9: Reflection and transmission probabilities R and T as a function of V_2/E_R. In this illustration $L = 8Å$ and $E_R = 3eV$. As the thickness increases, so does the sharpness of the transitions with respect to V_2 dependence. Thinner barriers transmit more than thicker barriers.

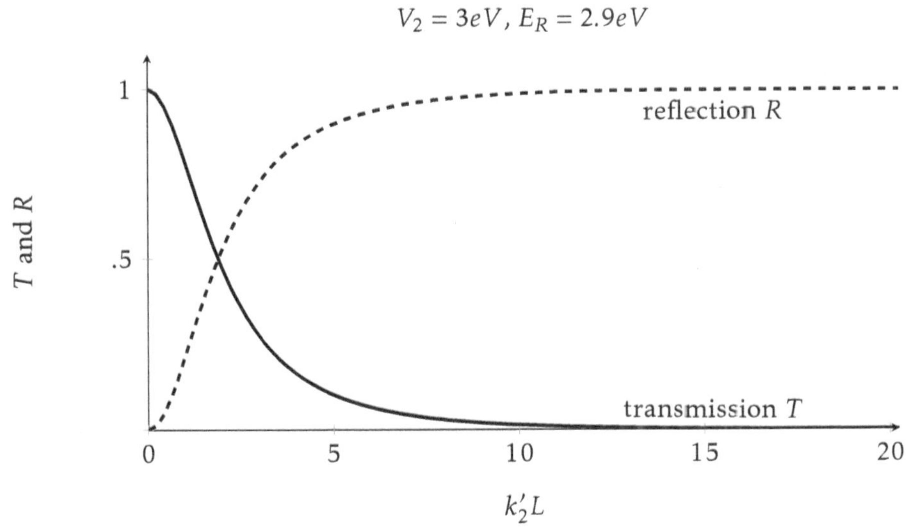

Figure 5.10: Reflection and transmission probabilities R and T as functions of barrier thickness L. Here, $V_2 = 3eV$, and $E_R = 2.9eV$, using the electron mass. For a given energy and barrier height, increasing thickness reduces the amount of transmission, and therefore, increases reflection.

lar barrier region. This phenomenon that we have found, indicating a *finite* transmission $T > 0$ in the condition where the energy E_R is strictly less than the potential barrier potential V_2, is known as *tunneling* or quantum tunneling. It is enabled by the existence of an evanescent wave function inside the region of the rectangular potential or region 2. Tunneling is more manifest in small or thin devices.

Tunneling is exploited in several technological devices including electronic transistors, widely used in computers for both processing and memory, and *magnetic tunnel junctions* utilized as the magnetic sensors.

You may have also noticed the nonmonotonic bahavior of T and R in the shaded region corresponding to $V_2/E_R \leq 1$. This unique behavior is related to a wave property known as *interference*. This is possible in any region where the wave function has the general form $\psi_x(x) = Ae^{ikx} + Be^{-ikx}$. In this case, as discussed, there is a boundary which reflects the incoming wave described by Ae^{ikx} resulting in the condition $B > 0$. So, let us consider the probabilities for this important case, namely, when $V_2 < E_R$.

5.3.2 Propagating Wave Barrier Solutions $V_2 < E_R$

If $V_2 < E_R$, it follows that $k_2 = \sqrt{2m_0(E_R - V_2)}/\hbar > 0$ or $k_2^2 > 0$. The processes of determining the solutions for the amplitude ratios B', C', D', and E' are the same as in the previous case. Since k_2 is real, the explicit form of B', C', D', and E' is given by Eqs. (5.22a), (5.22b), (5.22c), and (5.22d). Instead of evanescent waves described by $e^{-k_2' x}$ in region 2 (corresponding to 100% reflection at the first interface), the wave function contains purely propagating waves $e^{ik_2 x}$. For the transmission and reflection probability, the conjugates of E' and B' given by Eqs. (5.22c) and (5.22d), for this case, become

$$(E')^* = \frac{4k_1k_2e^{ik_3L}}{(k_1+k_2)(k_2+k_3)e^{ik_2L} + (k_1-k_2)(k_2-k_3)e^{-ik_2L}} \tag{5.29a}$$

$$(B')^* = \frac{(k_2+k_3)e^{ik_2L} + (k_2-k_3)e^{-ik_2L} - 2k_2e^{-ik_3L}}{2k_2e^{ik_3L}} \tag{5.29b}$$

$(E')^*E' = |E'|^2$ and $(B')^*B' = |B'|^2$ can be substituted into Eqs (5.23) and (5.24) to obtain R and T. Note that in obtaining $|E'|^2$ and $|B'|^2$, we have also used the relations $\cos x = (e^{ix} + e^{-ix})/2$ and $\cos^2 k_2L = 1 - \sin^2 k_2L$.

Then, we obtain the following result for the transmission probability $T = k_3|E'|^2/k_1$:

Barrier Transmission Probability ($k_2^2 > 0$)

$$T = \frac{4k_1k_2^2k_3}{(k_2^2-k_1^2)(k_2^2-k_3^2)\sin^2(k_2L)+(k_1k_2+k_2k_3)^2} \tag{5.30}$$

Similarly, the reflection probability $R = |B'|^2$ (c.Eq. 5.23) is given by

Barrier Reflection Probability ($k_2^2 > 0$)

$$R = \frac{(k_2^2-k_1^2)(k_2^2-k_3^2)\sin^2(k_2L)+(k_1k_2-k_2k_3)^2}{(k_2^2-k_1^2)(k_2^2-k_3^2)\sin^2(k_2L)+(k_1k_2+k_2k_3)^2} \tag{5.31}$$

In the simplified case of $k_1 = k_3$, Eqs. (5.30) and (5.31) become

$$T = \frac{4k_1^2k_2^2}{(k_2^2-k_1{}^2)^2\sin^2(k_2L)+4k_1^2k_2^2} \tag{5.32}$$

$$R = \frac{(k_2^2-k_1{}^2)^2\sin^2(k_2L)}{(k_2^2-k_1{}^2)^2\sin^2(k_2L)+4k_1^2k_2^2} \tag{5.33}$$

When $k_2^2 > 0$, we find that the solutions for T and R contain a $\sin^2(k_2L)$ term (as opposed to a $\sinh^2(k_2'L)$ term when $k_2^2 < 0$). This term, in particular, can lead to oscillations and/or undulations in T and R as functions of L and V_2. Examples of R and T from Eqs. (5.32) and (5.32) are illustrated in Figs. 5.14, 5.12, and 5.13. Both the transmission T and reflection R probabilities as a function of barrier thickness L are illustrated in Fig. 5.15. The potential barrier problem has different consequences depending on whether the system has more or less energy than the rectangular potential barrier V_0. One case leads to tunneling across the barrier, even though the system energy is less than that of the barrier. In the other case where the system energy is larger than that of the potential barrier, an interference phenomenon takes place for a wave propagating through a material with an interface. Fig. 5.16 illustrates the transmission for both cases $k_2^2 > 0$ and $k_2^2 < 0$, for direct

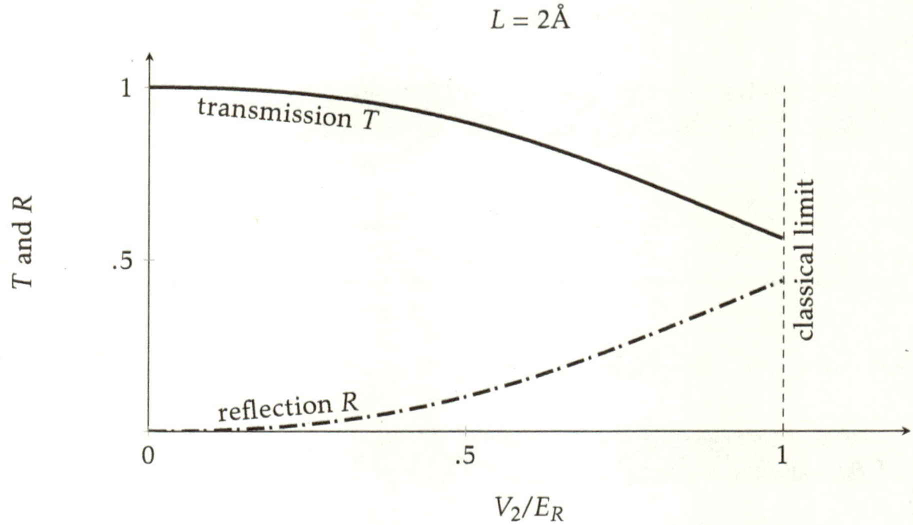

Figure 5.11: Illustration of the transmission and reflection probabilities R and T as functions of V_2/E_R, with $L = 2\text{Å}$ and $E_R = 3eV$.

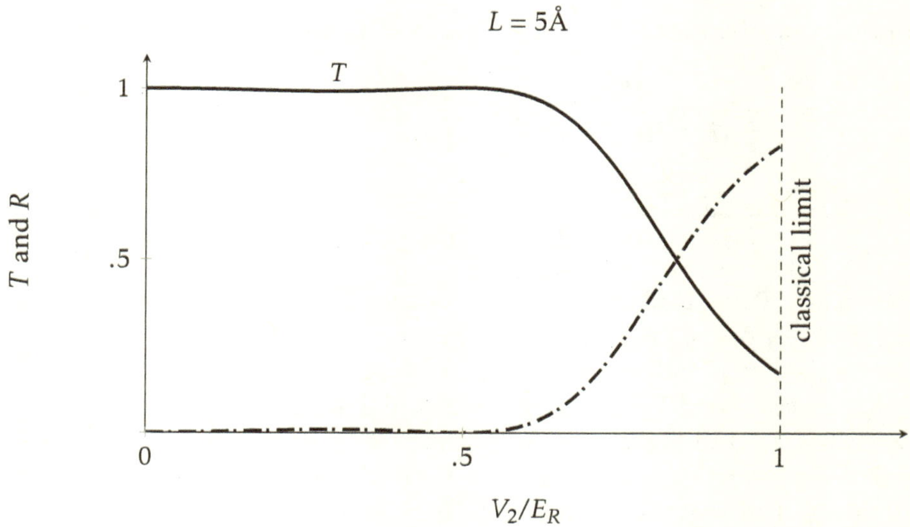

Figure 5.12: Illustration of the transmission and reflection probabilities R and T as functions of V_2/E_R, with $L = 5\text{Å}$ and $E_R = 3eV$.

comparison. The oscillatory nature of T and R is a *subtle* consequence of *wave function interference*. However, *the probability density does not reveal*

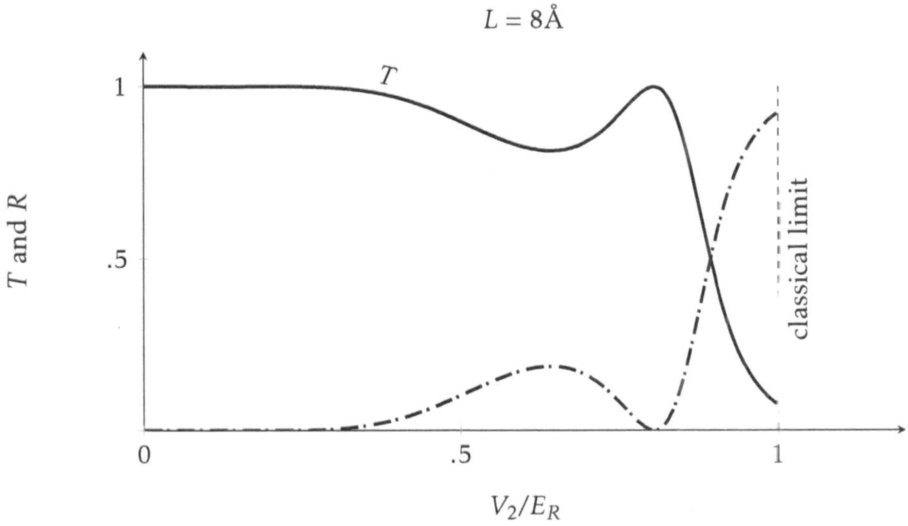

Figure 5.13: Illustration of the transmission and reflection probabilities R and T as functions of V_2/E_R, with $L = 8\text{Å}$ and $E_R = 3eV$.

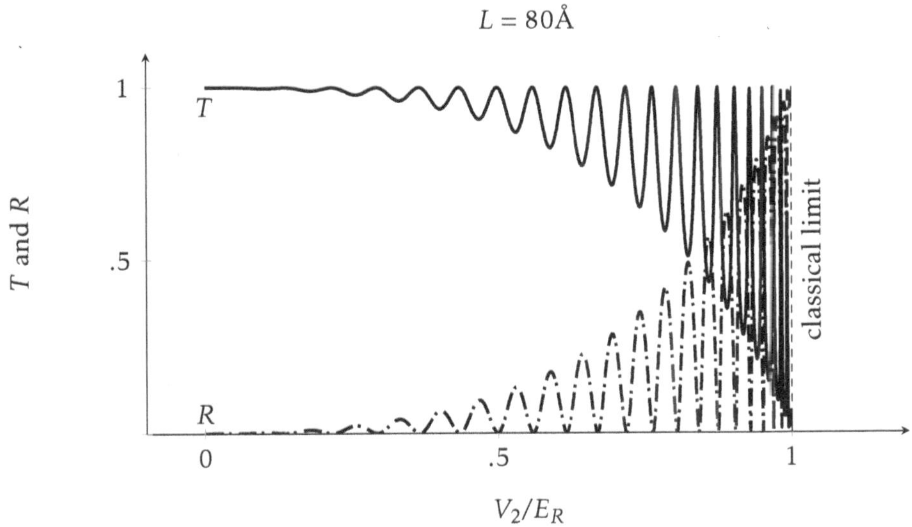

Figure 5.14: Illustration of the transmission and reflection probabilities R and T as functions of V_2/E_R, with $L = 80\text{Å}$ and $E_R = 3eV$.

interference. Let the wave function be given by

$$\psi_x(x) = Ce^{ikx} + De^{-ikx}$$

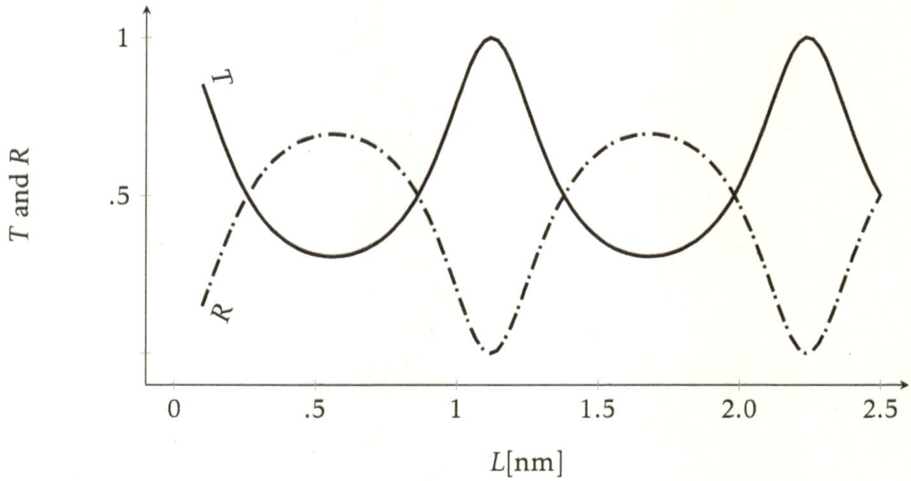

Figure 5.15: Illustration for the case $k_2^2 > 0$, R and T as a function of a barrier thickness L. The potential barrier height $V_2 = 3eV$, $E_R = 1.1V_2$, and electron mass m_e is also assumed. This behavior corresponds to interference of the electron wave function with itself.

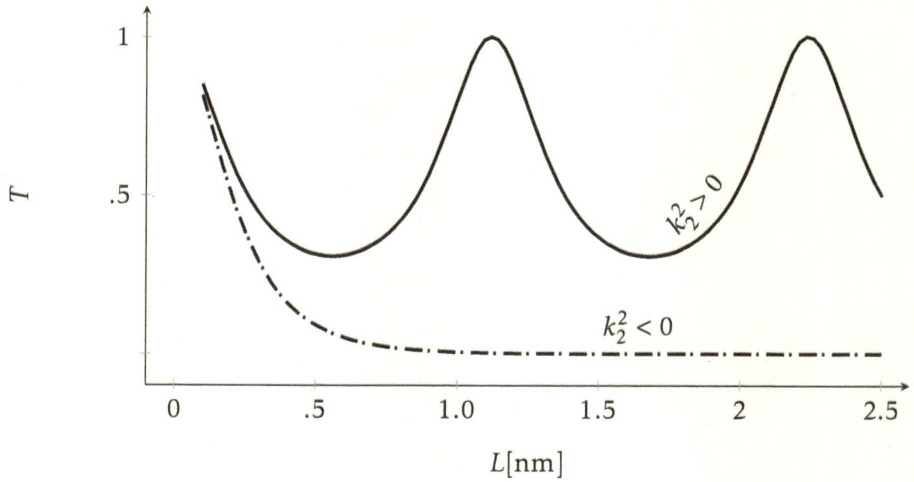

Figure 5.16: Illustration of the transmission probabilities T for both cases $k_2^2 > 0$ (solid black) and $k_2^2 < 0$, varying the rectangular potential barrier thickness L. For both cases, $V_2 = 3eV$, and in the first case $k_2^2 > 0$, $E_R = 1.1V_2$, while $E_R = 0.9V_2$ for the case $k_2^2 < 0$. The electron mass $m_0 = m_e$ is assumed. In one case, we have an oscillating transmission, while in the other case, we have a monotonically decaying transmission.

It can be shown that

$$\psi_x^*\psi_x(x) = |C|^2 + |D|^2 + (C^*D)_R\left[\cos^2 kx - \sin^2 kx\right] + (C^*D)_I \cos kx \sin kx \quad (5.34)$$

The subscripts R and I denotes the real and imaginary parts of that term, respectively. The second order trigonometric term $\cos^2 kx - \sin^2 kx$ is the term that conveys any interference phenomenon.

Both conditions $V_2 < E_R$ and $V_2 > E_R$ are relevant to the electrical conduction properties of real materials. As mentioned, this characteristic of an electron which back-scatters part of it's wave, superposes with all the other occurrences in the lattice (periodic or regular arrangement of atoms). Conditions are reached where the combined electron back-scattered wave cancels the incoming wave. When there is cancellation of this kind, standing waves form and this gives rise to *forbidden energy bands* where electrons no longer can propagate forward or backwards in the material. These bands essentially arise from the partial scattering of charge carriers having wave properties.

Next, we consider another problem with a potential energy function that is an *inversion* of the potential barrier just discussed. It is known as a *finite potential well.*

5.4 The Finite Potential Well Problem

In the *potential well problem*, the potential well has zero potential energy in the centered region (region 2). However, in the exterior regions one and three, there is a finite potential energy $V_1 = V_3 = V_0 > 0$. The potential dependence on x is an *inversion* of the rectangular potential barrier. Inverting the potential, however, also dictates that we start with the elementary particle located in region one, where $V = 0$.

The energy operator is, again, given by

$$\left[-\frac{\hbar^2}{2m_0}\frac{d^2}{dx^2} + V(x)\right]\psi_x = E_R\psi_x \Rightarrow \frac{d^2}{dx^2} + k^2\psi_x = 0 \quad (5.35)$$

where the corresponding wavenumbers $k_1 = k_3(V = V_0)$ and $k_2(V = 0)$ of the corresponding regions are given by

$$k_1^2 = k_3^2 = \frac{2m_0(V_0 - E_R)}{\hbar^2} \quad (5.36a)$$

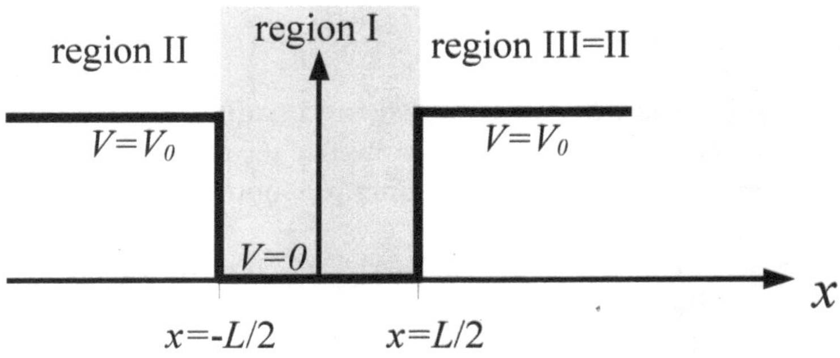

Figure 5.17: Illustration of a potential well. The origin is in the symmetric center of the potential well. The potential energy in region two is also taken to be the same as that of region three, but they are larger than in the well.

$$k_2^2 = \frac{2m_0 E_R}{\hbar^2} \quad (V = 0) \tag{5.36b}$$

In regions 1 and 3, if the eigenenergy is smaller than the potential $E_R < V_0$, an evanescence wave forms at the *exterior* side of the interfaces $(x = -L/2, L/2)$ of the potential well. This also comes with continual reflection of the particle in its quantum state *within* the well. When a quantum state is localized or bound in spatial coordinates, it is known as a *bound state*. It is the bound states within the potential well that are of primary interest in this problem.

Since the potential $V_1 = V_3 = V_0 < \infty$ at $x = -L/2, L/2$, $\psi_x(-L/2) = \psi_x(L/2) \neq 0$. Thus, we cannot explicitly specify *any* value of the wave function at $x = -L/2, L/2$, as it depends on $V(x) = V_0$ and E_R. Thus, like we did in the rectangular barrier problem, we use continuity of the wave functions and their first derivative at the interfaces at $x = -L/2, L/2$, while regions 1 and 3 wave functions must also vanish at $\mp\infty$, respectively.

To find the eigenenergy states $E_R = E_n$ assuming $E_R < V_1 = V_3$, we take the origin at the center point $x = 0$ (hence the interfaces are at $-L/2, L/2$, *etc.*). In Subsection 3.2.4, when we worked out the Hermite Polynomials for the harmonic oscillator, we found that the quantum states alternate between symmetric and anti-symmetric states as the quantum number n iterates from $n = 0$:symmetric, $n = 1$:anti-symmetric, $n = 2$:symmetric, *etc*. This is a property of bound quantum states. Thus, we can also take the eigensolutions to form a sequence of symmetric and anti-symmetric states. Each energy state is associated with either a symmetric or anti-symmetric wave function.

In the potential well of region 2, since $V = 0$ $V - E_R = -E_R$ and Eq. (5.35) becomes

$$\frac{d^2\psi_{2,x}}{dx^2} - k_2^2\psi_{2,x} = 0 \tag{5.37}$$

Eq. (5.37) has the general solution

$$\psi_{2,x} = A\cos k_2 x + B\sin k_2 x \tag{5.38}$$

Eq. (5.38) is a combination of a symmetric (cosine) and anti-symmetric (sine) function. In region 3 (and 1), the general solution is a combination of *evanescent* wave functions given by

$$\psi_{3,x} = Ce^{-k_3' x} + De^{k_3' x}$$

The normalization condition for $\psi_{3,x}$ (and $\psi_{1,x}$) requires that the wave functions vanish at ∞. This is true if $D = 0$. Otherwise, one of the terms diverge at ∞. Thus, $\psi_{3,x}$ becomes

$$\psi_{3,x} = Ce^{-k_3' x} \tag{5.39}$$

Any energy state E_n corresponds to *either* a symmetric or anti-symmetric solution, so let us consider each case separately. For the odd or *anti-symmetric* solutions, the wave function $\psi_{x,2}$ is given by

$$\psi_{x,2} = B\sin(k_2 x) \tag{5.40}$$

Using continuity at the boundaries for ψ_x and its first derivative at $x = L/2$, we have

$$B\sin\left(\frac{k_2 L}{2}\right) = Ce^{-\frac{k_3' L}{2}} \tag{5.41a}$$

$$k_2 B\cos\left(\frac{k_2 L}{2}\right) = -k_3' Ce^{-\frac{k_3' L}{2}} \tag{5.41b}$$

Taking the ratio of Eq. (5.41a) to Eq. (5.41b), for the anti-symmetric states, we obtain the relation

$$k_2\cot\left(\frac{k_2 L}{2}\right) = -k_3'$$

To express the above in a dimensionless form, let us multiply by $L/2$ so we have

$$\frac{k_2L}{2}\cot\left(\frac{k_2L}{2}\right) = -\frac{k_3'L}{2} \tag{5.42}$$

Eqs. (5.36a) can be used to express the following relation:

$$\frac{(k_2L)^2}{4} + \frac{(k_3'L)^2}{4} = \frac{2mV_0L^2}{4\hbar^2} \tag{5.43}$$

Using Eq. (5.43), we can define three parameters ν, β, and η as

$$\nu^2 + \beta^2 = \eta^2 \tag{5.44}$$

where Eq. (5.44) is just an alternative form of Eq. (5.43). Then, Eq. (5.42) can be written as

$$-\nu\cot(\nu) = \sqrt{\eta^2 - \nu^2} = \beta = \beta(\nu) \quad \text{(anti-symmetric)} \tag{5.45a}$$

For *symmetric* states, carrying out a similar process when $\psi_{x,2} = A\cos(k_2x)$, we obtain the relation

$$\nu\tan(\nu) = \sqrt{\eta^2 - \nu^2} = \beta(\nu) \quad \text{(symmetric)} \tag{5.45b}$$

For a chosen value of V_0 and L, Eq. (5.45a) is a function of the single variable E_R. Therefore, it can be solved for E_R. However, since both equations are nonlinear in ν, it can only be solved using numerical or computational methods, which will not be done here. Instead, we can infer the solutions by plotting both sides of Eqs. (5.45a) and (5.45b), then identifying where they overlap. Such points correspond to solutions. Fig. 5.18 shows both sets of symmetric (tangent) and anti-symmetric (cotangent) curves as functions of ν. Both the anti-symmetric negative cotangent curves and the symmetric tangent curves run mostly parallel to one another, separated by a distance of $\pi/2$. The dashed lines are $\nu\tan\nu$ curves (symmetric states), while the solid lines are $-\nu\cot\nu$ curves (anti-symmetric states). They intersect the black quarter circle lines, which are $\beta(\nu)$ or (ν, β) points, a finite number of times N_η, for any choice of η. The illustration also demonstates that for a given η, the solutions to the finite potential well problem correspond only to those discrete or quantized values (ν_n, β_n) at each intersection. The quantized solutions (ν_n,β_n) correspond to the open circles shown in Fig. 5.18.

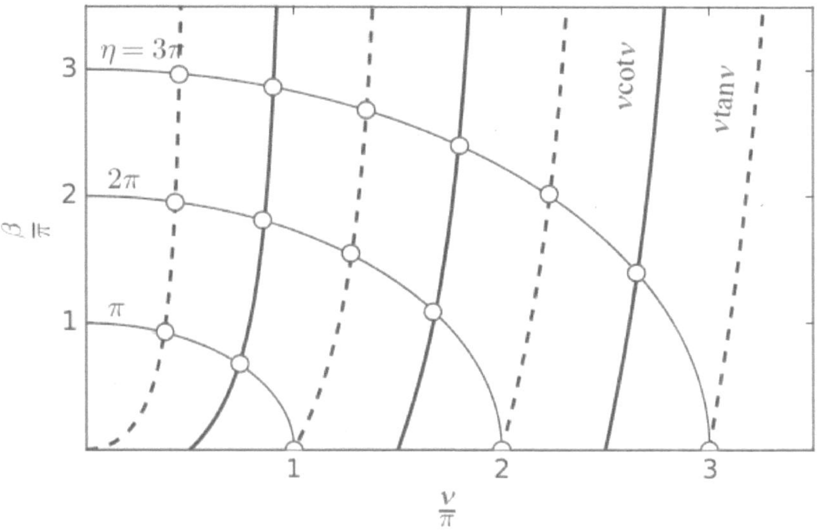

Figure 5.18: The potential well symmetric and anti-symmetric curves, along with the intersecting (ν,β) curve, and the discrete solutions corresponding to the solutions to the finite potential well problem.

The potential energy value V_0, as well as, the potential well length L influence η, which is the radius of the circles shown in Fig. 5.18. Increasing η, which increases the size of the circles, consequently increases the number of quantum states N_η. The number of quantum states is given by the number of intersections with the anti-symmetric cotangent and symmetric tangent curves. Determining these discrete values of ν_n provides the bound quantum energy state E_n, given by

Bound-state Energy Levels of 1D Potential Well

$$E_n = -\frac{2\hbar^2\nu_n^2}{m_0L^2} \quad : \quad \nu_n = \frac{k_{2,n}L}{2} \tag{5.46}$$

The number of quantum states N_η corresponding to a radius of η can be obtained by using the fact that each positive branch of *tan* and *cot* functions spans horizontally a width of $\pi/2$, and there is no overlap in their respective regions. This is because $\pi/2$ is the distance between the x-axis crossing and its asymptote to the immediate right. Since the first branch, which corresponds to a symmetric state, starts at the origin, for all circles of radius $\eta < \pi/2$, there will be only *one* symmetric state. From here, each addition to η of $\pi/2$ corresponds to an additional state. Thus, the number of discrete states N_η can be deduced as

$$N_\eta = \frac{\eta}{\pi/2} + 1$$

Substitution of the expression for η leads to

Number of Quantum States On A η Radius

$$N_\eta = \frac{L}{\pi\hbar}\sqrt{2m_0 V_0} + 1 \qquad (5.47)$$

For the values of η (π,2π, and 3π) illustrated in Fig. 5.18, Eq. (5.47) gives 3, 5, and 7, respectively, in agreement with the number of quantum states corresponding to number of open circles on that radius.

There is a wave function for each value of ν_n. For example, in the case of $\eta = 2\pi$ (the middle quarter circle), there are five discrete solutions, thus, there are five corresponding wave functions, all illustrated in Fig. 5.19. The potential well width L used in Fig. 5.19 is $L = 5$Å, and the corresponding energies for the eigenstates shown from bottom to top are $E_R = E_n$=1.1, 4.3, 9.8, 16.8, 24eV. From the bottom to the top wave function corresponds to going clockwise on the $\eta/\pi = 2$. The wave functions partly *leak* outside of the potential well, described by an evanescent wave function at the well exterior.

The solutions we have obtained can be contrasted to those of the so-called infinite potential ($V_0 \to \infty$) where all wave functions vanish exactly at the edges of the well. Fig. 5.20 illustrates the first five solutions for the infinite potential well $V_0 = \infty$.[4] The above problems provide good insight into some of the behavior of electrons being both in transport, but also the conditions that determine their ability to be in transport, or scatter, or localize etc. Using these concepts, let's discuss the technical meaning of electrical conduction, which should be easier to grasp now that we are familiar with some fundamental transport concepts and their basic energy conditions. This information should be useful when we return to applying 4D energy operators to electrons in the next Chapter.

[4]This is the only case where the wave functions can be set to 0 at $x = -L/2, L/2$.

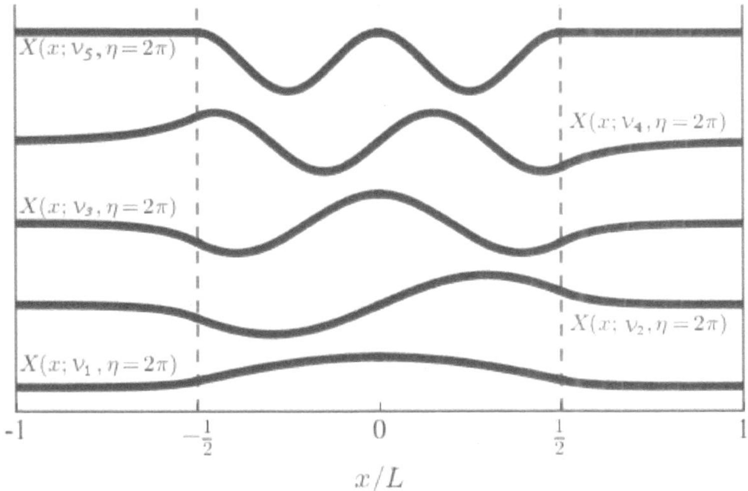

Figure 5.19: The five potential well symmetric and anti-symmetric wave functions for the case of $\eta = 2\pi$ (the second quarter circle in Fig. 5.18). The potential well width L used in this example is $L = 5\text{Å}$, and corresponding energies $E_R = E_n$=1.1, 4.3, 9.8, 16.8, 24eV.

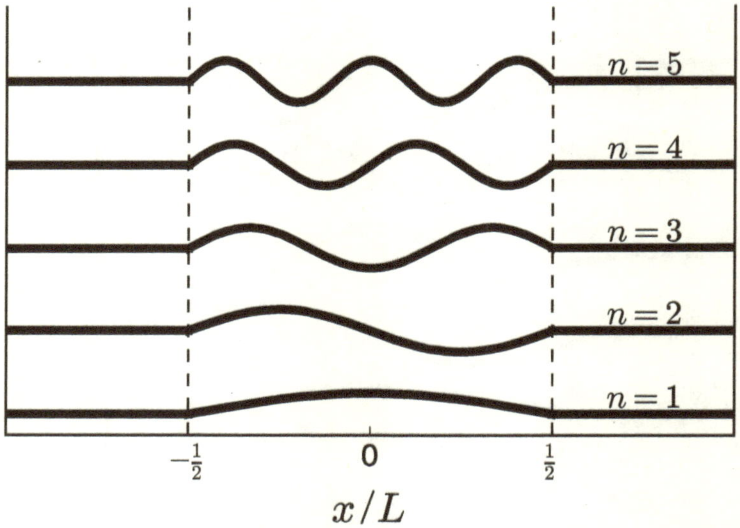

Figure 5.20: The first five ($n = 1, 2, 3, 4, 5$) wave functions for the *infinitely* deep potential well. Compare these to those shown in Fig. 5.19. In the finite potential well, the wave functions do not vanish at the boundaries, as with infinite potential. With finite potential, there is always finite thickness outside the body where wave decays.

5.5 The Conduction Electron

Real devices utilized in applications apply external voltages to support the process of electrical conduction. In a solid state material, an electron is ordinarily *bound* to an atom, resulting from an overall negative eigenenergy E_{bound}. However, *conducting electrons posses an energy above the bound state eigenenergy* ($E_R > E_{\text{bound}}$). Moreover, the additional externally driven energy is *directional*. The role of an external electric field, or voltage, is to raise the energy of an electron sufficiently to transform a bound electron, which has $E_{\text{bound}} < 0$ into an *itinerant electron*. Omitting the fourth dimension, we've seen that the description of the energy without an externally applied voltage is given by

$$\left[-\frac{\hbar^2}{2m_0}\frac{\partial^2}{\partial x^2} + V\right]\psi_x = E_R\psi_x \tag{5.48}$$

For an electron in an atom with two or more electrons, the potential $V = V_{n-e} + V_{e-e}$ generally includes two contributions given by

$$V_{n-e}(r) = -\frac{Ze^2}{4\pi\varepsilon_0|\mathbf{r}_1 - \mathbf{r}_2|} \quad \text{and} \quad V_{e-e}(r_i) = \sum_{j=1}^{Z-1}\frac{e^2}{4\pi\varepsilon_0|\mathbf{r}_i - \mathbf{r}_j|} \tag{5.49}$$

For an atom with a single electron with no electron-electron interactions ($V_{e-e} = 0$), solving the eigenoperator equation turns out to be a fairly sophisticated problem to solve. This is the problem of the hydrogen atom.[5] In the absence of a magnetic field, the bound state *eigenenergy* for the hydrogen electron are given by

$$E_{\text{bound}} = -\frac{Z^2 m^* e^4}{32\pi^2\hbar^2\varepsilon_0^2}\frac{1}{n^2} = -13.6\left(\frac{Z}{n}\right)^2 \tag{5.50}$$

n is called the *principal quantum number*. m^* is the *reduced mass* of a two-body (electron + nucleus) system given by

$$m^* = \frac{m_p m_e}{m_p + m_e} \tag{5.51}$$

where m_p is the hydrogen proton mass and m_e is the electron mass. For a multi-electron system, we must add V_{e-e} which acts as a *positive* potential energy between the electrons. Because $V_{e-e} > 0$, it *raises the*

[5] For a complete treatment of the hydrogen atom solution, see Chap2. 7 and 8, Quantum Mechanics With A SPIN On It (Eason)

energy of the atom/electrons. From Eq. (5.50), the larger the principal quantum number n, the smaller in amplitude is $|E_{\text{bound}}|$ and the closer to zero is the negative bound energy (i.e. as negative numbers, it's energy is higher in absolute units). Thus, $(Z/n)^2$ dictates how high the energy will be. For example, $Z/n = 79/6 = 13.2$ for gold (Au), while $Z/n = 29/4 = 7.25$ for copper (Cu). This suggests the bound state energy contribution for a valence electron in Cu might be more susceptible to an external electric field than those of Au. In general, the occupied electronic states having the highest energy (closest to zero, but negative in sign) are most susceptible to an external electric field E_{ext}, particularly when they can be excited to levels that are in proximity energetically.

In a system with a large number of electrons, when temperature $T = 0$, the highest occupied energy state is known as the *Fermi level*. All the filled states below this level occupy what is known as the *Fermi sea*. Note that the concept of the *Fermi level* applies to a 1D distribution of energies depending on a single wave vector component $k = k_x$. In 2D, the Fermi level becomes a *Fermi contour*, while in 3D, it becomes a *Fermi surface*, depending of $k = [k_x, k_y, k_z]^T$. The highest shell (which is determined by the principal quantum number n) below the *Fermi level* that houses the highest energy electrons is known as the *valence shell*. More generally, just the set of occupied states in this shell form the so-called **valence band**. The remaining nearest *available quantum states above the valence band* form the *conduction band*.

When there is a effectively small gap between the valence band and conduction band, as with metals, its easier for electrons to be exited into the conduction band states. However, these two bands are not necessarily strictly disjoint sets of energies. Valence and conduction bands can occupy the same shell, as in the case in metals. With this in mind, when an external electrical potential (*i.e.* voltage) U_{ext} is applied, there is an associated external electric field $\mathbf{E}_{ext}$ generated by the source that acts on electrons locally. The energy change due to the electric field is equal to the work done on the electron charge by E_{ext} over some small distance δx. This work is given by

$$W_{ext} = \int F_{ext} \cdot dx = \int qE_{ext} \cdot dx = -e \int E_{ext} dx = -eE_{ext}\left(x_f - x_i\right) > 0 \tag{5.52}$$

where F_{ext} is the electric force on a charge. Eq. (5.52) is *always positive* because for a positive electric field E_{ext}, say along $+x$, the force on the electron is along $-x$ due to a negative charge. This leads to $-e \times x_f - x_i$ which gives a $-\times-$ which is positive (+) in sign. Reversing the polarity of the external voltage swaps the sign of both E_{ext} and δx, which preserves the *positive* sign. Thus, with an externally applied voltage, the energy is *increased* or made *less* negative. So, the energy can be written as

$$E_R = qU_{ext} + E_{\text{bound}} + V_{e-e} \tag{5.53}$$

U_{ext} is the corresponding external electric potential.[6] The electric field not only raises the electronic energy, but also *kicks* the electron in a definite direction opposite the E–field axis ($F = qE$). These excited electrons are what we call *conduction electrons*, also known as *itinerant electrons*.. Itinerant electrons *are not free electrons*. Instead, they are in a class of their own. Freeing an electron *entirely* from a large collection of atoms is akin to providing enough energy for an object to escape the earth's atmosphere. It requires a much larger amount of energy for this, compared to the energy requirement for only *hopping* around on the earth as we do with aircrafts. Essentially, electrical conduction is the result of electrons being enabled to hop from one atom to another. This is the meaning of the word *itinerate* (*i.e.* to hop). The hopping process requires less energy because its obviously less work being done since its over a shorter distance. Freeing an electron requires moving the electron a distance of infinity, whereas with hopping, the electron only needs to move a distance on the order of the lattice constant, being finite compared to freeing the electron. When the electron *itinerates* to a nearby atom, there are two possibilities:

1. The electron can displace another electron in an arrival atom if the energy conditions are favorable. We can now understand the physics of this scenario based on the concepts discussed in the previous sections, considering external potentials versus bound state energies, *etc.* However, as the electron is excited into itineration, there still must be some available states below the excited electron energy for the electron to occupy in the arrival atom.

2. The other possibility is that the electron cannot displace another electron due to unfavorable energy conditions. This occurs when there is no proximate energy state to be occupied by the itinerant electron. For example, the available energy states are at a level

[6] U_{ext} is the amount of electric potential per unit charge to move the electron along the mean displacement δx.

element	valence	available states	no. avail. states
Au (P-shell)	$6s^1$	6s, 6p, 6d, 6f, 6g, 6h	72-1=71
Ag (O-shell)	$5s^1$	5s, 5p, 5d, 5f, 5g	50-1=49
Cu (N-shell)	$4s^1$	4s, 4p, 4d, 4f	32-1=31

sufficiently higher than the excited itinerant electronic energy. In this case, the electron scatters and the probability of conduction drops locally.

These two possibilities dictate that the process of electrical conduction is partly determined by the available energies *in the vicinity* of the valence electron energy levels, or near the Fermi edges (edge = general term for the level, contour, or surface). Thus, conduction properties depend on information relating to the Fermi edge, as well as the distribution of *available* energy states for the conducting material.

Good conductors have a single electron in the lowest possible energy of the highest shell. This leaves the largest number of available proximate energy states within the same shell. This is the case for the top three electrical conductors which are copper (Cu), silver (Ag), and gold (Au). Their properties are summarized in Table **??**. The proximate available higher energy states (beyond the Fermi edge) form the *conduction band,* as mentioned. Though itinerant electrons generally *still* have a negative energy,[7] they are energized just enough to *itinerate* from atom to atom, still more or less bound to the lattice of atoms. We must think of the (negative) electronic energy *relative* to a certain energy point. Only relative to this energy point, can the electron energy be considered *positive*. In transport, the appropriate *reference point* is the *Fermi edge* ϵ_F (1D).

Also related to transport, within a material lattice, which is a periodic arrangement of atoms, because of the combined effects from electron wave scattering at atoms, conditions are reached where some wavenumbers corresponding form standing waves in the lattice, resulting in no conduction for those particular wavenumber/energy pairs. This removes some $(\mathbf{k}, E)$ points from the distribution of probable energies. As mentioned earlier, these regions of non-permitted energy, created from scattering creating conditions of destructive interference, forms

[7] For this reason, they are always inclined to fall back into the Fermi sea.

the so-called *energy gaps*. The exact details must be computed for the specific lattice to determine the precise *band structure*. The net result is a further restriction of electron hopping utilizing, otherwise, available states. For the energy states in these gaps having some energy $E_j(k_j)$, now have a probability $f(k_j) \approx 0$. A general prescription can be outlined for how to determine a current density $\widehat{J} = g(E_{ext})$, which depends on the external field E_{ext}. We'll use an effective *conduction wave vector* k_{cond} *along the direction of the electric field* defined as

$$\frac{\hbar^2 k_{\text{cond}}^2}{2m_0} = \sum_j \left[E_R(E_{ext}) - E_j(k_j)\right] f(k_j) \tag{5.54}$$

$E_R(E_{ext})$ is the eigenenergy which depends on the external electric field. The sum is taken over the distribution of quantum states along with their corresponding occupation probability. $\mathbf{k}$ is varied over 1, 2, or 3D and the expectation of the corresponding energies is computed. In Eq. (5.54), for a large available energy $E_j > E_R$ in the conduction band, the sum adds a negative component to k_{cond}^2, while for small available energies where $E_j < E_R$, there is a positive contribution. From what we have considered in this chapter, when $k_{\text{cond}}^2 < 0$, the conduction wave function becomes evanescent (e.g. an insulator), while for $k_{\text{cond}}^2 > 0$, a propagating wave travels in the medium. This provides a mean conduction wave vector k_{cond}, which is a function of the external electric field. Thus, as a *quasi-free electron* using the conduction wave vector k_{cond}, the general eigenfunction becomes

$$\psi_{\text{cond}}(x) = Ae^{ik_{\text{cond}}x} + Be^{-ik_{\text{cond}}x} \tag{5.55}$$

For conduction of a quasi-free particle, the most general eigenfunction is given by

$$\Psi(x,t) = \frac{1}{\sqrt{2\pi\hbar}} \int_{-\infty}^{+\infty} F(p_{\text{cond}}) e^{-\frac{i}{\hbar}\left(p_{\text{cond}}x - \frac{p_{\text{cond}}^2}{2m}t\right)} dp_{\text{cond}} \tag{5.56}$$

$F(p_{\text{cond}})$ is normally the Fourier transform of the eigenfunction given by Eq. (5.55) and $p_{\text{cond}} = \hbar k_{\text{cond}}$. Since the largest contributions come from the eigenstates in the vicinity of the Fermi edge, this integral can be computed predominantly near the Fermi surface, where *available* states exist and the contributions from the band structure are largest (*i.e.* $f(k_j)$ are largest). The Fourier transform is given by

$$F(p) = \frac{1}{\sqrt{2\pi\hbar}} \int_{-\infty}^{+\infty} \psi(x')e^{-\frac{ip}{\hbar}x'} dx' \tag{5.57}$$

$F(p_{\text{cond}})$ is known as the probability amplitude of the momentum distribution. This yields a general wave function Ψ for the conduction electron which can be used to determine a current density $\widehat{J}(E_{ext})$ as[8]

$$\widehat{J}(E_{ext}) = \frac{i\hbar}{2m}\left(\Psi\frac{\partial\Psi^*}{\partial x} - \Psi^*\frac{\partial\Psi}{\partial x}\right)$$

Metals, generally, have smaller energy gaps in their band structure and can even overlap between the conduction and valence bands. At the opposite end of the spectrum, insulators have larger gaps so that electrons generally scatter more strongly leading less contribution to a current density.

In metals like Cu, Au, and Ag, *the single electron in the valence shell are also spin-down,* since spin-down electrons occupy orbitals before spin-up. This also means that good conductors (with less than half filled valence orbitals) tend to utilize spin-down electrons. It is this subtle detail that gets exposed when the electron experiences transport in a magnetic material. In the next chapter, using a 4D energy operator, we explore this behavior in more detail, analyzing a field known as *spintronics*.

5.6 Chapter Summary

In this chapter, we have discussed a relevant preliminary topic for any subject relating to electronic transport. We considered a few problems that serve to introduced some fundamental behavior in transport of quantum mechanical particles, their associated conditions, as well as the relationship between the two.

First we explored the evanescent wave and the conditions in which they form. These conditions liken to the ability of an electron to itinerate across atoms, versus more scattering as it conducts. We also introduced important concepts such as transmission and reflection probabilities an their utility in determination of current density J. The study of the finite well problem demonstrates in the simplest possible way, how electrons can go between localized (confined to the well), as opposed to excited enough to itinerate in the direction of an external field. The discussion of conduction brought us back to the true complexity of the topic and its relationship to quantum mechanical waves. As the transport considered

[8] conductivity σ can also be determined through the definition $J = \sigma E$

was purely charge-based, let's now move on to discuss another form of transport that involves the electron spin degree of freedom.

Some key equations are summarized in Table 5.1.

Table 5.1: Chapter 5 Summary Equations.

Name	Equation
Reflection definition	$R = \frac{\lvert J_R \rvert}{\lvert J_I \rvert}$
Transmission definition	$T = \frac{\lvert J_T \rvert}{\lvert J_I \rvert}$
Reflection ($k_2^2 < 0$)	$R = \frac{(k_1^2+k_2'^2)(k_2'^2+k_3^2)\sinh^2(k_2'L)+(k_1k_2'-k_2k_3)^2}{(k_1^2+k_2'^2)(k_2'^2+k_3^2)\sinh^2(k_2'L)+(k_1k_2'+k_2'k_3)^2}$
Transmission ($k_2^2 < 0$)	$T = \frac{4k_1k_2'^2k_3}{(k_1^2+k_2'^2)(k_2'^2+k_3^2)\sinh^2(k_2'L)+(k_1k_2'-k_2'k_3)^2}$
Reflection ($k_2^2 > 0$)	$R = \frac{(k_2^2-k_1^2)(k_2^2-k_3^2)\sin^2(k_2L)+(k_1k_2-k_2k_3)^2}{(k_2^2-k_1^2)(k_2^2-k_3^2)\sin^2(k_2L)+(k_1k_2+k_2k_3)^2}$
Transmission ($k_2^2 > 0$)	$T = \frac{4k_1k_2^2k_3}{(k_2^2-k_1^2)(k_2^2-k_3^2)\sin^2(k_2L)+(k_1k_2+k_2k_3)^2}$
Potential Well Energy	$E_n = -\frac{2\hbar^2\nu_n^2}{m_0L^2} \quad : \quad \nu_n = \frac{k_{2,n}L}{2}$

5.7 Chapter Problems

Problem 5.1 The wave number for an electron with kinetic energy and potential energy can sometimes be written as

$$k = \sqrt{\frac{2m(E_\lambda - V)}{\hbar^2}}$$

Calculate the wave number k for an electron with $E = 1$ eV in a potential of $V = +.5$ eV? Based on the result, is it a hill or hole? Will the wave slow down or speed up? Same for $V = -.5$ eV?

Problem 5.2 Show that if

$$k_2^2 < 0$$

That $R = 1$ and $T = 0$ using the definitions.

Problem 5.3 For two cases above, calculate the reflection and transmission probabilities using the appropriate (simplified)equations (e.g. Eqs. (5.28a) and (5.28b)) for R and T.

Problem 5.4 Earlier, in the problem of the rectangular barrier, the following transmission and reflection probabilities were obtained:

$$T = \frac{4k_1^2 k_2'^2}{(k_1^2 + k_2'^2)^2 \sinh^2(k_2' L) + 4k_1^2 k_2'^2}$$

and

$$R = \frac{(k_1^2 + k_2'^2)^2 \sinh^2(k_2' L)}{(k_1^2 + k_2'^2)^2 \sinh^2(k_2' L) + 4k_1^2 k_2'^2}$$

k_1 is the wave number in the first region and k_2' is the wave number in the barrier region of width L. The dependence on the barrier width L only appears in the sinh terms. Look at the asymptotic case of $L \rightarrow 0$, and show that $T = 1$ and $R = 0$. Then, take $L \rightarrow \infty$ and show that $T = 0$ and $R = 1$.

Problem 5.5 Use the above relations for reflection and transmission given by Eqs. (5.28a) and (5.28b) and show that

$$R + T = 1$$

Now show that the condition for $R = T = .5$ is given by

$$k_2' L = \sinh^{-1}\left[\frac{4k_1^2 k_2'^2}{k_1^2 + k_2'^2}\right]$$

Problem 5.6 Show the same for the more general reflection and transmission probabilities given by Eqs. (5.26) and (5.27).

Problem 5.7 Show that for any wave function of the form $X = Ae^{ikx} + Be^{-ikx}$, where A and B are the probability amplitudes, that $X = \overline{A}\cos kx +$

$\overline{B}\sin kx$ where $\overline{A} = A + B$ and $\overline{B} = i(A - B)$. *Hint*: Use the Euler identity $e^{i\theta} = \cos\theta + i\sin\theta$. Also show that

$$X^*X = |A|^2 + |B|^2 + (A^*B)_R\left[\cos^2 kx - \sin^2 kx\right] + (A^*B)_I \cos kx \sin kx$$

Discuss the conditions $B = 0$ and $A = B$.

Problem 5.8 Use Fig. 5.18 to estimate the energy of an electron in a potential well 1 angstrom wide, where $\alpha_n = 2\pi$.

For this case, also compute the ratio of wave numbers k_2/k_1 for a 2 angstrom wide potential well. Use one of the cot/tan trigonometric relations.

Problem 5.9 Show that for any wave function of the form $\psi = Ae^{ikx} + Be^{-ikx}$, that the current density is given by

$$J(x) = \frac{\hbar k}{m}\left(|A|^2 - |B|^2\right)$$

Comment on the meaning and conditions that lead to $J(x) = 0$?

Problem 5.10 For a propagating wave function $\psi_0 = Ae^{ikx}$, we have seen that when A is independent of space, ψ_0 is *non-normalizable*. To extricate the wave function from this dilemma, let

$$\psi(x) = f_c(x)\psi_0$$

$f_c(x)$ is any real valued compact (or square integrable) function of x, which satisfies the condition

$$\int_{-\infty}^{+\infty} |f_c(x')|^2 dx' = 1$$

Show that

$$\psi\frac{\partial\psi^*}{\partial x} - \psi^*\frac{\partial\psi}{\partial x} = -2ik\psi^*\psi$$

This result is identical in form to the result when $\psi = \psi_0$, however, the wave function we start with *is normalizable*. This represents the more correct form of an electron in motion, particularly, coming from a bound electronic state. The bound state energy information is, therefore, contained in the real-valued function $f_c(x)$.

CHAPTER 6

Spintronics In 4D

Non-Hermitian energy operators based on four dimensions, as we have seen, can be used to find non-equilibrium time-dependence of quantum mechanical systems. Let us continue along these lines by analyzing, in 4D, electron+spin transport in certain materials and material combinations. It turns out that there is distinct physics experienced by an electron flowing in copper (Cu) versus, for example, iron (Fe). Because iron is magnetic, the electron can couple to the internal energies of the magnetic atoms. When this happens, novel behavior can be observed that can be used in a variety of applications, from magnetic field sensors to magnetic-based memory providing a novel type of non-volatile memory (NVM).[1] The physics of *electron spin transport* and its applications is known as *spintronics*. In analyzing the quantum physics of spintronics, we consider problems involving electrons which are often flowing in and out of magnetic materials, as well as combinations with nonmagnetic materials. So, let's begin with the flow of electrons along with their spin through a single magnetic layer of material.

[1] non-volatile memory maintains its storage state even after external power is removed.

6.1 Asymmetric Spin Flow Of Itinerant Electrons

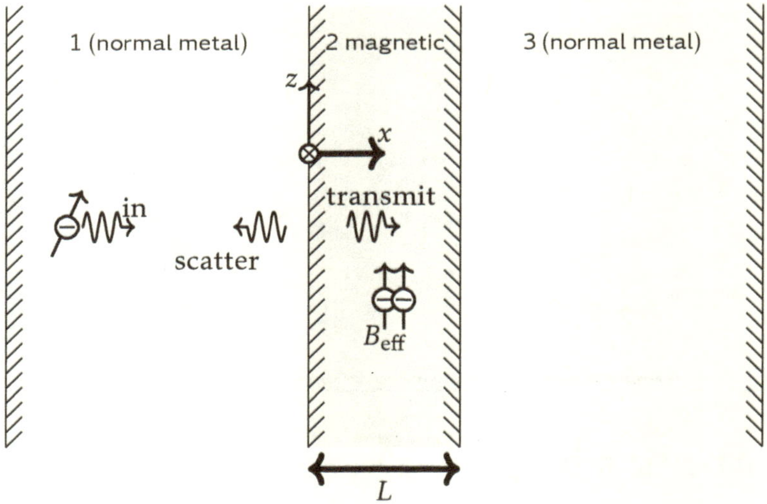

Figure 6.1: Itinerant electron+spin moving through a magnetic layer. When reaching the interface, some electrons can back-scatter or reflect while others enter the magnetic material more readily.

When electrons flow in a normal (nonmagnetic) conductor *e.g.* copper, both spin-up (↑) and spin-down (↓) electrons *see* equal conditions. The atoms of a normal metal provide equal conditions for both spins to flow. As a result, the expectation value of the spin angular momentum $\langle \mathbf{S} \rangle = 0$ for ordinary electrical current. Hence, we are left *only* with a flow of charges, though each electron carries both charge and spin. This changes when a charge carrier having *spin* flows through a material *e.g.* magnetic that *also* has some net spin by its localized electrons. In this case, related internal energies can couple to the itinerant electron-spin degree of freedom. This can lead to a break in spin-flow symmetry which can be useful for technological applications. The resulting physics, *spintronics,* is fundamental to any contemporary discussion of electronic transport in quantum mechanics. Using a 4D Hamiltonian, let's analyze the simplest example of the *breaking of symmetry* for electron spin carriers, arising from transport through a ferromagnet layer.

Fig. 6.1 illustrates a conducting *electron spin* having an effective kinetic energy conducting through a *magnetic* material of finite thickness L. Let us assume the localized spins in the magnetic layer are *fixed* in their orientation through internal mechanisms (*e.g.* in a ferromagnet with a 'strong' easy axis.), taken to be along the $+z$ direction of the diagram.

In a magnetic layer, the spin of the *itinerant* electron can couple to that of a *localized* electron spin much like it couples to a magnetic field. The strength of this coupling is known as *inter-electronic exchange coupling* often denoted by J[2]. Because itinerant electrons in metals come predominantly from s–orbitals and because magnetic materials have a net spin polarization from localized electrons predominantly in d–orbitals, this particular exchange coupling constant is often denoted by $J = J_{sd}$. We'll use this notation from here on out. There are at least two relevant energy couplings in a ferromagnetic layer, being the *inter-electron exchange* and *spin-orbit coupling* which creates the conditions of a preferred spin-alignment direction or an *easy axis*, e.g. $z-$ axis). [3]

Let the 2×1 spinor of electron wave functions containing the spin-up $\psi_\uparrow$ and spin-down $\psi_\downarrow$ eigenfunction be given by

$$\psi = \begin{bmatrix} \psi_\uparrow \\ \psi_\downarrow \end{bmatrix} \tag{6.1}$$

The 4D Hamiltonian for a flowing electron in a ferromagnetic layer becomes

$$\left[\frac{\widehat{\mathbf{p}}^2}{2m_0} + V_\sigma - i\hbar\frac{\partial}{\partial t}\right]\begin{bmatrix} \psi_\uparrow \\ \psi_\downarrow \end{bmatrix} = iE_I\begin{bmatrix} \psi_\uparrow \\ \psi_\downarrow \end{bmatrix} \tag{6.2}$$

V_σ denotes the spin-dependent potential energy of the system, *which can give rise to a quantum state transition.* It is Hermitian, or has real eigenvalues. This energy generally may contain more than one contribution. In the case of spintronics, it contains at least the inter-electronic exchange and spin-orbit interaction (c.Eq 2.132) given by

$$V_\sigma = -\frac{J_{sd}}{2}\widehat{\boldsymbol{\sigma}}_j \cdot \widehat{\boldsymbol{\sigma}}_k - \frac{\hbar}{m_0^2c^2}\widehat{\boldsymbol{\sigma}}_j \cdot (\widehat{\mathbf{p}} + e\mathbf{A}) \times \nabla V_e \tag{6.3}$$

where $\widehat{\sigma_j}$ is the spin operator of the itinerant electron, and $\widehat{\sigma_k}$ is that of the localized electron. In the spin-orbit interaction V_e *includes* an *electrical* potential energy due to the external electric field contributing to electrical conduction. For localized electrons, we can assume both the spin-orbit interaction (i.e. ∇V_e) and the localized ferromagnet spins σ_k are fixed in orientation. Then Eq. (6.5) can be expressed as

$$V_\sigma = -\widehat{\boldsymbol{\mu}}_j \cdot \mathbf{B}_{\text{eff}} \tag{6.4}$$

[2]For a detailed discussion and derivation of the exchange (integral) J, see Quantum Mechanics With A SPIN On It (Eason), Sec. 10.2-10.3.

[3]With exchange alone, spins only want to be parallel to one another. Spin-orbit coupling further provides a preferred direction for all the spins. *i.e.* an *easy axis*.

where the *effective magnetic field* $\mathbf{B}_{\text{eff}}$ can be defined as

$$\mathbf{B}_{\text{eff}} = -\frac{J_{sd}}{2\mu_B}\widehat{\sigma}_k - \frac{\hbar}{\mu_B m_0^2 c^2}(\widehat{\mathbf{p}} + e\mathbf{A}) \times \nabla V_e \tag{6.5}$$

$\mu_B = \gamma\hbar$ is the *Bohr magneton*.[4] In this way, both together have the form of an energy contribution due to the electron magnetic moment coupling to an external magnetic field *i.e.* $-\boldsymbol{\mu}\cdot\mathbf{B}$. Eq. (6.5) becomes

$$\left[\frac{\widehat{\mathbf{p}}^2}{2m_0} - \widehat{\boldsymbol{\mu}}_j \cdot \mathbf{B}_{\text{eff}} - i\hbar\frac{\partial}{\partial t}\right]\begin{bmatrix}\psi_\uparrow\\ \psi_\downarrow\end{bmatrix} = iE_I\begin{bmatrix}\psi_\uparrow\\ \psi_\downarrow\end{bmatrix} \tag{6.6}$$

Taking the imaginary eigenvalue $E_I = \alpha E_R$, Eq. (6.6) can be rearranged to give

$$i\hbar\frac{\partial}{\partial t}\begin{bmatrix}\psi_\uparrow\\ \psi_\downarrow\end{bmatrix} = \left[-\frac{\hbar^2}{2m_0}\frac{\partial^2}{\partial x^2} + (1 - i\alpha)V_\sigma\right]\begin{bmatrix}\psi_\uparrow\\ \psi_\downarrow\end{bmatrix} \tag{6.7}$$

We also require the *total energy operator* equation, given by

$$i\hbar\frac{\partial}{\partial t}\begin{bmatrix}\psi_\uparrow\\ \psi_\downarrow\end{bmatrix} = \begin{bmatrix}E_{R,\uparrow} - iE_I & 0\\ 0 & E_{R,\downarrow} - iE_I\end{bmatrix}\begin{bmatrix}\psi_\uparrow\\ \psi_\downarrow\end{bmatrix} \tag{6.8}$$

Because we have a spin-dependent energy operator, the eigenenergies E_R, generally, differ for both spin states. According to the spin-dependent operator given by Eq. (6.4), the two *distinct* energies corresponding to spin-up and spin-down states are described by

$$E_{R,\uparrow}\psi_\uparrow = -\frac{\hbar^2}{2m_0}\frac{d^2\psi_\uparrow}{dx^2} + \mu_B B_{\text{eff}}\psi_\uparrow \tag{6.9a}$$

$$E_{R,\downarrow}\psi_\downarrow = -\frac{\hbar^2}{2m_0}\frac{d^2\psi_\downarrow}{dx^2} - \mu_B B_{\text{eff}}\psi_\downarrow \tag{6.9b}$$

Eq. (6.9) corresponds to the real component of the RHS of Eq. (6.7).

We can obtain ψ by taking the spin wave functions as *separable* where $\psi_\uparrow = \psi_{t,\uparrow}\cdot\psi_{x,\uparrow}$ and $\psi_\downarrow = \psi_{t,\downarrow}\cdot\psi_{x,\downarrow}$. After substitution, we can divide by the respective wave function and equate components. This leads to the following four eigenequations:

$$E_{R,\uparrow}\psi_{x,\uparrow} = -\frac{\hbar^2}{2m_0}\frac{d^2\psi_{x,\uparrow}}{dx^2} + \mu_B B_{\text{eff}}\psi_{x,\uparrow} \tag{6.10a}$$

[4]Note that we have also used the relation $\boldsymbol{\mu} = -\mu_B\widehat{\boldsymbol{\sigma}}$.

$$i\hbar\frac{d\psi_{t,\uparrow}}{dt} = (1 - i\alpha)E_{R,\uparrow}\psi_{t,\uparrow} \tag{6.10b}$$

$$E_{R,\downarrow}\psi_{x,\downarrow} = -\frac{\hbar^2}{2m_0}\frac{d^2\psi_{x,\downarrow}}{dx^2} - \mu_B B_{\text{eff}}\psi_{x,\downarrow} \tag{6.10c}$$

$$i\hbar\frac{d\psi_{t,\downarrow}}{dt} = -(1 - i\alpha)E_{R,\uparrow}\psi_{t,\downarrow} \tag{6.10d}$$

If we consider the spin-up eigenequation given by Eq. (6.10a), it can be rearranged as

$$0 = \frac{d^2\psi_{x,\uparrow}}{dx^2} + \left[\frac{2m_0\left(E_{R,\uparrow} - \mu_B B_{\text{eff}}\right)}{\hbar^2}\right]\psi_{x,\uparrow} \tag{6.11a}$$

$$= \frac{d^2\psi_{x,\uparrow}}{dx^2} + k_\uparrow^2\psi_{x,\uparrow} \tag{6.11b}$$

Likewise, for the spin-down state eigenvalue equation, we have

$$0 = \frac{d^2\psi_{x,\downarrow}}{dx^2} + \left[\frac{2m_0\left(E_{R,\downarrow} + \mu_B B_{\text{eff}}\right)}{\hbar^2}\right]\psi_{x,\downarrow} \tag{6.12a}$$

$$= \frac{d^2\psi_{x,\downarrow}}{dx^2} + k_\downarrow^2\psi_{x,\downarrow} \tag{6.12b}$$

Eqs. (6.11) and (6.12) define the respective spin wavenumbers $k_\uparrow$ and $k_\downarrow$ as

Itinerant Electron Wavenumbers In A Magnet

$$k_\uparrow = \sqrt{\frac{2m_0\left(E_{R,\uparrow} - \mu_B B_{\text{eff}}\right)}{\hbar^2}}, \quad k_\downarrow = \sqrt{\frac{2m_C\left(E_{R,\downarrow} + \mu_B B_{\text{eff}}\right)}{\hbar^2}} \tag{6.13}$$

Eq. (6.13) indicates that the *spin-down wavenumber increases*, while *spin-up generally decreases.*

The spatial solutions described by Eqs. (6.11) and (6.12) are given by

$$\psi_{x,\uparrow} = A_\uparrow e^{ik_\uparrow x} + B_\uparrow e^{-ik_\uparrow x} \quad \text{(spin-up)} \tag{6.14a}$$

$$\psi_{x,\downarrow} = A_\downarrow e^{ik_\downarrow x} + B_\downarrow e^{-ik_\downarrow x} \quad \text{(spin-down)} \tag{6.14b}$$

The time-dependent equations given by Eqs. (6.10b) and (6.10d) can be rearranged as

$$0 = \frac{d\psi_{t,\uparrow}}{dt} + i\left[\frac{(1-i\alpha)E_{R,\uparrow}}{\hbar}\right]\psi_{t,\uparrow} \tag{6.15a}$$

and

$$0 = \frac{d\psi_{t,\downarrow}}{dt} - i\left[\frac{(1-i\alpha)E_{R,\downarrow}}{\hbar}\right]\psi_{t,\downarrow} \tag{6.16a}$$

The solutions to Eqs. (6.15) and (6.16) are given by

$$\psi_{t,\uparrow}(t) = C_\uparrow e^{-t/(\alpha E_{R,\uparrow})} e^{-iE_{R,\uparrow}t/\hbar} \quad , \quad \psi_{t,\downarrow}(t) = C_\downarrow e^{t/(\alpha E_{R,\downarrow})} e^{iE_{R,\downarrow}t/\hbar} \tag{6.17}$$

The total spin-state eigenenergies are given by

$$E_\uparrow = E_{R,\uparrow} + iE_{I,\uparrow} = \frac{\hbar^2 k_\uparrow^2}{2m_0} + \mu_B B_{\text{eff}} + i\alpha\mu_B B_{\text{eff}} \tag{6.18a}$$

$$E_\downarrow = E_{R,\downarrow} + iE_{I,\downarrow} = \frac{\hbar^2 k_\downarrow^2}{2m_0} - \mu_B B_{\text{eff}} - i\alpha\mu_B B_{\text{eff}} \tag{6.18b}$$

Combining the solutions for x and t, we have the *spin eigensolutions* $\psi(x,t) = \psi_{\uparrow,\downarrow}(0)e^{i(p_{\uparrow,\downarrow}x - E_{\uparrow,\downarrow}t)/\hbar}$, given by

$$\psi_\uparrow(x,t) = \left(A_\uparrow e^{ik_\uparrow x} + B_\uparrow e^{-ik_\uparrow x}\right) e^{-i\frac{\hbar^2 k_\uparrow^2}{2m_0}t} e^{-i\omega_{\text{eff}}t/2} e^{-\alpha\mu_B B_{\text{eff}}t/\hbar} \tag{6.19a}$$

$$\psi_\downarrow(x,t) = \left(A_\downarrow e^{ik_\downarrow x} + B_\downarrow e^{-ik_\downarrow x}\right) e^{-i\frac{\hbar^2 k_\downarrow^2}{2m_0}t} e^{+i\omega_{\text{eff}}t/2} e^{+\alpha\mu_B B_{\text{eff}}t/\hbar} \tag{6.19b}$$

ω_{eff} is the *precessional frequency* defined as

$$\omega_{\text{eff}} = 2\gamma B_{\text{eff}} \tag{6.20}$$

Thus, when spins enter via flow into a magnetic layer, the symmetry between electron spin states is 'broken', or the spin-up and spin-down energies are no-longer equal. For an assumed $+z$ internal field, the energy is biased towards spin-down electrons. Keep in mind that in a transport problem such as this, the boundary conditions of the layers involved should also be factored into the problem. However, general behavior can still be analyzed in the system to recognize distinct patterns that emerge.

One of the consequences of electrons flowing in a magnetic layer is to have bias between the spin-states. This is also known as *spin-splitting*. Let us discuss this concept further.

6.1.1 Spin-dependent Splitting In Wavenumbers $k_\uparrow, k_\downarrow$

When electrons flow in a magnetic, the spin wavenumbers are given by

$$k_\uparrow^2 = \frac{2m_0}{\hbar^2}\left(E_{R,\uparrow} - \mu_B B_{\text{eff}}\right) = \frac{2m_0 E_{R,\uparrow}}{\hbar^2}\left(1 - \frac{\mu_B B_{\text{eff}}}{E_{R,\uparrow}}\right) \tag{6.21a}$$

$$k_\downarrow^2 = \frac{2m_0}{\hbar^2}\left(E_{R,\uparrow} + \mu_B B_{\text{eff}}\right) = \frac{2m_0 E_{R,\uparrow}}{\hbar^2}\left(1 + \frac{\mu_B B_{\text{eff}}}{E_{R,\uparrow}}\right) \tag{6.21b}$$

The wavenumbers for an electron-spin flowing in a magnetic conductor *differ* for spin-up and spin-down states because of the coupling to the internal effective magnetic field in a magnetic layer. As mentioned, this corresponds to a spin splitting between spin-up and spin-down energy states.

According to Eq. (6.21a), mathematically, $k_\uparrow^2$, for example, can be either positive (e.g. a good conductor) or negative (e.g. a strong insulator). So, let us consider the consequences for these two extreme cases.

If $E_{R,\uparrow} > 0 \to 2m_0 E_{R,\uparrow}/\hbar^2 > 0$, then we have

$$\frac{k_\uparrow^2}{k_\downarrow^2} = \frac{1 - \frac{\mu_B B_{\text{eff}}}{E_{R,\uparrow}}}{1 + \frac{\mu_B B_{\text{eff}}}{E_{R,\uparrow}}} < 1 \tag{6.22}$$

From this relation, it follows that

$$k_\downarrow^2 > k_\uparrow^2 \quad \text{(spin down flows faster than spin up)} \tag{6.23}$$

Therefore, when spins enter a magnetic layer where the effective field is along a direction $+z$, the spin-down will flow more readily than spin up. Also recall from the discussion of electronic transport in Chap. 5, the spin-down equation of Eq. (6.13) is concomitant with scattering, while, to the contrary, it will be even easier for spin-down electrons to conduct in the magnetic layer. This is true even if $E_{R,\uparrow} < 0$, where we have

$$k_\uparrow^2 = \frac{-2m_0|E_{R,\uparrow}|}{\hbar^2}\left(1 + \frac{\mu_B B_{\text{eff}}}{|E_{R,\uparrow}|}\right) \tag{6.24}$$

$$= -|k_{R,\uparrow}^2|\left(1 + \frac{\mu_B B_{\text{eff}}}{|E_{R,\uparrow}|}\right) \tag{6.25}$$

Since this is a *negative-definite quantity,* the existing evanescent state resulting from $k^2 < 0$ only becomes *more pronounced,* or tunnels less efficiently. However, for spin-down, we have

$$k_\downarrow^2 = \frac{-2m_0|E_{R,\uparrow}|}{\hbar^2}\left(1 - \frac{\mu_B B_{\text{eff}}}{|E_{R,\uparrow}|}\right) \tag{6.26a}$$

$$= -|k_{R,\uparrow}^2|\left(1 - \frac{\mu_B B_{\text{eff}}}{|E_{R,\uparrow}|}\right) \tag{6.26b}$$

For the spin-down state, if $\mu_B B_{\text{eff}} < |E_{R,\uparrow}|$, then although $k_\downarrow^2 < 0$, it decreases in magnitude so that $\delta_\downarrow$ becomes larger, or the evanescence spreads farther at the interface for spin-down states. So, we have the general result:

(R) *Asymmetric spin conduction/scattering/tunneling*: When electrons flow through a magnetic material, the spin-down state, whose magnetic moment μ_S is parallel to that of the ferromagnet conducts more efficiently than the spin-up state, whose moments are anti-parallel. If evanescent conditions exist, then spin-down tunnels more efficiently.

Fig. 6.2 plots $E_\uparrow = \hbar^2 k_\uparrow^2/2m_0$ and $E_\downarrow = \hbar^2 k_\downarrow^2/2m_0$ as functions of $E_{R,\uparrow}$. The spin-dependent energies are proportional to kinetic energy $E_{R,\uparrow}$, however, there is a splitting or energy difference given by $E_\downarrow - E_\uparrow = 2\mu_B B_{\text{eff}}$. The corresponding plots of $|k_\uparrow|$ and $|k_\downarrow|$ are shown in Fig. 6.3. In the limit of $E_{R,\uparrow} \to \pm\infty$, both curves $k_{\uparrow,\downarrow}$ tend towards a single curve, so that increasing the applied voltage *beyond* the touch-down point, generally reduces splitting. Therefore, there is a tradeoff at higher voltages.

Using the above results, we can estimate length and energy scales for spin transport. First, let's use some estimates for Eq. (6.26) to see an example of the spin decay length δ.

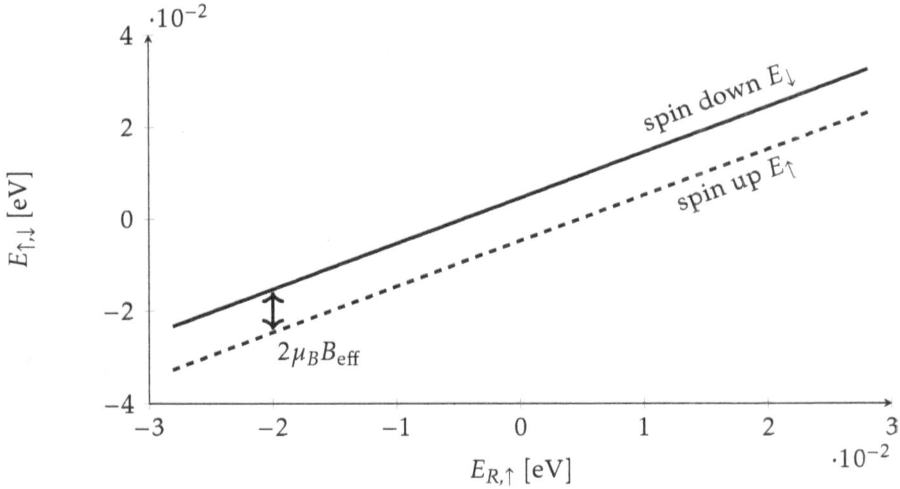

Figure 6.2: Spin-up and spin-down energies depending on the conduction kinetic energy $E^2_{R,\uparrow}$; a value of $\mu_B B_{\text{eff}} = 0.75 \times 10^{-21}$J is used.

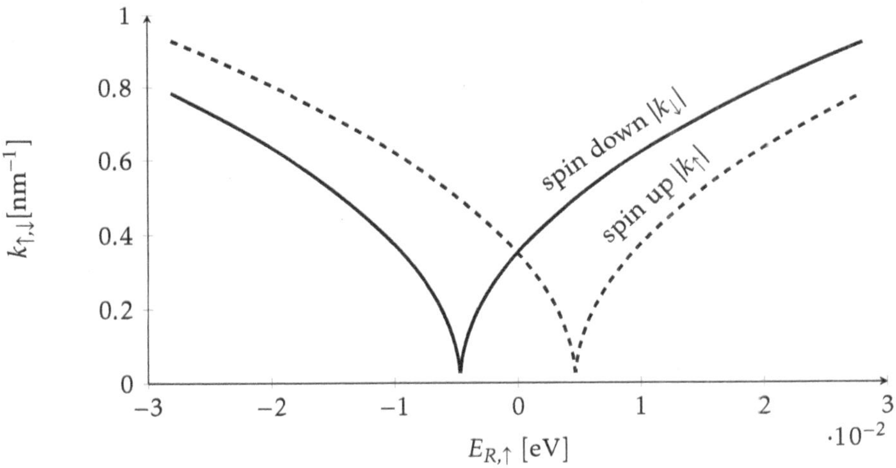

Figure 6.3: Corresponding wavenumbers for spin-up and spin-down depending on the ordinary conduction kinetic energy $E^2_{R,\uparrow}$. On the positive side, spin-down has larger momentum, while on the negative side (evanescence), spin-down has less rapid decay/scattering, *etc.*

Exercise 6.1 Estimate the spin decay length $\delta_{\downarrow,\uparrow} = 1/k_{\downarrow,\uparrow}$ using the relations

$$k^2_{\downarrow,\uparrow} = \frac{2m_0 E_{R,\downarrow,\uparrow}}{\hbar^2}\left(1 \pm \frac{\mu_B B_{\text{eff}}}{E_{R,\uparrow}}\right) \tag{6.27}$$

For an electron having a conduction kinetic energy of −0.15meV in a magnetic layer having an internal effective field strength $B_{\text{eff}} = 1.0$ Tesla, we have

$$|k^2_{\downarrow,\uparrow}| = \frac{2(9.109\times10^{-31})(.15\times10^{-3})}{(1.054\times10^{-34})^2}\left(1 \mp \frac{(9.274\times10^{-24})(1.0)}{(.15\times10^{-3})}\right)$$

$$= 3.9369\times10^{16}(1.0 \mp 0.38589)$$

$$= 2.417\times10^{15}\text{m}^{-2} \quad , \quad 5.456\times10^{15}\text{m}^{-2}$$

For the decay length, we can evaluate $\delta = 1/k$. Then, we have decay lengths given by

$$\delta_\downarrow = 20.338\times10^{-9}\text{m} \quad , \quad \delta_\downarrow = 13.538\times10^{-9}\text{m}$$

The length scales of decaying of spins are on the order of 10^{-9}m or nanometers. It is on this length scale that spin-dependent phenomena are usually observed and/or exploited.

■

In spin transport, as well as with localized spins in a magnetic material, we have been exploiting an effective magnetic field *in* the magnetic layer. This field contains an exchange interaction. To estimate magnitudes for the exchange field, let's consider two extreme cases corresponding to relatively large and, well, less-large exchange between two electrons.

6.1.2 Strengths of Electron Exchange Interaction J

One approximation for the magnitude of the exchange parameter can be made by considering the case of two electrons of *opposite spin* in the same atomic orbital. There will be an exchange interaction between the two electrons. For example, let us take two electronic states a and b as s-orbitals, defined by $\ell = 0$ (*i.e.* zero orbital angular momentum). The normalized single electron orbital state is given by[5]

$$\psi_{1s}(\mathbf{r}_j) = \left(\frac{Z^3}{a_0^3 \pi}\right)^{1/2} e^{-Zr_j/a_0} \quad \text{where} \quad a_0 = \frac{h^2 \epsilon_0}{m_e \pi e^2} \tag{6.28}$$

Z is the atomic number. For a multi-electron problem, given that the wave functions generally depend on the energy, self-consistent numerical methods become necessary to determine the precise energy and wave function simultaneously. Here, our aim to find reasonable approximations for order of magnitude estimates. The helium atom, which as 2 electrons in the $1s$–orbital has been studied extensively and the numerical analyses aligning with experiments with helium reveal that the form of the helium wave function, as well as lithium (Li) and beryllium (Be) wave functions, have a form of linear combinations of s–orbitals in the form of Eq. (6.28). The exchange interaction is obtained by evaluating the so-called *exchange integral* arising from multi-electron energy operators, given by[6]

$$J_{12} = \frac{e^2}{4\pi\epsilon_0} \int_V r_1^2 r_2^2 \frac{\psi_{1s}(\mathbf{r}_1)\psi_{1s}(\mathbf{r}_2)\psi_{1s}(\mathbf{r}_2)\psi_{1s}(\mathbf{r}_1)}{|\mathbf{r}_2 - \mathbf{r}_1|} dV \tag{6.29}$$

Then, setting the wave functions ψ_a and $\psi_b = \psi_{1s}$ from Eq. (6.28), and substituting into Eq. (6.29) using spherical coordinates gives

$$\begin{aligned} J_{12} &= \frac{Z^6 e^2}{4\pi^3 a_0^6 \epsilon_0} \int_0^\infty \int_0^\infty \int_0^{2\pi} \int_{\theta=0}^{\pi} r_1^2 r_2^2 \sin\theta \frac{e^{-2Zr_1/a_0} e^{-2Zr_2/a_0}}{|\mathbf{r}_2 - \mathbf{r}_1|} dr_1 dr_2 d\theta d\varphi \\ &= \frac{Z^6 e^2}{2\pi^2 a_0^6 \epsilon_0} \int_0^\infty \int_0^\infty \int_0^{\pi} r_1^2 r_2^2 \sin\theta \frac{e^{-2Zr_1/a_0} e^{-2Zr_2/a_0}}{|\mathbf{r}_2 - \mathbf{r}_1|} dr_1 dr_2 d\theta \end{aligned}$$

Because they occupy the same orbital, this corresponds to electrons that have opposite spins. The integral is nontrivial, however, it can be evaluated exactly and each step is provided. First, using a spherical coordinate system, take the angle between $\mathbf{r}_2$ and $\mathbf{r}_1$ as the polar angle θ. Then, we can use the law of cosines to write $|\mathbf{r}_2 - \mathbf{r}_1|$ as

$$|\mathbf{r}_2 - \mathbf{r}_1| = \left(r_1^2 + r_2^2 - 2r_1 r_2 \cos\theta\right)^{1/2} \tag{6.30}$$

[5] For a discussion and complete derivation of the atomic orbitals, see Quantum Mechanics With A SPIN On It (Eason), Chaps. 7 and 8.

[6] For a discussion and derivation of the exchange integral, see Chap. 10, Quantum Mechanics With A SPIN On It (Eason).

Then, the sub-integral with respect to *only* θ becomes

$$I_\theta = \int_0^\pi \frac{\sin\theta d\theta}{\left(r_1^2 + r_2^2 - 2r_1 r_2 \cos\theta\right)^{1/2}}$$

$$= -\int_{-1}^{1} \frac{dx}{\left(r_1^2 + r_2^2 - 2r_1 r_2 x\right)^{1/2}}$$

The minus sign appears because if $x = \cos\theta$, then $dx = -\sin\theta d\theta$. Using the integral formula $\int \left(\sqrt{ax+b}\right)^{-1} dx = 2a^{-1}\sqrt{ax+b}$, we have

$$I_\theta = -\frac{1}{2r_1 r_2} 2\sqrt{r_1^2 + r_2^2 - 2r_1 r_2 x}\bigg|_{-1}^{1}$$

$$= -\frac{1}{r_1 r_2}\left[\sqrt{r_1^2 + r_2^2 - 2r_1 r_2} - \sqrt{r_1^2 + r_2^2 + 2r_1 r_2}\right]$$

$$= -\frac{1}{r_1 r_2}\left(\sqrt{(r_1 - r_2)^2} - \sqrt{(r_1 + r_2)^2}\right)$$

From the last line, there are *two* solutions implied from the first radical on the RHS being $r_1 - r_2$ or $r_2 - r_1$. If $r_1 > r_2 \to r_1 - r_2 > 0$, then

$$(r_1 - r_2) - (r_1 + r_2) = -2r_2$$

while if $r_2 > r_1$, we have

$$(r_2 - r_1) - (r_1 + r_2) = -2r_1$$

The two conditions, therefore, allow the integral with respect to r_2 to be broken up such that

$$I_\theta = \frac{2}{r_1} \quad \text{if} \quad r_1 > r_2 \tag{6.31}$$

$$I_\theta = \frac{2}{r_2} \quad \text{if} \quad r_2 > r_1 \tag{6.32}$$

Using the above result, we have

$$J_{12} = \frac{Z^6 e^2}{\pi^2 a_0^6 \epsilon_0}\int_0^\infty dr_1 r_1^2 e^{-2Zr_1/a_0} \times$$

$$\left[\frac{1}{r_1}\int_0^{r_1} r_2^2 e^{-2Zr_2/a_0} dr_2 + \int_{r_1}^\infty r_2 e^{-2Zr_2/a_0} dr_2\right] \tag{6.33}$$

There are two known integral solutions we can use to solve Eq. (6.33), and they are given by

$$\int xe^{ax}dx = \left(\frac{x}{a} - \frac{1}{a^2}\right)e^{ax} \tag{6.34a}$$

$$\int x^2e^{ax}dx = \left(\frac{x^2}{a} - \frac{2x}{a} + \frac{2}{a^3}\right)e^{ax} \tag{6.34b}$$

With $a = -2Z/a_0$, the exchange integral in Eq. (6.33) works out to be

$$J_{12} = \frac{Z^3e^2}{4\pi^2a_0^3\epsilon_0}\int_0^\infty dr_1 r_1^2 e^{-2Zr_1/a_0} \times \left[\frac{1}{r_1} - e^{-2Zr_1/a_0}\left(\frac{1}{r_1} + \frac{Z}{a_0}\right)\right] \tag{6.35}$$

Evaluation of Eq. (6.35) using, again, the relations from Eqs. (6.34), we have the result

$$J_{12} = \frac{5Ze^2}{32\pi\epsilon_0 a_0} = 34.01\text{eV} \tag{6.36}$$

This turns out to be a relatively large exchange energy between the two electrons, leading to an effective magnetic field of order

$$B_{12} \approx \frac{J_{12}}{\mu_B} = \frac{34.01\cdot 1.6\times 10^{-19}\text{J}}{9.274\times 10^{-24}\text{J}\cdot\text{T}^{-1}} = 587,492\text{T}$$

You read it right. The exchange energy between two $1s$ orbit electrons is equivalent to an epic strength *unrealizable magnetic field*. Keep in mind that it will be reduced from this value when we also factor in temperature $T > 0$. The point is to demonstrate that significantly large energy exists between two electrons in the same orbit. This is part of the reason that filled shells (with pairs of electrons) are very stable (*e.g.* helium), while unpaired electrons can become available for transport. This type of exchange is called the *intra-atomic exchange*, originating within a single atom.

However, in magnetic materials, as well as, in spin transport, the two electrons do not wholly share a single atom. One of the two electrons is localized on a different atom, or the electron is conducting from another atom. When electrons originating from distinct atoms couple via exchange, it is known as *inter-atomic exchange*. To estimate the strength of this exchange, in particular, we can take an alternative approach and rely on well-known experimental data for a parameter that depends on the inter-atomic exchange parameter. For example, in a magnetic material with body-centered cubic crystalline structure, an experimental parameter known as the exchange stiffness parameter A_X is related to the *localized* interatomic exchange J (an upper bound for J_{sd}) by the relation

$$A_X = \frac{2JS^2}{a}$$

a is the lattice constant, S is the normalized spin magnitude, which is on the order of 1. In this case, the *spins are parallel.* For cobalt, for example, the lattice constant is $a \approx 0.286$ nm and $A_X \approx 10^{-11}$ J/m. This leads to

$$A_X = \frac{2JS^2}{a} \Rightarrow J = .5aA_x \approx 1.43 \times 10^{-21}\text{J} = 8.9\text{meV}$$

This value is about 4 orders of magnitude smaller compared to our first example evaluating the integral. It is, however, still large and is sufficient to flip an electron spin. In ferromagnetic materials, this is what keeps electron spins collectively aligned. The corresponding value for *sd* is smaller because the itinerant electron only spends *some* time close to the coupling atom's electron, and some time *between* atoms. The net effect is to weaken the interaction, however, it is sufficient either to flip a spin or scatter a spin of different orientation.

This is the key to the concept of spin polarization. Unlike conduction in a normal metal, with a splitting between the spin-up and spin-down energies and transport properties in a ferromagnet, it becomes possible to observe a net spin polarization in a normal metal or insulator adjacent to the ferromagnet. This net spin information can be found by evaluating the expectation value of the spin polarization (or magnetic moment) along the presumed +z-axis. So, let's do that for flowing spins that have passed through a magnetic layer.

6.1.3 Spin Polarization Of Itinerant Electrons

With three regions, we focus on the last region an electron flows into after passing through a magnetic layer. The wave function in the normal metal after the magnetic layer is given

$$\psi_{3,\uparrow,\downarrow}(x,t) = E'_{\uparrow,\downarrow}e^{ik_3x} \quad \text{(region 3/normal metal)} \tag{6.37}$$

We want to determine the expectation value at the *exit* of the device in region 3 using Eq. (6.37). In this region, the reflection amplitude is

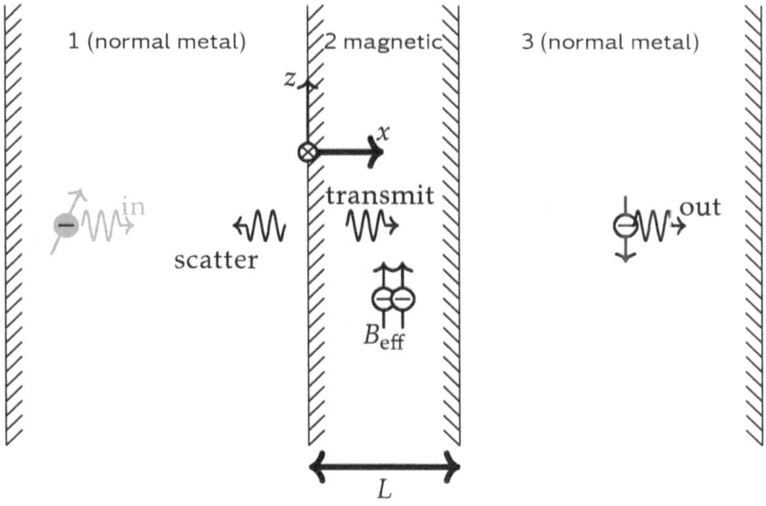

Figure 6.4: Electron spins flowing through a magnetic layer, sandwiched between two normal metallic layer. The spin polarization $\langle\mu_z\rangle$ is evaluated in the normal metal of region 3.

zero so the wave function takes the given form of $\psi_{3,\uparrow,\downarrow}(x,t)$. We already worked out the value we need for $E'_{\uparrow,\downarrow}$ (c.Eq. 5.22c). Using a *magnetic* layer for region 2, the coefficient $E'_{\uparrow,\downarrow}$ becomes

$$E'_{\uparrow,\downarrow} = \frac{4k_1 k_{\uparrow,\downarrow} e^{-ik_3 L}}{\left(k_1 + k_{\uparrow,\downarrow}\right)\left(k_{\uparrow,\downarrow} + k_3\right)e^{-ik_{\uparrow,\downarrow}L} + \left(k_1 - k_{\uparrow,\downarrow}\right)\left(k_{\uparrow,\downarrow} - k_3\right)e^{ik_{\uparrow,\downarrow}L}} \tag{6.38}$$

$k_{\uparrow,\downarrow}$ denotes the spin-up and spin-down wavenumbers, respectively. We can use Eq. (6.38) in Eq. (6.37) to compute the expectation value $\langle\widehat{\mu}_z\rangle$ given by

$$\langle\widehat{\mu}_z\rangle = \mu_B \cdot \frac{\int_V \begin{bmatrix}\psi_\uparrow^* & \psi_\downarrow^*\end{bmatrix}\begin{bmatrix}1 & 0\\ 0 & -1\end{bmatrix}\begin{bmatrix}\psi_\uparrow\\ \psi_\downarrow\end{bmatrix} dV}{\int_V \begin{bmatrix}\psi_\uparrow^* & \psi_\downarrow^*\end{bmatrix}\begin{bmatrix}\psi_\uparrow\\ \psi_\downarrow\end{bmatrix} dV} \tag{6.39}$$

The integration is carried out over the volume of region 3, which is *non*magnetic. Then, using the solution in region 3 for $\psi_{\uparrow,\downarrow}$, evaluation of Eq. (6.39) leads to

$$\langle\widehat{\mu}_z\rangle = \mu_B \cdot \frac{\int_V \psi_\uparrow^* \psi_\uparrow - \psi_\downarrow \psi_\downarrow^* dV}{\int_V \psi_\uparrow^* \psi_\uparrow + \psi_\downarrow^* \psi_\downarrow dV} \tag{6.40a}$$

$$= \mu_B \cdot \frac{\int_V |E'_\uparrow|^2 - |E'_\downarrow|^2 dV}{\int_V |E'_\uparrow|^2 + |E'_\downarrow|^2 dV} \tag{6.40b}$$

$$= \mu_B \cdot \frac{|E'_\uparrow|^2 - |E'_\downarrow|^2}{|E'_\uparrow|^2 + |E'_\downarrow|^2} \tag{6.40c}$$

The amplitudes $E'_{\uparrow,\downarrow}$ depend on *all three* wave numbers $k_1, k_2 = k_{\uparrow,\downarrow}, k_3$. They also depend on the thickness of the magnetic layer (see Fig. 6.1). According to Eq. (6.40c), the spin polarization ranges from $\langle\mu_z\rangle = 0$, when $E_\uparrow = E_\downarrow$ (normal metal) to $\langle\mu_z\rangle = \pm\mu_B$, when $E_\uparrow = 0$ or $E_\downarrow = 0$. When a material satisfies $\langle\widehat{\mu_z}\rangle = \pm\mu_B$, it conducts *only one of the two spin states*, while insulating against the other. Such a material is known as a *half-metal*.

Fig. 6.5 plots the itinerant spin polarization $\langle\widehat{\mu_z}\rangle$ in region 3, for both metallic and insulating conditions, using Eq. (6.40b).

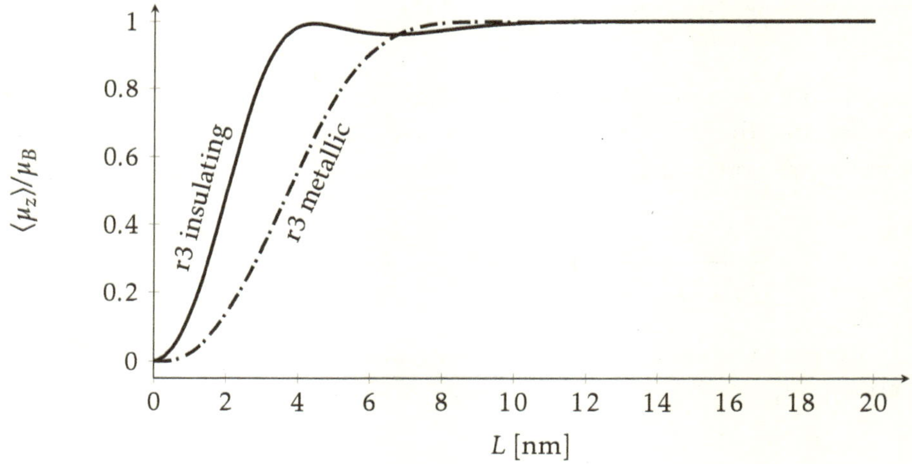

Figure 6.5: Expectation value of itinerant spin polarization given by Eq. (6.40c). The solid curve assumes insulating ($k_3 = ik'_3$) conditions, while the other assumes metallic (k_3 real) conditions in region 3.

$\langle\mu_z\rangle$ is known as the itinerant *spin polarization P*. The spin polarization, as Fig. 6.5 demonstrates, increases with increasing magnetic layer thickness up to a point of *saturation*. Between both curves shown, only k_3 differs. In one case, k_3 is set as a *real*-valued wave number, representing metallic conditions in region 3. The other case corresponds to have an imaginary wave number $k_3 = ik'_3$, which leads to insulating electrical conditions. Interestingly, we can see that insulating conditions (just after the polarizing magnet) can achieve more efficient spin polarization compared to a metallic conductor. This also demonstrates that the itinerant spin polarization is also sensitive to the layer adjacent to the magnetic polarizing layer.

In an insulator, the difference or splitting between spins is effectively enlarged due to much stronger scattering of spin up electrons. This enhances the asymmetric transport. Thus, placing an insulator after the magnet can create conditions somewhere between that of having a normal metal and a highly efficient *half-metal* in region 3.

The spin polarization $\langle \mu_z \rangle$ given by Eq. (6.40) can be expressed in another form by using the *spin-dependent* transmission probabilities $T_\uparrow$ and $T_\downarrow$. Ordinarily, without spin dependent energies in the Hamiltonian, the transmission probability T for the *non-magnetic* potential barrier is given by

$$T = \frac{J_T}{J_I} = \frac{k_3}{k_1}|E'|^2 \tag{6.41}$$

In this case, however, there are two distinct transmission probabilities corresponding to spin up $T_\uparrow = \frac{k_3}{k_1}|E'_\uparrow|^2$ and spin-down $T_\downarrow = \frac{k_3}{k_1}|E'_\downarrow|^2$. Using this relation, the *spin polarization* $\langle \mu_z \rangle = P$ can also be written as

Itinerant Electron Spin Polarization (fixed magnetic layer)

$$P = \mu_B \frac{T_\uparrow - T_\downarrow}{T_\uparrow + T_\downarrow} \tag{6.42}$$

Table 6.1 lists some measured spin polarization values for bulk (thick) magnetic metals.[7]

[7]Taken from Meservey and P.M. Tedrow 1994. See end of chapter references for more details.

Table 6.1: Measured itinerant spin polarizations P for some magnetic materials. From Table 2, pg. 204 of R. Meservey and P.M. Tedrow 1994 [8].

Material	bulk spin polarization P/μ_B
Fe	0.40
Co	0.35
Ni	0.23
Gd	0.14
Ho	0.075
Dy	0.070
Tb	0.0655
Er	0.055
Tm	0.027

In determining the itinerant spin polarization, we also find that *the post-adjacent layer to the magnetic layer can also influence the extent of the spin polarization*. It turns out that a magnetic layer and a carefully chosen post-adjacent layer leads to a number of useful applications for asymmetric spin transport. One common application is known as *magnetoresistance*. Let us discuss this topic in the next section.

6.2 Introduction To Magnetoresistance

In order to introduce the subject of magnetoresistance in a way that the reader can understand, we are going to approach it by examining the equations that lead to the effect. This will reveal how it arises in certain structures, and you will understand the fundamentals of how it can be calculated. However, the equations are treated in the most simple way so that the reader can also carry out the same analysis. Even in this approach, it is nontrivial. So, let's discuss where this effect comes from, and examine its effects, as well as applications.

With *asymmetric* transport of electron-spins in a magnetic layer, the distinct wavenumbers $k_\uparrow$ and $k_\downarrow$ for spin-up and spin-down gives rise to 2 unequal spin currents among a single ensemble of charge carriers. Changes in either spin-current can lead to changes in the electrical

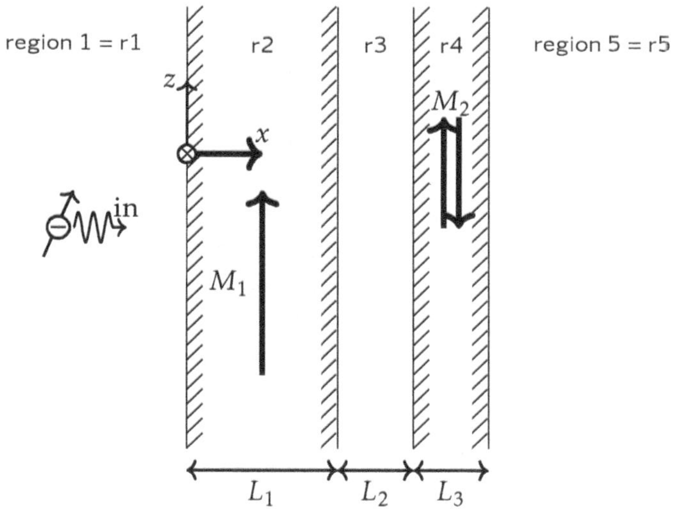

Figure 6.6: Illustration of a magnetoresistance (MR) device with two magnetic layers, a nonmagnetic layer between them, and two external contacts.

properties such as the resistance/conductivity. This effect can be used in a real device to provide information about, for example, a second magnetic layers in a system.

To analyze a device that can reveal these magneto-electrical changes, we can extend the ideas discussed in the polarization of itinerant electron spins by now considering *two magnetic layers sandwiching a nonmagnetic layer*, as illustrated in Fig. 6.6. In the diagram, the right-most magnetic layer, specifically, is designed so that the magnetization $\mathbf{M}_2$ can be changed by external influence, for example, by an external magnetic field $\mathbf{B}_{\text{ext}}$. In the diagram, there are *five layers* that must be taken into consideration. They include the *electrical contacts* at the left/right extremities of the device (regions 1 and 5). They are nonmagnetic. Howevever, between the electrical contacts, there are *three* additional layers forming a magnetic|nonmagnetic|magnetic multi-layer structure. Like the contacts, the middle-most layer is also nonmagnetic. When this layer is a normal metal, *e.g.* copper (Cu), the device is known as a *spin-valve*. However, if the middle layer is an *insulator*, the device is known as a *tunnel junction*.

To analyze this structure with the aim of revealing magnetoresitance, we must apply boundary conditions between all the layers/regions and utilize the appropriate wave functions in each respective region. As discussed in Chap. 5 with transport phenomena, the simplest general form for a wave function describing a quasi-free conducting electron is given by

$$\psi_x = Ae^{ikx} + Be^{-ikx} \tag{6.43}$$

where the coefficients A and B are independent of x. The first term is the *forward* propagating wave component, having $+ikx$ as an exponent. The second term is the backward or reflected wave component, having $-ikx$ as an exponent. Only one of the contacts, namely, at the exit (region 5), has no reflecting wave component as it is assumed there is no subsequent layer nearby to provide a scattering interface. Then, for the five layers, the respective wave functions become

$$\psi_{x,1} = Ae^{ikx} + Be^{-ikx} \quad \text{(layer 1-contact)} \tag{6.44a}$$

$$\psi_{x,2} = Ce^{ikx} + De^{-ikx} \quad \text{(layer 2-magnet)} \tag{6.44b}$$

$$\psi_{x,3} = Ee^{ikx} + Fe^{-ikx} \quad \text{(layer 3-nonmagnetic)} \tag{6.44c}$$

$$\psi_{x,4} = Ge^{ikx} + He^{-ikx} \quad \text{(layer 4-magnet)} \tag{6.44d}$$

$$\psi_{x,5} = Ie^{ikx} \quad \text{(layer 5-contact)} \tag{6.44e}$$

The objective is to use the wave functions to determine the transmission probability T of the *whole* device structure shown in Fig. 6.6. The observed value of T is determined at the exit in region 5 using the wave function in the contact, or $\psi_{x,5}$. We can set corresponding wave functions *as well as* their first derivatives equal at the interfaces between adjacent layers. From Fig. 6.6, the positions where the boundary conditions must be applied are at $x = 0, L_1, L_1 + L_2$, and $L_1 + L_2 + L_3$ (there are four interfaces for five layers). In the above equations, there are *nine unknown amplitudes* A, B, C, D, E, F, G, H and I. However, as mentioned, we can only express eight equations at four interfaces.[8]. To ensure a unique solution, we can divide all nine equations by the incoming wave amplitude A. This reduces the number of unknowns from nine to eight. When this is done, the eight boundary condition equations, setting wave functions and their derivatives equal, become

$$1 + B' = C' + D' \tag{6.45a}$$

$$1 - B' = \frac{k_2}{k_1}[C' - D'] \tag{6.45b}$$

[8]Eight equations for nine unknowns

$$C'e^{ik_2L_1} + D'e^{-ik_2L_1} = E'e^{ik_3L_1} + F'e^{-ik_3L_1} \tag{6.45c}$$

$$C'e^{ik_2L_1} - D'e^{-ik_2L_1} = \frac{k_3}{k_2}\left[E'e^{ik_3L_1} - F'e^{-ik_3L_1}\right] \tag{6.45d}$$

$$E'e^{ik_3(L_1+L_2)} + F'e^{-ik_3(L_1+L_2)} = G'e^{ik_4(L_1+L_2)} + H'e^{-ik_4L_1} \tag{6.45e}$$

$$E'e^{ik_3(L_1+L_2)} - F'e^{-ik_3(L_1+L_2)} = \frac{k_4}{k_3}\left[G'e^{ik_4(L_1+L_2)} - H'e^{-ik_4(L_1+L_2)}\right] \tag{6.45f}$$

$$G'e^{ik_4(L_1+L_2+L_3)} + H'e^{-ik_4(L_1+L_2+L_3)} = I'e^{ik_5(L_1+L_2+L_3)} \tag{6.45g}$$

$$G'e^{ik_4(L_1+L_2+L_3)} - H'e^{-ik_4(L_1+L_2+L_3)} = \frac{k_5}{k_4}I'e^{ik_4(L_1+L_2)} \tag{6.45h}$$

The prime symbol $'$ denotes the amplitude has been divided by the incoming wave function amplitude $A \neq 0$. Though exponential terms are present, the equations are still *linear* in the normalized amplitudes. To solve them, we can take certain combinations of Eqs. 6.45 to obtain the equations/solutions for the amplitudes $B', C', D', E', F', G', H'$, and ultimately I'. Specifically, we can use the following combinations:

$$\begin{aligned}
(6.45g) + (6.45h) &\Rightarrow \quad G'(I') \\
(6.45g) - (6.45h) &\Rightarrow \quad H'(I') \\
(6.45e) + (6.45f) &\Rightarrow \quad E'(I') \\
(6.45e) - (6.45f) &\Rightarrow \quad F'(I') \\
(6.45c) + (6.45d) &\Rightarrow \quad C'(I') \\
(6.45c) - (6.45d) &\Rightarrow \quad D'(I') \\
(6.45a) - (6.45b) &\Rightarrow \quad B'(I') \\
(6.45a) + (6.45b) &\Rightarrow \quad I'
\end{aligned}$$

Calculating the above combinations, the relative amplitudes are found to be

$$G' = \frac{1}{2}\left(1 + \frac{k_5}{k_4}\right)I'e^{ik_5(L_1+L_2+L_3)}e^{-ik_4(L_1+L_2+L_3)} \tag{6.46a}$$

$$H' = \frac{1}{2}\left(1 - \frac{k_5}{k_4}\right)I'e^{ik_5(L_1+L_2+L_3)}e^{ik_4(L_1+L_2+L_3)} \tag{6.46b}$$

$$E' = \frac{1}{2}\left[\left(1+\frac{k_4}{k_3}\right)G'e^{ik_4(L_1+L_2)}+\left(1-\frac{k_4}{k_3}\right)H'e^{-ik_4(L_1+L_2)}\right]e^{-ik_3(L_1+L_2)} \tag{6.46c}$$

$$F' = \frac{1}{2}\left[\left(1-\frac{k_4}{k_3}\right)G'e^{ik_4(L_1+L_2)}+\left(1+\frac{k_4}{k_3}\right)H'e^{-ik_4(L_1+L_2)}\right]e^{ik_3(L_1+L_2)} \tag{6.46d}$$

$$C' = \frac{1}{2}\left[\left(1+\frac{k_3}{k_2}\right)E'e^{ik_3L_1}+\left(1-\frac{k_3}{k_2}\right)F'e^{-ik_3L_1}\right]e^{-ik_2L_1} \tag{6.46e}$$

$$D' = \frac{1}{2}\left[\left(1-\frac{k_3}{k_2}\right)E'e^{ik_3L_1}+\left(1+\frac{k_3}{k_2}\right)F'e^{-ik_3L_1}\right]e^{ik_2L_1} \tag{6.46f}$$

$$B' = \frac{1}{2}\left[\left(1-\frac{k_2}{k_1}\right)C'+\left(1+\frac{k_2}{k_1}\right)D'\right] \tag{6.46g}$$

Starting with Eqs. (6.46a) and (6.46b), since G' and H' are proportional to I', finding their relation solely to I', then, substituting into subsequent equations leads to the other amplitudes E', F', C', D', and B' as functions of I'. Then, any pair *e.g.* Eqs. (6.46a) and (6.46b), can be used to determine I', which can be used to determine the transmission probability T given by

$$T = \hbar k_5|I'|^2/m_0 \quad \text{(in contact/region 5)} \tag{6.47}$$

Since there are *two* unique spin states, there are also *two transmission probabilities* corresponding to spin-up (↑) and spin-down (↓). Resolving Eq. (6.47) down to the spin states gives

$$T_\uparrow = \hbar k_5|I'_\uparrow|^2/m_0 \quad \text{and} \quad T_\downarrow = \hbar k_5|I'_\downarrow|^2/m_0 \tag{6.48}$$

Although I' is the relative amplitude of the wave function in a nonmagnetic layer, because it depends on the amplitudes of all the previous layers, *there is propagation of the spin splitting behavior* into region 5. This is probably the simplest calculation that can be done that allows visibility of such subtle behavior. Because of this propagation, the amplitude I' have a corresponding $I'_\uparrow$ and $I'_\downarrow$. Consequently, in the contact region $r5$, there are also spin-up and spin-down wave functions $\psi_{x,5,\uparrow} = I'_\uparrow e^{ik_5x}$ and $\psi_{x,5,\downarrow} = I'_\downarrow e^{ik_5x}$, respectively.

In an experiment with a 5-layered device like that shown in Fig. 6.6, an output voltage and/or current can be measured between the metallic contacts of regions 1 and 5. If a *fixed* external voltage is applied to the device, then the total *charge+spin* current (proportional to the current density) across the device can be calculated as the expectation value of the *total* current density $\langle J \rangle$ through the device as a function of the free magnetic layer (region 4) orientations. The expectation value of the *total current density* through the device is given by

$$\langle J \rangle = \frac{J_\uparrow |\psi'_{x,5,\uparrow}|^2 + J_\downarrow |\psi'_{x,5,\downarrow}|^2}{|\psi'_{x,5,\uparrow}|^2 + |\psi'_{x,5,\downarrow}|^2} = \frac{J_\uparrow |I'_\uparrow|^2 + J_\downarrow |I'_\downarrow|^2}{|I'_\uparrow|^2 + |I'_\downarrow|^2} \tag{6.49}$$

The prime used alongside the wave functions in Eq. (6.49) denotes the use of the normalized coefficients, rather than the absolute. Because we have a ratio, the quantities using either form are identical. The expectation value of the current density $\langle J \rangle$, which used in the *magnetoresistance*, can be computed for *two* particular cases of interest: (1) *parallel* magnetization ($\mathbf{M}_2 \parallel +\mathbf{M}_1$) $\rightarrow \langle J_p \rangle$ and (2) *anti-parallel* magnetization ($\mathbf{M}_2 \parallel -\mathbf{M}_1$) $\rightarrow \langle J_{ap} \rangle$. The two current densities can then be used to define a *dimensionless magnetoresistance fraction* as

$$\text{MR Fraction} = \frac{\langle J_p \rangle - \langle J_{ap} \rangle}{\langle J_{ap} \rangle} = \frac{G_p - G_{ap}}{G_{ap}} \tag{6.50}$$

G is the bulk conductance of the device. In expressing the MR, the convention is to use the relation $G = 1/R$, where R is the corresponding resistance, so the *MR Fraction* becomes

MR Fraction

$$\text{MR Fraction} = \frac{\langle R_{ap} \rangle - \langle R_p \rangle}{\langle R_p \rangle} \tag{6.51}$$

It is often multiplied by 100 to give what is known as the *magnetoresistance ratio* (MR ratio) defined as

MR Ratio

$$\text{MR} = \frac{\langle R_{ap} \rangle - \langle R_p \rangle}{\langle R_p \rangle} \times 100 \tag{6.52}$$

In Subsec. 6.1.3, when we analyzed the polarization of itinerant spins passing through a magnetic layer, we observed that the subsequent layer (just after the polarizing magnetic layer) can have a significant effect on the resulting spin polarization. This effect can be explored in the 5-layered structure, as well, by changing the potential energy V_3 in the layer. This parameter affects the wave number in this layer given by

$$k_3^2 = \frac{2m_0(E_\lambda - V_3)}{\hbar^2} \Rightarrow k_3$$

E_λ is the total energy eigenvalue in region 3 or r3. Fig. 6.7 plots the calculated *magnetoresistance fraction* (c.Eq. 6.50) as a function of V_3/E_λ. From Fig. 6.7, it can be seen that when the potential energy in region 3

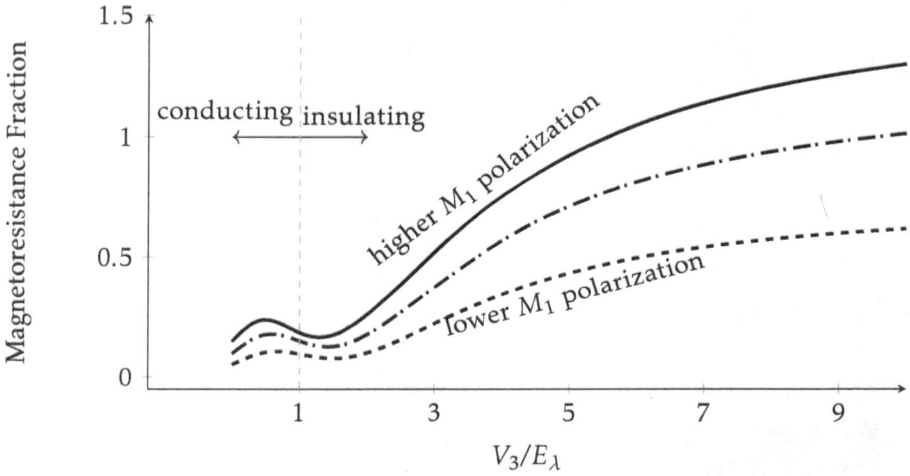

Figure 6.7: Calculated magnetoresistance (MR) fraction as a function of the potential energy of layer/region 3. The value spans from conducting conditions ($V_3 < E_\lambda$) to insulator conditions ($V_3 > E_\lambda$). The dashed gray line at $V_3/E_\lambda = 1$ marks the transition from conducting to insulating. The curves labeled 'higher' and 'lower polarization' respectively have higher and lower values for B_{eff} in the first magnet (region 2).

V_3 exceeds the incoming conduction energy E_λ, there is a transition of sorts accompanied by the GMR generally becoming significantly larger, even exceeding the fractional value of 1.0 or 100%. This corresponds to a transition between conducting and insulating. When $V_3 > E_\lambda$, the wavenumber in this region becomes

$$k_3^2 = \frac{2m_0(E_\lambda - V_3)}{\hbar^2} \Rightarrow k_3 = i\sqrt{2m_0(E_\lambda - V_3)}/\hbar$$

Since the wave function is $\psi_{x,3} \propto e^{ik_3x}$, an *evanescent wave* condition is created. Whenever transport takes place across a region where evanescent wave conditions exist, this type of transport is distinct from conventional electronic transport, and is known as *tunneling*. For this reason, when a magneto-resistance device utilizes a material between the two magnets that is an insulating barrier, the magneto-resistance is called *tunneling magneto-resistance* (TMR).

The thickness L_2 of the spacer layer also affects spin-transport. Fig. 6.8 illustrates the *magnetoresistance fraction* (c.Eq. 6.50), as a function of region three layer thickness L_2. One of the first insulating (barrier) mate-

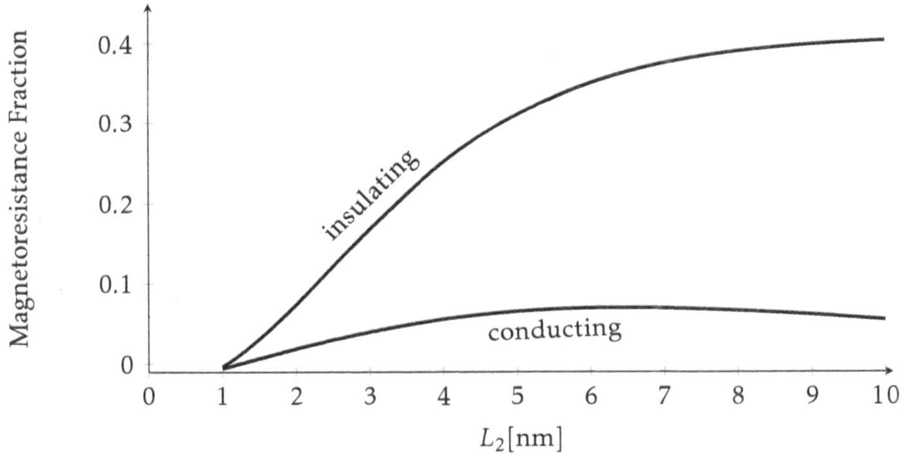

Figure 6.8: Computed magnetoresistance in an MR device as a function of region 3 thickness (material between magnetic layers), labeled as L_2. Additional parameters used: $L_1 = L_3 = 3$nm; $E_\lambda = 1.5meV$.

rials investigated in MTJs was alumina or aluminum oxide (Al_2O_3). This compound is abundant and very stable. This is one of the reasons it is very widely used in many applications. This property, however, becomes somewhat of a detriment in this application, because its conduction properties do not alter easily.

The most effective insulating barrier used in an MTJ is magnesia or magnesium oxide MgO. Just as in the case of Al_2O_3, it is also an insulator, so there is an insulating gap between the *valence* and *conduction* bands. This imposes an effectively large potential energy that must be overcome by any valence band electrons. However, MgO readily undergoes a unique modification to its valence shell due to an interplay between magnesium and oxygen. The valence electrons for Mg ordinarily are $3s^2$, while *singly* oxygen has valence electrons that are $2p^4$ (see Table 6.4). In MgO, the two electrons in Mg can lower their energy by occupying a 2p orbital available in oxygen (known as charge transfer). Consequently, oxygen places a single spin-down from 2p into 3s. As a single spin in 3s, the spin-down conduction is enhanced, while spin-up scatters more in the barrier due to the larger gap. This leads to relatively larger amounts of spin-down electrons tunneling through MgO, and even greater disparity between the spin-up and down-electrons. In Al_2O_3, the combining effect is very different. As electrons transfer to oxygen, the $p-$ orbitals only fill more, without placing singly electrons in 3s. Because of this difference, when MgO is sandwiched between two magnetic layers, the TMR is enhanced further in comparison.

In order to calculate the MR, the *boundary conditions are necessary*. This is especially true for thin film devices because the device layers involved can be very thin, *e.g.* $\approx 1nm$. Consequently, the interfaces can play a significant role in the behavior. Their chemical makeup and the quality of processing during their fabrication, thus, become important to understand. In TMR devices, similar spin-splitting effects can occur at the interface. For example, if the magnetic layers are Fe(Co), then Fe-O (Co-O) tends to form at the interfaces with an oxide barrier. Fe has two valence electrons in the 4s shell (Mg has two in 3s), thus, very similar conditions are created at the interfaces (see Table 6.4). Ultimately, it is the combination of interface and volume/bulk MgO properties that leads to enhanced TMR ratios. A detailed analysis of the enhanced attributes of MgO were first predicted *c.*2000 by William Butler *et. al* while at Oak Ridge National Laboratory in Tennessee, USA. These were revealed by a numerical study of the transport properties in a magnetic tunnel junction having the structure Fe|MgO|Fe. Experimental demonstrations with MgO verifying this fact with MgO followed in *c.*2004. For a specific barrier material, taking account of the interfaces require a relatively complex calculation. Thus, keep in mind that only the simplest formulation possible for an insulating nonmagnetic layer is illustrated here.

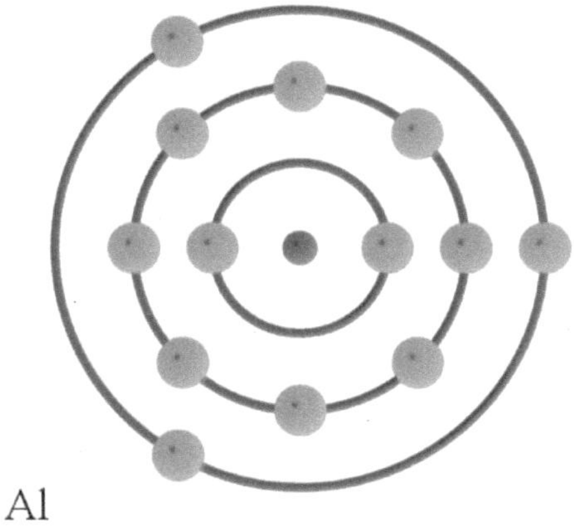

(a) Aluminum

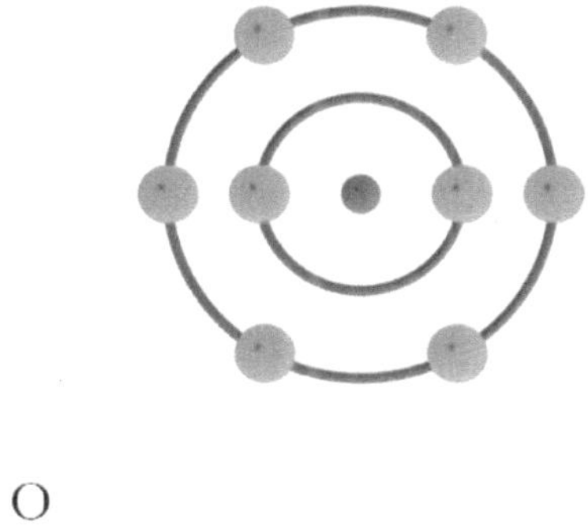

(b) Oxygen

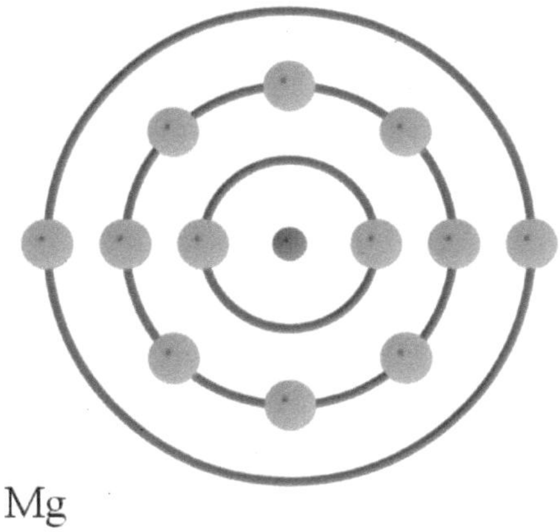

(c) Magnesium

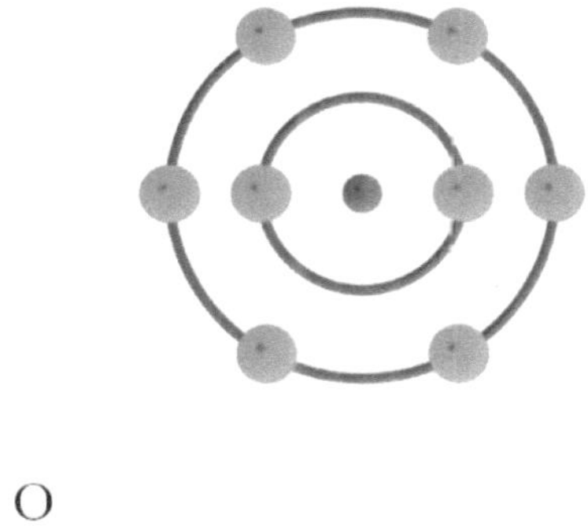

(d) Oxygen

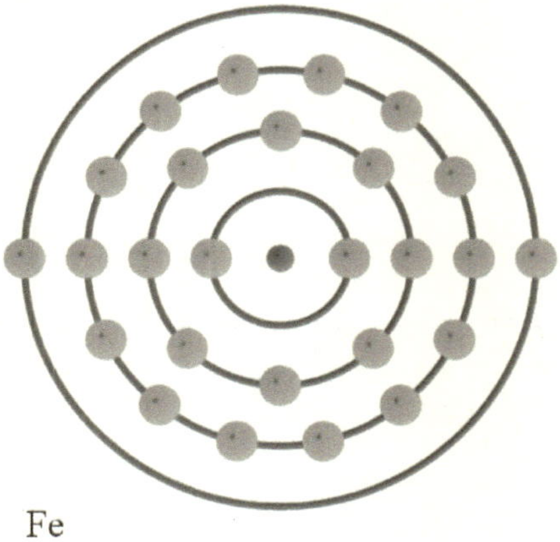

(e) Iron

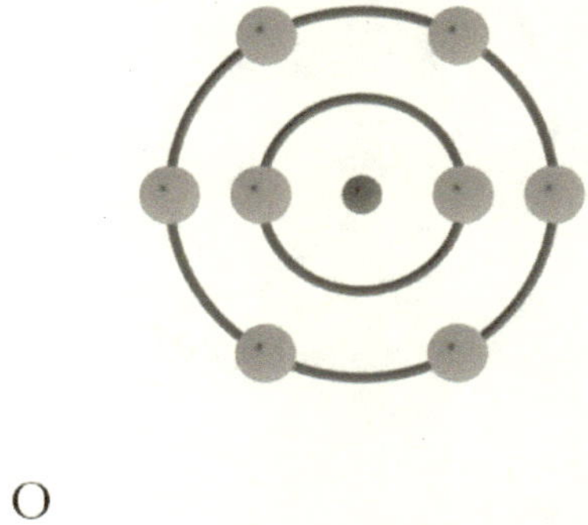

(f) Oxygen

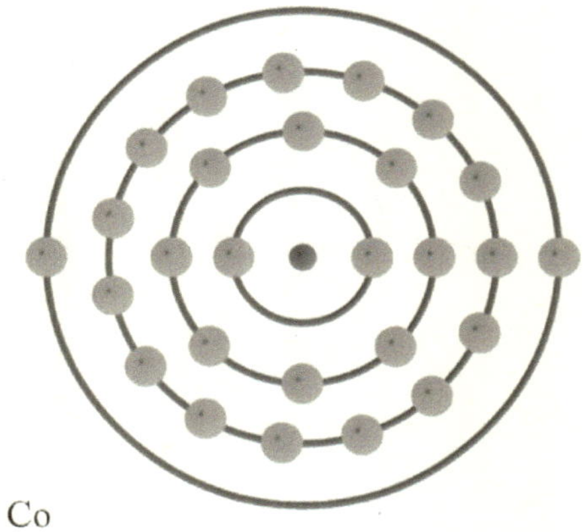

(g) Cobalt

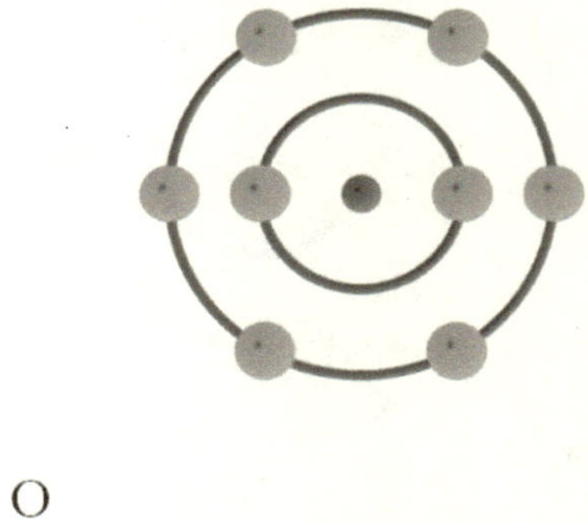

(h) Oxygen

Table 6.3: Atoms of the elements typical of magnetic tunnel junctions. In the left column, notice the similarities between all but Aluminum (Al). Except for Al, the electrons from filled s-orbitals move to oxygen to create singly spin-down electrons in an s-orbital.

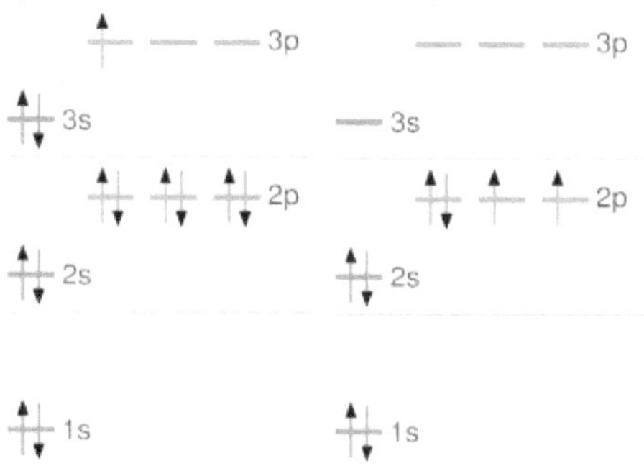

(a) Aluminum (left) Oxygen (right)

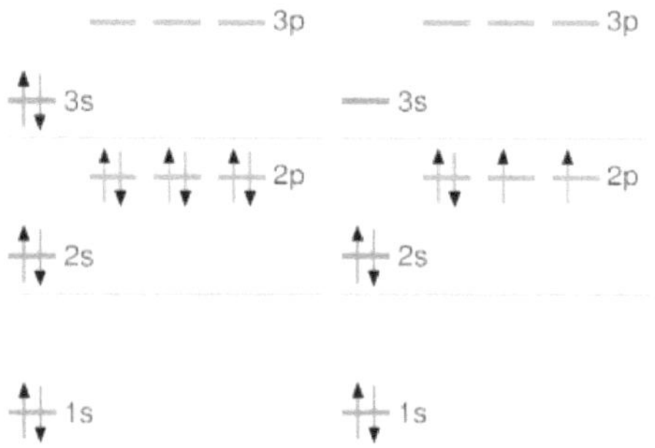

(b) Magnesium (left) Oxygen (right)

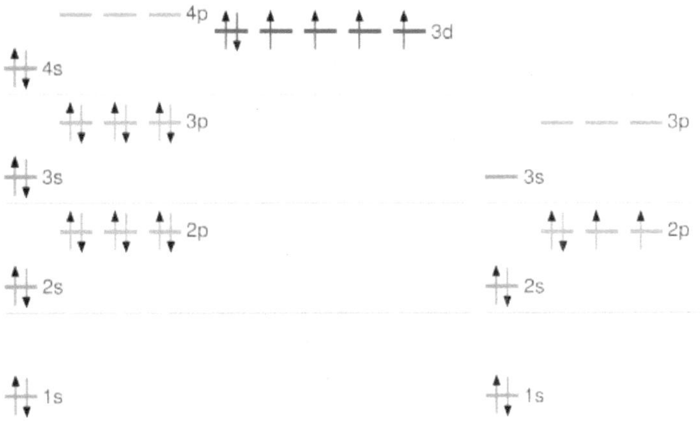

(c) Iron (left) Oxygen (right)

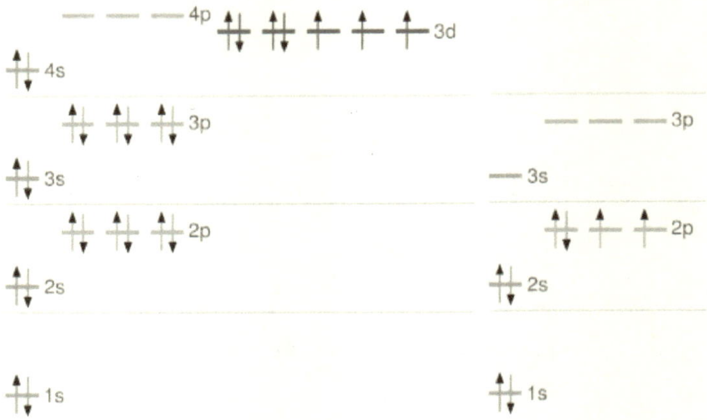

(d) Cobalt (left) Oxygen (right)

Table 6.4: Corresponding electron configuration for some typical elements of a magnetic tunnel junctions. Except in Al, the electrons from filled s-orbitals in the metal move to oxygen to create singly spin-down electrons in an s-orbital.

Summarizing, we find that the spin-splitting can lead to observable changes in the electrical transport properties. In a structure with two magnetic layers, when the second magnetic layer changes its magnetization, it shows up in the electrical current, which is composed of both the spin-up and spin-down currents. We, again, found that when an insulating barrier is used (between both magnets), the effect can be further enhanced.

For what kind of practical device application might this be useful? There are several, but let's discuss one of the more widely used applications, which uses TMR/GMR to measure external magnetic fields.

6.2.1 Magnetoresistance For External Field Measurements

Fig. 6.9 illustrates a MR device as used in an application for external magnetic field measurements. In this application, the field must *not* rotate or affect M_1. In practice, this is achieved by designing M_1 to

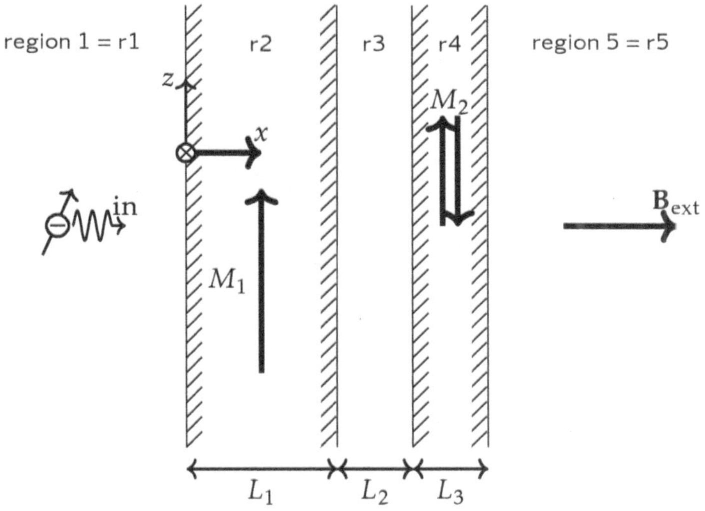

Figure 6.9: Illustration of a magnetoresistance (MR) device with two magnetic layers, exposed to an external magnetic field $\mathbf{B}_{ext}$. The field rotates M_2, reducing the net polarization along the $+z$ axis.

have sufficiently large uniaxial anisotropy K_u. It should only rotate M_2. Because the field is not oriented along the $+z$ direction, it will tend to rotate the polarization of M_2 *away* from the $+z$ direction. This leads to a *diminution* of the component of M_2 along this direction. As you might have guessed, this affects the spin currents created by M_1.

For the calculations discussed here, they are carried out the same way as the calculations of the magnetoresistance above, using Eq. (6.51). However, here, we are concerned with effects due to changes in the polarization M_2. The polarization is contained in the effective magnetic field B_{eff} in that layer. Fig. 6.10 illustrates the calculated magnetoresistance fraction for a device like that shown in Fig. 6.9, as a function of the polarization strength in the second magnet *i.e.* $\mathbf{M}_2$. Three curves are shown, where each one corresponds to different M_1 polarization strengths along the $+z$ axis. Because the energy is axis-sensitive because the spins are polarized, for example, along the $+z$ or $-z$ polarization direction. Fig. 6.10 shows that as the $+z$ component of polarization decreases, so does the TMR. Though it is not exactly linear, for a small enough range, the MR has an approximately linear response, going right through the origin so that no response is observed when there is no z component of polarization in the second magnet M_2.

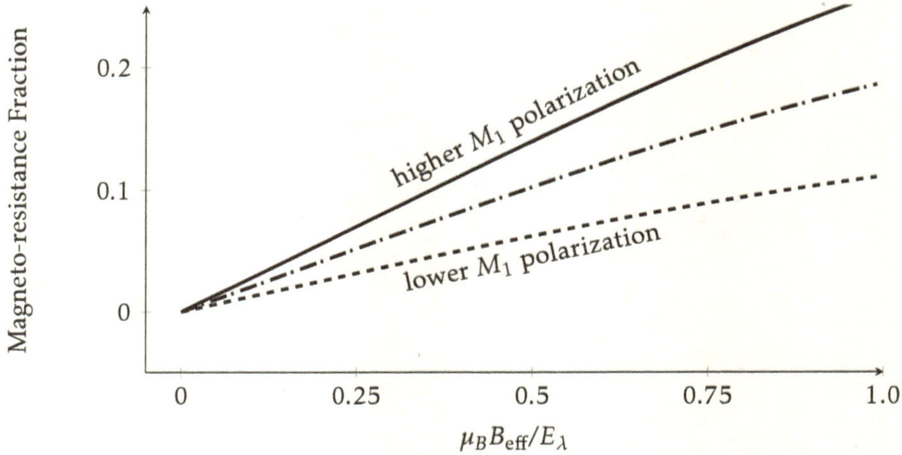

Figure 6.10: Calculated magnetoresistance (MR) fraction as a function of magnetic polarization of layer/region 4 in insulator conditions ($V_3 > E_\lambda$). The curves labeled 'higher' and 'lower polarization' respectively have higher and lower values for B_{eff} in the first magnet (region 2) *i.e.* $\mathbf{M}_1$. Additional parameters used: $L_1 = L_3 = 2$nm;$L_2 = 1.5$nm; $E_\lambda = 1.5meV$.

Let's take this one step further. If the second magnetization layer is reoriented *away* from the $+z$ direction by an external magnetic field $\mathbf{B}_{ext}$, the result will be to diminish the net polarization M_2 along the $+z$ direction. Then, one would observe an effect like that shown in Fig. 6.10. Thus, it can be used for the detection of an external magnetic field in any direction that changes the spin polarization direction of $\mathbf{M}_2$ away from that of M_1. In an experiment, one can apply an external magnetic field to vary the orientation of M_2, and measure the MR as a function of the external field along the $+x$ direction. This experiment can be simulated, as well.

The component of equilibrium magnetic moment along x can be approximated using the *Brillouin function* as [9]

$$M_{x,2} = B_J(\beta) = \frac{1}{J}\left[\left(J + \frac{1}{2}\right)\coth\left(\frac{2J+1}{2J}\cdot\beta\right) - \frac{1}{2}\coth\left(\frac{1}{2J}\cdot\beta\right)\right] \quad (6.53)$$

The parameter $\beta = V_4/k_B T$ (energy of M_2 layer/region 4) where k_B is Boltzmann's constant and T is the temperature. J is the total angular momentum quantum number. This value is known for many materials.

[9] For a detailed discussion and derivation of the Brillouin function, see Sec. 10.12 of Quantum Mechanics With A SPIN On It (Eason)

For our calculations, we'll use the value for iron (Fe) $J = 5/2$. Then, the normalized component of magnetic moment along the z axis can be approximated as

$$M_{z,2} = \sqrt{1 - M_{x,2}^2} \tag{6.54}$$

Fig. 6.11 shows examples of the kind of experimental curves one obtains when measuring MR as a function of B_{ext} which is along the $+x$ direction.

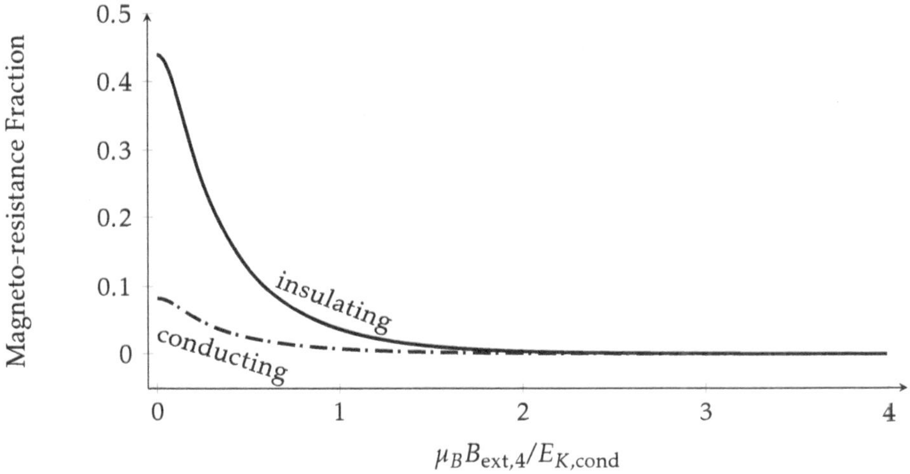

Figure 6.11: Calculated magnetoresistance (MR) fraction as a function of the externally applied field along $+x$. Additional parameters used: $L_1 = L_3 =$ 2nm;$L_2 = 1.5$nm; $E_{K,\text{cond}} = 10.5meV$, $T = 13K$. The ratio is maximum when there is no field as the layer polarization is along $+z$. As the field increases, the $+z$ component is reduced due to the rotation towards $+x$.

GMR was discovered experimentally *c.*1988 by Albert Fert and Peter Grünberg. Fig. 6.12 shows experimental data formatted from one of their historic publications. The MR fraction is measured as a function of the external field which is perpendicular to the films/interfaces. Note that the device structure used in the experiments was a so-called *superlattice structure* (repeats of *e.g.* Fe|Cr) with in-plane magnetization ($+z$) and an external field perpendicular to the superlattice layers, *i.e.* consistent with Fig. 6.9.

The behavior illustrated in Fig. 6.9 is experimentally demonstrated in Fig. 6.12. There is, generally, a reduction in the magnetoresistance as

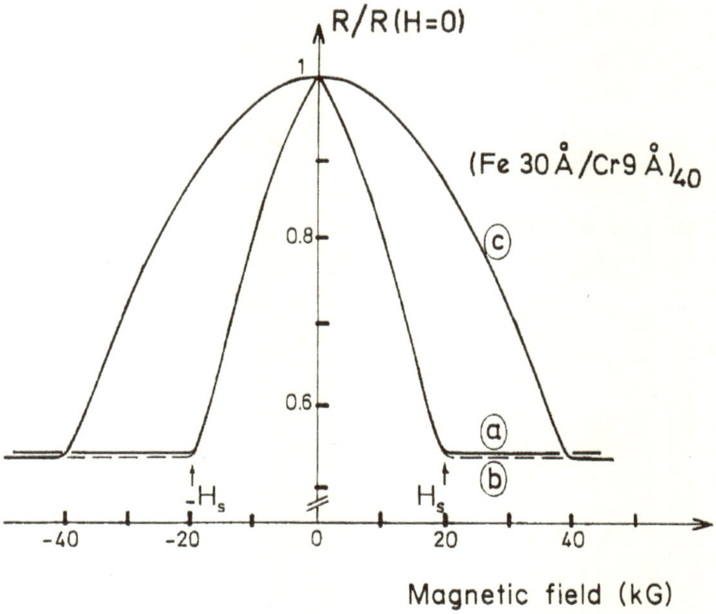

Figure 6.12: Experimental magnetoresistance plots from A. Fert's team in Orsay in 1988, constituting the discovery of the giant magnetoresistance effect. Curve c corresponds to a perpendicular magnetic field B_{ext}. They used a superlattice structure (repeats of *e.g.* Fe|Cr) with in- plane magnetization (+z) and an external field perpendicular to the superlattice layers.[Phys. Rev. Lett., Vol. 61, No. 21, November 1988]

the magnetic field along +x increases. Note that curve c corresponds to an external field perpendicular to the film. The other curves correspond to rotations in the plane of the film. Beyond the saturation along +x, no more difference between the two spin-currents can be observed. Since the field is rotating the M_2, it leads to a reduction or change in the device's polarization component along +z. This effect is used in magnetic field sensors for a variety of applications from hard disk drives to GPS systems.

For general geometries as well as materials, it is useful to have a set of differential equations describing the physics of asymmetric spin transport and magnetoresistance. Also, it can be generalized to more types of spin-polarizing materials than magnets. For example, in materials with appreciable spin-orbit interaction (see Sec. 2.5), there is also coupling to the spin degree of freedom. By including the fourth dimension, the corresponding wave functions can be determined and the resulting quantum mechanical spin-transport equations can be formulated. The resulting 4D equations will be applicable to any system providing a complete description in space as well as time. Let's find these differential equations using the 4D framework.

6.3 Spin Diffusion Equations

Spin transport and diffusion equations can be obtained directly from the quantum mechanical continuity equation. In Sec. 4.2, a scalar form of the *continuity equation* was derived for the case of a *non-Hermitian* operator. However, since spin energy operators are *not* scalars, but 2×2 operators, we must generalize the result further so that is is useful for 2×2 operators. So, let the system be described by the following 4D energy operator:

$$\widehat{H}^4 = \left[\widehat{H}_\sigma - i\hbar\frac{\partial}{\partial t}\right]\psi = (iE_I)\psi \tag{6.55}$$

The Hermitian Hamiltonian contribution is given by

$$\widehat{H}_\sigma = -\frac{\hbar^2}{2m_0}\nabla^2 + V_\sigma \tag{6.56}$$

Substituting Eq. (6.56) into Eq. (6.55), and introducing $E_I = \alpha E_R$, leads to

$$i\hbar\frac{\partial}{\partial t}\begin{bmatrix}\psi_\uparrow \\ \psi_\downarrow\end{bmatrix} = \left[-\frac{\hbar^2}{2m_0}\nabla^2 + (1 - i\alpha)V_\sigma\right]\begin{bmatrix}\psi_\uparrow \\ \psi_\downarrow\end{bmatrix} \tag{6.57a}$$

The *conjugate-transpose* (or *dagger*) of Eq. (6.57a) becomes

$$-i\hbar\frac{\partial}{\partial t}\begin{bmatrix}\psi_\uparrow^* & \psi_\downarrow^*\end{bmatrix} = \left[-\frac{\hbar^2}{2m_0}\nabla^2 + (1 + i\alpha)V_\sigma\right]\begin{bmatrix}\psi_\uparrow^* & \psi_\downarrow^*\end{bmatrix} \tag{6.57b}$$

Let us define the 2×2 *spin density matrix* Υ as

$$\Upsilon = \psi\psi^\dagger = \begin{bmatrix}\psi_\uparrow \\ \psi_\downarrow\end{bmatrix}\begin{bmatrix}\psi_\uparrow^* & \psi_\downarrow^*\end{bmatrix} \tag{6.58}$$

Evaluating the spinor product, Υ leads to

2 × 2 Spin Density Matrix

$$\Upsilon = \begin{bmatrix} \psi_\uparrow \psi_\uparrow^* & \psi_\uparrow \psi_\downarrow^* \\ \psi_\downarrow \psi_\uparrow^* & \psi_\downarrow \psi_\downarrow^* \end{bmatrix} \tag{6.59}$$

Right-multiplying Eq. (6.57a) by $\psi^\dagger = \begin{bmatrix} \psi_\uparrow^* & \psi_\downarrow^* \end{bmatrix}$, we have

$$i\hbar \frac{\partial}{\partial t} \begin{bmatrix} \psi_\uparrow \\ \psi_\downarrow \end{bmatrix} \begin{bmatrix} \psi_\uparrow^* & \psi_\downarrow^* \end{bmatrix} = \left[-\frac{\hbar^2}{2m_0} \nabla^2 + (1 - i\alpha) V_\sigma \right] \begin{bmatrix} \psi_\uparrow \\ \psi_\downarrow \end{bmatrix} \begin{bmatrix} \psi_\uparrow^* & \psi_\downarrow^* \end{bmatrix} \tag{6.60}$$

Likewise, if we *left*-multiply Eq. (6.57b) by $\psi = \begin{bmatrix} \psi_\uparrow & \psi_\downarrow \end{bmatrix}^T$, we have

$$-i\hbar \begin{bmatrix} \psi_\uparrow \\ \psi_\downarrow \end{bmatrix} \frac{\partial}{\partial t} \begin{bmatrix} \psi_\uparrow^* & \psi_\downarrow^* \end{bmatrix} = \begin{bmatrix} \psi_\uparrow \\ \psi_\downarrow \end{bmatrix} \left[-\frac{\hbar^2}{2m_0} \nabla^2 + (1 + i\alpha) V_\sigma \right] \begin{bmatrix} \psi_\uparrow^* & \psi_\downarrow^* \end{bmatrix} \tag{6.61}$$

Taking the *difference* between Eqs. (6.60) and (6.61) gives

$$i\hbar \left(\frac{\partial}{\partial t} \begin{bmatrix} \psi_\uparrow \\ \psi_\downarrow \end{bmatrix} \begin{bmatrix} \psi_\uparrow^* & \psi_\downarrow^* \end{bmatrix} + \begin{bmatrix} \psi_\uparrow \\ \psi_\downarrow \end{bmatrix} \frac{\partial}{\partial t} \begin{bmatrix} \psi_\uparrow^* & \psi_\downarrow^* \end{bmatrix} \right) = \cdots$$

$$-\frac{\hbar^2}{2m_0} \nabla^2 \begin{bmatrix} \psi_\uparrow \\ \psi_\downarrow \end{bmatrix} \begin{bmatrix} \psi_\uparrow^* & \psi_\downarrow^* \end{bmatrix} + \begin{bmatrix} \psi_\uparrow \\ \psi_\downarrow \end{bmatrix} \frac{\hbar^2}{2m_0} \nabla^2 \begin{bmatrix} \psi_\uparrow^* & \psi_\downarrow^* \end{bmatrix} - i2\alpha V_\sigma \Upsilon \tag{6.62}$$

Using the definition of the *spin density matrix* in Eq. (6.58), the LHS of Eq. (6.62) can be written as

$$i\hbar \left(\frac{\partial}{\partial t} \begin{bmatrix} \psi_\uparrow \\ \psi_\downarrow \end{bmatrix} \begin{bmatrix} \psi_\uparrow^* & \psi_\downarrow^* \end{bmatrix} + \begin{bmatrix} \psi_\uparrow \\ \psi_\downarrow \end{bmatrix} \frac{\partial}{\partial t} \begin{bmatrix} \psi_\uparrow^* & \psi_\downarrow^* \end{bmatrix} \right) = i\hbar \frac{d\Upsilon}{dt} \tag{6.63}$$

Using this result, Eq. (6.64a) becomes

$$i\hbar \frac{d\Upsilon}{dt} = -\frac{\hbar^2}{2m_0} \left(\nabla^2 \begin{bmatrix} \psi_\uparrow \\ \psi_\downarrow \end{bmatrix} \begin{bmatrix} \psi_\uparrow^* & \psi_\downarrow^* \end{bmatrix} - \begin{bmatrix} \psi_\uparrow \\ \psi_\downarrow \end{bmatrix} \nabla^2 \begin{bmatrix} \psi_\uparrow^* & \psi_\downarrow^* \end{bmatrix} \right) - i2\alpha V_\sigma \Upsilon \tag{6.64a}$$

$$= -\frac{\hbar^2}{2m_0} \begin{bmatrix} e_{11} & e_{12} \\ e_{21} & e_{22} \end{bmatrix} - i2\alpha V_\sigma \Upsilon \tag{6.64b}$$

Next, we want to modify the RHS. The four elements of our matrix which are the differences between the two kinetic energy operator terms, are given by

$$e_{11} = \psi_\uparrow^* \nabla^2 \psi_\uparrow - \psi_\uparrow \nabla^2 \psi_\uparrow^* \tag{6.65a}$$

$$e_{12} = \psi_\downarrow^* \nabla^2 \psi_\uparrow - \psi_\uparrow \nabla^2 \psi_\downarrow^* \tag{6.65b}$$

$$e_{21} = \psi_\uparrow^* \nabla^2 \psi_\downarrow - \psi_\downarrow \nabla^2 \psi_\uparrow^* \tag{6.65c}$$

$$e_{22} = \psi_\downarrow^* \nabla^2 \psi_\downarrow - \psi_\downarrow \nabla^2 \psi_\downarrow^* \tag{6.65d}$$

The matrix elements e_{ij} can be rewritten by using the relation

$$\nabla \cdot (\psi \nabla \psi^*) = \psi \nabla \cdot (\nabla \psi^*) + \nabla \psi \cdot \nabla \psi^* \tag{6.66}$$

Rearranging, the Laplacian term we need is given by

$$\psi \nabla^2 \psi^* = \nabla \cdot (\psi \nabla \psi^*) - \nabla \psi \cdot \nabla \psi^* \tag{6.67}$$

Writing out all eight of the terms we'll need (2 for each difference) using Eq. (6.67), we have

$$\psi_\uparrow^* \nabla^2 \psi_\uparrow = \nabla \cdot (\psi_\uparrow^* \nabla \psi_\uparrow) + \nabla \psi_\uparrow^* \cdot \nabla \psi_\uparrow \qquad (e_{11}) \tag{6.68a}$$

$$\psi_\uparrow \nabla^2 \psi_\uparrow^* = \nabla \cdot (\psi_\uparrow \nabla \psi_\uparrow^*) + \nabla \psi_\uparrow \cdot \nabla \psi_\uparrow^* \qquad (e_{11}) \tag{6.68b}$$

$$\psi_\downarrow^* \nabla^2 \psi_\uparrow = \nabla \cdot (\psi_\downarrow^* \nabla \psi_\uparrow) + \nabla \psi_\downarrow^* \cdot \nabla \psi_\uparrow \qquad (e_{12}) \tag{6.68c}$$

$$\psi_\uparrow \nabla^2 \psi_\downarrow^* = \nabla \cdot (\psi_\uparrow \nabla \psi_\downarrow^*) + \nabla \psi_\uparrow \cdot \nabla \psi_\downarrow^* \qquad (e_{12}) \tag{6.68d}$$

$$\psi_\uparrow^* \nabla^2 \psi_\downarrow = \nabla \cdot (\psi_\uparrow^* \nabla \psi_\downarrow) + \nabla \psi_\uparrow^* \cdot \nabla \psi_\downarrow \qquad (e_{21}) \tag{6.68e}$$

$$\psi_\downarrow \nabla^2 \psi_\uparrow^* = \nabla \cdot (\psi_\downarrow \nabla \psi_\uparrow^*) + \nabla \psi_\downarrow \cdot \nabla \psi_\uparrow^* \qquad (e_{21}) \tag{6.68f}$$

$$\psi_\downarrow^* \nabla^2 \psi_\downarrow = \nabla \cdot (\psi_\downarrow^* \nabla \psi_\downarrow) + \nabla \psi_\downarrow^* \cdot \nabla \psi_\downarrow \qquad (e_{22}) \tag{6.68g}$$

$$\psi_\downarrow \nabla^2 \psi_\downarrow^* = \nabla \cdot (\psi_\downarrow \nabla \psi_\downarrow^*) + \nabla \psi_\downarrow \cdot \nabla \psi_\downarrow^* \qquad (e_{22}) \tag{6.68h}$$

Notice that terms belonging to the same element e_{ij} of the form $\nabla \psi_i \cdot \nabla \psi_j$ are equal. Because of this, they will cancel each other in taking the difference. Using $\psi \nabla \cdot (\nabla \psi^*) = \psi \psi^* \nabla^2$, the matrix elements then become

$$e_{11} = \nabla \cdot \left(\psi_\uparrow^* \nabla \psi_\uparrow\right) - \nabla \cdot \left(\psi_\uparrow \nabla \psi_\uparrow^*\right) = \nabla \cdot \left(\psi_\uparrow^* \nabla \psi_\uparrow - \psi_\uparrow \nabla \psi_\uparrow^*\right) \tag{6.69a}$$

$$e_{12} = \nabla \cdot \left(\psi_\downarrow^* \nabla \psi_\uparrow\right) - \nabla \cdot \left(\psi_\uparrow \nabla \psi_\downarrow^*\right) = \nabla \cdot \left(\psi_\downarrow^* \nabla \psi_\uparrow - \psi_\uparrow \nabla \psi_\downarrow^*\right) \tag{6.69b}$$

$$e_{21} = \nabla \cdot \left(\psi_\uparrow^* \nabla \psi_\downarrow\right) - \nabla \cdot \left(\psi_\downarrow \nabla \psi_\uparrow^*\right) = \nabla \cdot \left(\psi_\uparrow^* \nabla \psi_\downarrow - \psi_\downarrow \nabla \psi_\uparrow^*\right) \tag{6.69c}$$

$$e_{22} = \nabla \cdot \left(\psi_\downarrow^* \nabla \psi_\downarrow\right) - \nabla \cdot \left(\psi_\downarrow \nabla \psi_\downarrow^*\right) = \nabla \cdot \left(\psi_\downarrow^* \nabla \psi_\downarrow - \psi_\downarrow \nabla \psi_\downarrow^*\right) \tag{6.69d}$$

These are just alternative forms of the quantum mechanical current density, which is given by

$$J = \frac{i\hbar}{2m_0}\left(\psi\nabla\psi^* - \psi^*\nabla\psi\right) \tag{6.70}$$

Given the form of the elements e_{ij}, we can define a 2×2 *spin current density matrix* as

2 × 2 Spin-Current Density Matrix

$$[\mathrm{J}] = \frac{i\hbar}{2m_0}\begin{bmatrix} \psi_\uparrow\nabla\psi_\uparrow^* - \psi_\uparrow^*\nabla\psi_\uparrow & \psi_\uparrow\nabla\psi_\downarrow^* - \psi_\downarrow^*\nabla\psi_\uparrow \\ \psi_\downarrow\nabla\psi_\uparrow^* - \psi_\uparrow^*\nabla\psi_\downarrow & \psi_\downarrow\nabla\psi_\downarrow^* - \psi_\downarrow^*\nabla\psi_\downarrow \end{bmatrix} \tag{6.71}$$

There are *four* effective current densities varying in space and time, all contained in the spin-current density matrix, or

$$[\mathrm{J}] = \begin{bmatrix} J_\uparrow & J_{\uparrow,\downarrow} \\ J_{\downarrow,\uparrow} & J_\downarrow \end{bmatrix} \tag{6.72}$$

The off-diagonal elements $J_{\uparrow,\downarrow}$ and $J_{\uparrow,\downarrow}$ are known as *spin-mixing currents*, and are generally nonzero, at least during nonequilibrium. Dividing Eq. (6.64a) by $i\hbar$, and using the *spin current density matrix* from Eq. (6.71), we have a *spin continuity equation* given by

2 × 2 Spin Continuity Equation

$$\frac{\partial \Upsilon}{\partial t} = -\begin{bmatrix} \nabla\cdot J_\uparrow & \nabla\cdot J_{\uparrow,\downarrow} \\ \nabla\cdot J_{\downarrow,\uparrow} & \nabla\cdot J_\downarrow \end{bmatrix} - \frac{2\alpha}{\hbar}V_\sigma\Upsilon \tag{6.73}$$

Eq. (6.73) is a differential spin-transport equation of size 2×2. In defining the current density elements J_{ij}, they are expressed as $-1\times$ the elements of Eqs. (6.69) because dividing by $i\hbar$ introduces a factor of -1 since $1/i = -i$.

It is possible to express Eq. (6.73) in the form of a classical diffusion equation. Let's see how the continuity equation can be expressed in this alternative form.

6.3.1 Spin Diffusion And Diffusion Length

Recall that the *ordinary* electrical current density **J** is given by

$$\mathbf{J} = \sigma \mathbf{E} \tag{6.74}$$

σ is the electrical conductivity and **E** is the electric field driving the conduction. The electric field is defined as

$$\mathbf{E} = -\nabla \mu$$

∇ is the spatial gradient operator and μ is the *electric potential*, which has units of volts, or energy per unit charge. Because there are generally up to four spin currents, this relation requires some resolving beyond charge, down to the level of the electron spin. Adapting Eq. (6.74) to four spin currents, let us write each spin current in the following form:

$$\mathbf{J}_\uparrow = \sigma_\uparrow \mathbf{E}_\uparrow = -\sigma_\uparrow \nabla \mu_\uparrow \quad , \quad \mathbf{J}_\downarrow = \sigma_\downarrow \mathbf{E}_\downarrow = -\sigma_\downarrow \nabla \mu_\downarrow \tag{6.75a}$$

$$\mathbf{J}_{\uparrow,\downarrow} = \sigma_{\uparrow,\downarrow} \mathbf{E}_{\uparrow,\downarrow} = -\sigma_{\uparrow,\downarrow} \nabla \mu_{\uparrow,\downarrow} \quad , \quad \mathbf{J}_{\downarrow,\uparrow} = \sigma_{\downarrow,\uparrow} \mathbf{E}_{\downarrow,\uparrow} = -\sigma_{\downarrow,\uparrow} \nabla \mu_{\downarrow,\uparrow} \tag{6.75b}$$

$\sigma_{\uparrow,\downarrow}$ are the respective spin-dependent electrical conductivities. From both states, the total current density J can be found from the relation

$$\mathbf{J} = \frac{\mathbf{J}_\uparrow |\psi_\uparrow|^2 + \mathbf{J}_\downarrow |\psi_\downarrow|^2}{|\psi_\uparrow|^2 + |\psi_\downarrow|^2}$$

Using Eqs. (6.75), we can define electrical *spin potentials* $\mu_{\uparrow,\downarrow}$ as

Definition of Spin Potentials

$$\mu_\uparrow = -\frac{1}{\sigma_\uparrow} \int \mathbf{J}_\uparrow(\mathbf{r}) \cdot d\mathbf{r} \quad , \quad \mu_\downarrow = -\frac{1}{\sigma_\downarrow} \int \mathbf{J}_\downarrow(\mathbf{r}) \cdot d\mathbf{r}$$

$$\mu_{\uparrow,\downarrow} = -\frac{1}{\sigma_{\uparrow,\downarrow}} \int \mathbf{J}_{\uparrow,\downarrow}(\mathbf{r}) \cdot d\mathbf{r} \quad , \quad \mu_{\downarrow,\uparrow} = -\frac{1}{\sigma_{\downarrow,\uparrow}} \int \mathbf{J}_{\downarrow,\uparrow}(\mathbf{r}) \cdot d\mathbf{r} \tag{6.76}$$

Using the spin-potential definitions given by Eq. (6.76), the *spin continuity equation* becomes

$$\frac{\partial \Upsilon}{\partial t} = \nabla \cdot \begin{bmatrix} \sigma_\uparrow \nabla \mu_\uparrow & \sigma_{\uparrow,\downarrow} \nabla \mu_{\uparrow,\downarrow} \\ \sigma_{\downarrow,\uparrow} \nabla \mu_{\downarrow,\uparrow} & \sigma_\downarrow \nabla \mu_\downarrow \end{bmatrix} - \frac{2\alpha}{\hbar} V_\sigma \Upsilon \tag{6.77}$$

Since

$$\nabla \cdot \nabla \psi = \nabla^2 \psi$$

if the spin-dependent conductivities are assumed to be uniform in space, then we have a so-called *spin diffusion equation* given by

2 × 2 Spin Diffusion Equation

$$\frac{\partial \Upsilon}{\partial t} = \nabla^2 \begin{bmatrix} \sigma_\uparrow \mu_\uparrow & \sigma_{\uparrow,\downarrow} \mu_{\uparrow,\downarrow} \\ \sigma_{\downarrow,\uparrow} \mu_{\downarrow,\uparrow} & \sigma_\downarrow \mu_\downarrow \end{bmatrix} - \frac{2\alpha}{\hbar} V_\sigma \Upsilon \tag{6.78}$$

Both the spin continuity equation and the spin diffusion equation describes *four* variables depending on space and time. In the continuity equations, although the total spin-current computes using $J_\uparrow$ and $J_\downarrow$, they are both coupled to the *mixing* terms. In both equations, the itinerant spin couples to the internal energy of the material. For example, in Sec. 6.1, the spin-dependent potential operator V_σ was given by

$$V_\sigma = -\boldsymbol{\mu}_j \cdot \mathbf{B}_{\text{eff}}$$

where $\mathbf{B}_{\text{eff}}$ contains local spin dependent energy operators.

6.3.2 Spin-Diffusion Length

In evanescent wave conditions, we have discussed the decay length $\delta = 1/k$ in Chap. 5. However, it is only relevant in insulating conditions. In the case of a metal, for example, there can be a wave contribution of the form Ae^{ikx}. Even in this case, there is still a decay in the amplitude, however, with the inclusion of the fourth dimension, this is a time-dependent decay. This type of amplitude decay is known as *diffusion,* and is often recognized in the form of a *spatial* variant, though it originates from time-dependent changes.

Because there is generally an *unequal* flow of spin-up and spin-down electrons, the electron motion is sometimes imagined as having two *parallel* currents; a spin-up electrical current $J_\uparrow$ and a spin-down current $J_\downarrow$. The traditional semi-classical picture sometimes assumes there is no mixing or correlation between the two spin states. However, when the fourth dimension is included, we've see that this is, generally, not the case. However, it is still useful to introduce some of the concepts and their connection to quantum mechanics.

Let the eigensolutions for the spin-up and spin-down states be given by

$$\psi_\uparrow = \left(A_\uparrow e^{ik_\uparrow x} + B_\uparrow e^{-ik_\uparrow x}\right) e^{-iE_{R,\uparrow} t} e^{-\alpha\omega_{\text{eff}} t/2} \tag{6.79a}$$

$$\psi_\downarrow = \left(A_\downarrow e^{ik_\downarrow x} + B_\downarrow e^{-ik_\downarrow x}\right) e^{+iE_{R,\downarrow} t} e^{+\alpha\omega_{\text{eff}} t/2} \tag{6.79b}$$

Evaluating the spin currents using the definitions from Eq. (6.71), we find $J_\uparrow$ and $J_\downarrow$ are given by

$$J_\uparrow = \frac{\hbar k_\uparrow}{m_0}\left(|A_\uparrow|^2 - |B_\uparrow|^2\right) e^{+\alpha\omega_{\text{eff}} t} = K_\uparrow e^{+\alpha\omega_{\text{eff}} t} \tag{6.80a}$$

$$J_\downarrow = \frac{\hbar k_\downarrow}{m_0}\left(|A_\downarrow|^2 - |B_\downarrow|^2\right) e^{-\alpha\omega_{\text{eff}} t} = K_\downarrow e^{-\alpha\omega_{\text{eff}} t} \tag{6.80b}$$

Eq. (6.80) describe a current with *decaying* amplitude. Since there is motion, this can be expressed in an equivalent form using the relation between velocity and time. The itinerant electrons have motion in an electrical current proportional to $k_\uparrow$ and $k_\downarrow$, so there is a velocity v which permits a relation to time t given by

$$x = vt = \frac{\hbar\langle k\rangle}{m_0} t \Rightarrow t = \frac{m_0}{\hbar k} x \tag{6.81}$$

$\langle k\rangle$ is the expectation value of the wavenumber. Substitution of t into Eqs. (6.80) gives the relations

$$J_\uparrow = K_\uparrow e^{+\alpha\omega_{\text{eff}} m_0 x/\hbar\langle k\rangle} = K_\uparrow e^{x/\ell_{sd}} \tag{6.82a}$$

$$J_\downarrow = K_\downarrow e^{-\alpha\omega_{\text{eff}} m_0 x/\hbar\langle k\rangle} = K_\downarrow e^{-x/\ell_{sd}} \tag{6.82b}$$

Eqs. (6.82a) and (6.82b) define a so-called *spin-diffusion length* ℓ_{sd} as

Spin-Diffusion Length

$$\ell_{sd} = \frac{\hbar\langle k\rangle}{\alpha m_0 \omega_{\text{eff}}} \tag{6.83}$$

where

$$\omega_{\text{eff}} = 2\gamma B_{\text{eff}} \quad \text{and} \quad \gamma = \frac{e}{2m_0}$$

The *spin diffusion length* ℓ_{sd} is a measure of the length scale over which electron spin currents decay or amplify, respectively along the direction of conduction.

In most of the analyses discussed so far in topics of spintronics, we have assumed the localized spins are fixed in orientation when there is spin-transport. However, if an itinerant spin can be rotated by a local spin, it stands to reason that a local spin may be rotated by a spin in transport. For this more general case, we must add another equation to describe the dynamical changes to the local spins in, say, a magnetic layer. The resulting coupled dynamical response of a magnetic layer spins to the itinerant spin polarization is known as *spin-transfer torque*. In the next section, we'll examine this part of the spin transport picture, which generally couples bidirectionally to the itinerant spin dynamics.

6.4 Spin Transfer Torque

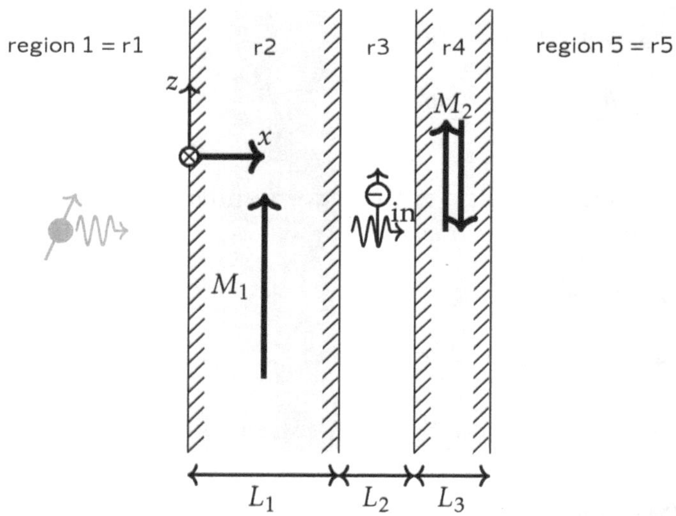

Figure 6.13: Illustration of a spin which has been polarized by the first magnetic layer M_1, then subsequently injected into a second magnetic layer M_2.

With the creation of a *persistent* spin-polarized current, it becomes possible to utilize the moving spin to impart rotation or torque to the localized spins of a magnet. First, we analyze the resulting physics using the 4D energy operator to obtain the electron spin wave functions. Then, the wave functions are used to evaluate time-dependent expectation values. Finally, we'll work out the corresponding dynamical equation that generally describes the expectation values found.

Let us consider the situation illustrated in Fig. 6.13, which is very similar to the MR device. However, in this case, the polarized spin passes into a *free* magnetic layer which is *not fixed* in orientation. We are interested in the dynamics resulting from interactions with the polarized itinerant spin. As Fig. 6.13 shows, the itinerant electrons coupling to the localized spins of M_2 are spin-polarized by the first magnetic layer (M_1 or region 2). Then, they pass into a nonmagnetic material between both magnetic layers. This layer presumably does not deteriorate the spin polarization significantly. In this sort of application, this nonmagnetic layer is also *necessary to decouple the two magnetic layers* from one another. Otherwise, if the magnetization of one moves, the other may be affected. The design aims to enable independent manipulation of the magnetization of the second magnetic layer M_2.

Describing the energy of the localized spins in M_2, the 4d Hamiltonian is given by

$$\widehat{H}^4 = \left[\widehat{H}_\sigma - i\hbar\frac{\partial}{\partial t}\right]\psi = (iE_I)\psi \tag{6.84}$$

The total energy eigenequation is given by

$$i\hbar\frac{\partial\psi}{\partial t} = (E_R - iE_I)\psi \tag{6.85}$$

Using the electron's gyromagnetic ratio $-\gamma = -e/2m_0$ along the magnetic moment $\mu = \gamma\hbar$, the Hamiltonian $\widehat{H}_\sigma$ is given by

$$\widehat{H}_\sigma = V_\sigma = -\boldsymbol{\mu}\cdot\mathbf{B}_{\text{eff}} = -(-\gamma\hbar)\widehat{\boldsymbol{\sigma}}\cdot\mathbf{B}_p = \gamma\hbar\widehat{\boldsymbol{\sigma}}\cdot\mathbf{B}_{\text{eff}} \tag{6.86}$$

$\mathbf{B}_{\text{eff}}$ is the *polarizing magnetic field* including the exchange field from the spin-transport electron. The localized electrons also experience the spin-orbit interaction, so the total effective magnetic field becomes

$$\mathbf{B}_{\text{eff}} = -\frac{J_{sd}}{2\mu_B}\widehat{\boldsymbol{\sigma}}_j - \frac{\hbar}{\mu_B m_0^2 c^2}(\widehat{\mathbf{p}} + e\mathbf{A}) \times \nabla V \tag{6.87}$$

$\mu_B = \gamma\hbar$ is the Bohr magneton. $\widehat{\boldsymbol{\sigma}}_j$ is the itinerant spin polarization (*e.g.* from the continuity equation). The spin continuity equation describes the density *e.g.* $\rho_\uparrow$, having units of m^{-3}. From ρ, a *spin-density* or *moment-density* can also be expressed using

$$\rho_\uparrow \to s_\uparrow = \frac{\hbar\rho_\uparrow}{2} \quad \text{(spin-angular momentum density)}$$

$$\rho_\uparrow \to \mu_\uparrow = \hbar\gamma\rho_\uparrow \quad \text{(magnetic moment density)}$$

The exact contributions are not important for our purposes since only $\mathbf{B}_{\text{eff}}$ is used, however, in practice, the relevant contributions should be understood and included in a proper calculation.

Substituting Eq. (6.86) into Eq. (6.84) and introducing $E_I = \alpha E_R$, leads to

$$i\hbar\frac{\partial}{\partial t}\begin{bmatrix}\psi_\uparrow\\ \psi_\downarrow\end{bmatrix} = -(1-i\alpha)\boldsymbol{\mu}\cdot\mathbf{B}_{\text{eff}}\begin{bmatrix}\psi_\uparrow\\ \psi_\downarrow\end{bmatrix} \tag{6.88a}$$

$$= (1-i\alpha)\gamma\hbar\widehat{\boldsymbol{\sigma}}\cdot\mathbf{B}_{\text{eff}}\begin{bmatrix}\psi_\uparrow\\ \psi_\downarrow\end{bmatrix} \tag{6.88b}$$

If $\mathbf{B}_{\text{eff}}$ is taken along the $+z$ axis, then

$$\widehat{\boldsymbol{\sigma}}\cdot\mathbf{B}_{\text{eff}} = B_{\text{eff}}\widehat{\sigma_z} \tag{6.89}$$

where $\widehat{\sigma_z}$ is the spin matrix along the z axis given by

$$\widehat{\sigma_z} = \begin{bmatrix}1 & 0\\ 0 & -1\end{bmatrix} \tag{6.90}$$

Dividing Eq. (6.88) by $i\hbar$, we have

$$\frac{\partial}{\partial t}\begin{bmatrix}\psi_\uparrow\\ \psi_\downarrow\end{bmatrix} = \frac{(1-i\alpha)\gamma\hbar B_{\text{eff}}}{i\hbar}\widehat{\sigma_z}\begin{bmatrix}\psi_\uparrow\\ \psi_\downarrow\end{bmatrix} \tag{6.91a}$$

$$= (-i-\alpha)\gamma B_{\text{eff}}\widehat{\sigma_z} \tag{6.91b}$$

Eq. (6.91) resolves into two scalar equations describing $\psi_\uparrow$ and $\psi_\downarrow$ given by

$$\frac{d\psi_\uparrow}{dt} = (-i-\alpha)\gamma\hbar B_{\text{eff}}\psi_\uparrow = -(i\omega_{\text{eff}} + \alpha\omega_{\text{eff}})\psi_\uparrow \tag{6.92a}$$

$$\frac{d\psi_\downarrow}{dt} = (i+\alpha)\gamma\hbar B_{\text{eff}}\psi_\downarrow = (i\omega_{\text{eff}} + \alpha\omega_{\text{eff}})\psi_\downarrow \tag{6.92b}$$

Then, the solutions for $\psi_\uparrow$ and $\psi_\downarrow$ become

$$\psi_\uparrow = \psi_\uparrow(0)e^{-i\omega_{\text{eff}}\cdot t}e^{-\alpha\omega_{\text{eff}}\cdot t} \tag{6.93a}$$

$$\psi_\downarrow = \psi_\uparrow(0)e^{i\omega_{\text{eff}}\cdot t}e^{\alpha\omega_{\text{eff}}\cdot t} \tag{6.93b}$$

where we have defined an *effective field frequency* ω_{eff} as

$$\omega_{\text{eff}} = \gamma B_{\text{eff}} \tag{6.94}$$

To determine the dynamical behavior of the spin, the spin-wave functions of Eq. (6.93) can be used to calculate the expectation value **S** or equivalently, the electron magnetic moment μ. Let's do this next.

6.4.1 Time-dependent Expectation Values of Spin

In order to use the solutions given by Eqs. (6.93), we must *also* specify the initial conditions $\psi_\uparrow(0)$ and $\psi_\downarrow(0)$ of the spin states. Their values are *not* arbitrary. Because we have taken the magnetic field along $+z$, the relevant spinor *initial condition* is given by[10]

$$\psi(0) = \begin{bmatrix}\psi_\uparrow(0) & \psi_\downarrow(0)\end{bmatrix}^T = \begin{bmatrix}\cos\frac{\theta}{2}e^{-i\frac{\varphi}{2}} & \sin\frac{\theta}{2}e^{i\frac{\varphi}{2}}\end{bmatrix}^T \tag{6.95}$$

The generalized spinors to be used for the initial condition are found from working out the eigenvector of a generalized spin matrix $\widehat{\sigma}$ for an arbitrary field direction. They are given in terms of two angles θ and φ. Both are used to specify an initial condition for a *spinor*. θ is the polar angle measured from the $+z$ axis, while φ is the azimuthal angle measured from the $+x$ axis.

With the specified initial state for the spinor, the wave functions can be used to determine the time-dependent expectation values of the components of the intrinsic angular momentum $\widehat{S_x}$, $\widehat{S_y}$, and $\widehat{S_z}$, or equivalently, the magnetic moment $\mu = g\gamma\mathbf{S}$. The first component of the spin angular momentum $\langle\widehat{S_x}\rangle$ is defined as

$$\langle\widehat{S_x}\rangle = \frac{\hbar}{2}\frac{\begin{bmatrix}\psi_\uparrow^* & \psi_\downarrow^*\end{bmatrix}\widehat{\sigma_x}\begin{bmatrix}\psi_\uparrow \\ \psi_\downarrow\end{bmatrix}}{\begin{bmatrix}\psi_\uparrow^* & \psi_\downarrow^*\end{bmatrix}\begin{bmatrix}\psi_\uparrow \\ \psi_\downarrow\end{bmatrix}} = \frac{\hbar}{2}\frac{\psi_\uparrow^*\psi_\downarrow + \psi_\downarrow^*\psi_\uparrow}{\psi_\uparrow^*\psi_\uparrow + \psi_\downarrow^*\psi_\downarrow} \tag{6.96}$$

[10] For a discussion and derivation of spinor initial states, see Sec. 9.4 & 9.5 of Quantum Mechanics With A SPIN On It (Eason).

In solving for the expectation values, to simply some expressions, we will also make use of the following trigonometric identities:

$$\cos x = \frac{e^{ix}+e^{-ix}}{2} \qquad \cos^2\frac{\theta}{2} = \frac{1+\cos\theta}{2} \qquad \sin^2\frac{\theta}{2} = \frac{1-\cos\theta}{2}$$

$$\cosh x = \frac{e^{x}+e^{-x}}{2} \qquad \sinh x = \frac{e^{x}-e^{-x}}{2} \qquad \sin x\cos x = \frac{\sin(2x)}{2}$$

Substituting the initial condition from Eq. (6.95) with the solutions given by Eqs. (6.93a)-(6.93b), $\langle\widehat{S}_x\rangle$ becomes

$$\begin{aligned}
\langle\widehat{S}_x\rangle &= \frac{\hbar}{2}\frac{\cos\frac{\theta}{2}\sin\frac{\theta}{2}\left(e^{i2\omega_{\text{eff}}t}e^{i\varphi}+e^{-i2\omega_{\text{eff}}t}e^{-i\varphi}\right)}{\left(\cos^2\frac{\theta}{2}e^{-2\alpha\omega_{\text{eff}}t}+\sin^2\frac{\theta}{2}e^{2\alpha\omega_{\text{eff}}t}\right)} \\
&= \frac{\hbar}{2}\frac{\cos\frac{\theta}{2}\sin\frac{\theta}{2}\left(e^{i2\omega_{\text{eff}}t}e^{i\varphi}+e^{-i2\omega_{\text{eff}}t}e^{-i\varphi}\right)}{\left(\left[\frac{1+\cos\theta}{2}\right]e^{-2\alpha\omega_{\text{eff}}t}+\left[\frac{1-\cos\theta}{2}\right]e^{2\alpha\omega_{\text{eff}}t}\right)} \\
&= \frac{\hbar}{2}\frac{\cos\frac{\theta}{2}\sin\frac{\theta}{2}(e^{i2\omega_{\text{eff}}t}e^{i\varphi}+e^{-i2\omega_{\text{eff}}t}e^{-i\varphi})}{\frac{\left(e^{-2\alpha\omega_{\text{eff}}t}+e^{2\alpha\omega_{\text{eff}}t}\right)}{2}+\cos\theta\frac{\left(e^{-2\alpha\omega_{\text{eff}}t}-e^{2\alpha\omega_{p}t}\right)}{2}} \\
&= \frac{\hbar}{2}\frac{(\sin\theta/2)(e^{i2\omega_{\text{eff}}t}e^{i\varphi}+e^{-i2\omega_{\text{eff}}t}e^{-i\varphi})}{\frac{\left(e^{-2\alpha\omega_{\text{eff}}t}+e^{2\alpha\omega_{\text{eff}}t}\right)}{2}+\cos\theta\frac{\left(e^{-2\alpha\omega_{\text{eff}}t}-e^{2\alpha\omega_{\text{eff}}t}\right)}{2}} \\
&= \frac{\hbar}{2}\frac{\sin\theta\cos(2\omega_{\text{eff}}t+\varphi)}{\cosh(2\alpha\omega_{\text{eff}}t)-\cos\theta\sinh(2\alpha\omega_{\text{eff}}t)}
\end{aligned}$$

Since the denominator of $\langle\widehat{S}_x\rangle$ is identical for all three expectation values $\langle S_x\rangle$, $\langle S_y\rangle$, and $\langle S_z\rangle$, we won't repeat the calculation (of the denominator) for the remaining two components since we can just use the above result.

The initial spinor angles (θ,φ) are dictated by the energy operator where $\boldsymbol{\mu}\cdot\mathbf{B}_{\text{eff}}$ leads to the spinor angles $\theta=\pi/2$, while $\varphi=\varphi_i$ is arbitrary.[11] Using these initial values, $\langle S_x\rangle$ becomes

$$\langle\widehat{S}_x\rangle(t) = \frac{\hbar}{2}\frac{\cos(2\omega_{\text{eff}}t+\varphi)}{\cosh(2\alpha\omega_{\text{eff}}t)} \tag{6.97}$$

Fig. 6.14 illustrates $\langle\widehat{S}_x\rangle(t)$ for three different values of $\alpha=0.01$, $\alpha=0.03$, and $\alpha=0.05$.

---

[11] For a detailed discussion on why the values dictated by $\boldsymbol{\mu}\cdot\mathbf{B}_{\text{eff}}$, see Sec. 7.2 Quantum Mechanics With A Spin On It.

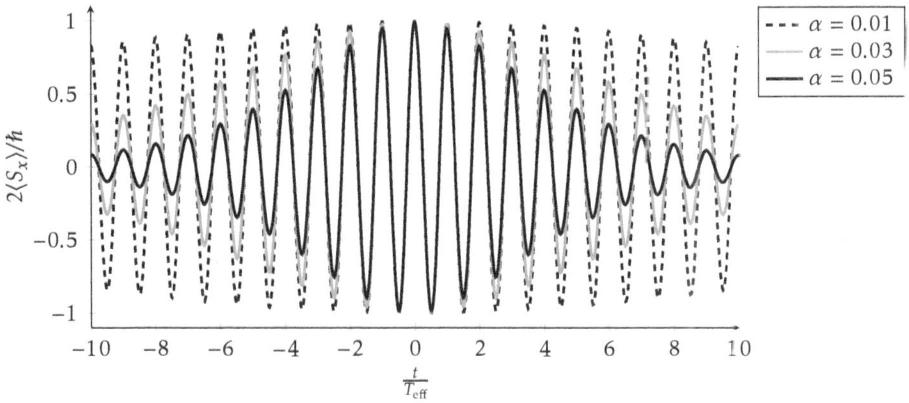

Figure 6.14: Plot of the x-component of the electron spin momentum $\langle S_x\rangle(t)$. $\langle S_x\rangle$ oscillates with frequency $\omega_{\text{eff}} = 2\gamma B_{\text{eff}} = 2\pi/T_{\text{eff}}$, with damping leading to decaying away from zero. The larger the damping parameter α, the faster the damping in the damped oscillations.

A similar evaluation for $\langle S_y\rangle$ can be done using

$$\langle \widehat{S}_y\rangle(t) = \frac{\hbar}{2}\frac{\begin{bmatrix}\psi_\uparrow^* & \psi_\downarrow^*\end{bmatrix}\begin{bmatrix}0 & -i\\ i & 0\end{bmatrix}\begin{bmatrix}\psi_\uparrow\\ \psi_\downarrow\end{bmatrix}}{\begin{bmatrix}\psi_\uparrow^* & \psi_\downarrow^*\end{bmatrix}\begin{bmatrix}\psi_\uparrow\\ \psi_\downarrow\end{bmatrix}} = \frac{i\hbar}{2}\frac{\psi_\downarrow^*\psi_\uparrow - \psi_\uparrow^*\psi_\downarrow}{\psi_\uparrow^*\psi_\uparrow + \psi_\downarrow^*\psi_\downarrow} \tag{6.98}$$

Then, substitution of the wave functions from Eq. (6.93) into Eq. (6.98) leads to

$$\langle \widehat{S}_y\rangle = \frac{i\hbar}{2}\frac{\cos\frac{\theta}{2}\sin\frac{\theta}{2}\left(e^{-i2\omega_{\text{eff}}t}e^{-i\varphi} - e^{i2\omega_{\text{eff}}t}e^{i\varphi}\right)}{\frac{\left(e^{-2\alpha\omega_{\text{eff}}t}+e^{2\alpha\omega_{\text{eff}}t}\right)}{2} + \cos\theta\frac{\left(e^{-2\alpha\omega_{\text{eff}}t}-e^{2\alpha\omega_{\text{eff}}t}\right)}{2}}$$

$$= \frac{\hbar}{2}\frac{\sin\theta\sin(2\omega_{\text{eff}}t+\varphi)}{\cosh(2\alpha\omega_{\text{eff}}t) - \cos\theta\sinh(2\alpha\omega_{\text{eff}}t)}$$

Substitution of the initial boundary condition $(\theta,\varphi) = (\pi/2,\varphi_i)$ into $\langle \widehat{S}_y\rangle$ leads to

$$\langle \widehat{S}_y\rangle = \frac{\hbar}{2}\frac{\sin(2\omega_{\text{eff}}t+\varphi)}{\cosh(2\alpha\omega_{\text{eff}}t)} \tag{6.99}$$

Fig. 6.15 illustrates $\langle \widehat{S}_y\rangle$, also for the values of α = .01,.03, and .05. Lastly, $\langle \widehat{S}_z\rangle$ is given by

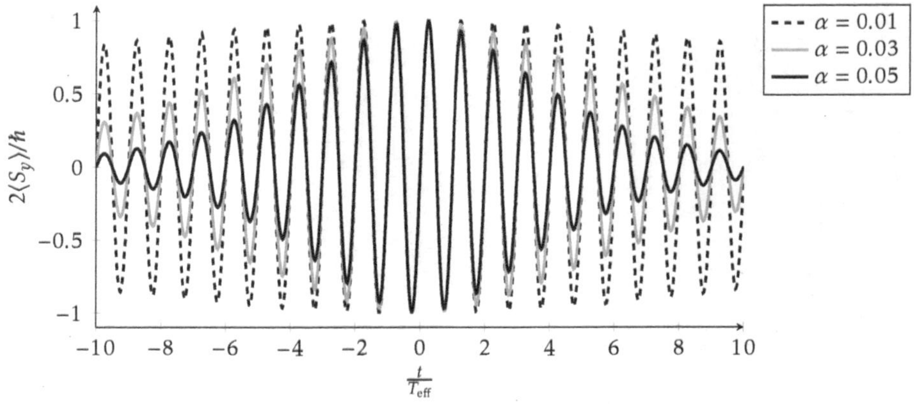

Figure 6.15: Plot of the y-component of the electron spin momentum. Like $\langle\widehat{S}_x\rangle$, $\langle\widehat{S}_y\rangle$ oscillates with frequency $\omega_{\text{eff}} = 2\pi/T_{\text{eff}}$, with larger parameter α leading to more rapidly decaying oscillations.

$$\langle\widehat{S}_z\rangle(t) = \frac{\hbar}{2}\frac{\begin{bmatrix}\psi_\uparrow^* & \psi_\downarrow^*\end{bmatrix}\begin{bmatrix}1 & 0\\0 & -1\end{bmatrix}\begin{bmatrix}\psi_\uparrow\\\psi_\downarrow\end{bmatrix}}{\begin{bmatrix}\psi_\uparrow^* & \psi_\downarrow^*\end{bmatrix}\begin{bmatrix}\psi_\uparrow\\\psi_\downarrow\end{bmatrix}} = \frac{\hbar}{2}\frac{\psi_\uparrow^*\psi_\uparrow - \psi_\downarrow^*\psi_\downarrow}{\psi_\uparrow^*\psi_\uparrow + \psi_\downarrow^*\psi_\downarrow}$$

Substitution of the wave functions from Eqs. (3.118) gives

$$\begin{aligned}\langle\widehat{S}_z\rangle &= \frac{\hbar}{2}\frac{\left(e^{-2\alpha\omega_{\text{eff}}t} - e^{2\alpha\omega_{\text{eff}}t}\right) + \cos\theta\left(e^{-2\alpha\omega_{\text{eff}}t} + e^{2\alpha\omega_{\text{eff}}t}\right)}{\left(e^{-2\alpha\omega_{\text{eff}}t} + e^{2\alpha\omega_{\text{eff}}t}\right) + \cos\theta\left(e^{-2\alpha\omega_{\text{eff}}t} - e^{2\alpha\omega_{\text{eff}}t}\right)}\\ &= \frac{\hbar}{2}\frac{-\sinh(2\alpha\omega_{\text{eff}}t) + \cos\theta\cosh(2\alpha\omega_{\text{eff}}t)}{\cosh(2\alpha\omega_{\text{eff}}t) - \cos\theta\sinh(2\alpha\omega_{\text{eff}}t)}\end{aligned}$$

Substituting the initial spinor condition into $\langle S_z\rangle$, we have

$$\langle\widehat{S}_z\rangle = -\frac{\hbar}{2}\frac{\sinh(2\alpha\omega_{\text{eff}}t)}{\cosh(2\alpha\omega_{\text{eff}}t)} = -\frac{\hbar}{2}\tanh(2\alpha\omega_{\text{eff}}t) \tag{6.100}$$

Fig. 6.16 shows $\langle\widehat{S}_z\rangle(t)$ for values of $\alpha = 0.01, 0.03, 0.05$. The component of the spin angular momentum along the z–axis (or field axis) aligns *anti-parallel* with the effective magnetic field. Since z is the presumed axis of the field, we find that the localized electron in M_2 undergoes a dynamical transition between the two spin states. Additionally, taking $\mathbf{B}_{\text{eff}}$ along this axis, we are provided a time-scale for the *quantum transition* τ_S, given by

$$\tau_S = \frac{1}{2\alpha\omega_{\text{eff}}} = \frac{1}{2\alpha\gamma B_{\text{eff}}} \tag{6.101}$$

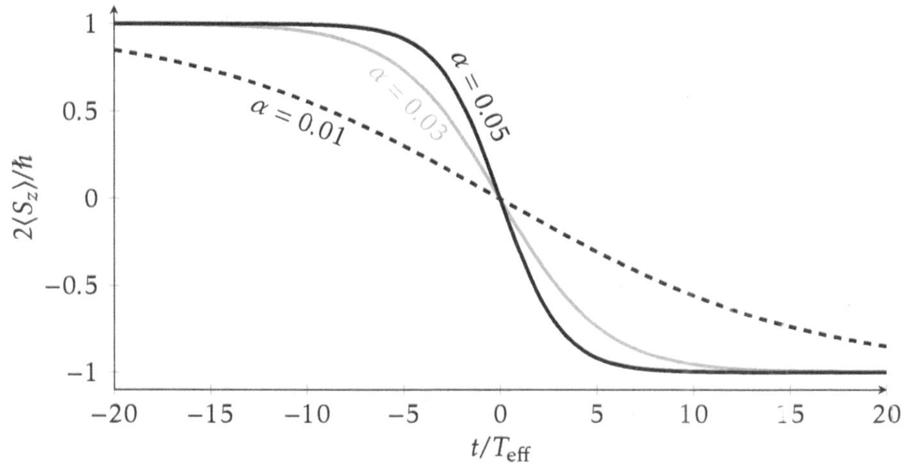

Figure 6.16: Plot of the z-component of the electron spin momentum, which is the component along the field axis. $T_{\text{eff}} = \omega_{\text{eff}}/2\pi$. As with any magnetic field, the spin is anti-parallel with B_{eff} (the magnetic moment becomes parallel to B_{eff}). The larger the parameter α, the faster the dynamical transition.

while the two components perpendicular to this axis *precess* about the field axis at a rate $2\omega_{\text{eff}}$.

Therefore, we find that by applying a 4D Hamiltonian to a *localized* electron coupling to an effective magnetic field, the three resulting components of the time-dependent spin angular momentum are given by

x–Spin Angular Momentum Expectation Value

$$\langle\widehat{S}_x\rangle = \frac{\hbar}{2}\frac{\cos(2\omega_{\text{eff}}t + \varphi)}{\cosh(\alpha\omega_{\text{eff}}t)} \tag{6.102a}$$

y–Spin Angular Momentum Expectation Value

$$\langle\widehat{S}_y\rangle = \frac{\hbar}{2}\frac{\sin(2\omega_{\text{eff}}t + \varphi)}{\cosh(\alpha\omega_{\text{eff}}t)} \tag{6.102b}$$

z–Spin Angular Momentum Expectation Value

$$\langle\widehat{S_z}\rangle = -\frac{\hbar}{2}\tanh(2\alpha\omega_{\text{eff}}t) \tag{6.102c}$$

As mentioned, the two components $\langle\widehat{S_x}\rangle$ and $\langle\widehat{S_y}\rangle$, being *perpendicular* to $\mathbf{B}_{\text{eff}}$, oscillate about it at a frequency $2\omega_{\text{eff}}$, which is the *electron spin precessional frequency*.

The above expectation values can be used to examine the value of the spin angular momentum amplitude $|S|$ in time t. Let us consider the amplitude of $\langle S\rangle(t)$ while undergoing a dynamic transition in the effective magnetic field. The quantity $\langle\widehat{S}\rangle^2$ becomes

$$\langle\widehat{S}\rangle^2 = \langle\widehat{S_x}\rangle^2 + \langle\widehat{S_y}\rangle^2 + \langle\widehat{S_z}\rangle^2 =$$

$$\left[\frac{\hbar}{2}\frac{\cos(2\omega_{\text{eff}}t+\varphi)}{\cosh(2\alpha\omega_{\text{eff}}t)}\right]^2 + \left[\frac{\hbar}{2}\frac{\sin(2\omega_{\text{eff}}t+\varphi)}{\cosh(2\alpha\omega_{\text{eff}}t)}\right]^2 + \left[-\frac{\hbar}{2}\tanh(2\alpha\omega_{\text{eff}}t)\right]^2 =$$

$$\frac{\hbar^2}{4}\left[\frac{\cos^2(2\omega_{\text{eff}}t+\varphi)+\sin^2(2\omega_{\text{eff}}t+\varphi)+\sinh^2(2\alpha\omega_{\text{eff}}t)}{\cosh^2(2\alpha\omega_{\text{eff}}t)}\right] =$$

$$\frac{\hbar^2}{4}\left[\frac{1+\sinh^2(2\alpha\omega_{\text{eff}}t)}{\cosh^2(2\alpha\omega_{\text{eff}}t)}\right] = \frac{\hbar^2}{4}\left[\frac{\cosh^2(2\alpha\omega_{\text{eff}}t)}{\cosh^2(2\alpha\omega_{\text{eff}}t)}\right] = \frac{\hbar^2}{4} \tag{6.103}$$

According to Eq. (6.103), the length of the expectation value of $\mathbf{S}$ *is conserved* across quantum dynamical transitions, or

$$|\langle\widehat{\mathbf{S}}\rangle(t)| = \frac{\hbar}{2} \tag{6.104}$$

Since the electron spin magnetic moment $\boldsymbol{\mu}_S$ is proportional to the spin angular momentum as $\mathbf{S}$

$$\widehat{\boldsymbol{\mu}_S} = -\frac{e}{2m_0}\widehat{\mathbf{S}} \tag{6.105}$$

The corresponding magnetic moment components $\langle\mu_x\rangle(t)$ become

$$\langle\mu_x\rangle(t) = -\gamma\hbar\frac{\cos(2\omega_{\text{eff}}t+\varphi)}{\cosh(2\alpha\omega_{\text{eff}}t)} \tag{6.106a}$$

$$\langle\mu_y\rangle(t) = -\gamma\hbar\frac{\sin(2\omega_{\text{eff}}t+\varphi)}{\cosh(\alpha\omega_{\text{eff}}t)} \tag{6.106b}$$

$$\langle\mu_z\rangle(t) = \gamma\hbar\tanh(2\alpha\omega_{\text{eff}}t) \tag{6.106c}$$

If we use the Bohr magneton $\mu_B = \gamma\hbar$, it follows that

$$|\boldsymbol{\mu}(t)| = \mu(t) = \mu_B \tag{6.107}$$

The transition observed here between the spin states has been enabled by inclusion of the fourth dimension. This leads to a time-dependence in the eigenfunctions and gives rise to time-dependent expectation values which can describe transitions between quantum states $+\hbar/2$ to $-\hbar/2$.

It is also useful to have a differential equation that contains the time-dependent behavior of the expectation values. This would be a classical equation, and it can be obtained by using the *Most General Ehrenfest Theorem* derived and given by Eq. (4.21). So, lets find out what kind of differential equation describes the vector motion given by Eq. (6.106).

6.4.2 Dynamical Torque Equation

The *Most General Ehrenfest Theorem* is obtained from generalizing the Ehrenfest Theorem for a non-Hermitian operator. The result is given by

$$\frac{d\langle\widehat{\boldsymbol{\mu}}\rangle}{dt} = \frac{i}{\hbar}\left[\langle\widehat{H}^{\dagger}\widehat{\boldsymbol{\mu}}\rangle - \langle\widehat{\boldsymbol{\mu}}\widehat{H}\rangle\right] - \frac{i}{\hbar}\langle\widehat{\boldsymbol{\mu}}\rangle\left[\langle\widehat{H}^{\dagger}\rangle - \langle\widehat{H}\rangle\right] \tag{6.108}$$

The energy operator $\widehat{H}$ needed for Eq. (6.108) was given by Eq. (6.88) as

$$\widehat{H} = (1 - i\alpha)V_\sigma = (1 - i\alpha)\widehat{\boldsymbol{\mu}}\cdot\mathbf{B}_{\text{eff}} \tag{6.109}$$

Since the energy operator is the sum of two terms, let us evaluate each of the terms *separately*, then combine the result at the end. The first term of $\widehat{H} = (1 - i\alpha)V_\sigma$ is just V_σ, which is *Hermitian*. We'll denote this contribution using the superscript A. In this case, the second term on the RHS of Eq. (6.108) vanishes,[12] so we have

$$\frac{d\widehat{\boldsymbol{\mu}}^A}{dt} = \frac{i}{\hbar}\left[\widehat{H}_\sigma\widehat{\boldsymbol{\mu}} - \widehat{\boldsymbol{\mu}}\widehat{H}_\sigma\right] = \frac{i}{\hbar}\left[\widehat{H}_\sigma, \widehat{\boldsymbol{\mu}}\right] \tag{6.110}$$

Substituting $\widehat{\boldsymbol{\mu}} = (-\gamma\hbar)\widehat{\boldsymbol{\sigma}}$ into the above along with the spin matrices $\widehat{\sigma}_x, \widehat{\sigma}_y, \widehat{\sigma}_z$, we have

$$\frac{d\widehat{\boldsymbol{\mu}}^A}{dt} = -i\gamma^2\hbar\left[(\widehat{\boldsymbol{\sigma}}\cdot\mathbf{B}_{\text{eff}})\widehat{\boldsymbol{\sigma}} - \widehat{\boldsymbol{\sigma}}(\widehat{\boldsymbol{\sigma}}\cdot\mathbf{B}_{\text{eff}})\right]$$

[12]They correspond to real eigenvalues, $\widehat{H}^{\dagger} = \widehat{H}$.

Expanding the dot products, we then have

$$\frac{d\widehat{\mu}^A}{dt} = -i\gamma^2\hbar\left[(\widehat{\sigma}_x B_x + \widehat{\sigma}_y B_y + \widehat{\sigma}_z B_z)\widehat{\boldsymbol{\sigma}} - \widehat{\boldsymbol{\sigma}}(\widehat{\sigma}_x B_x + \widehat{\sigma}_y B_y + \widehat{\sigma}_z B_z)\right]$$

$$= -i\gamma^2\hbar\begin{bmatrix}(\widehat{\sigma}_y\widehat{\sigma}_x - \widehat{\sigma}_x\widehat{\sigma}_y)B_y + (\widehat{\sigma}_z\widehat{\sigma}_x - \widehat{\sigma}_x\widehat{\sigma}_z)B_z\\ (\widehat{\sigma}_x\widehat{\sigma}_y - \widehat{\sigma}_y\widehat{\sigma}_x)B_x + (\widehat{\sigma}_z\widehat{\sigma}_y - \widehat{\sigma}_y\widehat{\sigma}_z)B_z\\ (\widehat{\sigma}_x\widehat{\sigma}_z - \widehat{\sigma}_z\widehat{\sigma}_x)B_x + (\widehat{\sigma}_y\widehat{\sigma}_z - \widehat{\sigma}_z\widehat{\sigma}_y)B_z\end{bmatrix}$$

Using the spin matrix relation $\widehat{\sigma}_i\widehat{\sigma}_j - \widehat{\sigma}_j\widehat{\sigma}_i = 2i\widehat{\sigma}_k$, we then have

$$\frac{d\widehat{\mu}^A}{dt} = -i\gamma^2\hbar\begin{bmatrix}-2i\widehat{\sigma}_z B_y + 2i\widehat{\sigma}_y B_z\\ 2i\widehat{\sigma}_z B_x - 2i\widehat{\sigma}_x B_z\\ -2i\widehat{\sigma}_y B_x + 2i\widehat{\sigma}_x B_z\end{bmatrix} = 2\gamma^2\hbar\widehat{\boldsymbol{\sigma}}\times\mathbf{B}_{\text{eff}} = -2\gamma\widehat{\boldsymbol{\mu}}\times\mathbf{B}_{\text{eff}}$$

This contribution is known as the *precessional torque*, and it contains the motion of the components of $\boldsymbol{\mu}$ *perpendicular* to the effective field $\mathbf{B}_{\text{eff}}$. This result accounts for *one of the two contributions* to the dynamic equation for $\widehat{\boldsymbol{\mu}}$.

For the second contribution from the imaginary part of the energy operator $-i\alpha V_\sigma$, both terms on the RHS of Eq. (6.108) must be included. We denote this contribution using the superscript B. The first part for this second contribution becomes

$$\frac{d\widehat{\mu}^B}{dt} = \frac{i}{\hbar}\left[\widehat{H}_\sigma^\dagger\widehat{\boldsymbol{\mu}} - \widehat{\boldsymbol{\mu}}\widehat{H}_\sigma\right] = \frac{i}{\hbar}\left[i\alpha\widehat{H}_\sigma\widehat{\boldsymbol{\mu}} + i\alpha\widehat{\boldsymbol{\mu}}\widehat{H}_\sigma\right]$$

$$= \frac{-\alpha}{\hbar}\left(\widehat{H}_\sigma\widehat{\boldsymbol{\mu}} + \widehat{\boldsymbol{\mu}}\widehat{H}_\sigma\right) = \alpha\gamma^2\hbar(\widehat{\boldsymbol{\sigma}}\cdot\mathbf{B}\widehat{\boldsymbol{\sigma}} + \widehat{\boldsymbol{\sigma}}\widehat{\boldsymbol{\sigma}}\cdot\mathbf{B})$$

$$= \alpha\gamma^2\hbar\begin{bmatrix}2\widehat{\sigma}_x^2 B_x + (\widehat{\sigma}_y\widehat{\sigma}_x + \widehat{\sigma}_x\widehat{\sigma}_y)B_y + (\widehat{\sigma}_z\widehat{\sigma}_x + \widehat{\sigma}_x\widehat{\sigma}_z)B_z\\ 2\widehat{\sigma}_y^2 B_y + (\widehat{\sigma}_x\widehat{\sigma}_y + \widehat{\sigma}_y\widehat{\sigma}_x)B_x + (\widehat{\sigma}_z\widehat{\sigma}_y + \widehat{\sigma}_y\widehat{\sigma}_z)B_z\\ 2\widehat{\sigma}_z^2 B_z + (\widehat{\sigma}_x\widehat{\sigma}_z + \widehat{\sigma}_z\widehat{\sigma}_x)B_x + (\widehat{\sigma}_y\widehat{\sigma}_z + \widehat{\sigma}_z\widehat{\sigma}_y)B_z\end{bmatrix}$$

Using the relations $\widehat{\sigma}_i^2 = \mathbf{I}(2)$ along with the spin matrix anti-commutation relation $\widehat{\sigma_i}\widehat{\sigma_j} + \widehat{\sigma_j}\widehat{\sigma_i} = 0$, the above becomes

$$\frac{d\widehat{\boldsymbol{\mu}}^B}{dt} = \alpha\gamma^2\hbar \begin{bmatrix} 2B_x \\ 2B_y \\ 2B_z \end{bmatrix} = 2\alpha\gamma^2\hbar\mathbf{B}_{\text{eff}} \tag{6.111}$$

For the second term, we have

$$\frac{-i}{\hbar}\langle\boldsymbol{\mu}\rangle\left[\widehat{H}_\sigma^\dagger - \widehat{H}_\sigma\right] = \frac{-i}{\hbar}\langle\boldsymbol{\mu}\rangle\left[i\alpha\gamma\hbar + i\alpha\gamma\hbar\right](\boldsymbol{\sigma}\cdot\mathbf{B}) \tag{6.112}$$

$$= \frac{2\alpha}{\hbar}\langle\boldsymbol{\mu}\rangle\left[\gamma\hbar\right](\boldsymbol{\sigma}\cdot\mathbf{B}_{\text{eff}}) \tag{6.113}$$

$$= -\frac{2\alpha}{\hbar}\left(\boldsymbol{\mu}\cdot\mathbf{B}_{\text{eff}}\right)\boldsymbol{\mu} \tag{6.114}$$

In order to more compactly tie our results together, we write the effective magnetic field $\mathbf{B}_{\text{eff}}$ in another way, using a vector identity which states that for a 3D unit vector $\mathbf{u} = \boldsymbol{\mu}/\mu$ ($|\mathbf{u}| = 1$), any vector $\mathbf{B}_{\text{eff}}$ can be decomposed using the unit vector as

$$\mathbf{B}_{\text{eff}} = \frac{1}{\mu^2}\left[(\boldsymbol{\mu}\cdot\mathbf{B}_{\text{eff}})\boldsymbol{\mu} - \boldsymbol{\mu}\times\boldsymbol{\mu}\times\mathbf{B}_{\text{eff}}\right] \tag{6.115}$$

This idea is illustrated in Fig. 6.17. According to Eq. (6.107), for an

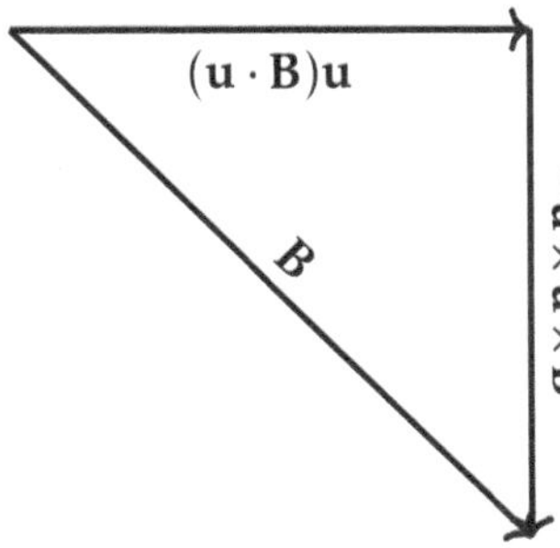

Figure 6.17: Illustration of the vector relations between $\mathbf{B}_{\text{eff}}$, $\mathbf{u}\times\mathbf{u}\times\mathbf{B}$, and $(\mathbf{u}\cdot\mathbf{B}_{\text{eff}})\mathbf{u}$.

electron spin, the moment μ is conserved and given by

$$|\boldsymbol{\mu}| = \mu_B = \gamma\hbar \tag{6.116}$$

Substitution of these relations into Eqs. (6.111) gives

$$\mathbf{B}_{\text{eff}} = \frac{2\alpha}{\hbar}\left[(\boldsymbol{\mu}\cdot\mathbf{B}_{\text{eff}})\boldsymbol{\mu} - \boldsymbol{\mu}\times\boldsymbol{\mu}\times\mathbf{B}_{\text{eff}}\right] \tag{6.117}$$

The first term in Eq. (6.117) *cancels* the identical term from Eq. (6.114). Then, we have the following for the time-derivative of the operator μ:

$$\begin{aligned}\frac{d\langle\mu\rangle}{dt} &= \frac{d\widehat{\mu}^A}{dt} + \frac{d\widehat{\mu}^B}{dt} \\ &= -2\gamma\mu \times \mathbf{B}_{\text{eff}} - \frac{2\alpha}{\hbar}\mu \times \mu \times \mathbf{B}_{\text{eff}} \\ &= -2\gamma\mu \times \mathbf{B}_{\text{eff}} - \frac{2\alpha\gamma}{\mu}\mu \times \mu \times \mathbf{B}_{\text{eff}}\end{aligned}$$

Using the definition for the electron g-factor $g = 2$, the components of the electron magnetic moment μ satisfy an equation of motion given by

Electron Magnetic Moment Dynamical Equation

$$\frac{d\mu}{dt} = \underbrace{-g\gamma\mu \times \mathbf{B}_{\text{eff}}}_{\text{precession}} - \underbrace{\frac{\alpha g\gamma}{\mu}\mu \times \mu \times \mathbf{B}_{\text{eff}}}_{\text{transverse damping}} \tag{6.118}$$

As mentioned, the first term in Eq. (6.118) describes *precessional* motion, involving the components of μ *perpendicular* to $\mathbf{B}_{\text{eff}}$. The second term is called the *transverse torque* corresponding to pure rotation of the electron magnetic moment vector μ *towards* the field $\mathbf{B}_{\text{eff}}$.

Eq. (6.118) is general, and can be solved together with a well-posed spin transport equation like that given by Eq. (6.73). In our treatment, we have mentioned two contributions to the effective magnetic field $\mathbf{B}_{\text{eff}}$, being the *sd*-exchange coupling from the spin-current and the spin-orbit interaction. Ordinarily, in a magnetic material, the spin-orbit interaction manifests in the form of an *easy axes*. However, there are additional contributions that can arise which are also important in spintronics. One contribution is known as the spin Hall effect, and we discuss it's properties next.

6.5 Spin-Hall Spin-Orbit Effects

The spin-Hall effect is an active area of research within the scope of *spintronics*. To examine the spin-Hall effect in a way that is relevant to spintronics, we must look at momentum values associated with the spin-orbit interaction, particularly when an electric field is applied. We can make use of the Generalized Ehrenfest theorem, or Eq. (4.21). It involves commuting the energy operator or Hamiltonian with the operator whose derivative we are after. For example, with an energy operator given by the kinetic energy $-\hbar^2\widehat{\mathbf{p}}^2/2m_0$, one can obtain dp/dt which becomes the charge-based Lorentz force[13]. We can use the same approach to determine how the expectation value of the extended momentum $\widehat{\mathbf{P}} = \widehat{\mathbf{p}} + e\mathbf{A}$ changes in time when the energy operator is the *spin-orbit interaction*. In these kinds of devices, there is current flowing, say along x, in a film structure as illustrated in Fig. 6.18.

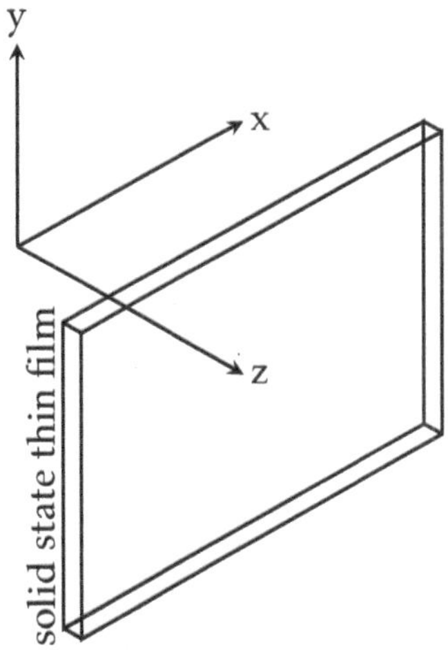

Figure 6.18: A thin film in a defined coordinate system. Solid state thin films of special materials are often used to exploit spin-orbit interactions. In the examples we consider, current flows along the longitudinal axis x.

Before we find the time-derivative of the momentum $\widehat{\mathbf{P}}$, let us start with something *simpler* like the position $\mathbf{x}$. For this, we have

$$\frac{d\langle \mathbf{x} \rangle}{dt} = \frac{i}{\hbar}\langle [\widehat{H}_{SO}, \mathbf{x}] \rangle \tag{6.119}$$

[13]For an introduction to quantum commutation operators, see Quantum Mechanics With A SPIN On It (Eason), Secs. 7.8 & 7.9

For convenience, we will omit the expectation brackets $e.g.\langle x\rangle \rightarrow x$. Then, substituting the spin-orbit operator from Eq. (2.132), we have

$$\frac{d\mathbf{x}}{dt} = \left[-\frac{i}{m_0^2c^2}\widehat{\sigma}\cdot(\widehat{\mathbf{p}}+e\mathbf{A})\times\nabla V\right]\mathbf{x}+\mathbf{x}\left[\frac{i}{m_0^2c^2}\widehat{\sigma}\cdot(\widehat{\mathbf{p}}+e\mathbf{A})\times\nabla V\right] \quad (6.120)$$

Let us take the gradient of the potential energy ∇V as spatially uniform. ∇V corresponds to a uniform external electric field $\mathbf{E}_0$ driving charges along the x–axis of the wire/film. Consequently, it *does not operate* on the wave function. Then, using the vector identity $\mathbf{a}\cdot\mathbf{b}\times\mathbf{c} = \mathbf{b}\cdot\mathbf{c}\times\mathbf{a}$, we can rewrite the dot-cross product as

$$\widehat{\sigma}\cdot\widehat{\mathbf{P}}\times\nabla V = \widehat{\mathbf{P}}\cdot(\nabla V\times\widehat{\sigma}) \quad (6.121)$$

This is permitted since both ∇V and $\widehat{\sigma}$ are *non-spatial* operators on the wave function. Then, using this result in Eq. (6.120), we have

$$\frac{d\mathbf{x}}{dt} = \frac{i}{m_0^2c^2}\left[\mathbf{x}\widehat{\mathbf{P}}\cdot(\nabla V\times\widehat{\sigma}) - \widehat{\mathbf{P}}\cdot(\nabla V\times\widehat{\sigma})\mathbf{x}\right] \quad (6.122)$$

Expanding Eq. (6.122), we have

$$\frac{m_0^2c^2}{i}\frac{d\mathbf{x}}{dt} = \left[\mathbf{x}\widehat{\mathbf{P}}\cdot(\nabla V\times\widehat{\sigma}) - \widehat{\mathbf{P}}\cdot(\nabla V\times\widehat{\sigma})\mathbf{x}\right] \quad (6.123a)$$

$$= \mathbf{x}\left(\widehat{P_x}(\nabla V\times\widehat{\sigma})_x + \widehat{P_y}(\nabla V\times\widehat{\sigma})_y + \widehat{P_z}(\nabla V\times\widehat{\sigma})_z\right) \quad (6.123b)$$

$$-\left((\nabla V\times\widehat{\sigma})_x\widehat{P_x} + (\nabla V\times\widehat{\sigma})_y\widehat{P_y} + (\nabla V\times\widehat{\sigma})_z\widehat{P_z}\right)\mathbf{x} \quad (6.123c)$$

Rearranging the RHS leads to each row of the vector equation having the form:

$$RHS = x_i\left[\left(\widehat{P_x}(\nabla V\times\widehat{\sigma})_x - (\nabla V\times\widehat{\sigma})_x\widehat{P_x}\right) + \left(\widehat{P_y}(\nabla V\times\widehat{\sigma})_y - (\nabla V\times\widehat{\sigma})_y\widehat{P_y}\right)\right.$$
$$\left. + \left(\widehat{P_z}(\nabla V\times\widehat{\sigma})_z - (\nabla V\times\widehat{\sigma})_z\widehat{P_z}\right)\right]x_i \quad (6.124)$$

In the above, $x_i = x, y, z$. Eq. (6.122) can be evaluated using the commutation relation $[x_i, \widehat{p_i}] = i\hbar$.[14] There are three nonzero commutations involving $[x, \widehat{P_x}]$, $[y, \widehat{P_y}]$, and $[z, \widehat{P_z}]$. For all the commutations with different components (e.g. z and $\widehat{P_x}$), they vanish since they commute. Then,

[14]This commutation relation is derived in the Appendix (*c*.Eq B.168).

the non-commuting components can be replaced by their eigenvalue $i\hbar$. This gives the result

$$\frac{m_0^2c^2}{i}\frac{d\mathbf{x}}{dt} = \begin{bmatrix} \left(x\widehat{P}_x - \widehat{P}_x x\right)(\nabla V \times \widehat{\boldsymbol{\sigma}})_x \\ \left(y\widehat{P}_y - \widehat{P}_y y\right)(\nabla V \times \widehat{\boldsymbol{\sigma}})_y \\ \left(z\widehat{P}_z - \widehat{P}_z z\right)(\nabla V \times \widehat{\boldsymbol{\sigma}})_z \end{bmatrix} = i\hbar(\nabla V \times \widehat{\boldsymbol{\sigma}}) \tag{6.125}$$

So we have a contribution to the electron's momentum as a consequence of the spin-orbit interaction, under an externally applied electric field. Then, using $\widehat{\mathbf{P}}_{SO} = m_0 d\langle\mathbf{x}\rangle/dt$, we have an additional momentum given by

$$\langle\widehat{\mathbf{P}}_{SO}\rangle = m_0\frac{d\langle\mathbf{x}\rangle}{dt} = -\frac{\hbar}{m_0c^2}\nabla V \times \langle\widehat{\boldsymbol{\sigma}}\rangle \tag{6.126}$$

In terms of the spin angular momentum $\widehat{\mathbf{S}}$, the *spin-orbit transverse momentum* becomes

Spin-Orbit Transverse Momentum

$$\langle\widehat{\mathbf{P}}_{SO}\rangle = -\frac{2}{m_0c^2}\langle\widehat{S}\rangle \times \nabla V \tag{6.127}$$

This momentum should be added to the two existing contributions so the total momentum $\widehat{\mathbf{P}}_T$ with spin-orbit interaction becomes

$$\widehat{\mathbf{P}}_T = \widehat{\mathbf{p}} + e\mathbf{A} + \widehat{\mathbf{P}}_{SO} \tag{6.128}$$

The first term in Eq. (6.128) is a conserved momentum, for example, due to an electric field $\mathbf{E}_0 = -\nabla V/e$ associated with the electrical current. The second term gives rise to the ordinary *charge-based Hall effect*. The additional spin-orbit term, however, is a *spin-dependent momentum* and will differentiate between different spin states. This spin-dependent momentum corresponds to a spin-dependent *force* on a charged particle *carrying spin*.

We can determine the spin-dependent force using the same approach. Analogous to what was done with x, we can evaluate

$$\frac{d\langle\mathbf{P}\rangle}{dt} = \frac{i}{\hbar}\langle[\widehat{H}_{SO}, \mathbf{P}]\rangle \tag{6.129}$$

Again, omitting the expectation value brackets for convenience, we have

$$\frac{d\widehat{\mathbf{P}}_{SO}}{dt} = \left[-\frac{i}{m_0^2c^2}\widehat{\mathbf{P}}\cdot(\nabla V\times\widehat{\boldsymbol{\sigma}})\right]\mathbf{P}+\mathbf{P}\left[\frac{i}{m_0^2c^2}\widehat{\mathbf{P}}\cdot(\nabla V\times\widehat{\boldsymbol{\sigma}})\right] \tag{6.130}$$

or

$$\begin{aligned}\frac{m_0^2c^2}{i}\frac{d\widehat{\mathbf{P}}_{SO}}{dt} &= \left[\widehat{\mathbf{P}}\widehat{\mathbf{P}}\cdot(\nabla V\times\widehat{\boldsymbol{\sigma}})-\widehat{\mathbf{P}}\cdot(\nabla V\times\widehat{\boldsymbol{\sigma}})\widehat{\mathbf{P}}\right]\\ &= \widehat{\mathbf{P}}\left(\widehat{P_x}\cdot(\nabla V\times\widehat{\sigma})_x+\widehat{P_y}\cdot(\nabla V\times\widehat{\sigma})_y+\widehat{P_z}\cdot(\nabla V\times\widehat{\sigma})_z\right)\\ &\quad -\left((\nabla V\times\widehat{\sigma})_x\widehat{P_x}+(\nabla V\times\widehat{\sigma})_y\widehat{P_y}+(\nabla V\times\widehat{\sigma})_z\widehat{P_z}\right)\widehat{\mathbf{P}}\end{aligned}$$

The result above is identical in form to Eq. (6.123). Here, instead of using the commutation relation $[x_i, P_i] = i\hbar$, we can use the commutation relation

$$\left[\widehat{P_i},\widehat{P_j}\right] = -ie\hbar\varepsilon_{ijk}B_k \tag{6.131}$$

ε_{ijk} is the Levi-Civita symbol.[15] Substituting this commutation relation, we end up with all the terms of the form P_iP_i canceling because they commute. Terms of the form P_iP_j can deploy Eq. (6.131). Evaluation then leads to the following time-derivative of the extended momentum operator:

$$\frac{m_0^2c^2}{i}\frac{d\widehat{\mathbf{P}}_{SO}}{dt} = -ie\hbar(\nabla V\times\widehat{\boldsymbol{\sigma}})\times\mathbf{B}\Rightarrow\langle\widehat{\boldsymbol{F}}\rangle_{SO}$$

Rearranging and solving for $dPSO/dt$, we have

Spin-Orbit Force

$$\langle\widehat{\boldsymbol{F}}\rangle_{SO} = -\frac{e\hbar}{m_0^2c^2}(\nabla V\times\langle\widehat{\boldsymbol{\sigma}}\rangle)\times\mathbf{B} \tag{6.132}$$

Eq. (6.132) is a *spin-dependent force* arising from the spin-orbit interaction.

[15] Eq. (6.131) is derived in the Appendix (c.Eq. (B.187)).

This unique force can drive a spin-carrier in a direction normal to a magnetic field **B** and momentum **P**. It is a *spin-dependent Lorentz force* because Eq. (6.132) can be written as

$$\langle\widehat{F}\rangle_{SO} = -\frac{e}{m_0}\widehat{\mathbf{P}}_{SO} \times \mathbf{B} = -e\langle\mathbf{v}_{SO} \times \mathbf{B}\rangle \tag{6.133}$$

This is the familiar Lorentz force form of $q\mathbf{v} \times \mathbf{B}$.

So, if there is an electric field $E \propto \nabla V_e$ (∇V_e is a contribution to ∇V) pointing along the x–axis, providing sustained charge motion, with spin-orbit interaction, the result will be that spins pointing along $+z$, for example, will develop spin-orbit momentum laterally along y. When a magnetic field **B** is applied along x, these spins will be *forced* 'upwards/downwards' because $(\widehat{\mathbf{x}} \times \pm\widehat{\mathbf{z}}) \times \widehat{\mathbf{x}} = \pm\widehat{\mathbf{z}}$. Therefore, *the spins end up moving towards the top and/or bottom surfaces* of the film, perpendicular to the z–axis. *This motion by spin-force is a form of spin-current.* This effect is generally referred to as the *spin-Hall effect.* As we learned in the previous section, a spin-current is equivalent to a magnetic field (dynamical), and can excite dynamics in a magnetic layer.

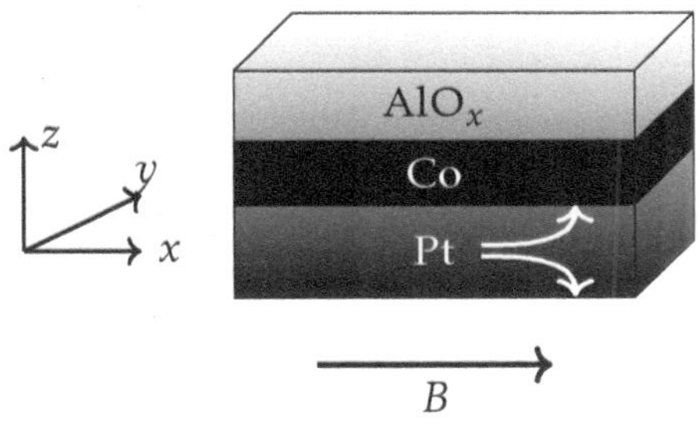

Figure 6.19: Illustration of a thin film device utilizing the *spin-Hall effect.* The platinum layer provides the large spin-orbit interaction, injecting polarized spins (along z) into the Co layer. The force on the spins is proportional to $(\widehat{\mathbf{x}} \times \pm\widehat{\mathbf{z}}) \times \widehat{\mathbf{x}} = \pm\widehat{\mathbf{z}}$.

Observations of the *spin-Hall* effect are a well-established. It has been used to drive the magnetization of a thin magnetic film. A well-known example of the *spin-Hall effect* used to control the switching of a magnetic layer was carried out by Liu *et. al.* The electrical current was injected into a device structure like that shown in Fig. 6.19. It was composed of a bottom layer of platinum (Pt), a cobalt (Co) layer above the Pt, and an aluminum oxide (AlO_x) layer on top. The magnetic field was along the same axis as current flow or the x axis. The resulting force created in the Pt *generated a spin current injected vertically* into the Co layer. Through its spin-current, the spin-Hall effect was able to reverse the magnetization of the Co layer.

In this example, the bottom platinum (Pt) layer is *essential* for the device physics because it provides the spin-orbit interaction where the overall spin-current effect originates. As electrical current initially flows longitudinally along x (*e.g.*$\sim$ 1mA), the spin-orbit interaction gives rise to spin-orbit momentum $\widehat{\mathbf{P}}_{SO}$ along the $\pm y$–axis. However, a magnetic field *e.g.* $B \sim 10$mT along x causes a *spin-orbit force* along z, which is what *injects* the spins of uniform polarization along $\pm z$ into the Co layer. As mentioned, the process of injecting spin-current into a magnetic layer can generate a *spin-transfer torque* causing it to reverse or switch its magnetization direction (also along z).

An important consideration with the spin-Hall effect is that *too large a magnetic field can be adverse* because beyond a certain magnetic field strength, the z–polarized spins will tend to rotate towards the magnetic field axis, which is perpendicular to z. Moreover, *without* a magnetic field, the magnetization of the Co layer can be rotated, but it cannot be switched.

6.6 Chapter Summary

In this chapter, the four dimensional energy operator developed in Chap. 2 was applied to the problem of electrical current flowing through structures including a magnetic layer. An early conclusion was a break in symmetry between spins of opposite direction. Having different wave numbers, this leads to the creation of a *spin polarized current* created by the magnetic layer through its coupling to the electron spin.

We considered examples of physical device structures used to create spin currents, extending from a single magnetic layer to a structure with a second (free) magnetic layer. In this kind of structure, magneto-electrical effects such as the magnetoresistance effect was derived. This is the change in overall device resistance when there is a change in the magnetization direction of the second magnetic layer. Related to this concept, we defined the *magnetoresistance ratio* (MR ratio), which is often used as a metric. With two magnetic layers, there must be a nonmagnetic layer between them to both decouple them, and provide a layer whose properties can be tailored to enhance the spin current and/or magnetoresistance. We found that an insulating barrier layer can generate larger magnetoresistance than a metallic layer by widening the split of the spin flow. We also discussed how *magnetoresistance* provides a way to measure an external magnetic field. Finally, we discussed concepts related to the torques that spin currents can impart to the free magnetic layer of a device. This is known as *spin-transfer torque*, examples of which can be created by the *spin-Hall* spin-orbit effect. We also derived a set of 4D space and time-dependent transport equations via the continuity equation,

Table 6.5 summarizes some of the key results from this chapter.

Table 6.5: Chapter 6 Summary Equations.

Name	Equation
Spin Polarization	$P = \mu_B \frac{T_\uparrow - T_\downarrow}{T_\uparrow + T_\downarrow}$
Magnetoresistance Fraction	$\text{MRF} = \frac{\langle R_{ap}\rangle - \langle R_p\rangle}{\langle R_p\rangle}$
Spin Density Matrix	$\Upsilon = \begin{bmatrix} \psi_\uparrow\psi_\uparrow^* & \psi_\uparrow\psi_\downarrow^* \\ \psi_\downarrow\psi_\uparrow^* & \psi_\downarrow\psi_\downarrow^* \end{bmatrix}$
Spin Cont. Equation	$\frac{\partial\Upsilon}{\partial t} = -\begin{bmatrix} \nabla\cdot J_\uparrow & \nabla\cdot J_{\uparrow,\downarrow} \\ \nabla\cdot J_{\downarrow,\uparrow} & \nabla\cdot J_\downarrow \end{bmatrix} - \frac{2\alpha}{\hbar} V_\sigma \Upsilon$
Spin-orbit Momentum	$\langle\widehat{\mathbf{P}}_{SO}\rangle = -\frac{2}{m_0 c^2}\langle\widehat{\boldsymbol{S}}\rangle \times \nabla V$
Spin-orbit Force	$\langle\widehat{\boldsymbol{F}}\rangle_{SO} = -\frac{e\hbar}{m_0^2 c^2}(\nabla V \times \langle\widehat{\boldsymbol{\sigma}}\rangle) \times \mathbf{B}$
Spin-diffusion Length (↓)	$\ell_{sd} = \frac{\hbar k}{2\alpha m_0 \omega_{\text{eff}}}$

6.7 Chapter Problems

Problem 6.1 A dynamical equation of motion, known as the spin-transfer torque equation, was found to be (*c*.Eq. (6.118))

$$\frac{d\boldsymbol{\mu}}{dt} = \underbrace{-g\gamma\boldsymbol{\mu}\times\mathbf{B}_{\text{eff}}}_{\text{precession}} - \underbrace{\frac{\alpha g\gamma}{\mu}\boldsymbol{\mu}\times\boldsymbol{\mu}\times\mathbf{B}_{\text{eff}}}_{\text{transverse rotation}}$$

Considering the STT equation, can a design be devised for a spin-transfer torque system using a static external magnetic field that has *zero precessional motion*? Considering a non-static magnetic field, is it possible? If so, how can this be achieved?

Problem 6.2 Similarly, based on the torque terms in spin-transfer torque, can a design be devised for a spin-transfer torque system using any external magnetic field having *pure* precessional motion, and no transverse rotation?

Problem 6.3 Based on this equation, is the following statement *true* or *false*?

$$\frac{\partial}{\partial t}(\boldsymbol{\mu}_S\cdot\boldsymbol{\mu}_S) = 0$$

Explain why you have chosen your answer.

Problem 6.4 Estimate the strength of the spin-orbit force (per spin) for an electron flowing through along 1cm sample, driven by 1 V, in an external magnetic field of 1 Tesla, using the relation (*c*.Eq.6.132)

$$\langle\widehat{\boldsymbol{F}}\rangle_{SO} = \frac{e\hbar}{m_0^2c^2}(\nabla V\times\langle\widehat{\boldsymbol{\sigma}}\rangle)\times\mathbf{B}$$

Problem 6.5 Let the number of charge carriers per unit volume be $n = 10^{28}\text{m}^{-3}$, with an applied voltage of 1V over a sample 1cm in length L. Using an electrical conductivity of $\sigma = 1.0\times10^7\text{S/m}$, first, estimate the average momentum per charge carrier using the relation

$$nev = \frac{nep}{m_0} = \sigma E \Rightarrow p \approx \frac{\sigma E m_0}{ne}$$

E is the electric field, which can be estimated by $E \approx V/L$. Use the result to estimate the maximum strength of spin-orbit interaction using the relation

$$SOI \approx \frac{\hbar p E}{m_0^2 c^2}$$

Problem 6.6 Earlier, we obtained a spin-diffusion length (*c*.6.83) given by

$$\ell_{sd} = \frac{\hbar k}{2\alpha m_0 \omega_{\text{eff}}} = \frac{p}{2\alpha m_0 \omega_{\text{eff}}}$$

Using $p = \hbar k$ with the value of p from Prob. 6.5, calculate the spin-diffusion length for the following parameters:

$$\alpha = 0.01, B_{\text{eff}} = 0.1 \text{ Tesla}$$

$$\alpha = 0.02, B_{\text{eff}} = 0.4 \text{ Tesla}$$

$$\alpha = 0.05, B_{\text{eff}} = 0.5 \text{ Tesla}$$

Problem 6.7 Estimate the *precessional frequency* for an electron spin coupled to an exchange energy of value $J = 1 \times 10^{-21}$ Joules. What is the frequency in GHz? Also determine the effective exchange field $B_{exch} = J/2\mu_B$.

Problem 6.8 Prove that a magnetic potential energy of the form given by Eq. (4.48) leads to the relation given by

$$\frac{\partial V_M}{\partial \boldsymbol{\mu}} = -g|\gamma|\mathbf{B} - \frac{-g|\gamma|\alpha}{2\mu}\boldsymbol{\mu} \times \mathbf{B} + \frac{g|\gamma|\alpha B}{\mu}\boldsymbol{\mu}$$

Problem 6.9 Show that the expectation value of the eigenenergy is given by $\langle E_\lambda(t)\rangle = \frac{g|\gamma|\hbar B}{2}\tanh\left(\alpha\omega_p t\right)$.

Problem 6.10 Let the wave function $\psi(x,t)$ for a Hamiltonian be given by

$$\psi(x,t) = Ae^{i\left(k_r x - \frac{\alpha V}{\hbar}t\right)}e^{-k_i x + \frac{\alpha V}{\hbar}t}$$

$k = k_r + ik_i$ is the wave number, α is the dissipation parameter, while V is a real valued potential. Use the *dissipative continuity equation* given by

$$\frac{\partial \rho}{\partial t} + \nabla \cdot \mathbf{J} = -\frac{2\alpha}{\hbar}V\rho$$

and show that it leads to the following condition which must be satisfied:

$$\hbar^2 k_i k_r = 2\alpha m_0 V$$

Problem 6.11 For a pair of 3D vectors **A** and **B**, show that

$$\mathbf{B} = \frac{1}{|A|^2}\left[(\mathbf{A}\cdot\mathbf{B})\mathbf{A} - \mathbf{A}\times\mathbf{A}\times\mathbf{B}\right]$$

6.8 Suggested Readings & References

[1] K. Eason, *Quantum Mechanics With A SPIN On It*, Sym. Pub. (2019).

[2] J. E. Hirsch, *Spin Hall Effect*, **117**, pp. 18341837, Phys. Rev. Lett., 83 (9) (1999).

[3] W. Butler, et. al. *Spin-dependent tunneling conductance of Fe|MgO|Fe sandwiches*, Phys. Rev. B., **63**, 054416 (2001).

[4] S. Yuasa, et. al. *Giant room-temperature magnetoresistance in single-crystal Fe|MgO|Fe magnetic tunnel junctions*, Nature Mat., **3**, pp. 868-871 (2004).

[5] T. Valet and A. Fert, *Theory of the perpendicular magnetoresistance in magnetic multilayers*, Phys. Rev. B., **63**, 054416 (1993).

[6] L. Berger, *Precession of Conduction-Electron Spins Near an Interface Between Normal and Magnetic Metals*, IEEE T. Mag., **31**, pp. 3871-3873 (1995).

[7] J. Slonczewski, *Current-driven excitation of magnetic multilayers*, **159**, pp. L1-L7, J. Mag. Magn. Mater. (1996).

[8] R. Meservey and P.M. Tedrow, *Spin-polarized electron tunneling*, PHYSICS REPORTS (Review Section of Physics Letters) **238**, No. 4 pp. 173243 (1994).

Chapter 7

Relativistic Treatment Of Superconductors

Superconductors are recognized as exotic and interesting materials. They are of high interest in the scientific world due to their intriguing possibilities for novel low-energy consumption applications. However, there is a price. Still, since the discovery of *superconductivity* c.1911, numerous applications have still been realized from medical technologies to highly sensitive magnetic field detectors. More recently, superconductors have even found their way into a *potentially* novel compute-processor technology known as a *quantum computer*. Some of these systems rely on quantum circuits that aim to exploit superconductors. At this time, two of the front-runners include *ion trapping* and those that use quantum circuits based on *superconductingquantum bits*, or *qubits*. In quantum computers using superconductivity, a special type of junction is used which is the heart of the quantum circuit. It is somewhat analogous to how the *transistor* is a building block for today's computer processors and some memory. This junction, exploiting superconductivity, is called a *Josephson junction*, which we'll be discussing in this chapter, also. However, before we can understand how a Josephson junction works, we should examine what *relativistic* quantum mechanics has to say about superconductors at a more fundamental level.

In this chapter, we reach into the bag of relativistic quantum mechanics

Sym	T_c[K]	$B_c(0)$[mT]	Valence	L	S
Nb	9.50	198.0	$\cdots 3d^{10}\,4s^2p^6\mathbf{d}^4\,5s^1$	2	5/2
Tc	7.77	141.0	$\cdots 3d^{10}\,4s^2p^6\mathbf{d}^5\,5s^2$	0	5/2
Pb	7.19	80.3	$\cdots 4s^2p^6d^{10}f^{14}\,5s^2p^6d^{10}6s^2\mathbf{p}^2$	1	2/2
La	6.00	110.0	$\cdots 4s^2p^6d^{10}\,5s^2p^6\mathbf{d}^1\,6s2$	2	1/2
V	5.38	142.0	$1s^2\,2s^2p^6\,3s^2p^6\mathbf{d}^3\,4s^2$	3	3/2
Ta	4.48	83.0	$\cdots 4s^2p^6d^{10}f^{14}\,5s^2p^6\mathbf{d}^3 6s^2$	3	3/2
Hg*	4.15	41.2	$\cdots 4s^2p^6d^{10}f^{14}\,5s^2p^6\mathbf{d}^{10}\,6s^2$	0	0
Sn	3.72	30.9	$\cdots 3d^{10}\,4s^2p^6d^{10}\,5s^2\mathbf{p}^2$	1	2/2
In	3.40	29.3	$\cdots 3d^{10}\,4s^2p^6d^{10}\,5s^2\mathbf{p}^1$	1	1/2

Table 7.1: List of some superconducting elements (free atom) with corresponding critical temperature T_c and field B_c. The third and four columns contain (principal quantum number)(L=total orbital quantum number) and S, being the total atomic spin number. Bold indicates the highest energy state, which conduction electrons are most likely to come from.

to analyze superconductors. To do this, we must discuss the role of higher order energy operators like those worked out in Chap. 2. Together with quantum transport concepts and statistical thermodynamics, superconductors *can* be explained satisfactorily. An important question to answer in this discussion is towards the origins of Type I and Type II superconductors. The relativistic quantum mechanical treatment of superconductors is capable of providing this explanation, and more.

7.1 Energy Operators For Superconductivity

Mercury (Hg) was the first superconducting materials discovered experimentally by physicist Heike Onnes and his research group at University of Leiden, in Netherlands *c.*1911. For superconductors, it is important to consider the experimental conditions in which superconductivity is observed. Generally, it is only observed in *certain* materials taken to a sufficiently *low temperature*. These conditions have implications on the energy operator contributions that should be taken into consideration.

In quantum mechanics, there are infinitely many energy contributions which belong to the relativistic kinetic energy. Here are some relevant quantum mechanical energy operators plucked from the relativistic energy operator:

$$\widehat{E}_K = \frac{(\widehat{\mathbf{p}} + e\mathbf{A})^2}{2m_0} \quad \text{(kinetic energy)} \tag{7.1a}$$

$$\widehat{E}_Z = -\widehat{\mu}_L \cdot \mathbf{B} \quad \text{(Zeeman energy)} \tag{7.1b}$$

$$\widehat{E}'_Z = -\widehat{\mu}_S \cdot \mathbf{B} \quad \text{(anomolous Zeeman energy)} \tag{7.1c}$$

$$\widehat{E}'_K = -\frac{(\widehat{\mathbf{p}} + e\mathbf{A})^4}{2m_0^3c^2} \quad \text{(higher order kinetic energy)} \tag{7.1d}$$

$$\widehat{E}'_K = -\frac{\hbar}{m_0^2c^2}\widehat{\boldsymbol{\sigma}}' \cdot (\widehat{\mathbf{p}} + e\mathbf{A}) \times \nabla V \quad \text{(spin-orbit energy)} \tag{7.1e}$$

For our discussion of superconductors, any energy contribution considered, with the exception of the external electric field to drive the electrical conduction, is among those listed in Eqs. (7.1).

The last two in the list are examples of higher order energy operators, going with $1/c^2$. As such, they are expected to be smaller in strength. In contrast, the first three are of the order $1/c^0 = 1$. The contributions of $1/c^2$, however, should be next in line in terms of their contribution. Also, in metals, because conduction is, generally, *more* efficient with reduced temperature, momentum *increases*. The higher order operators are momentum dependent. This is another reason why we need to include some higher order terms *e.g.* Eq. (7.1d) in our operators for superconductors, particularly, since they are materials in reduced temperatures.

Since the *conduction* properties, in particular, do not require an external magnetic field, to understand the transition from ordinary conduction to *super*, we'll turn it off. This also means we can ignore the Zeeman energy of Eq. (7.1b), as well as Eq. (7.1c).

We'll consider two types of transition conditions: from bound to itinerant or quasi-itinerant electrons and second: from itinerant electrons back to localized. Throughout both, there is a non-zero electric field. Relating to the first type of transition, it is useful to examine the behavior of the spin-orbit interaction given by Eq. (7.1e) of a localized electron.

7.2 Localized Spin-Orbit And Temperature Dependence

There are two contributions we'll consider towards the enlargement of the localized spin-orbit interaction. Their combined effect takes place when *reducing* the temperature. For the first contribution, let us assume that the potential gradient in the spin-orbit interaction is due to the nucleus having a charge of Ze where Z is the atomic number and e is the elementary charge. The electrical potential energy gradient due to the nuclear charge source is given by

$$\nabla V = \nabla\left(\frac{Ze^2}{4\pi\epsilon_0}\frac{1}{r}\right) = -\frac{Ze^2}{4\pi\epsilon_0}\frac{\mathbf{r}}{r^3} \tag{7.2}$$

$\mathbf{r}$ is the *relative* position vector with respect to the origin, located at the nucleus (or source charge) of the atom. Substituting the potential energy gradient from Eq. (7.2) into the spin-orbit interaction of Eq. (7.1e), we have

$$\widehat{H}_{SO} = \frac{Ze^2\hbar}{4\pi\epsilon_0 m_0^2 c^2}\widehat{\sigma}\cdot(\widehat{\mathbf{p}}+e\mathbf{A})\times\frac{\mathbf{r}}{r^3}$$

There is a sign-change since ∇V is negative and we already had a negative sign in the spin-orbit interaction.

Since a magnetic field is *not* required to observe the unique electrical properties of a superconductor, let's first consider the case of no magnetic field field or $\mathbf{A} = 0$. Then, we have

$$\widehat{H}_{SO} = \frac{Ze^2\hbar}{4\pi\epsilon_0 m_0^2 c^2}\widehat{\sigma}\cdot\widehat{\mathbf{p}}\times\frac{\mathbf{r}}{r^3} \tag{7.3}$$

The operator given by $\widehat{\mathbf{p}}\times\frac{\mathbf{r}}{r^3}$ can be expanded into an explicit vector form given by

$$\begin{vmatrix} i & j & k \\ \widehat{p}_x & \widehat{p}_y & \widehat{p}_z \\ \frac{r_x}{r^3} & \frac{r_y}{r^3} & \frac{r_z}{r^3} \end{vmatrix} = \begin{bmatrix} i\left(\widehat{p}_y\frac{r_z}{r^3} - \widehat{p}_z\frac{r_y}{r^3}\right) \\ -j\left(\widehat{p}_x\frac{r_z}{r^3} - \widehat{p}_z\frac{r_x}{r^3}\right) \\ k\left(\widehat{p}_x\frac{r_y}{r^3} - \widehat{p}_y\frac{r_x}{r^3}\right) \end{bmatrix} \tag{7.4}$$

Evaluating the *first component* of the vector operator in Eq. (7.4), we have

$$\widehat{p}_y\frac{r_z}{r^3} - \widehat{p}_z\frac{r_y}{r^3} = -i\hbar\left(\frac{\partial}{\partial y}\frac{r_z}{r^3} - \frac{\partial}{\partial z}\frac{r_y}{r^3}\right) \tag{7.5}$$

Using the relation $\mathbf{r} = [r_x, r_y, r_z]^T = [x^2, y^2, z^2]^{1/2}$, the first term in parentheses gives[1]

$$\frac{\partial}{\partial y}\frac{r_z}{r^3} = r_z \cdot \frac{\partial}{\partial y}\left(x^2+y^2+z^2\right)^{-3/2} + \frac{r_z}{r^3}\frac{\partial}{\partial y} \tag{7.6a}$$

$$= -\frac{3r_z(2r_y)}{2}\left(x^2+y^2+z^2\right)^{-5/2} + \frac{r_z}{r^3}\frac{\partial}{\partial y} \tag{7.6b}$$

$$= -\frac{3r_z r_y}{r^5} + \frac{r_z}{r^3}\frac{\partial}{\partial y} \tag{7.6c}$$

Similarly, evaluating the second term in parentheses of Eq. (7.5) gives

$$\frac{\partial}{\partial z}\frac{r_y}{r^3} = r_y \cdot \frac{\partial}{\partial z}\left(x^2+y^2+z^2\right)^{-3/2} + \frac{r_y}{r^3}\frac{\partial}{\partial z} \tag{7.7a}$$

$$= -\frac{3r_y(2r_z)}{2}\left(x^2+y^2+z^2\right)^{-5/2} + \frac{r_y}{r^3}\frac{\partial}{\partial z} \tag{7.7b}$$

$$= -\frac{3r_y r_z}{r^5} + \frac{r_y}{r^3}\frac{\partial}{\partial z} \tag{7.7c}$$

The first term

$$\text{common term} = -\frac{3r_z r_y}{r^5}$$

in both evaluations is *common*. Thus, in taking the difference between them, this term falls out and we have

$$\widehat{p}_y\frac{r_z}{r^3} - \widehat{p}_z\frac{r_y}{r^3} = -i\hbar\left(\frac{r_z}{r^3}\frac{\partial}{\partial y} - \frac{r_y}{r^3}\frac{\partial}{\partial z}\right) = \frac{1}{r^3}\left(\widehat{p}_y r_z - \widehat{p}_z r_y\right) \tag{7.8}$$

Carrying out the evaluations for the second and third components provides the remaining elements of $\widehat{\mathbf{p}} \times \mathbf{r}$. Then, we have that

$$\widehat{\mathbf{p}} \times \frac{\mathbf{r}}{r^3} = \frac{1}{r^3}\left(\widehat{\mathbf{p}} \times \mathbf{r}\right) \tag{7.9}$$

Substitution into Eq. (7.3) and using the operator relation $\widehat{\mathbf{L}} = \mathbf{r} \times \widehat{\mathbf{p}} = -\widehat{\mathbf{p}} \times \mathbf{r}$, we have

$$\widehat{H}_{SO} = \frac{Ze^2\hbar}{4\pi\epsilon_0 m_0^2 c^2 r^3}\widehat{\boldsymbol{\sigma}} \cdot \widehat{\mathbf{p}} \times \mathbf{r} \tag{7.10a}$$

---

[1] Keep in mind $x = r_x, y = r_y$, and $z = r_z$.

$$= -\frac{Ze^2}{2\pi\epsilon_0 m_0^2 c^2 r^3}\widehat{\mathbf{S}}\cdot\widehat{\mathbf{L}} \tag{7.10b}$$

From Eq. (7.10b), we find that a localized electron spin having spin-orbit interaction can effectively couple to the orbital angular momentum, analogous to how both can couple to an external magnetic field. These are known as Zeeman energies. So, the spin-orbit interaction is also given by

Localized Spin-Orbit Interaction

$$\widehat{H}_{SO} = -\frac{Ze^2}{2\pi\epsilon_0 m_0^2 c^2 r^3}\widehat{\mathbf{S}}\cdot\widehat{\mathbf{L}} \tag{7.11}$$

We can use Eq. (7.11) to estimate the *magnitude* of the spin-orbit coupling for bound electrons in the atoms of superconducting materials such as Niobium (Nb), Technetium (Tc), Lead (Pb), and similar materials to those listed in Table 7.1. Fig. 7.1 plots the coefficient of Eq. (7.11) for atoms with nuclei charge $Z = 41, 43, 82$, corresponding to Nb, Tc, Pb. Note that for $\widehat{\mathbf{S}}\cdot\widehat{\mathbf{L}}$, an order of magnitude value of $\hbar^2$ has been used.

Fig. 7.1 demonstrates the following for the spin-orbit interaction:

1. The interaction is on the order of meV, particularly, in the conditions and size range of superconducting metallic atoms
2. The interaction increases with *any* effect that reduces atomic size.

For a metal, atomic size reduction *turns* out to be one of the consequences of reduced temperature. As pointed out in Sec. 7.4, reducing temperature in metals causes the occupation of quantum states by bound electrons to *squeeze* into lower and lower energies near the Fermi edge according to a Fermi population.[2] The distribution is governed by the so-called Fermi-Dirac distribution given by[3]

$$f_{F-D}(E_\lambda) = \frac{1}{e^{E_\lambda/k_B T} + 1} \tag{7.12}$$

[2]Here, the electrons are the fermions, which have a unique set of rules for how to occupy quantum states.

[3]For a detailed discussion and derivation of the Fermi-Dirac distribution, see Chap. 3, Quantum Mechanics With A SPIN On It (Eason).

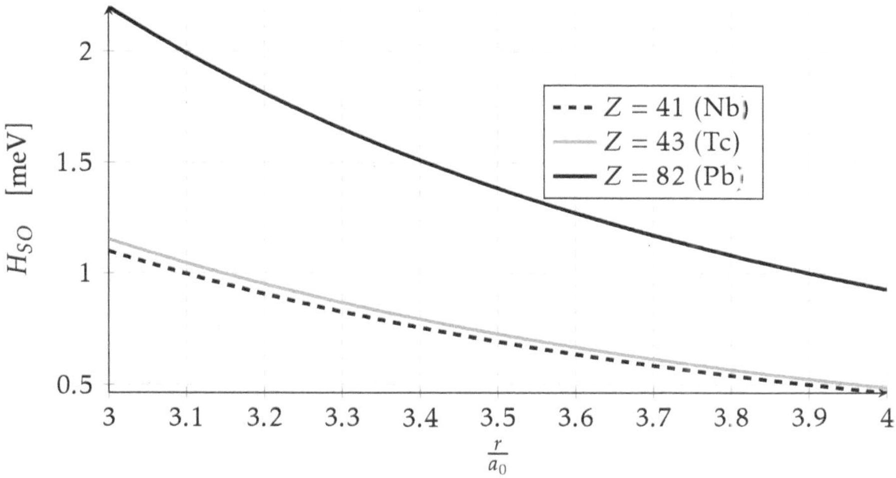

Figure 7.1: Calculated *internal* atomic spin-orbit coupling experienced by a localized electron using Eq. (7.11). The x-axis is measured in units of Bohr radius $a_0 = .529\text{Å}$. The elements shown correspond to Nb ($Z = 41$)-atomic radius ($r_0 = 2.08\text{Å}$), Tc ($Z = 43$)-atomic radius ($r_0 = 1.95\text{Å}$), and Pb ($Z = 82$)-atomic radius ($r_0 = 1.81\text{Å}$). In this context, r_0 denotes the known atomic radius for the element.

$k_B = 1.38065 \times 10^{-23}$ J-K^{-1} is the Boltzmann constant and T is the *thermodynamic equilibrium temperature*. The energy variation *width* is more directly observed in the width of f_{F-D}/dE_λ, given by

$$\frac{df_{F-D}}{dE_\lambda} = -\frac{1}{k_B T} \frac{e^{E_\lambda/k_B T}}{e^{E_\lambda/k_B T} + 1} \tag{7.13}$$

Both distributions are illustrated in Fig. 7.2 for $T = 4, 6,$ and 8K .

We can see from Fermi-Dirac distribution that the scale of the energy variations for this temperature range are on the order of meV. This can be seen when we evaluate the quantity

$$10k_B = 10 \times 1.3806505 \times 10^{-23} = 1.3806505 \times 10^{-22}\text{J} = 0.86183\text{meV}$$

For temperatures 10K, the energies correspond to being on the order of 1meV. And, as the temperature is reduced, the distributions can be seen to *squeeze in* closer towards the Fermi edge ϵ_F. Because of this effect from temperature, the atomic size is reduced as it is also known that lower and *lower bound atomic energy states have smaller spatial extents*.

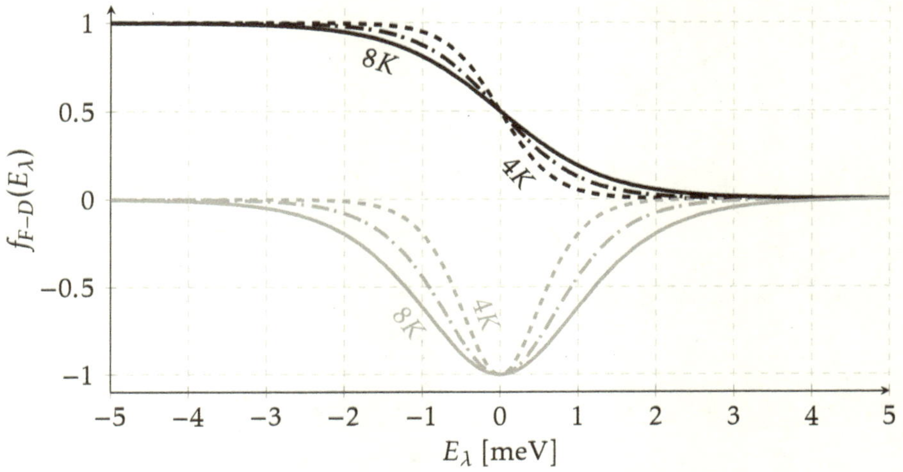

Figure 7.2: Plots of the Fermi-Dirac distribution $f_{F-D}(E_\lambda)$ which determines the occupation of states for electrons or fermions. Each curve corresponds to a different temperature $T = 4, 6$, and/or $8K$. $\epsilon_F = 0$ meV corresponds to the Fermi edge located near the 'edge' of an atom. As temperature decreases, the width narrows as electrons occupy tighter and tighter spaces (or lower energies) within the atom.

Therefore, if the effective atomic size is reduced, based on Eq. (7.11), we can also expect an increase in the spin-orbit interaction for localized electrons. The question becomes in what manner will it increase with a reduced temperature? In this case, the spin-orbit coupling rises as $1/r^2$ because $L/r^3 \propto rp/r^3 \propto p/r^2$. r is the mean distance between the valence electron and the nucleus/origin.

Any energy contribution on this scale can, therefore, potentially impact the state of an electron in the vicinity of the Fermi edge. In the context of conduction, the spin-orbit is more relevant acting on valence electrons lying in non-full p and d orbitals, as they have non-zero spin-orbit interaction. This is the case for the elements of Table 7.1, being mostly metallic superconductors, particularly after an external electric field has been introduced.

We can use the following relations to express the spin-orbit interaction in the form of a Zeeman energy (*i.e.* $H_{SO} = -\boldsymbol{\mu} \cdot \mathbf{B}$):

$$\mu_S = -\frac{ge}{2m_0}\widehat{\mathbf{S}} = -\frac{2e}{2m_0}\widehat{\mathbf{S}} \tag{7.14a}$$

$$\mu_L = -\frac{e}{2m_0}\widehat{\mathbf{L}} \tag{7.14b}$$

$$c^2 = \frac{1}{\epsilon_0\mu_0} \tag{7.14c}$$

$\mu_0 = 4\pi 10^{-7}\ N/A^2$(units of Newtons per Ampere squared) is the *magnetic permeability in vacuum*. Then, we can write Eq. (7.11) as

$$\widehat{H}_{SO} = -\frac{Ze^2}{2\pi\epsilon_0 m_0^2 c^2 r_0^3}\widehat{\mathbf{S}}\cdot\widehat{\mathbf{L}} \tag{7.15a}$$

$$= -\frac{Z}{\pi\epsilon_0 c^2 r_0^3}\left[\frac{-2e}{2m_0}\widehat{\mathbf{S}}\right]\cdot\left[\frac{-e}{2m_0}\widehat{\mathbf{L}}\right] \tag{7.15b}$$

$$= -\frac{Z}{\pi\epsilon_0 c^2 r_0^3}\mu_S\cdot\mu_L \tag{7.15c}$$

$$= -\frac{Z\mu_0}{\pi r_0^3}\mu_S\cdot\mu_L \tag{7.15d}$$

$$= -\mu_S\cdot\widehat{\mathbf{B}}_{SO} \tag{7.15e}$$

where the *spin-orbit field operator* can be defined as

$$\widehat{\mathbf{B}}_{SO} = \frac{Z\mu_0}{\pi r_0^3}\widehat{\mu}_L \tag{7.16}$$

The spin-orbit interaction is, thus, analagous to a Zeeman energy whose energy is minimized when **S** and **L** or μ_S and μ_L are *parallel*. In parallel conditions, *the energy is also negative in sign.*

In physical materials, the orbital angular momentum state is more dependent on the entire atomic structure, and thus, more energy is required to alter the states of orbital angular momentum. When conditions are reached that can alter **L**, the spin states can follow suit according to the spin-orbit interaction. This kind of nonlinear behavior can be observed in magnetic materials as the so-called *hysteresis* phenomenon.

Along with the *exchange interaction*, the spin-orbit interaction is also the origin of the *easy axes* because for moments aligned 'downward', a positively applied external field has to overcome $\mathbf{B}_{SO} + \mathbf{B}_{ex}$ *before* the collective moments μ will rotate towards the opposite direction. Fig. 7.3 shows an example of a magnetic hysteresis curve, which is the nonlinear magnetic response to externally applied magnetic field. .

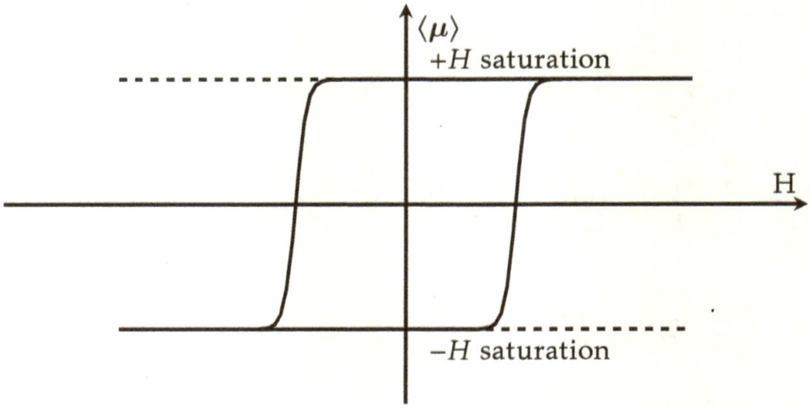

Figure 7.3: Hysteresis loop example. To achieve reversal, the applied magnetic field (x-axis) *must* overcome the internal effective (spin-orbit + exchange) field, given approximately by the crossing value along the H (or x) axis.

Thus, the spin-orbit interaction is largely dominated by the characteristics of the orbital angular momentum. Another relevant property is the symmetry of **L** states. The eigenstates or quantum states of $\widehat{\mathbf{L}}$ have symmetric qualities. For example, for a state given by

$$\mathbf{L} = \sqrt{\ell(\ell+1)}\hbar\mathbf{e}_z$$

there exists a corresponding quantum state of

$$\mathbf{L} = -\sqrt{\ell(\ell+1)}\hbar\mathbf{e}_z$$

Because of this kind of symmetry, the spin-orbit interaction will inherit symmetric properties, as well. This is evident in it's form. For example, the energy is equally minimized whether both spin and orbital states point 'up' or 'down'. These symmetry properties can be used to find an equilibrium temperature dependence of the spin-orbit interaction.

Summarizing, metallic elements like those shown in Table 7.1 have net *spin-down* electrons that generally possess a non-zero spin-orbit interaction. For the localized electron, this energy contribution is *negative in sign*. Let us take the properties of the spin-orbit interaction to find the equilibrium temperature dependence.

7.2.1 Equilibrium Temperature Dependence Of Spin-Orbit

To introduce finite temperature, we can evaluate an appropriate thermodynamic expectation value of the electron spin, as well as the spin-orbit interaction. By this, we can obtain a temperature dependence under conditions of *thermodynamic equilibrium* or constant temperature. This is consistent with most experiments with superconductors. For a localized electron, the spin-orbit interaction is expected to be negative in sign. Therefore, in computing the average, *this sign must also be taken into account.*

Taking the spin-orbit energy as symmetric, we can use the *Boltzmann probability distribution* where the probability of a potential energy state E_i is given by

$$f_B(E_i) = \frac{e^{-E_i/k_B T}}{\sum_{m_j=-j}^{m_j=+j} e^{-E_i/k_B T}} \tag{7.17}$$

Let us define the energy V_0 for a single electron according to Eq. (7.11) as

$$V_0 = -\mu_S \cdot \widehat{\mathbf{B}}_{SO} = -\frac{Ze^2}{2\pi\epsilon_0 m_0^2 c^2 r^3}\widehat{\mathbf{S}} \cdot \widehat{\mathbf{L}} \tag{7.18}$$

For a multi-electron system, the angular momentum operators given by

$$\widehat{L^2}\psi = \ell(\ell+1)\hbar^2\psi \qquad \widehat{L}\psi = \sqrt{\ell(\ell+1)}\hbar\psi \tag{7.19}$$

generalizes to

$$\widehat{J^2}\psi = j(j+1)\hbar^2\psi \qquad \widehat{J}\psi = m_j\hbar\psi \tag{7.20}$$

m_j is the total magnetic quantum number inclusive of both spin and orbital angular momentum, taking positive or negative values of j. Then, the expectation value of the spin-orbit interaction can be written as

$$\langle V_S\rangle(T) = \sum_{m_j=-j}^{m_j=+j} f_B(E_{m_j})E_{m_j} \tag{7.21}$$

which can be written as

$$\langle V_S\rangle = N\sum_{m_j=-j}^{j} m_j V_0 f_B(E_{m_j})$$

$$= N \sum_{m_j=-j}^{j} m_j V_0 \frac{e^{(m_j V_0/k_B T)}}{\sum_{m_j=-j}^{j} e^{(m_j V_0/k_B T)}}$$

Let $x = V_0/(k_B T)$. Then, we can write

$$\langle V_S \rangle = N \frac{\sum_{m_j=-j}^{j} m_j V_0 e^{m_j x}}{\sum_{m_j=-j}^{j} e^{m_j x}} \tag{7.22}$$

Using the relation $d(e^{m_j x})/dx = m_j e^{m_j x}$, Eq. (7.22) becomes

$$\langle V_S \rangle = N V_0 \frac{\sum_{m=-j}^{j} \frac{d}{dx} e^{mx}}{\sum_{m=-j}^{j} e^{mx}}$$

We have used the fact that V_0 does not depend on m_j. If we let $u = \sum_{m_j=-j}^{j} e^{m_j x}$, we can write

$$\langle V_S \rangle = N V_0 \frac{1}{u} \frac{du}{dx} = N V_0 \frac{d}{dx} \ln u$$

It is useful to write u in the following form:

$$u = \sum_{m_j=-j}^{j} e^{m_j x} = e^{-jx} + e^{(-j+1)x} + \cdots + e^{(j-1)x} + e^{jx}$$

$$= e^{-jx}(1 + e^x + e^{2x} + \cdots + e^{2jx})$$

$$= e^{-jx}(1 + r + r^2 + \cdots + r^{2j})$$

If we let $r = e^x$, the summation in parentheses is in the form of a *finite geometric series*, which has the form

$$a + ar + ar^2 + \cdots + ar^n = a \frac{1 - r^{n+1}}{1 - r}$$

In this case, $a = 1$, so u can be written as

$$u = e^{-jx} \frac{1 - r^{2j+1}}{1 - r} = \frac{e^{-jx} - e^{(j+1)x}}{1 - e^x}$$

This leads to the following form:

$$\langle V_S \rangle = N V_0 \frac{d}{dx} \ln \left[\frac{e^{-jx} - e^{(j+1)x}}{1 - e^x} \right]$$

$$= NV_0 \frac{d}{dx} \ln \left[\frac{e^{-(j+\frac{1}{2})x} - e^{(j+\frac{1}{2})x}}{e^{-\frac{x}{2}} - e^{\frac{x}{2}}} \right]$$

$$= NV_0 \frac{d}{dx} \ln \left[\frac{\sinh\left(j+\frac{1}{2}\right)x}{\sinh\frac{x}{2}} \right]$$

$$= NV_0 \frac{d}{dx} \left[\ln\left(\sinh\left(j+\frac{1}{2}\right)x\right] - \ln\left(\sinh\frac{x}{2}\right) \right]$$

We have multiplied the log argument by $1 = e^{-x/2}/e^{-x/2}$. Using the derivative of the ln(sinh u) term which is $du/dx \cdot (\cosh u/\sinh u)$, we obtain the result

$$\langle V_S \rangle = NV_0 \left[\left(j+\frac{1}{2}\right) \coth\left[\left(j+\frac{1}{2}\right)x\right] - \frac{1}{2}\coth\left(\frac{x}{2}\right) \right]$$

Let us examine the asymptotic limit of $\langle V_S \rangle / NV_0$ as $x \to \infty$. The hyperbolic cotangent function is defined as

$$\coth x = \frac{e^x + e^{-x}}{e^x - e^{-x}}$$

The limit of Eq. (**??**) is given by

$$\lim_{x\to\infty} \coth x = \frac{e^{\infty} + e^{-\infty}}{e^{\infty} - e^{-\infty}} = \frac{e^{\infty}}{e^{\infty}} = 1$$

Thus, $\langle V_S \rangle / NV_0$ has a limit given by

$$\lim_{x\to\infty} \frac{\langle V_S \rangle}{NV_0} = \left(j+\frac{1}{2}\right) - \frac{1}{2} = j$$

Therefore, the *Brillouin function* can be normalized by defining it as

$$B_J(x) = \frac{1}{j}\left[\left(j+\frac{1}{2}\right)\coth\left(j+\frac{1}{2}\cdot x\right) - \frac{1}{2}\coth\left(\frac{1}{2}\cdot x\right) \right] \tag{7.23}$$

Defining a modified $x' = (V_0 j)/(k_B T)$, then B_J can be written as

Brillouin Function

$$B_J(x) = \frac{1}{j}\left[\left(j+\frac{1}{2}\right)\coth\left(\frac{2j+1}{2j}\cdot x'\right) - \frac{1}{2}\coth\left(\frac{1}{2j}\cdot x'\right) \right] \tag{7.24}$$

Then, the temperature dependent expectation value of the spin-orbit interaction can be written as

Spin-Orbit Expectation Value

$$\langle V_S \rangle(T) = -NjV_0 B_J(x') \tag{7.25}$$

Examples of Brillouin functions are shown in Fig. 7.4 for different values of the total magnetic quantum number j. The smaller the value of j, the

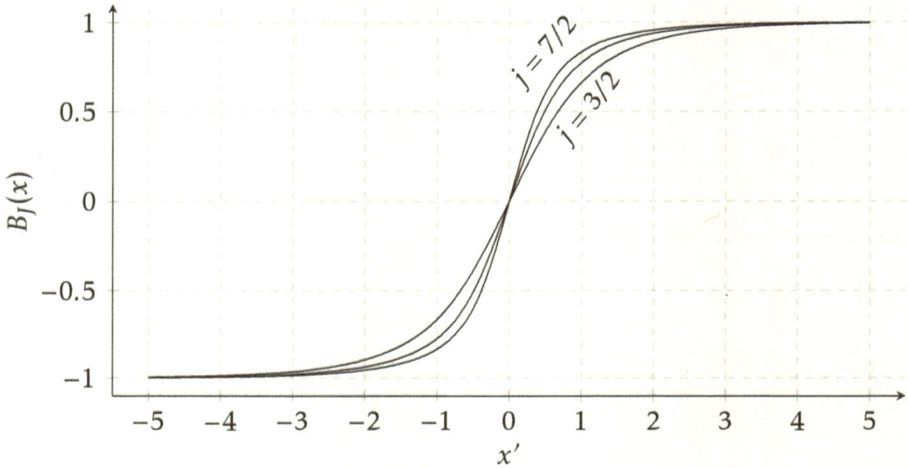

Figure 7.4: Brillouin function for $j = 3/2$, $5/2$, and $7/2$. The smaller is j, the more shallow B_J transitions.

more shallow B_J becomes. The Brillouin function in the limit of $x \to \infty$, becomes the *Langevin function* $L(x')$, given by

$$L(x') = \coth(x') - \frac{1}{x'} \tag{7.26}$$

To simplify some of our subsequent analysis, we can use the fact that the Langevin function $L(x')$ can be approximated by a hyperbolic tangent function tanh. Using this approximation, the expectation value for the spin-orbit interaction becomes

Spin-Orbit Interaction Expectation Value

$$\langle V_S \rangle(T) \approx -NjV_0 \cdot \tanh\left(\frac{jV_0}{k_B T}\right) \tag{7.27}$$

Fig. 7.5 shows example expectation values computed using the approximation for the *spin-orbit* interaction with $jV_0 = 1, 1.5, 2, 2.5$ meV.

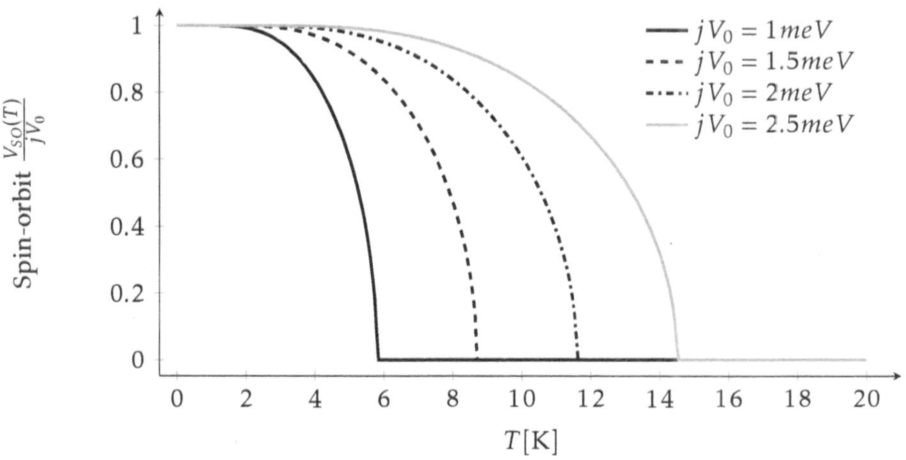

Figure 7.5: Solutions for spin-orbit interaction energy as a function of temperature T for four different values of $jV_0 = 1, 1.5, 2.0$, and 2.5 meV.

There are two noteworthy features that can be pointed out from Fig. 7.5:

(1) a *discontinuity* appears at a specific temperature, *below* which the energy rises relatively quickly from zero. This temperature corresponds to the *superconducting transition temperature* T_c or the transition temperature.

(2) the transition temperature T_c is in proportion to the energy strength at jV_0, so that materials with weaker spin-orbit interaction have lower transition temperatures T_c compared to a material with larger interaction.

Based on thermodynamic expectation values, the spin-orbit interaction tends to contribute additional energy, *negative* in sign, that enlarges in strength with reduced temperature. However, the behavior of the interaction energy *turns on discontinuously* at a unique temperature, depending on the strength of the spin-orbit interaction and/or the energy state.

7.2.2 Critical Temperature In Superconductors

We can *estimate* the relation of proportionality between the strength jV_0 and the temperature T_c by using an expansion of the Langevin function in conditions where $L(x') \approx 0$, corresponding to the definition for T_c. This is where the expectation value vanishes. However, since $V_{SO} \approx 0$, yet, $jV_0 \neq 0$, it follows that the Langevin function must vanish. This is where the approximation for the Langevin function $L(x')$ can be used. For small x, this is given by

$$L(x') \approx \frac{x}{3} - \frac{x^3}{45} \tag{7.28}$$

Substitution of Eq. (7.28) into the expectation relation for the spin-orbit interaction, along with the condition that $\langle V_{SO} \rangle = 0$, we have

$$0 \approx NjV_0 L\left(\frac{jV_0}{k_B T}\right) \approx NjV_0\left(\frac{jV_0}{3k_B T} - \frac{j^3 V_0^3}{45k_B^3 T^3}\right) \tag{7.29}$$

or

$$0 \approx \frac{jV_0}{3k_B T} - \frac{j^3 V_0^3}{45k_B^3 T^3} \tag{7.30}$$

Divide both sides by $jV_0 3/k_B T$ leads to

$$0 = 1 - \frac{j^2 V_{SO}^2}{15k_B^2 T^2} \tag{7.31}$$

Solving for the temperature T, the transition temperature T_c is found to be

Critical Temperature

$$T_c = \frac{jV_0}{\sqrt{15}k_B} \tag{7.32}$$

(R) *Spin-Orbit Strength and Critical Temperature*: The critical temperature T_c is proportional to the zero temperature value of the spin-orbit energy $V_{so}(0)$.

Fig. 7.6 illustrates the normalized solutions to V_{SO} as a function of temperature, *dividing the temperature by* $T_c = jV_0/k_B$. For comparison, included are normalized experimental data of Tin (Sn) and Tantalum (Ta). As the experimental data does, the calculated values demonstrate the vanishing point occurs when $k_BT/jV_0 = 1$.

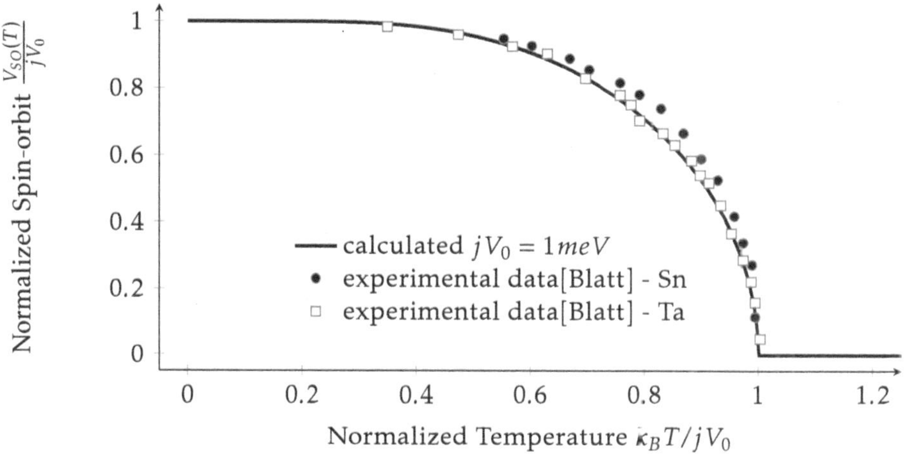

Figure 7.6: Calculated solution for spin-orbit interaction energy as a function of temperature k_BT/jV_0 using $|jV_0 = 1$ meV. Absolute temperature or Kelvin was used in the calculations illustrating that the vanishing point is at $k_BT/jV_0 = 1$. For comparisons, experimental data from Blatt, Frank J., Modern Physics is also shown.

From Fig. 7.6, we see a sort-of universal curve given by the dimensionless form of the temperature-dependent energy. In this case, they collapse onto a single curve.

Let us consider more of the possible energy contributions that are relevant to superconductors. These contributions are based purely on relativistic quantum mechanics. From these combinations, we can further analyze the behaviors found in these unique materials known as superconductors.

7.3 Solving Eigen-Operators For Superconductors

We have examined how the spin-orbit interaction at the relevant temperatures of superconductors can give rise to a discontinuity in the electronic state. This discontinuity has correlation with the critical temperature T_c.

Another question we may consider is what can happen to an electron that gets excited out of the valence band *somehow* and is introduced to an external magnetic field $\mathbf{B} = \nabla \times \mathbf{A}$. This is a very relevant question to superconductors. To analyze these conditions, let's start with an 4D energy operator given by

$$\left[E_{K,\text{cond}} + \frac{(\widehat{\mathbf{p}} + e\mathbf{A})^2}{2m_0} + \widehat{V_s} - i\hbar\frac{\partial}{\partial t}\right]\begin{bmatrix}\psi_\uparrow \\ \psi_\downarrow\end{bmatrix} = (iE_I)\begin{bmatrix}\psi_\uparrow \\ \psi_\downarrow\end{bmatrix} \tag{7.33}$$

$E_{K,\text{cond}}$ is the kinetic energy of ordinary conduction due to the externally applied electric potential/voltage. This contribution is *always* present in the analysis of superconductors, but can be treated as a constant. $\widehat{V_s}$ denotes *any* additional contributions in a superconductor. From our discussions so far, the spin-orbit interaction is an example in the quasi-bound state.

If we *only* consider conditions of *forces moving charges around, i.e.* no quantum state excitations, then we can ignore the 4th dimension's contribution in Eq. (7.33). Then, we have

$$\left[E_{K,\text{cond}} + \frac{(\widehat{\mathbf{p}} + e\mathbf{A})^2}{2m_0} + \widehat{V_s}\right]\begin{bmatrix}\psi_\uparrow \\ \psi_\downarrow\end{bmatrix} = E_\lambda\begin{bmatrix}\psi_\uparrow \\ \psi_\downarrow\end{bmatrix} \tag{7.34}$$

Because the magnetic field is applied *after* a transition into the superconducting state, we are describing quasi-free electrons conducting in an effectively negative potential energy *i.e.* in electronic holes.[4] Therefore, from the list of relevant operators given by Eq. (7.1), let us add the following contribution given by

$$\widehat{V_s} = -\frac{(\widehat{\mathbf{p}} + e\mathbf{A})^4}{2m_0^3c^2} \qquad \text{(after transition to sc)} \tag{7.35}$$

so we have an energy operator given by

$$\left[E_{K,\text{cond}} + \frac{(\widehat{\mathbf{p}} + e\mathbf{A})^2}{2m_0} - \frac{(\widehat{\mathbf{p}} + e\mathbf{A})^4}{2m_0^3c^2} + \widehat{V_s}'\right]\begin{bmatrix}\psi_\uparrow \\ \psi_\downarrow\end{bmatrix} = E_\lambda\begin{bmatrix}\psi_\uparrow \\ \psi_\downarrow\end{bmatrix} \tag{7.36}$$

Because of the spin-orbit interaction of the atoms, when an external magnetic field is introduced, we know a spin-orbit force is also generated.

[4] In the unique case of a *nonzero* magnetic field, the electrons are actually *not* itinerant or hopping if the spin-orbit-force if present (*c*.Sec. 6.5).

As a force on spins, it is equivalent to an electric field, where it also increases the kinetic energy of the electrons. In Sec. 6.5, we found the *spin-dependent force* to be

$$\langle \widehat{\boldsymbol{F}} \rangle_{SO} = -\frac{e\hbar}{m_0^2 c^2} (\nabla V \times \langle \widehat{\boldsymbol{\sigma}} \rangle) \times \mathbf{B} \tag{7.37}$$

Consider if the external magnetic field is along the axis of a cylindrical superconductor, say the $+z$ direction. Then, the spin-orbit force will tend to push radially toward the surface of the superconductor. Because this contribution is positive, it opposes the localized spin-orbit interaction contributing to electronic holes. However, by pushing electrons *away* from their atoms and/or toward the surface, ordinary electron drift behavior is reduced. We want to examine this *intermediary condition* between the extreme optimal condition for electronic holes, corresponding to $\mathbf{B} = 0$, and the ordinary electrical conduction state at some temperature below T_C. As mentioned, with the magnetic field held constant, this force is equivalent to a static electric field which is also a force per unit charge. Thus, the energy operator can then be written as

$$\left[E_{K,\text{cond}} + E_{K,\text{sof}} + \frac{(\widehat{\mathbf{p}} + e\mathbf{A})^2}{2m_0} - \frac{(\widehat{\mathbf{p}} + e\mathbf{A})^4}{2m_0^3 c^2} + V_{\text{barrier}} \right] \begin{bmatrix} \psi_\uparrow \\ \psi_\downarrow \end{bmatrix} = E_\lambda \begin{bmatrix} \psi_\uparrow \\ \psi_\downarrow \end{bmatrix} \tag{7.38}$$

where *sof* denotes the spin-orbit force energy. V_{barrier} is new and includes any potential energy barrier that exists between the valence band and conduction band. This energy is acquired in the excitation process. Taking the conduction, spin-orbit force, and barrier potential to be constant, we can just write

$$\left[\frac{(\widehat{\mathbf{p}} + e\mathbf{A})^2}{2m_0} - \frac{(\widehat{\mathbf{p}} + e\mathbf{A})^4}{2m_0^3 c^2} + V_{\text{free}} \right] \begin{bmatrix} \psi_\uparrow \\ \psi_\downarrow \end{bmatrix} = E_\lambda \begin{bmatrix} \psi_\uparrow \\ \psi_\downarrow \end{bmatrix} \tag{7.39}$$

where V_{free} lumps all the *constant* contributions to the electron motion. Other than the constant term V_{free}, there is only one unique operator in Eq. (7.39). This means that we can solve the simplest eigenvalue equation for $\widehat{\mathbf{p}} + e\mathbf{A}$, which is given by the second order kinetic energy operator. This is also the largest contribution. The more general eigen-energy corresponding to Eq. (7.39) can be determined from this eigenvalue. Then, we are *first* interested in solving

$$\frac{(\widehat{\mathbf{p}} + e\mathbf{A})^2}{2m_0} \psi(x,y,z) = E_\lambda \psi(x,y,z) \tag{7.40}$$

As mentioned, by solving Eq. (7.40), the eigenenergy of Eq. (7.39) is also determined for any static magnetic/electric field. Later, in Sec. 7.9, all these terms will be required for analysis. But, for now, let us solve Eq. (7.40).

7.3.1 Quantized Vortex Solutions

For the magnetic vector potential **A**, let us use the *symmetric gauge* given by

$$\mathbf{A} = (1/2)[-B \cdot y, B \cdot x, 0]^T \tag{7.41}$$

Then, expanding the square in Eq. (7.40), we have

$$\frac{1}{2m_0}\left[\widehat{\mathbf{p}}^2 + e\widehat{\mathbf{p}} \cdot \mathbf{A} + e\mathbf{A} \cdot \widehat{\mathbf{p}} + e^2\mathbf{A}^2\right]\psi = E_\lambda \psi \tag{7.42}$$

$\widehat{\mathbf{p}}^2$ is given by $-\hbar^2\nabla^2\psi$. Let us consider the second term $e\widehat{\mathbf{p}}\cdot\mathbf{A}$ on the RHS. Since operators are applied to the wave function ψ, we can make use of a vector-calculus identity for the *gradient of a (vector times a scalar)*.[5] Applying it leads to

$$e\widehat{\mathbf{p}} \cdot (\mathbf{A}\psi) = e\left[(\widehat{\mathbf{p}} \cdot \mathbf{A})\psi + \mathbf{A} \cdot (\widehat{\mathbf{p}}\psi)\right] \tag{7.43}$$

The first term on the RHS is proportional to the divergence of **A** or $\nabla \cdot \mathbf{A}$. Because of the form of **A** given by Eq. (7.41), an inner product with the momentum operator (or divergence) is identically zero, so we have

$$\frac{1}{2m_0}\left[\widehat{\mathbf{p}}^2 + 2e\mathbf{A} \cdot \widehat{\mathbf{p}} + e^2\mathbf{A}^2\right]\psi = E_\lambda \psi \tag{7.44}$$

With rotational motion in mind, it is better to use a coordinate system that describes rotation more naturally. Therefore, to obtain the solutions to Eq. (7.40), we'll use a cylindrical coordinate system (r, φ, z). To transform Eq. (7.41), we can use the relations

$$y = r\sin\varphi$$

$$x = r\cos\varphi$$

$$\mathbf{e}_\varphi = -\sin\varphi\widehat{\mathbf{x}} + \cos\varphi\widehat{\mathbf{y}}$$

[5]It is the generalization of the product rule for differentiation

Comparing the azimuthal unit vector above to the symmetric gauge in Eq. (7.41), it follows that the magnetic vector potential **A** is oriented *azimuthally*, or it circles around the z–axis. Then, using the above relations in the symmetric gauge, we can write the magnetic vector potential **A** as

$$\mathbf{A} = [0 \quad rB_z/2 \quad 0]^T \tag{7.45}$$

In *cylindrical* coordinates, the momentum operator is given by

$$\widehat{\mathbf{p}} = -i\hbar\nabla = -i\hbar\left[\frac{\partial}{\partial r} \quad \frac{1}{r}\frac{\partial}{\partial\varphi} \quad \frac{\partial}{\partial z}\right]^T$$

Then, the energy operator given by Eq. (7.44), in cylindrical coordinates, becomes

$$E_\lambda\psi = \frac{1}{2m_0}\left[-\hbar^2\nabla^2\psi - ier\hbar B\mathbf{e}_\varphi\cdot\nabla\psi + \frac{e^2B^2}{4}r^2\psi\right] \tag{7.46a}$$

$$= \frac{1}{2m_0}\left[-\hbar^2\left(\frac{1}{r}\frac{\partial}{\partial r}\left(r\frac{\partial\psi}{\partial r}\right) + \frac{1}{r^2}\frac{\partial^2\psi}{\partial\varphi^2} + \frac{\partial^2\psi}{\partial z^2}\right) - ie\hbar B\frac{\partial\psi}{\partial\varphi}\right. \tag{7.46b}$$

$$\left. + \frac{e^2B^2}{4}r^2\psi\right] \tag{7.46c}$$

Let ψ be *separable* so that it is given by

$$\psi = \phi_r(r)\phi_\varphi(\varphi)\psi_z(z) \quad \text{and} \quad E_\lambda = E_r + E_\varphi + E_z \tag{7.47}$$

Substituting Eq. (7.47) into Eq. (7.46b), we can separate the operator equation into two distinct equations, one for z, and the other for the combination (r, φ). For the z dependence, $\psi_z(z)$ is described by

$$-\frac{\hbar^2}{2m_0}\frac{d^2\psi_z}{dz^2} = E_z\psi_z(z) \Rightarrow \frac{d^2\psi_z}{dz^2} + k_z^2 Z = 0 \tag{7.48}$$

The solution to Eq. (7.48) is given by

$\psi_z(z)$ Solution Of $\psi(x, y, z)$

$$\psi_z(z) = \psi_z(0)e^{ik_z z} \quad \text{where} \quad k_z^2 = \frac{2m_0E_z}{\hbar^2} \tag{7.49}$$

Substituting Eq. (7.49) into Eq. (7.48), it can be verified as a solution.

The remaining part of Eq. (7.46b) depends on r and φ. The equation describing functions $\psi_r(r)$ and $\psi_\Phi(\varphi)$ is given by

$$\frac{1}{2m_0}\left[-\hbar^2\left(\psi_\varphi\frac{1}{r}\frac{\partial}{\partial r}\left(r\frac{\partial\psi_r}{\partial r}\right)+\psi_r\frac{1}{r^2}\frac{\partial^2\psi_\varphi}{\partial\varphi^2}\right)-ie\hbar B\psi_r\frac{\partial\psi_\varphi}{\partial\varphi}+\right.$$
$$\left.\frac{e^2B^2}{4}r^2\psi_r\psi_\varphi\right]=(E_r+E_\varphi)\psi_r\psi_\varphi \quad (7.50)$$

It turns out that we can still use *separability* for Eq. (7.50), however, there is a slight trick we should use. Since it can be shown that if $\psi_\varphi(\varphi)=C_\varphi e^{ik_\varphi r\varphi}$, then both terms in Eq. (7.50) depending on φ sum to zero. This also means $E_\varphi=0$, so that there is *no* energy associated with motion azimuthally! Because the time-dependence dictated by the 4th dimension ($E_I=\alpha E_R$) is in proportion to the real eigenvalue, we have that pure motion along this direction has no associated amplitude change, *i.e.* no loss.

To determine the value of k_φ, we sum all the terms depending on φ, which gives

$$-\frac{\hbar^2}{2m_0}\left(\frac{1}{r^2}\frac{\partial^2\psi_\varphi}{\partial\varphi^2}\right)-i\frac{e\hbar B}{2m_0}\frac{\partial\psi_\varphi}{\partial\varphi}=E_\varphi\psi_\varphi \quad (7.51)$$

Let the solution be written as $\psi_\varphi(\varphi)=C_\varphi e^{ik_\varphi r\varphi}$. Evaluating the derivatives of Φ leads to

$$\frac{\partial\psi_\varphi}{\partial\varphi}=C_\varphi ik_\varphi re^{ik_\varphi r\varphi}=ik_\varphi r\psi_\varphi \quad (7.52)$$

$$\frac{\partial^2\psi_\varphi}{\partial\varphi^2}=C_\varphi(ik_\varphi r)^2e^{ik_\varphi r\varphi}=-k_\varphi^2r^2\psi_\varphi \quad (7.53)$$

Then, substitution of these results into the LHS of Eq. (**??**) gives

$$-\frac{\hbar^2}{2m_0}\left(\frac{1}{r^2}\frac{\partial^2\psi_\varphi}{\partial\varphi^2}\right)-i\frac{e\hbar B}{2m_0}\frac{\partial\psi_\varphi}{\partial\varphi}=-\frac{\hbar^2}{2m_0}\frac{1}{r^2}\left(-k_\varphi^2r^2\psi_\varphi\right)-i\frac{e\hbar B}{2m_0}\left(ik_\varphi r\psi_\varphi\right) \quad (7.54a)$$

$$=\left[\frac{\hbar^2k_\varphi^2}{2m_0}+\frac{e\hbar Bk_\varphi r}{2m_0}\right]\psi_\varphi \quad (7.54b)$$

We want the value of k_φ that makes Eq. (7.54) vanish satisfying the condition

$$0 = \hbar k_\varphi + eB_z r$$

Solving for k_φ, we have

Azimuthal Wavenumber k_φ

$$k_\varphi = -\frac{eBr}{\hbar} \tag{7.55}$$

Since the function $\psi_\varphi(\varphi) = \psi_\varphi(r,\varphi) = C_\varphi e^{ik_\varphi r\varphi}$, we also have for the azimuthal solution $\psi_\varphi(r,\varphi)$

Azimuthal Solution $\psi_\varphi(\varphi)$

$$\psi_\varphi(r,\varphi) = C_\varphi e^{-i\frac{eBr^2}{\hbar}\varphi} \tag{7.56}$$

The momentum in the azimuthal direction is given by

$$p_\varphi = \hbar k_\varphi = -eBr \tag{7.57}$$

Equivalently, for a particle with mass m_0 where $p \approx m_0 v$, we have

$$v_\varphi = \frac{p_\varphi}{m_0} = -\frac{eBr}{m_0} = -\omega_0 r \tag{7.58}$$

From Eq. (7.58), we can define a rotational frequency ω_0 as

Rotational Frequency ω_0

$$\omega_0 = \frac{eB}{m_0} \tag{7.59}$$

Eq. (7.59) is also known as the *cyclotron frequency*. It is the rotational frequency of a mass m_0 having charge e, moving along φ, due to the axial field of strength $B = B_z$. Because we have found the solution for ψ_φ, this also means that $E_\varphi = 0$. As mentioned, it follows there is not loss with motion along this direction.

> R *Lossless motion along azimuthal axis*: In a superconductor, the only direction capable of demonstrating lossless electron motion is along the azimuthal direction. They are known as *persisting currents*.

C_φ is the *normalization factor* which can be found from the *normalization condition*, or by evaluating the following integral in cylindrical coordinates

$$\begin{aligned} 1 &= \int_0^{2\pi} C_\varphi^* C_\varphi e^{i\frac{e}{\hbar}Br^2\varphi} e^{-i\frac{e}{\hbar}Br^2\varphi} d\varphi \\ &= |C_\varphi|^2 \int_0^{2\pi} d\varphi \\ &= |C_\varphi|^2 \cdot 2\pi \end{aligned}$$

This leads to the **azimuthal normalization factor** being given by

Normalization Factor For $\psi_\varphi(\varphi)$

$$|C_\varphi| = \frac{1}{\sqrt{2\pi}} \tag{7.60}$$

So, we have obtained the solutions $\psi_z(z)$ and $\psi_\varphi(r,\varphi)$. From here, we only have to solve for $\psi_r(r)$. The remaining terms, being only radial, are given by[6]

$$\frac{1}{2m_0}\left[-\hbar^2 \frac{1}{r}\frac{d}{dr}\left(r\frac{d\psi_r}{dr}\right) + \frac{e^2B^2}{4}r^2\psi_r\right] = E_r\psi_r \tag{7.61}$$

Solving for the radial part $\psi_r(r)$ is the more challenging in obtaining the separate solutions. To solve Eq. (7.61), we must express it in an alternative form, as then, it becomes amenable to the powerful method

[6] since ψ_r is a strict function of r, partial derivatives become total derivatives.

known as the *Method of Frobenius.* , which finds a solution in the form of an *infinite series* expansion. It can be used to find an exact solution to Eq. (7.61).

First, expanding the derivative term and rearranging, we have

$$\frac{d^2\psi_r}{dr^2} + \frac{1}{r}\frac{d\psi_r}{dr} + \left(\frac{2m_0E_r}{\hbar^2} - \frac{e^2B^2}{4\hbar^2}r^2\right)\psi_r = 0 \tag{7.62}$$

We can express Eq. (7.62) in terms of another variable ζ defined as

$$\sqrt{\zeta} = \sqrt{\frac{eB}{2\hbar}}r = \nu r \Rightarrow \underline{\zeta = \nu^2 r^2} \tag{7.63}$$

From the definition of ζ, we can use the following useful relations:

$$d\zeta = 2\nu^2 r dr \quad \text{or} \quad \frac{d\zeta}{dr} = 2\nu^2 r$$

$$\frac{d\psi_r}{dr} = \frac{d\psi_r}{d\zeta}\frac{d\zeta}{dr} = 2\nu^2 r\frac{d\psi_r}{d\zeta}$$

$$\frac{d^2\psi_r}{dr^2} = \frac{d}{dr}\left(2\nu^2 r\frac{d\psi_r}{d\zeta}\right) = 2\nu^2\frac{d\psi_r}{d\zeta} + 2\nu^2 r\frac{d\zeta}{dr}\frac{d}{d\zeta}\left(\frac{d\psi_r}{d\zeta}\right)$$

$$= 2\nu^2\frac{d\psi_r}{d\zeta} + 4\nu^4 r^2\frac{d^2\psi_r}{d\zeta^2} = 2\nu^2\frac{d\psi_r}{d\zeta} + 4\nu^2\zeta\frac{d^2\psi_r}{d\zeta^2}$$

Substitution of these relations into Eq. (7.62) leads to the following equation in terms of ζ:

$$4\nu^2\zeta\frac{d^2\psi_r}{d\zeta^2} + 4\nu^2\frac{d\psi_r}{d\zeta} + \left(\frac{2m_0E_r}{\hbar^2} - \nu^2\zeta\right)\psi_r(\zeta) = 0 \tag{7.64}$$

or

$$4\zeta\frac{d^2\psi_r}{d\zeta^2} + 4\frac{d\psi_r}{d\zeta} + \left(\frac{2m_0E_r}{\nu^2\hbar^2} - \zeta\right)\psi_r(\zeta) = 0 \tag{7.65}$$

So, let us now express the wave function ψ_r in the form $\psi_r = \psi_{r,\infty}(\zeta)\psi_{r,1}(\zeta)$, where $\psi_{r,\infty}(\zeta)$ is an *asymptotic* solution we can find first. For this, we consider the asymptotic limit $\zeta \to \infty$, where Eq. (7.65) becomes

$$4\frac{d^2\psi_{r,\infty}}{d\zeta^2} - \psi_{r,\infty}(\zeta) = 0 \Rightarrow \frac{d^2\psi_{r,\infty}}{d\zeta^2} - \frac{1}{4}\psi_{r,\infty}(\zeta) = 0 \tag{7.66}$$

Eq. (7.66) has a solution given by

$$\psi_{r,\infty}(\zeta) = C_\infty e^{-\frac{\zeta}{2}} + D_\infty e^{+\frac{\zeta}{2}} \Rightarrow \psi_{r,\infty}(\zeta) = C_\infty e^{-\frac{eB}{4\hbar}r^2} \tag{7.67}$$

The result follows because the other solution $\psi_{r,\infty}(\zeta) = D_\infty e^{+\frac{\zeta}{2}}$ goes to infinity as $\zeta \to \infty$. This *does not satisfy the boundary condition* where $\psi_r(r = \infty) = 0$. Therefore, the coefficient $D_\infty = 0$. Because $\psi_{r,\infty}(\zeta)$ is in the form of a *Gaussian*, now we can obtain an eigenvalue equation *purely* in terms of $\psi_{r,1}$. If $\psi_r = \psi_{r,\infty}\psi_{r,1}$, we have the following derivatives of $\psi_r(\zeta)$ in terms of $\psi_{r,1}$:

$$\frac{d\psi_r}{d\zeta} = \frac{d}{d\zeta}(\psi_{r,\infty}\psi_{r,1}) = \left[\frac{d\psi_{r,1}}{d\zeta} - \frac{1}{2}\psi_{r,1}\right]e^{-\frac{1}{2}\zeta}$$

$$\frac{d^2\psi_r}{d\zeta^2} = \left[\left(\frac{d^2\psi_{r,1}}{d\zeta^2} - \frac{1}{2}\frac{d\psi_{r,1}}{d\zeta}\right) + \left(\frac{d\psi_{r,1}}{d\zeta} - \frac{1}{2}\psi_{r,1}\right)\left(-\frac{1}{2}\right)\right]e^{-\frac{1}{2}\zeta}$$

Using these relations in Eq. (7.65), we obtain the following equation for $\psi_{r,1}$:

$$4\zeta\left[\left(\frac{d^2\psi_{r,1}}{d\zeta^2} - \frac{1}{2}\frac{d\psi_{r,1}}{d\zeta}\right) + \left(\frac{d\psi_{r,1}}{d\zeta} - \frac{1}{2}\psi_{r,1}\right)\left(-\frac{1}{2}\right)\right]e^{-\frac{1}{2}\zeta}$$
$$+ 4\left[\frac{d\psi_{r,1}}{d\zeta} - \frac{1}{2}\psi_{r,1}\right]e^{-\frac{1}{2}\zeta} + \left[\frac{2m_0E_r}{\nu^2\hbar^2} - \frac{1}{2}\zeta\right]\psi_{r,1}e^{-\frac{1}{2}\zeta} = 0 \tag{7.68}$$

The Gaussian factors cancel, and we can combine respective terms as

$$4\zeta\frac{d^2\psi_{r,1}}{d\zeta^2} + 4(1-\zeta)\frac{d\psi_{r,1}}{d\zeta} + \left[\frac{2m_0E_r}{\nu^2\hbar^2} - 2\right]\psi_{r,1} = 0$$

or

$$\zeta\frac{d^2\psi_{r,1}}{d\zeta^2} + (1-\zeta)\frac{d\psi_{r,1}}{d\zeta} - \left[\frac{2 - \frac{2m_0E_r}{\nu^2\hbar^2}}{4}\right]\psi_{r,1} = 0 \tag{7.69}$$

Defining Ω as

$$\Omega = \frac{2m_0E_r}{\nu^2\hbar^2} \tag{7.70}$$

we have

$$\zeta\frac{d^2\psi_{r,1}}{d\zeta^2} + (1-\zeta)\frac{d\psi_{r,1}}{d\zeta} - \left(\frac{2-\Omega}{4}\right)\psi_{r,1} = 0 \tag{7.71}$$

Eq. (7.71) is in the form of the so-called *confluent hypergeometric equation*(CHE) which is given by

$$x\frac{d^2y}{dx^2}+(c-x)\frac{dy}{dx}-ay(x)=0$$

The solution to this equation is given by the *confluent hypergeometric series* (CHS).[7] The CHS is given by

$$\text{CHS}(\zeta;a,c)={}_1F_1(\zeta;a,c)=1+\frac{1}{1!}\frac{a}{c}\zeta+\frac{1}{2!}\frac{a(a+1)}{c(c+1)}\zeta^2+$$

$$\frac{1}{3!}\frac{a(a+1)(a+2)}{c(c+1)(c+2)}\zeta^3+\frac{1}{4!}\frac{a(a+1)(a+2)(a+3)}{c(c+1)(c+2)(c+3)}\zeta^4+$$

$$\frac{1}{5!}\frac{a(a+1)(a+2)(a+3)(a+4)}{c(c+1)(c+2)(c+3)(c+4)}\zeta^5+$$

$$\frac{1}{6!}\frac{a(a+1)(a+2)(a+3)(a+4)(a+5)}{c(c+1)(c+2)(c+3)(c+4)(c+5)}\zeta^6+\ldots$$

From Eq. (7.71), we have

$$c=1\quad\text{and}\quad a=\frac{2-\Omega}{4}=\frac{2-\frac{2m_0E_r}{\hbar^2\nu^2}}{4}\tag{7.72}$$

As pointed out earlier, to be a *valid* quantum mechanical wave function solution, it must satisfy the boundary conditions including $\psi(r=\infty)=0$. For this to happen, it turns out that the infinite series ${}_1F_1(\zeta;a,c)$ has to be truncated, and this is done if the parameter *a only takes negative integer values*. So, we have a compulsory condition given by

$$a=\frac{2-\Omega}{4}=-n\tag{7.73}$$

This condition given by Eq. (7.73) also determines the allowed values for E_r because

$$\frac{2-\frac{2m_0E_r}{\hbar^2\nu^2}}{4}=-n\Rightarrow-\frac{2m_0E_r}{\hbar^2\nu^2}=-4n-2\tag{7.74}$$

Substitution of $\nu^2=eB/2\hbar$, and solving for E_r, we have

$$E_r=\hbar\omega_0\left(n+\frac{1}{2}\right)\tag{7.75}$$

or

[7] For an in-depth discussion and derivation of the CH series, see Chap. 5, Quantum Mechanics With A SPIN On It (Eason)

Radial Eigenenergy

$$E_r = \frac{\hbar e B}{m_0}\left(n + \frac{1}{2}\right) \tag{7.76}$$

where $n = 0, 1, 2, 3, \ldots$ ω_0 is the rotational frequency from Eq. (2.156). The corresponding *radial eigenfunctions* are

Radial Eigensolution

$$\psi_r(r) = C_0 e^{-\frac{eB}{4\hbar}r^2}{}_1F_1\left(\frac{eB}{2\hbar}\cdot r^2; -n, 1\right) \tag{7.77}$$

Combining our results for z, φ, and r, we have a *complete* eigensolution given by

Free Electron Kinetic Energy Equation Eigensolution

$$\psi(r, \varphi, z) = C_n \cdot {}_1F_1\left(\frac{eB}{2\hbar}\cdot r^2; -n, 1\right) e^{-\frac{eB}{4\hbar}r^2} e^{i\frac{eBr^2}{\hbar}\varphi} e^{ik_z z} \tag{7.78}$$

Because the radial functions including the Gaussian function in the solution strongly determines the decaying length scale, a length-scale of variation for a quantized vortex can be found. Since $\nu_0 r^2 = eBr^2/2\hbar$ is dimensionless, it follows that $\nu_0^2 = \frac{2\hbar}{eB}$ has units of length^{-2}. Then, ν_0 becomes

Superconducting Coherence Length

$$\nu_0 = \sqrt{\frac{2\hbar}{eB}} \tag{7.79}$$

Alternatively, since the energy E_r is proportional to eB, we can solve for it in terms of E_r as follows

$$E_0 = \frac{\hbar e B}{m_0}\left(n + \frac{1}{2}\right) \Rightarrow eB = \frac{2m_0 E_0}{\hbar}$$

Then, substituting eB for the ground state, an equivalent length-scale for the decay-length of the quantized vortex along r becomes

$$\nu_0 = \sqrt{\frac{\hbar^2}{m_0 E_0}}$$

Eq. (7.79) is also known as the *superconducting coherence length.* It provides a maximum length scale for the variation of the charge density in the radial direction, or in the direction normal to the superconductor surface. Because this dimension is quantized, there will be variation in vortices along this direction.

The solution given by Eq. (7.78) describes a *quantized vortex.* C_n is the normalization constant obtained by evaluating the integral

$$1 = \int \psi^* \psi dV \tag{7.80}$$

$$= \int |C_n|^2 e^{-\frac{eB}{2\hbar}r^2} {}_1F_1\left(\frac{eB}{2\hbar}r^2;-n,1\right) {}_1F_1\left(\frac{eB}{2\hbar}r^2;-n,1\right) rdrd\varphi dz \tag{7.81}$$

We can rewrite the integral in a form so we can use a lookup table to find the anti-derivative.[8]. For this, lets write $\zeta = eBr^2/2\hbar = \nu^2 r^2$. Then $d\zeta = 2\nu^2 rdr$. Then, the integral from Eq. (7.80) becomes

$$1 = \frac{1}{2\nu^2} \int |C_n|^2 e^{-\zeta} {}_1F_1(\zeta;-n,1) {}_1F_1(\zeta;-n,1) d\zeta d\varphi dz \tag{7.82}$$

Eq. (7.82) has a form that can be found in *Factorization Method In Quantum Mechanics,* where we have

$$1 = \frac{1}{2\nu^2} \left(|C_n|^2 2\pi L_z \frac{n!\Gamma(1)^2}{\Gamma(n+1)} \right) \tag{7.83}$$

$\Gamma()$ is the *Gamma* function, L_z is the thickness of the domain along the z direction. Using $\Gamma(1) = 1$, and solving for $|C_n|$, we have a normalization constant given by

Normalization Constant

[8] **Factorization Method In Quantum Mechanics, Shi-Hai Dong, 2007.**

$$|C_n| = \sqrt{\frac{eB\Gamma(n+1)}{2\pi\hbar n! L_z}} \tag{7.84}$$

Fig. 7.7 shows one of the earliest examples of direct observation of a lattice of quantized vortices, made in 1967 by Essmann and Trauble. The lattice of quantized vortices is also known as an *Abrikosov lattice.*

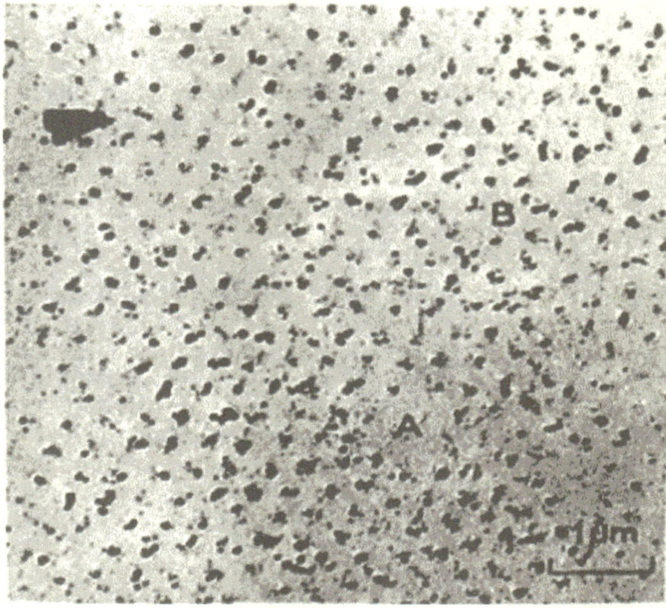

Figure 7.7: One of the first direct observations of the **Abrikosov lattice**, in Pb_4In reported in the historic paper from Essmann and Trauble (1967).

Some examples of the quantized vortex solution given by Eq. (7.78) are shown in Table 2.1 for a variety of quantum numbers n and magnetic field strengths B.

$(a)\ n = 0, B = 0.5T$

$(b)\ n = 0, B = 1.5T$

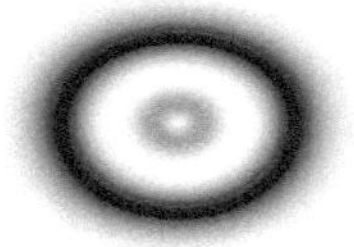

$(c)\ n = 1, B = 0.5T$

$(d)\ n = 1, B = 1.5T$

$(e)\ n = 2, B = 0.5T$

$(f)\ n = 2, B = 1.5T$

$(g)\ n = 3, B = 0.5T$

$(h)\ n = 3, B = 1.5T$

Table 7.2: n–vortices for different values of $n = 0, 1, 2, 3$ and external magnetic field $B = 0.5, 1.5$T. As pure radial functions, they have azimuthal symmetry in the x and y plane, and only motion in the azimuthal direction. Note that the probability density functions are *not* normalized here. All figures shown are on the same length scale spanning ±200nm, or 400nm.

The quantized vortices shown in Table 7.2 are all on the same length scale and range spanning ±200nm. The electron is predominantly confined to an *orbital exterior ring*. Note that this excludes the spin-orbit force. The orbital motion about the magnetic field form orbits that are *quantized*, being examples of quantized vortices.

Therefore, summarizing observations made from the solution, we find the following:

1. Superconductors form quantized vortices around magnetic fields which can sustain motion azimuthally *without loss*.
2. The general effect of the magnetic field strength on a quantized vortex is to further confine the orbital ring radially.

Now that we have an understanding of the energies at play in superconductors, let's use some of these ideas in discussing the unique discontinuities that are observed in superconductors.

7.4 General Effect on Resistance With Temperature

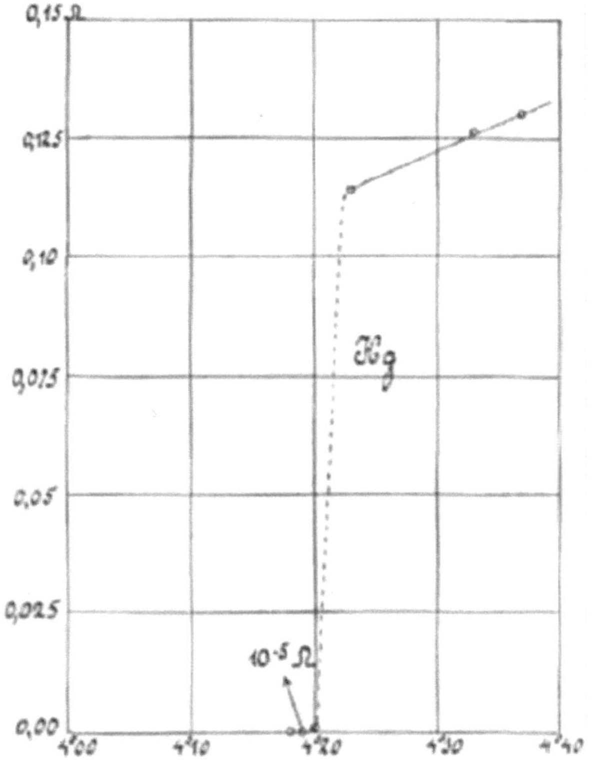

Figure 7.8: The historical plot reported by Onnes' team after their discovery of superconductivity in mercury (Hg) in 1911. The x–axis shows temperature in Kelvin, while the y–axis measures the resistance in Ohms.

Fig. 7.8 demonstrates a drop in the electrical resistance of mercury (Hg) near it's *superconducting transition temperature* T_c. Even in the absence of the superconducting transition, the resistance can be seen to still change with temperature, increasing approximately linearly. This dependence has nothing to do with superconductivity. So, the more pronounced effects in the electrical properties in superconductors are *in addition to* the more general electronic properties. The origin of the increase in resistance with temperature in metals like mercury can be understood by considering Fig. 7.9.

The Fermi-Dirac distribution affects the distribution of electrons in their available energy states near the Fermi edge. It is a function of tempera-

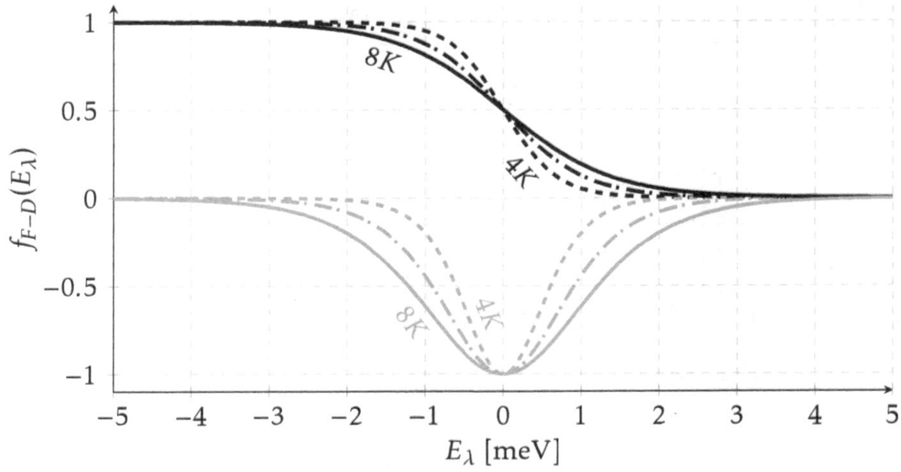

Figure 7.9: Fermi-Dirac distribution $f_{F-D}(E_\lambda)$ for temperatures $T = 4, 6$, and $8K$. $\epsilon_F = 0$ meV corresponds to the Fermi edge located near the 'edge' of an atom. As temperature decreases, the Fermi-Dirac width narrows indicating that electrons occupy tighter and tighter spaces (=lower energies) within the atom.

ture, as well.[9] For any $T = 4,\ 6,\ 8$K, it turns out that the lower bound for available states does not change. Instead, electrons tend to occupy lower energy states with *decreasing* temperature. Lower energy quantum states occupy less physical space, so electron distribution effectively get smaller. This causes electrons in conduction to see a positive electric potential *sooner* with increasing temperature. Thus, increasing temperature effectively introduces more and more of a potential barrier to a conducting electron. Hence, conduction is hindered or resistance increases with temperature in metals, in particular.

In insulators, however, the effect can manifest very differently. At best, the effect is much weaker, so if there is an increase in resistance, it is much slower to increase. The other situation in insulators is that the electrons get excited, but they do *not* enter into the conduction band due to the existing large potential gap. Because of this, the atomic size is much less affected if at all. The result can be to make the job of the electric field easier to conduct. This corresponds to a decrease in resistance in some insulators. Experiments will demonstrate all these effects together. However, we can now separate the general increase in

[9] For a detailed discussion on density of states, Fermi-Dirac distributions, etc., see Chaps. 3 and 4, Quantum Mechanics With A SPIN On It (Eason).

resistance from changes due to superconductivity.

7.5 Discontinuities In Electronic Properties Near T_c

A well-known behavior of a superconductor is the demonstration of a unique *discontinuity* in the resistance $R(T)$ and/or resistivity $\rho(T)$ in its temperature dependence. One of the first examples observing this discontinuity in Mercury (Hg) is shown in Fig. **??**. Another property that demonstrates a discontinuity at the same temperature is the heat capacity $C(T)$. This is a measure of the materials change in internal energy relative to a temperature change.

With what has been discussed so far, we are in a position to examine how these properties demonstrate discontinuities. Let's start with the resistance.

7.5.1 Discontinuity is Resistance

In a solid-state material such as a superconductor, the discontinuity in electrical conductivity $\sigma_s = 1/\rho_s$ takes place the vicinity of the so-called *transition temperature* T_c or *critical temperature*. ρ_s is the bulk resistivity of the superconducting material (proportional to the resistance R). If we assume that over the thermoequilibrium temperature variation, the applied external electric field is held constant, then variation with temperature is contained in the current density $\widehat{\mathbf{J}}_c$. For conduction along the axis of a superconducting cylinder ($z - axis$), the charge current density is given by

$$\begin{aligned}
\widehat{\mathbf{J}}_z &= \frac{ie\hbar}{2m_0}\left(\psi\nabla\psi^* - \psi^*\nabla\psi\right) \\
&= \frac{ie\hbar}{2m_0}\left[(-ik_z)e^{ik_z z}e^{-ik_z z} - ik_z C_n^* e^{-ik_z z}e^{ik_z z}\right]\psi_\varphi^*\psi_\varphi\psi_r^*\psi_r \\
&= \frac{ie\hbar k_z}{2m_0}\left(-2i\psi_r^2\right) \\
&= \frac{e\hbar k_z}{m_0}\psi_r^2
\end{aligned}$$

Above, we have used the fact that $\psi_\varphi^* \psi_\varphi = e^{-ik_\varphi d\ell} e^{ik_\varphi d\ell} = 1$. ψ_r is the radial wave function describing the *quantized vortices* obtained in Sec. 7.3.1. Thus, to examine temperature dependent conduction properties, k_z is necessary. To find k_z, we can return to the eigenenergy of a charge carrier given by

$$E_\lambda = E^+_{K,\text{cond}} + \langle V_{so} \rangle \tag{7.85a}$$

$$= \frac{\hbar^2 k_z^2}{2m_0} - V_{so}(0)\tanh\left(\frac{V_{so}}{k_B T}\right) \tag{7.85b}$$

using the temperature-dependent expectation value of the spin-orbit interaction strength, which was found to be

$$V_{so}(T) = V_{so}(0)\tanh\left(\frac{V_{so}(T)}{k_B T}\right) \tag{7.86}$$

$E^+_{K,\text{cond}}$ is the kinetic energy contribution including the electric external field existing *before* the transition into superconductivity. Solving for k_z gives

$$k_z = \sqrt{\frac{2m_0}{\hbar^2}\left[V_{so}(0)\tanh\left(\frac{V_{so}}{k_B T}\right) + E_\lambda\right]} \tag{7.87}$$

Since $J_z \propto k_z$, and $J_z \propto \sigma_s E$, it follows that $k_z(T) \propto \sigma_s$, and thus $\rho(T) \propto 1/k_z(T)$. To obtain k_z we must remember that Eq. (7.86) is a *nonlinear* equation in $V_{so}(T)$. To obtain the temperature dependence of $V_{so}(T)$, it must be solved numerically. Fig. 7.10 illustrates an example of the temperature dependence obtained by solving Eq. (7.86).

We are interested to examine what happens *very close* to the critical temperature T_c, as the temperature is reduced. The temperature dependence is contained in the solution to the nonlinear equation for k_z. Let us denote the energy just above T_c by E^+_λ. Then, we can rewrite Eq. (7.87) as follows:

$$k_z = \sqrt{\frac{2m_0}{\hbar^2}\left[V_{so}(0)\tanh\left(\frac{V_{so}}{k_B T}\right) + E^+_\lambda\right]}$$

$$= \frac{\sqrt{2m_0 E^+_\lambda}}{\hbar}\sqrt{\frac{V_{so}(0)}{E^+_\lambda}\tanh\left(\frac{V_{so}}{k_B T}\right) + 1}$$

Defining the dimensionless ratio $\hbar k_z/\sqrt{2m_0 E^+_\lambda} = k_z(T)/k_z^+(T_c^+)$, we obtain a dimensionless wavenumber given by

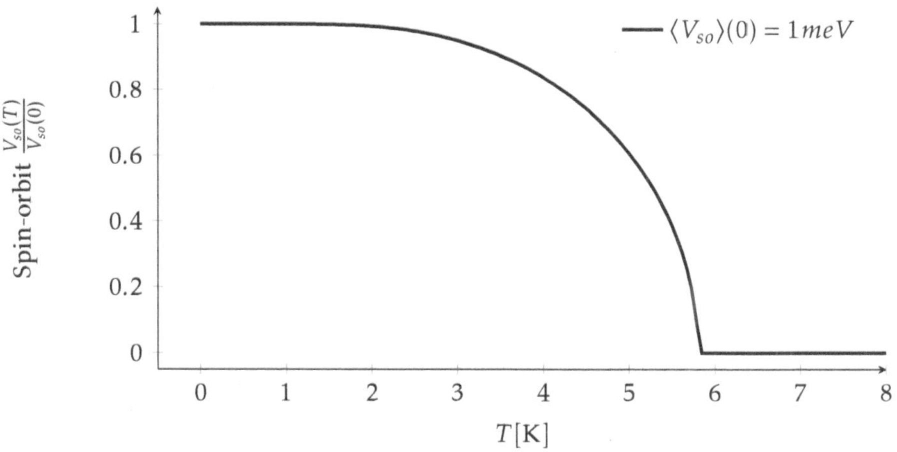

Figure 7.10: Example of expectation value of spin-orbit interaction energy as a function of temperature T for $V_{so}(0) = 1$ meV.

Discontinuous Conductivity Factor

$$\frac{k_z(T)}{k_z^+} = \sqrt{\frac{V_{so}(0)}{E_\lambda^+}\tanh\left(\frac{V_{so}}{k_B T}\right)+1} \tag{7.88}$$

It is the *inverse* of Eq. (7.88) which is proportional to the temperature-dependent resistivity. Fig. 7.11 plots examples of $k_z^+(T_c^+)/k_z(T)$ for different ratios of $V_{so}(0)/E_\lambda^+ = 1000, 500$, and 100. From our discussion of the spin-orbit interaction expectation value and its dependence on temperature, we know that the discontinuity at T_c follows from the discontinuous, sharp potential rise of V_{so} at T_c. Because of the dependence of the resistance on the kinetic energy of the electron, the discontinuity propagates into the electrical properties. Notice, also, that the discontinuities in Fig. 7.11 are more pronounced for systems where the kinetic energy before the transition is *smallest*. Thus, relatively poor conduction above T_c can express stronger discontinuities in resistivity.

As mentioned, the discontinuity from the symmetric spin-orbit interaction also arises in another physical property of a superconductor. Let's examine this property next.

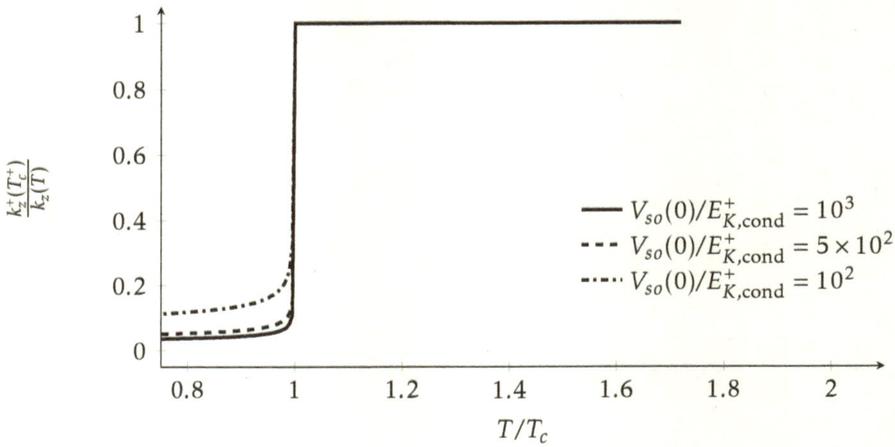

Figure 7.11: Discontinuities in $k_z^+/k_z(T)$ *near* critical temperature T_c, which is proportional to the resistivity, for three ratios $V_{so}(0)/E^+_{K,\text{cond}}$ 1000, 500, and 100.

7.5.2 Discontinuity In Heat Capacity

The *electronic heat capacity* C_e is a measure of the *amount of change in electronic energy of a system per unit change in temperature.* We have already seen that the energy of electrons is affected by the spin-orbit interaction. The *electronic heat capacity* is a function of the kinetic energy $E_K = \hbar^2 k_z^2/2m_0$. Substituting k_z^2 from Eq. (7.87), we have

$$C_e = \frac{\partial\langle E_K\rangle}{\partial T} = \frac{\partial}{\partial T}\left(E_\lambda + V_{so}(0)\tanh\left(\frac{V_{so}}{k_B T}\right)\right) = C_{ne} + C_{se} \tag{7.89}$$

There are two contributions. C_{ne} can be regarded as the heat capacity due to normal conductivity, while C_{se} is the heat capacity due to the superconducting charge carriers. For the superconductivity contribution, the spin-orbit energy responsible is for the addition conduction electrons. We are primarily interested in this contributions as it is the primary one in the vicinity of the transition or discontinuity. Then, we have

$$C_{se} = \frac{\partial\langle V_{so}\rangle}{\partial T} = \frac{\partial}{\partial T}\left[V_{so}(0)\tanh\left(\frac{V_{so}}{k_B T}\right)\right] \tag{7.90}$$

In what follows, we use the following hyperbolic trigonometric function relations:

$$\frac{d}{dx}\tanh x = \text{sech}^2 x$$

$$\text{sech}^2 x + \tanh^2 x = 1$$

Then, evaluating the above derivative, we have

$$C_{se} = \frac{\partial V_{so}}{\partial T} = \frac{\partial}{\partial T}\left[V_{so}(0)\tanh\left(\frac{V_{so}}{k_B T}\right)\right]$$

$$= V_{so}(0)\left[1 - \tanh^2\left(\frac{V_{so}}{k_B T}\right)\right]\left[\frac{1}{k_B T}\frac{\partial V_{so}}{\partial T} + \frac{V_{so}}{k_B}\left(\frac{-1}{T^2}\right)\right]$$

The first term on the RHS, in square brackets, is the derivative of $\tanh u$. The second term comes from the temperature dependent argument of the $\tanh u$ function or du/dT. This gives

$$\frac{\partial V_{so}}{\partial T} = \frac{V_{so}(0)}{k_B T}\left[1 - \tanh^2\left(\frac{V_{so}}{k_B T}\right)\right]\frac{\partial V_{so}}{\partial T} + \frac{V_{so}}{k_B T^2}\frac{V_{so}(0)}{k_B}\left[1 - \tanh^2\left(\frac{V_{so}}{k_B T}\right)\right] \quad (7.91)$$

We must recognize that we have $\partial V_{so}/\partial T$ on *both* the LHS and RHS. So, rearranging to solve for $\partial V_{so}/\partial T$, we have

$$\frac{\partial V_{so}}{\partial T} = \frac{\frac{V_{so}V_{so}(0)}{k_B T^2}\left[1 - \tanh^2\left(\frac{V_{so}}{k_B T}\right)\right]}{1 - \frac{V_{so}(0)}{k_B T}\left[1 - \tanh^2\left(\frac{V_{so}}{k_B T}\right)\right]} \quad (7.92)$$

Because the energy in a superconductor *increases with decreasing temperature,* Eq. (7.92) is negative in sign. However, the heat capacity is defined as a positive, so we take the magnitude of Eq. (7.92). Then, multiplying the above numerator and denominator by $-k_B T^2$, and using the hyberbolic relations gives an equal expression for the *superconductor electronic heat capacity* $C_{se}(T)$ as

Electronic Heat Capacity C_{se} Of Superconductor

$$C_{se} = \frac{V_{so}V_{so}(0)\text{sech}^2\left(\frac{V_{so}}{k_B T}\right)}{k_B T^2 - V_{so}(0)T\text{sech}^2\left(\frac{V_{so}}{k_B T}\right)} \quad (7.93)$$

Eq. (7.93) can be evaluated simultaneously when solving V_{so} given by Eq. (7.27), to obtain both as functions of temperature T. Fig. 7.12 shows a solution of Eq. (7.93) for the case of $V_{so}(0) = 1.5$meV. The heat capacity of a superconductor also exhibits an exponential rise accompanied by a discontinuity at T_c.

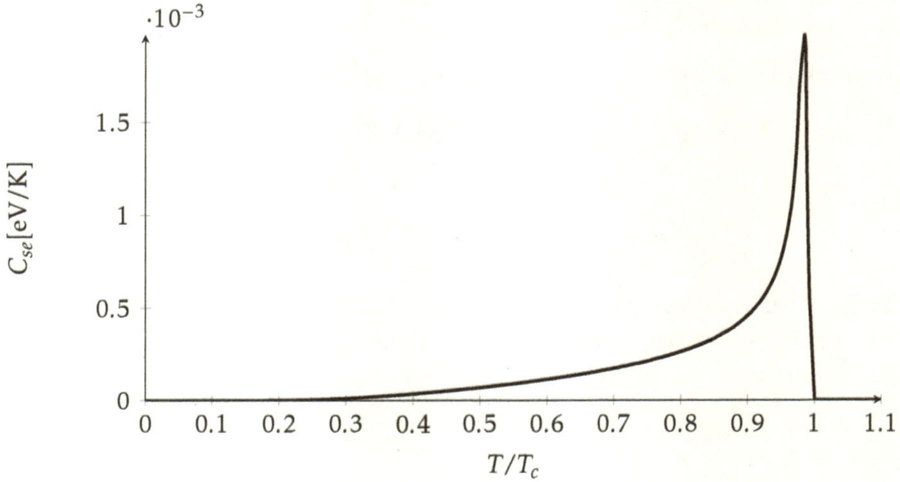

Figure 7.12: Temperature dependence of electronic heat capacity of a superconductor approaching the critical temperature T_c.

7.6 Magnetic Field Effect On T_c

To examine the effect of an external magnetic field near the transition temperature T_c, we can apply satisfying the condition where the free charge-carrier energy should be close to the localized energy in quantity or

$$E_{\text{free}} = E_{\text{local}} \tag{7.94}$$

In the superconductor, the free energy includes the eigenenergy of the quantized vortices, while the localized energy includes the symmetric potential which gives rise to the transition. This conditions leads to

$$\hbar\omega_s\left(n+\frac{1}{2}\right)+E_{\text{common}} = \Delta_c(0)\tanh\left(\frac{\Delta_c(T)}{2k_BT}\right)+E_{\text{common}} \tag{7.95}$$

E_{common} denotes the terms common to both free and ordinary conduction electrons. For example, this includes the energy due to the external electric field, as well as the spin-orbit force.

$$\Delta_c(0)\tanh\left(\frac{\Delta_c(T)}{2k_BT}\right)-\hbar\omega_s\left(n+\frac{1}{2}\right)=0$$

In the vicinity of the transition, this is equivalent to a temperature and field-dependent potential $\Delta_c(T,B)$ given by

Energy Near T_c With External Magnetic Field

$$\Delta_c(T,B) = \Delta_c(0,B)\tanh\left(\frac{\Delta_c(T)}{2k_BT}\right) - \frac{\hbar q_s B}{m_s}\left(n+\frac{1}{2}\right) \tag{7.96}$$

Eq. (7.96) corresponds to downshifted potential function. Fig. 7.13 shows the temperature dependent potential energy for several values of external field B based on Eq. (7.96). Fig. **??** shows more directly, the

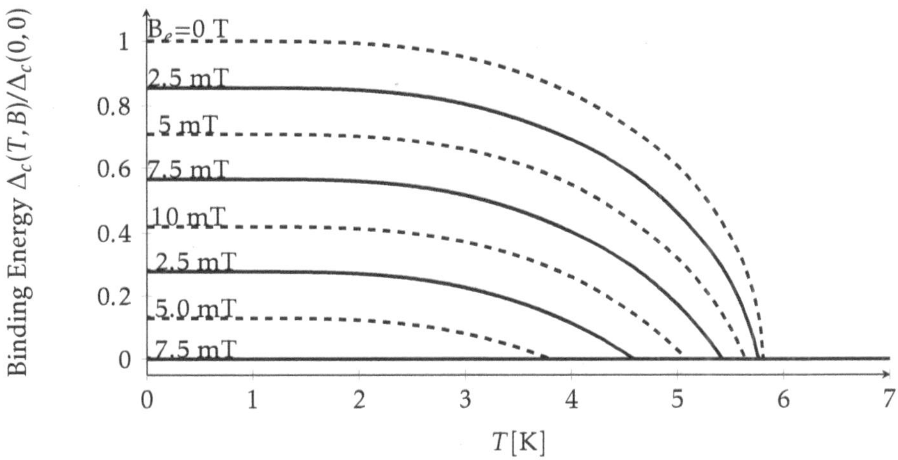

Figure 7.13: Potential energy as a function of temperature T and external magnetic field B for V_0 = 1meV. As more field is introduced, the transition temperature shifts to the left, or to lower values.

change in T_c with the external magnetic field. Because the curve shifts downwards, the slope at $T = 0$ remains unchanged. However, the slope at $T = T_c$ is reduced for larger and larger magnetic fields. Generally, introducing an external magnetic field in superconductors (consistent with the traditional configuration) reduces the critical temperature.

Let's move on and discuss an important magnetic property that we'll also need in future discussions. Specifically, let us evaluate the magnetic flux created by the quantized vortices of a superconductor.

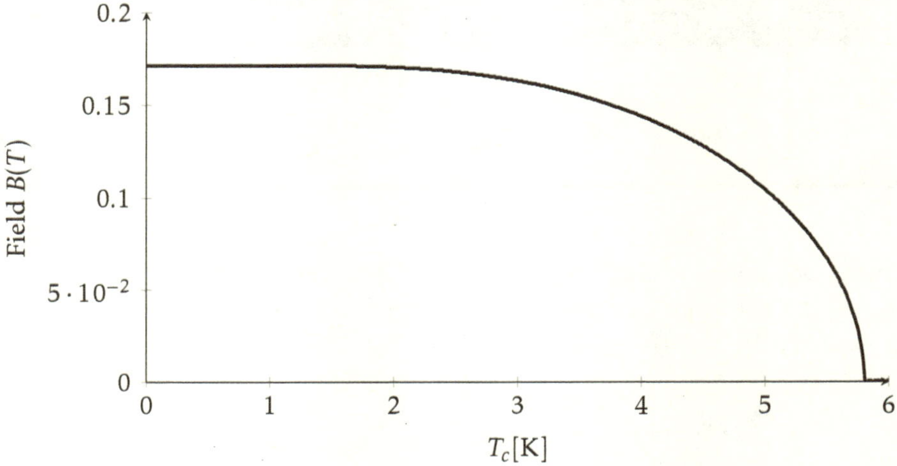

Figure 7.14: Effect of an external magnetic field on the critical temperature T_c using Eq. (7.96), and corresponding to Fig. 7.13. The calculations assume the ground state energy where $n = 0$.

7.7 Quantum Of Magnetic Flux

The solution given by Eq. (7.78) leads to another important property when we consider the following: if the wave function is forming a loop around the magnetic field axis, then for the continuity of the wave function in the φ direction, one complete loop around must return to the same starting point. So, we have a boundary condition along the azimuthal direction of $\mathbf{e}_\varphi$. With the relevant part of the solution given by $\psi_\varphi = C_\varphi e^{ik_\varphi r\varphi}$, the exponent contains the phase ϕ of the carrier around the loop. Then, for a continuity condition for a general number of loops, we have

$$\cos(k_\varphi 2\pi rn) = \cos(k_\varphi 2\pi r(n+1)) \tag{7.97}$$

$$= \cos(k_\varphi 2\pi rn + k_\varphi 2\pi r) \tag{7.98}$$

On the RHS, we can use the trigonometric identity given by

$$\cos(\alpha+\beta) = \cos(\alpha)\cos(\beta) - \sin(\alpha)\sin(\beta) \tag{7.99}$$

After use of Eq. (7.99) in Eq. (7.97) along with the condition that sine terms at $2\pi n$ are zero, we get

$$\cos(k_\varphi 2\pi rn) = \cos(k_\varphi 2\pi r(n+1)) \tag{7.100}$$

$$= \cos(k_\varphi 2\pi r n)\cos(k_\varphi 2\pi r) \tag{7.101}$$

Because we have non-zero terms, we can divide by the cosine term of the LHS to get

$$1 = \cos(k_\varphi 2\pi r) \tag{7.102}$$

Of course, this is true if and only if the argument is a multiple of 2π, so we have

$$k_\varphi 2\pi r = 2\pi n \Rightarrow k_\varphi r = n \tag{7.103}$$

Using $k_\varphi = -eBr/\hbar$ from Eq. (7.55) and $\hbar = h/2\pi$, we have

$$-\frac{eBr^2}{\hbar} = -\frac{eBr^2}{h/2\pi} = -\frac{2eB\pi r^2}{h} = n \tag{7.104}$$

For a quantized vortex of flux area A, the flux is $\Phi_s = BA = B\pi r^2$, so Eq. (7.104) becomes

$$B\pi r^2 = B \cdot A = -\frac{nh}{2e} \tag{7.105}$$

Since Eq. (7.105) describes an integer multiplied by a fixed amount, we can define that fixed amount as a *unit of quantized flux* given by

Unit Of Quantized Flux

$$\Phi_{\text{unit}} = \frac{h}{2e} \tag{7.106}$$

Then, the flux given by Eq. (7.105) becomes

Superconductor Quantized Flux

$$\Phi_n = -n\Phi_0 \tag{7.107}$$

The reciprocal of Eq. (7.106) is also known as the *Josephson constant*. If we use the values for Planck's constant h and the elementary charge e, we have

$$\frac{h}{2e} = \frac{6.6261 \times \times 10^{-34}\text{J} \cdot s}{2 \times 1.6022 \times 10^{-19}\,\text{C}} = 2.06783 \times 10^{-15}\text{Tesla} \cdot \text{m}^2 \tag{7.108}$$

The above result is the amount of flux change $\Phi_{n+1} - \Phi_n$ created by quantized vortices. This gives an indication of how small a magnetic field can be detected by device using a superconducting element. For a device with area $\approx 1\text{m}^2$, a magnetic field of $B = 2.06783 \times 10^{-15}$Tesla can be detected, which is remarkably small. For this reason, magnetic sensors based on superconductors remain one of the most sensitive magnetometers in use for the measurement and detection of external magnetic fields. Fig. 7.15 demonstrates the discrete changes in the voltage due to the quantum steps in flux in a superconductor.

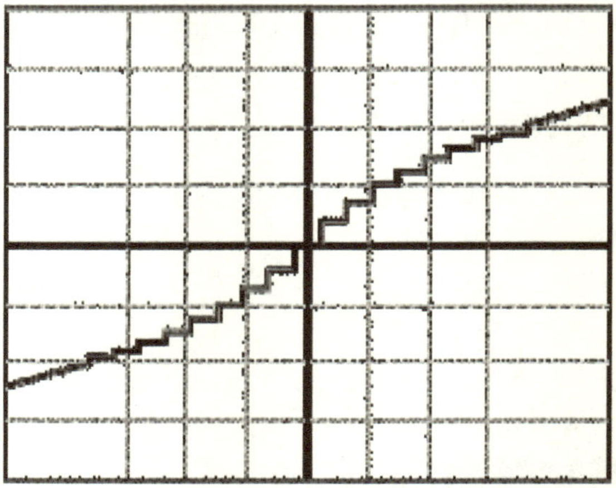

Figure 7.15: Example of quantum flux change demonstrated by a Josephson junction, using an ac + dc input voltage. If a constant biasing voltage/current is maintained, the measured voltage oscillates with the changes in wave-function phase $kx = (p/\hbar)x$. This phase change is proportional to the change in the magnetic flux $\Delta\Phi_n$. Because the unit is known, simply counting the discrete changes from a measurement provides the flux change.

Thus, because the vortices in superconductors are quantized, it turns out that the amount of flux they create is *also* quantized, and Eq. (7.106) provides the unit amount by which the flux can change. It is also known as the *flux quantum* and we see that it arises from satisfying the continuity of the quantum mechanical wave function *azimuthally*. The quantized nature of the flux in superconductors was first measured around 1961 with Tin cylinders in well-controlled magnetic fields in which the superconducting cylinders were moved cyclically along the axis of the magnetic field. This allowed determination of the fluxes. They observed that the fluxes were indeed quantized by an amount given by Eq. (7.106).

Due to quantization of the vortices, the radius of the circulating currents changes discretely in the presence of an external magnetic field. As we observed, increasing the external field reduces the radius of the vortex. If we consider the case of vortices whose area is the size of the wire, it can be shown that these conditions, having maximum surface current, can lead to *perfect diamagnetic* behavior. This is the subject of the next section.

7.8 Diamagnetism From Circulating Surface Currents

The presence of a magnetic field leads to circulating charge carriers *about* the magnetic field axis. According to the solutions we obtained in Sec. 7.3, increasing the external magnetic field *reduces* the radius of carrier circulation. This suggests that under a sufficiently weak magnetic field, we can take the charge carriers to be on the surface of a fixed size cylindrical body with radius r. Then, loop currents can be described in terms of a magnetic moment $\mu = IA$. Magnetic moments associated with solid state materials have *three distinct origins.* They are

1. Localized electron spin-associated angular momentum
2. Localized electron orbital-associated angular momentum
3. Change in orbital motion of 'free' electrons caused by an external magnetic field

Diamagnetism, in particular, arises from the third mechanism in the list.. It is a distinct form of magnetism. Similar to conventional paramagnetism, *it is only present or created when an external field is applied.*

Ⓡ *Diamagnetic material*: A diamagnetic material is defined as a material having negative magnetic susceptibility χ.

For most materials, it causes a very weak magnetic moment that *opposes an externally applied magnetic field.* In the case of weak diamagnetism, it arises from either small numbers of charge carriers participating, circulating in smaller current loops, or because of the presence of other forms of magnetism which tend to align with an external field, and this competes to ultimately reduce the net strength. However, for any structure, *if other forms of opposing magnetism are absent and/or sufficient weak e.g.* non-magnetic, and a sufficiently large number of charge carriers per unit volume circulate at the exterior surface, conditions can be created for diamagnetism to be sufficiently large.

Here, we investigate how a circulating current, created by *freed* charges like those found in superconductors in sufficiently weak magnetic fields, inherently leads to diagmagnetism as a result of the external magnetic field. This behavior applies to the charge carriers on the surface of a superconductor. One key distinction is in the fact that *diamagnetism* arises from *variations* in the external magnetic field. The variation in the field causes a corresponding *change* in magnetic moment. Let us define the magnetic moment of a superconductor as

$$\mu_s = I_s\mathbf{A} \tag{7.109}$$

The current for a single electron in a circular orbit relates to the azimuthal velocity as

$$I_s = -\frac{e}{T_s} = -\frac{e|v_\varphi|}{2\pi r} \tag{7.110}$$

The negative sign assumes *the charge of a carrier is negative,* or is due to electrons. $|v_\varphi|$ is the magnitude of the velocity around the loop $v_\varphi = -r\omega_s$ from Eq. (**??**). The diamagnetic moment relation $\mu_s = I_s A$ then becomes

$$\mu_s = -\frac{e|v_\varphi|r}{2} \tag{7.111}$$

The *diamagnetic moment is maximal in amplitude at the largest radii.* Hence, surface currents correspond to the maximum possible magnetic moment of diamagnetic material. This leads to a variation relation given by

$$\Delta\mu_s = -\frac{e\Delta v_\varphi r}{2} \tag{7.112}$$

We need to determine the relationship between Δv_φ and ΔB. This will allow us to obtain the magnetic susceptibility $\Delta\mu_s/\Delta B$. Fortunately, this is straightforward because we already have the result

$$|v_\varphi| = r\omega_s = r\frac{eB}{m_0} \Rightarrow \Delta v_\varphi = r\omega_s = r\frac{e\Delta B}{m_0} \tag{7.113}$$

Substitution of Eq. (7.113) into Eq. (7.112) gives the result

$$\Delta\mu_s = -\frac{e^2r^2}{2m_0}\Delta B = -\frac{\mu_0 e^2 r^2}{2m_0}\Delta H \tag{7.114}$$

H is given by B/μ_0 and $\mu_0 = 4\pi 10^{-7} H/m$ is the magnetic permeability in vacuum. Eq. (7.114) contains the *diamagnetic susceptibility* $\chi_s = \Delta\mu_s/\Delta H$. So, we have

Diagmagnetic Susceptibility Per Charge Carrier

$$\chi_s = -\frac{\mu_0 e^2 r^2}{2m_0} < 0 \tag{7.115}$$

From Eq. (7.111), the unit magnetic moment of a charge carrier in a surface current then becomes

Magnetic Moment Per Charge Carrier

$$\mu_s = -\frac{\mu_0 e^2 H r^2}{2m_0} \tag{7.116}$$

The SI unit of the magnetic moment μ_s is $A - m^2$. To express in the form of a *moment density* or magnetization $\mathbf{M}_s$, Eq. (7.116) can be scaled by the charge carrier density $\langle n_s \rangle$ (units of m^{-3}), so we have

$$\langle n_s \rangle \mu_s = \mathbf{M}_s = -\frac{\langle n_s \rangle \mu_0 e^2 r^2}{2m_0}\mathbf{H} \quad \text{units of A/m} \tag{7.117}$$

Fig. 7.16 shows an example of an ideal response of the magnetic moment, based on Eq. (7.117).

The word 'ideal' is used in Fig. 7.16 because, as has already been mentioned, this behavior is limited to a certain strength of magnetic field, before the quantized vortices will prefer a reduced size, which affects other properties of the superconductor. We'll discuss the details of this limitation in Sec. 7.9, which relates to the change in the magnetization response after this condition is reached.

We can also express the magnetization in units of Tesla if we multiply by μ_0, so we have

$$\mu_0 \mathbf{M}_s = -\frac{\mu_0^2 \langle n_s \rangle e^2 r^2}{2m_0}\mathbf{H} \tag{7.118}$$

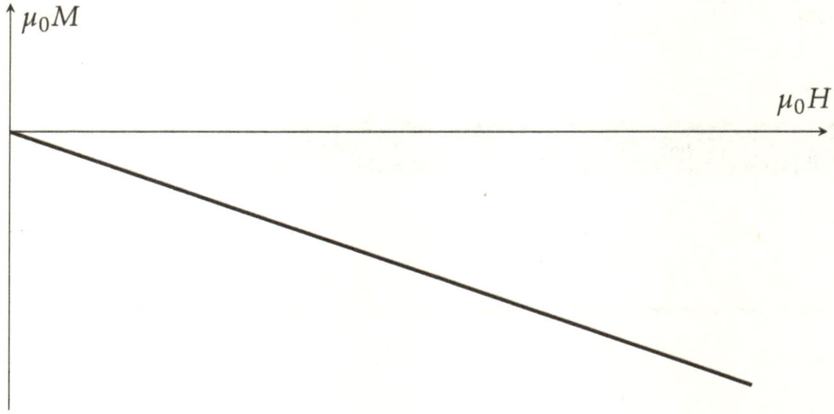

Figure 7.16: Ideal response of the magnetic moment created by an external field in a superconductor.

This also provides a way to write a magnetic induction in the superconductor $\mathbf{B}_s = \mu_0 \mathbf{M}_s + \mu_0 \mathbf{H}$ as

$$\mathbf{B}_s = -\frac{\langle n_s \rangle \mu_0^2 e^2 r^2}{2m_0}\mathbf{H} + \mu_0 \mathbf{H} = \left(1 - \frac{\langle n_s \rangle \mu_0 e^2 r^2}{2m_0}\right)\mu_0 \mathbf{H} \tag{7.119}$$

Therefore, we find the *superconductor relative magnetic permeability* to be

Superconductor Relative Magnetic Permeability

$$\mu_r = 1 - \frac{\langle n_s \rangle \mu_0 e^2 r^2}{2m_0} \tag{7.120}$$

$\langle n_s \rangle$ is the number of charge carriers per unit volume, and is specific to each material.

Eq. (7.120) satisfies the condition $0 < \mu_r < 1$, for any finite amount of circulating current. Weak diamagnetism corresponds to a value just below unity or $\mu_r \approx 1$, while strong diamagnetism, with maximum circulating current, corresponds to the condition $\mu_r \approx 0$. The distinctions are dictated by the value of $\langle n_s \rangle$.

Let's consider an example problem demonstrating how to estimate the number of charge carriers per unit volume in a superconductor.

The geometrical parameters we'll use are taken from experiments carried out in 1961, by Deaver and Fairbank on cylinders of tin (Sn) of approximately 0.8cm in length having a radius of $r = 2.33{\times}10^{-3}$cm.

Exercise 7.1 Use the following condition to estimate the number of charge carriers per unit volume $\langle n_s \rangle$ in a tin superconducting cylinder of radius $r = 2.33 \times 10^{-3}$cm, using unit charge and mass e, m_e to be that of the electron:

$$-\frac{\langle n_s \rangle \mu_0 e^2 r^2}{2m_s} + 1 \approx 0$$

Solving for $\langle n_s \rangle$ from the above condition gives

$$\begin{aligned}\langle n_s \rangle &\approx \frac{2m_s}{\mu_0 e^2 r^2} \\ &= \frac{2 \cdot \left(9.10938356 \times 10^{-31} kg\right)}{4\pi 10^{-7} H \cdot \mathrm{m}^{-1} \left(1.60217662 \times 10^{-19} \mathrm{C}\right)^2 \left(2.33 \times 10^{-5} \mathrm{m}\right)^2} \\ &= 1.04 \times 10^{23} \mathrm{m}^{-3}\end{aligned}$$

Therefore, there are on the order of $\langle n_s \rangle \approx 10^{23}$ charge carriers per unit volume in a cylindrical superconductor of size $r = 2.33{\times}10^{-3}$cm.

■

Summarizing, we've found in superconductors that the electrons freed by the combined action of the external electric field and the increase in potential energy, with reduced temperature, leads to the freeing of electrons from being localized in the atoms. Moreover, the spin-orbit force begins to push these electrons to the surface, particularly, in the presence of a magnetic field. Because the magnetic field will cause these electrons to orbit, we have found that in conditions where these electrons are confined to the exterior surface, maximum diamagnetism is created, but only up to a certain magnetic field value, because we also know that an increasing field will eventually cause the quantized vorices to reduce their radii. While on the surface, we have a relationship between the magnetic moment of the superconductor $\mathbf{M}_s$ and the external magnetic field.

The next question to discuss is what happens when conditions are reached that will no longer support the quantized vortices strictly confined to the outer surface of the superconductor.

7.9 Origins Of Type I & Type II Superconductors

In Subsection 8.1.1, two distinct types of superconductors were described, being called *Type I* and *Type II*. Both are distinguished by their magnetic field response, like those shown in Fig. 7.17.

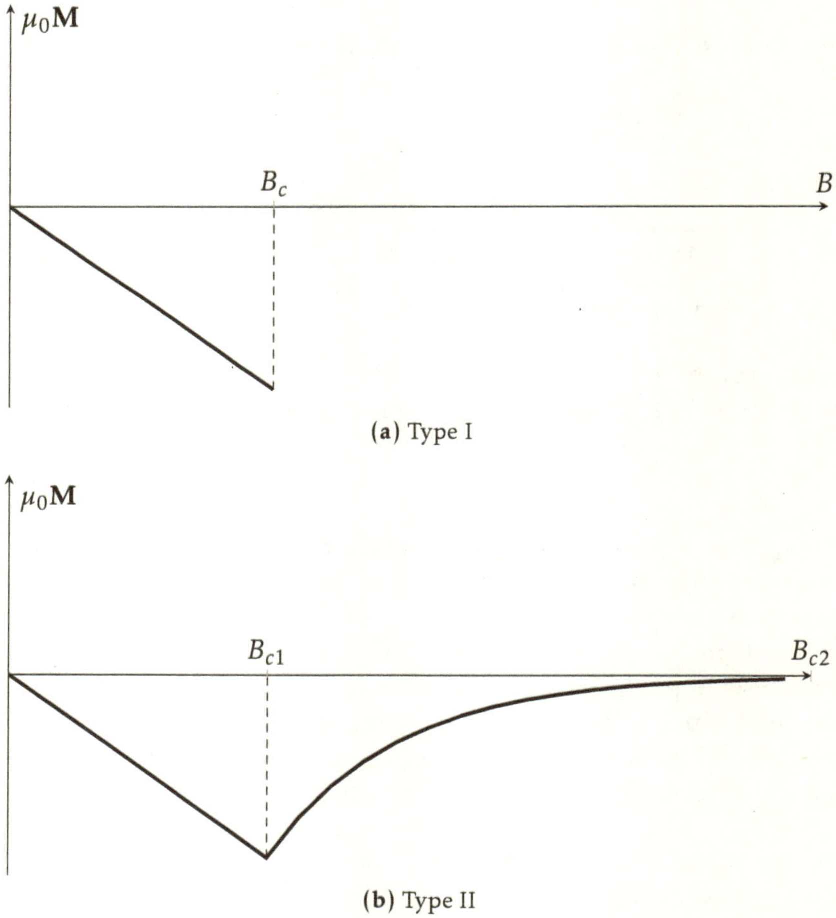

Figure 7.17: Illustration of magnetic responses from Type I and Type II superconductors.

Each type has their respective magnetization response which is important to understand, especially for the purpose of developing supercon-ducting materials, or other diamagnetic materials. Here, we'll discuss the origins of both types and why we observe *only two types* of magnetization responses.

Let us return to the energy operator from Eq. (7.38), where we have

$$E_\lambda \begin{bmatrix} \psi_\uparrow \\ \psi_\downarrow \end{bmatrix} = \left[E_{K,\text{cond}} + E_{K,\text{sof}} + \frac{(\widehat{\mathbf{p}} + e\mathbf{A})^2}{2m_0} - \frac{(\widehat{\mathbf{p}} + e\mathbf{A})^4}{2m_0^3 c^2} + V_{\text{gap}} \right] \begin{bmatrix} \psi_\uparrow \\ \psi_\downarrow \end{bmatrix} \tag{7.121}$$

$$= \left[\frac{(\widehat{\mathbf{p}} + e\mathbf{A})^2}{2m_0} - \frac{(\widehat{\mathbf{p}} + e\mathbf{A})^4}{2m_0^3 c^2} + V_{\text{super}} \right] \begin{bmatrix} \psi_\uparrow \\ \psi_\downarrow \end{bmatrix} \tag{7.122}$$

where it is most probable that

$$V_{\text{super}} > 0$$

$E_{K,\text{cond}}$ is the ordinary conduction due to the external electric field. $E_{K,\text{sof}}$ is the spin-orbit force pushing the charge carriers toward the surface. V_{gap} is the additional potential energy from any existing band gap between the valence band and the conduction band. We can use the above eigenoperator equation to evaluate the condition on the vortex quantum number n near zero energy or $E_s = 0$.

7.9.1 Nonzero Quantum Number At Zero Energy

In Sec. 7.1, we found the kinetic energy eigenvalue contribution for the operator given by $[(\widehat{\mathbf{p}} + e\mathbf{A})^2/2m_0]\psi$ to be

$$E_K = \frac{\hbar e B}{m_0}\left(n + \frac{1}{2}\right) = \hbar\omega_s\left(n + \frac{1}{2}\right) \tag{7.123}$$

Replacing the operator with its eigenvalue from Eq. (7.123), we have

$$E_s = -\frac{e^2\hbar^2 B^2}{2m_0^3 c^2}\left(n + \frac{1}{2}\right)^2 + \frac{e\hbar B}{m_0}\left(n + \frac{1}{2}\right) + V_{\text{super}} \tag{7.124a}$$

$$= -\frac{\hbar^2\omega_s^2}{2m_0 c^2}\left(n + \frac{1}{2}\right) + \hbar\omega_s\left(n + \frac{1}{2}\right) + V_{\text{super}} \tag{7.124b}$$

$$= A\alpha_n^2 + B\alpha_n + C \tag{7.124c}$$

where the polynomial coefficients and parameter α_n introduced are defined as

$$\alpha_n = n + \frac{1}{2} \tag{7.125a}$$

$$A = -\frac{\hbar^2 \omega_s^2}{2m_0 c^2} < 0 \tag{7.125b}$$

$$B = \hbar \omega_s > 0 \tag{7.125c}$$

$$C = V_{\text{super}} > 0 \tag{7.125d}$$

Without the higher order correction, we have only the non-relativistic kinetic energy term. However, adding the higher order kinetic energy term turns out to make a significant difference in the energy dependence on the external field B. Fig. (7.18) illustrates the eigenenergy of Eqs. (7.124) for the two cases being (1) *without* the higher order kinetic energy correction so $E_s = E_K$ and (2) *with* the correction E_c where $E_s = E_K + E_c$.

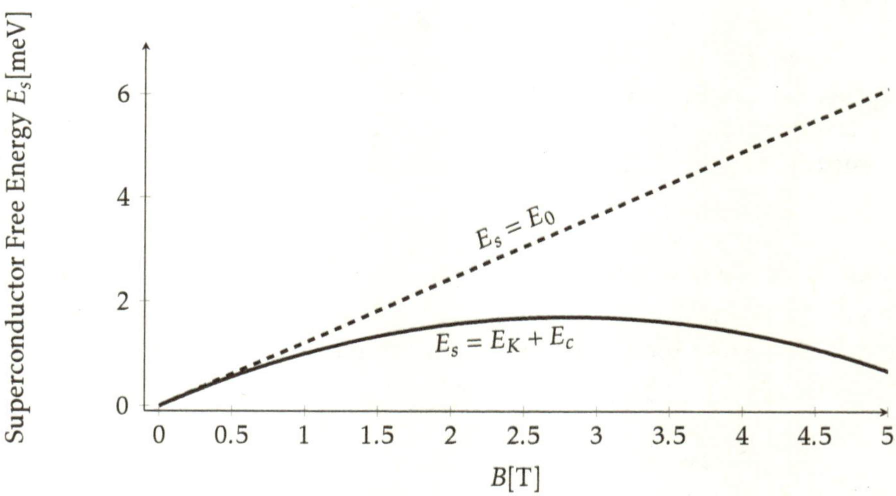

Figure 7.18: Eigenenergy E_s versus the external magnetic field, with and without the relativistic correction. A quantum number of $n = 10$ is used. The correction eventually begins to dominate, and thus, reduces the energy E_s.

By adding the additional kinetic energy contribution, we see that the magnetic field eventually drives the energy back down to zero over a finite range of magnetic field. We can examine the conditions of the quantum number n in the vicinity where $E_\lambda = 0$. To do this, we can just set Eq. (7.124c) equal to zero. Beyond this point, the kinetic energy become negative, and we know that localization is generally indicated by a negative eigenenergy, at least relative to the Fermi energy of the system. As the energy drops below the Fermi edge, the conduction electron returns back to the valence band, or near the ordinary conduction band comprised of additional mechanisms such as electronic drift. This effectively *removes* holes from the atoms in the superconductor rendering the system back towards *ordinary* conduction.

For the roots, α_n^*, they are given by

$$\alpha_n^* = n^* + \frac{1}{2} = \frac{-B \pm \sqrt{B^2 - 4AC}}{2A} \tag{7.126}$$

Using the coefficient values from Eq. (7.125), we have

$$\alpha_n^* = \frac{-\hbar\omega_s \pm \sqrt{\hbar^2\omega_s^2 - 4\left(-\frac{\hbar^2\omega_s^2}{2m_0c^2}\right)(V_{\text{super}})}}{2 \times \left(-\frac{\hbar^2\omega_s^2}{2m_0c^2}\right)} \tag{7.127a}$$

$$= +\frac{m_0c^2}{\hbar\omega_s} \mp \frac{m_0c^2}{\hbar^2\omega_s^2}\sqrt{\hbar^2\omega_s^2 - 4\left(-\frac{\hbar^2\omega_s^2}{2m_0c^2}\right)(V_{\text{super}})} \tag{7.127b}$$

$$= +\frac{m_0c^2}{\hbar\omega_s} \mp \frac{m_0c^2}{\hbar\omega_s}\sqrt{1 + \left(\frac{V_{\text{super}}}{m_0c^2/2}\right)} \tag{7.127c}$$

$$= +\frac{m_0^2c^2}{\hbar eB}\left[1 \mp \sqrt{1 + \left(\frac{V_{\text{super}}}{m_0c^2/2}\right)}\right] \tag{7.127d}$$

Because energy contributions like spin-orbit interaction are higher order relativistic energy operators, being on the order of $1/c^2$, the term in parentheses under the radical is expected to be very small *i.e.* $V_{\text{super}} \lll m_0c^2/2$. In fact, this term is on the order of $1/c^4$. Consequently, it indicates the term in brackets essentially takes on the binary values of

$$\alpha_n^1 \approx +\frac{m_0^2c^2}{\hbar eB} \cdot (1-1) = 0 \tag{7.128a}$$

$$\alpha_n^2 \approx +\frac{m_0^2c^2}{\hbar eB} \cdot (1+1) = +\frac{2m_0^2c^2}{\hbar eB} \tag{7.128b}$$

In the case where $\alpha_n = \alpha_n^2$, $n \gg 1$ so that even though the superconductor energy reaches near zero, there is still a quantized vortex state in this condition. Because the system will be driven into the superconducting state, it will already possess a nonzero quantum number prior to reaching the condition $E_\lambda = 0$.

Moreover, we find that the root α_n^* *diminishes* with *increasing* magnetic field so the quantum number will tend towards smaller values as the field increases, with $1/B$. This also indicates that the vortex quantum number values n *prior* to reaching E_λ satisfy the condition

$$n \geq n^* \quad \text{for} \quad B \leq B^*$$

where B^* is the magnetic field value at the crossover point $E_\lambda = 0$. The combination of a nonzero quantum number (i.e. an existing quantum state) along with the relationship between $\alpha \propto 1/B$ can be observed in the magnetization curve. Recall that the circulating currents of the superconductor give rise to a *diagmagnetic* effect having a *negative* susceptibility given by

$$\chi_s = -\langle n_s \rangle \frac{\mu_0 e^2 r^2}{2m_0} < 0 \tag{7.129}$$

where $\langle n_s \rangle$ is the number of charge carriers per unit volume in the superconductor *contributing* to the net magnetic moment. The charge carrier density correlates with the quantum number n. For example, when the vortices have large n, this corresponds to maximum charge carriers, while $n = 0$ corresponds to the ground state from the system can be driven readily out of the vortex state. So, smaller $n \rightarrow$ smaller $\langle n_s \rangle$, while larger $n \rightarrow$ larger $\langle n_s \rangle$ etc. Therefore, if n decreases with increasing magnetic field, χ_s will also diminish in amplitude from its negative value towards zero. So, we have

$$(\text{inceasing} \quad \text{B}) \rightarrow (\text{decreasing} \quad n) \rightarrow (\text{decreasing} \quad |\chi_s| \text{ as} as \ 1/B) \tag{7.130}$$

There is still the important question as to when can the $1/B$ reduction in χ_s response be observed in superconductors. For this, we must consider the energy contribution V_{super}.

7.9.2 Effects Of A Band Gap

Fig. 7.18 showed that by adding the relativistic correction to the kinetic energy, the ordinarily linear field dependence becomes a nonmonotonic/downward-parabolic energy. This has an impact on the behavior of superconductors, particularly, when we consider the energy V_{super} of the system. As the energy of 'free' electrons, we can write the energy as an effective kinetic energy, where

$$\frac{\hbar^2 k_{\text{super}}^2}{2m_0} = -\frac{\hbar^2 \omega_s^2}{2m_0 c^2}\left(n+\frac{1}{2}\right) + \hbar\omega_s\left(n+\frac{1}{2}\right) + V_{\text{super}} \tag{7.131}$$

Recall that $V_{\text{super}} = E_{K,\text{cond}} + E_{K,\text{sof}} + V_{\text{gap}}$ where $E_{K,\text{cond}}$ is the ordinary conduction due to an external electric field. $E_{K,\text{sof}}$ is unique conduction of the spin-orbit force pushing the charge carriers toward the surface. V_{gap} is the additional potential energy charge carriers will pickup from leaping any existing band gap between the valence band and the conduction band.

One effect of V_{super} is to shift the $E_s(B)$ upwards, as illustrated in Fig. 7.19.

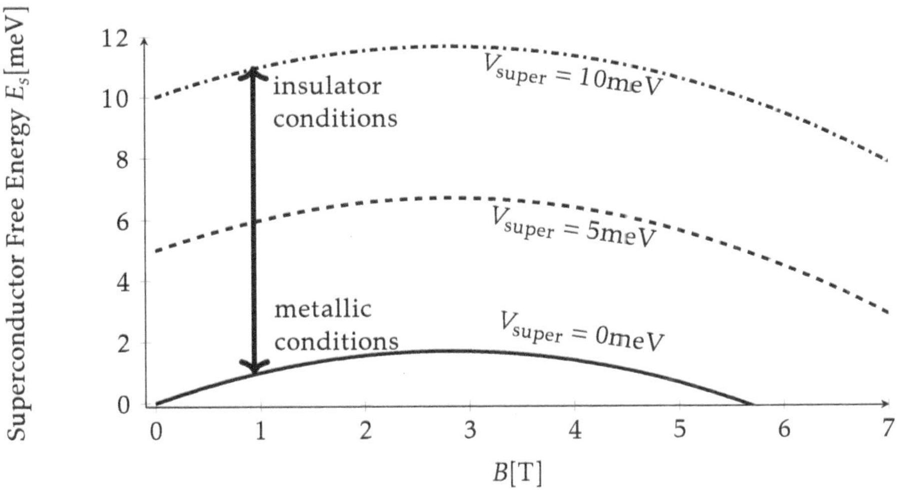

Figure 7.19: Eigenenergy E_s versus the external magnetic field for different values of the energy $V_{\text{super}} = 0, 5, 10\text{meV}$.

Curves toward the bottom correspond to small values of V_{super}. This is more like a metal, with relatively small band gaps. In this case, the curve reaches a crossover, and according to our discussion of the quantum number at this point, the system goes from $n \gg 1$ directly to the condition where $E_\lambda < 0$. Therefore, the response goes immediately from quantized vortices to bound electrons. This behavior corresponds to a Type I superconductor, and they are all metallic.

However, when V_{super} is relatively large and the curve shifts upward, electrons are ejected from the unique superconducting state, however, they still have positive kinetic energy, or $E_\lambda > 0$. This condition allows the external magnetic field to continue, but following the relation $1/B$. Therefore, from the point of ejection from the pure superconducting states, the magnetization curve changes shape from the linear response, to the response where χ_s reduces towards zero as B increases. This behavior corresponds to a *Type II superconductor*. Type II superconductors all have insulator properties which effectively enhances V_{super}.

Now, we have a combination of conditions that explain the origins of Type I and Type II superconductor magnetization curves. For Type I, their energies crossover directly, and they transition immediately to bound state electrons. This produces a magnetization response given by.

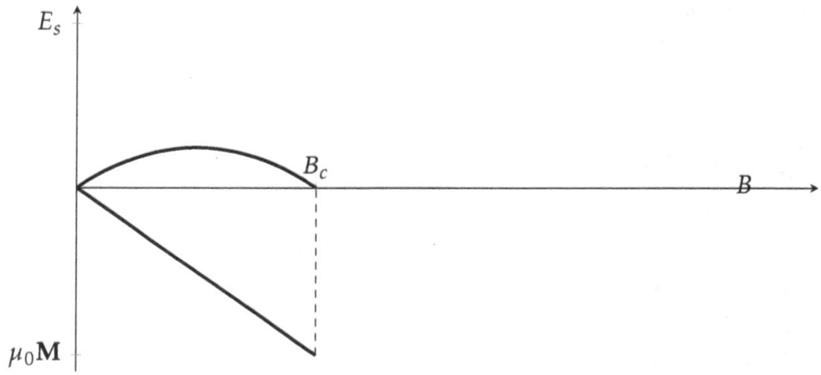

Figure 7.20: Type I superconductor magnetization response and corresponding energy dependence on an external magnetic field B. In this type, because the carriers escape only near $E_\lambda = 0$, they immediately return to ordinary electrical conduction between interior atoms.

In Type II, when electrons escape the conduction band, the situation differs from metals because they are still in a state of *strictly positive energy*. This results in a different response because the increase in magnetic field begins to reduce the quantum number n, which consequently reduces the magnetic charge carrier concentration $\langle n_s \rangle$ so that $\chi_s \to 0$. The relationship between the magnetization response and the free energy curve for Type II superconductors is illustrated in Fig. 7.21.

In superconductors, we find that the band gap between the valence and superconduction band can help to significantly improve the range of magnetic field from which a response can be obtained. Because of this,

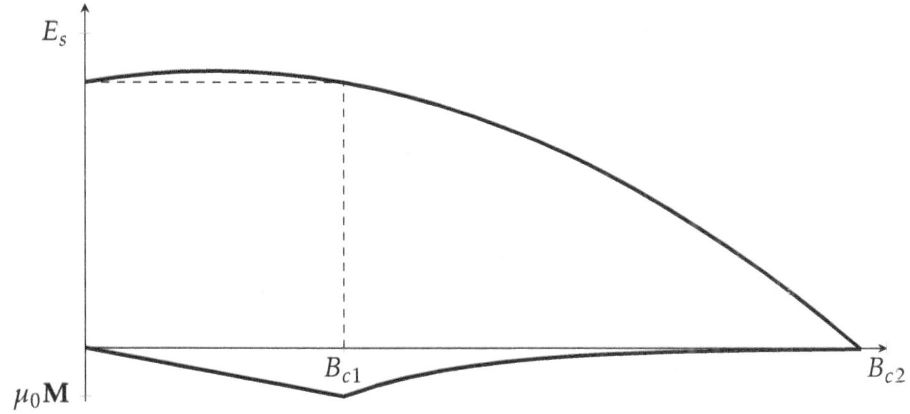

Figure 7.21: Type II superconductor magnetization response and corresponding energy dependence on an external magnetic field B. In Type II, the carriers reach a condition to exit the superconduction band near B_c1. Due to the band gap, they still have a significant amount of kinetic energy. From B_c1 onwards, the electrons are no longer held near the surface by the spin-orbit force, and follow the $1/B$ dependence by the quantum number n.

many applications using superconductors will have some form of barrier potential introduced into the material.

It is remarkable that uncovering the origins of Type I and II has only been possible by adding the kinetic energy correction term. In low temperature physics, higher order operators like these can become surprisingly indispensable. In this case, we find cogent evidence that at low temperatures, the higher order kinetic energy contribution becomes significant and even compulsory.

Let's now discuss an example of an application of superconductors using in superconductor|insulator|superconductor junction, often used to measure large and small magnetic fields. It is known as the Josephson junction.

7.10 The Josephson Junction

Because of their unique properties, superconductors are used in a number of applications. A Josephson junction is a three-layer device whose operation is based on the *principles of superconductivity*. The most basic

Josephson junction device is composed of two superconducting layers sandwiching an insulator material, e.g. Al|Al_2O_3|Al (see Table ?? for aluminum critical temperature). The wave function in the insulator

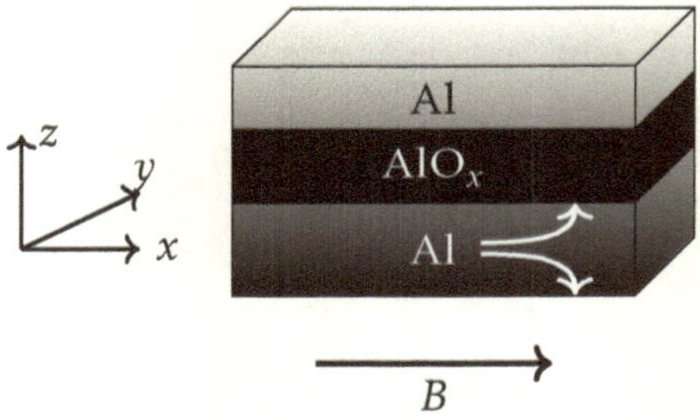

Figure 7.22: Illustration of a *Josephson Junction* device composed of two superconducting layers sandwiching an insulator. Superconductivity in the contacts, of course, is designed through low temperature settings, etc.

region between both superconducting contacts is generally given by

$$\psi_c = Ce^{ik_2x} + De^{-ik_2x} \tag{7.132}$$

The wavenumber of the barrier layer, denoted by k_2, is

$$k_2^2 = \frac{2m}{\hbar^2}(E - V_2) \tag{7.133}$$

7.11 Josephson Junction Voltage Relation

Fig. 7.23 shows an example of an application of the Josephson junction in a *superconducting quantum interference device* or *SQUID*. One common application of a SQUID is to measure and/or detect external magnetic fields. The illustrated SQUID consists of two Josephson junctions in parallel forming a loop-like structure. The loop structure is designed to be conducive to a quantized vortex. By understanding properties of superconductors, much of the behavior of Josephson junctions can also be understood. In Sec. 7.9, we learned that insulating conditions can extend the magnetic range of superconducting behavior, so that its properties can be exploited more robustly using its response to an external magnetic field.

In Sec. 7.3, when we obtained the wave function ψ for the quantized vortex, we also found that for the motion around the vortex loop, it's

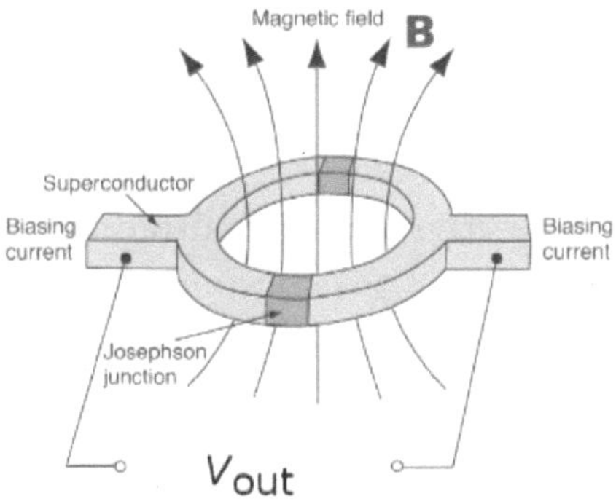

Figure 7.23: Illustration of a device using *two* Josephson junctions in parallel (junctions are center of top and bottom arms). It is known as a superconducting-quantum-interference-device (SQUID), used as a magnetometer or to detect magnetic fields.

eigenenergy $E_\varphi = 0$. Then, for *pure* circular motion, the wave function $\psi = \psi_r \psi_\varphi$ becomes

$$\psi_\varphi = \psi_r(r)\psi_\varphi(0)e^{i\left(\frac{p_\varphi}{\hbar}\cdot r d\varphi - \frac{E_\varphi}{\hbar}t\right)} \tag{7.134a}$$

$$= \psi_r(r)\psi_\varphi(0)e^{i\frac{p_\varphi}{\hbar}\cdot r d\varphi} \tag{7.134b}$$

$$= \psi_r(r)\psi_\varphi(0)e^{ik_\varphi \cdot d\ell} = \psi_r(r)\psi_\varphi(0)e^{i\theta} \tag{7.134c}$$

In this situation, even a constant magnetic field along the axis of the loop generates a nonzero voltage drop around it.[10] Using Faraday's law and a few other relations, an explicit relationship between the nonconservative voltage change around the loop, the quantum flux change, and the rate at which the phase change takes place can be found.

Faraday's electromagnetic law of induction is given by

$$\nabla \times \mathbf{E} = -\frac{\partial \mathbf{B}}{\partial t} \tag{7.135}$$

The electric field $\mathbf{E}$ is given by

$$\mathbf{E} = \mathbf{E}_c + \mathbf{E}_n = -\nabla V_c - \frac{\partial \mathbf{A}}{\partial t} \tag{7.136}$$

[10]Because the motion around the loop is circular $\partial \mathbf{p}_\varphi / \partial t \propto \partial \mathbf{A}/\partial t$ changes, and is therefore nonzero.

$\mathbf{E}_c$ denotes the conservative contribution to the electric field and $\mathbf{E}_n$ denotes the *nonconservative* contribution. V_c is the corresponding conservative electric potential.

Integrating Faraday's law over the loop surface area S, we get

$$\int \nabla \times \mathbf{E} \cdot d\mathbf{S} = -\int \frac{\partial \mathbf{B}}{\partial t} \cdot d\mathbf{S} = -\frac{\partial}{\partial t}\left(\int \mathbf{B} \cdot d\mathbf{S}\right) \tag{7.137}$$

The RHS is the loop magnetic flux $\Phi = \mathbf{B} \cdot \mathbf{S}$, so Eq. (7.137) becomes

$$\int \nabla \times \mathbf{E} \cdot d\mathbf{S} = -\frac{\partial \Phi}{\partial t} \tag{7.138}$$

Let us rewrite the LHS, as well. In Sec. 4.6, the quantum mechanical operators for Faraday's law of induction where derived, and for the additional potential degree of freedom $\phi(\mathbf{r}, t)$ included in the usual potential V owing to the 4th dimension, we found (*c*.Eq. 4.64)

$$\nabla \phi = \frac{\partial \mathbf{p}_B}{\partial t} = e\frac{\partial \mathbf{A}}{\partial t} \tag{7.139}$$

$\mathbf{p}_B = \mathbf{p}_\varphi$ is the momentum of a charge carrier around the loop, arising from a magnetic field B. Then, we have

$$-\frac{\partial \Phi}{\partial t} = -\oint_S \nabla \times \left(\mathbf{E}_c + \frac{\partial \mathbf{A}}{\partial t}\right) \cdot d\mathbf{S} \tag{7.140a}$$

$$= -\oint_S \nabla \times \mathbf{E}_c \cdot d\mathbf{S} - \frac{\partial}{\partial t}\left(\oint_S \nabla \times \mathbf{A} \cdot d\mathbf{S}\right) \tag{7.140b}$$

$$= -\oint_S \mathbf{E}_c \cdot d\ell - \frac{\partial}{\partial t}\oint_S \mathbf{A} \cdot d\ell \tag{7.140c}$$

S is the surface area defining the flux $\mathbf{B} \cdot \mathbf{S}$, and have used the following vector-calculus identity for a vector field $\mathbf{F}$:

$$\int \nabla \times \mathbf{F} \cdot dS = \oint_S \mathbf{F} \cdot d\ell \tag{7.141}$$

Because $\mathbf{E}_c$ is the conservative contribution to the electric field, the first integral on the RHS is zero. Therefore, we have

$$\frac{\partial \Phi}{\partial t} = \frac{\partial}{\partial t}\oint_S \mathbf{A} \cdot d\ell \tag{7.142a}$$

$$= \oint_S \frac{\partial \mathbf{A}}{\partial t} \cdot d\ell \tag{7.142b}$$

$$= \frac{1}{e} \oint_S \nabla \phi_n \cdot d\ell \tag{7.142c}$$

$$= \frac{\Delta \phi_n}{e} \tag{7.142d}$$

Let us define the *nonconserved electrical voltage* around the loop as

$$\Delta V_n = \frac{\Delta \phi_n}{e} \tag{7.143}$$

Then, for the RHS of Eq. (7.138), we can write

$$\frac{\partial \Phi}{\partial t} = \frac{\partial \Phi}{\partial n} \frac{\partial n}{\partial t} \tag{7.144}$$

Recall that we derived the flux relation to be (*c*.Eq. 7.107)

$$\Phi = \frac{nh}{2e} \Rightarrow \frac{\partial \Phi}{\partial n} = \frac{h}{2e} \tag{7.145}$$

In this context, n counts the number of revolutions around the flux area. Substituting this result into Eq. (7.144), we have

$$\frac{\partial \Phi}{\partial t} = \frac{h}{2e} \frac{\partial n}{\partial t} \tag{7.146}$$

Here, the quantity dn/dt is a constant given by the frequency of a unit loop voltage change. Thus,

$$\frac{\partial n}{\partial t} = \frac{1}{T_p} = f_s = \frac{\omega_s}{2\pi} = \frac{d\theta}{dt} \tag{7.147}$$

Therefore, we have the voltage change around the loop is given by

$$\Delta V_n = \frac{h}{2e} \cdot f_s \tag{7.148}$$

or

Superconductor Nonconserving Loop Frequency

$$f_s = \frac{2e\Delta V_n}{h} \tag{7.149}$$

This also means that the wave function phase θ can be related, as well, because

$$\frac{d\theta}{dt} = 2\pi f_s = \frac{2\pi}{h} \cdot 2e\Delta V_n = \frac{2e\Delta V_n}{\hbar} \tag{7.150}$$

so

Superconductor Phase Evolution Equation

$$\frac{d\theta}{dt} = \frac{2e\Delta V_n}{\hbar} \tag{7.151}$$

Biasing a SQUID device with a constant current greater than the sum of the Josephson critical currents of the two junctions, the output voltage across the Josephson junctions is a non linear periodic function of the magnetic flux through the superconductive loop having the period equal to a flux quantum [2.07x10-15 Wb].

7.12 SQUID As A Super RLC Circuit

An example of a *superconducting-quantum-interference device*, known as a SQUID, is illustrated in Fig. 7.23. The SQUID is an electronic circuit that utilizes two identical *Josephson junctions*, which are superconductor|barrier|superconductor junctions. As shown, the junctions are arranged in parallel, where a current which always includes a steady or dc component, enters from the left. We can consider an ordinary circuit first, to understand the effect of such a design on the electrical response.

We'll limit the discussion of how the SQUID operates to the so-called *dc SQUID*, which injects a current containing a steady or dc current I into the device having a topology as shown in Fig. 7.23. The current then splits up between both arms of the SQUID with I_a and I_b, respectively, so that

$$I = I_a + I_b \tag{7.152a}$$

$$= \frac{dQ_a}{dt} + \frac{dQ_b}{dt} \tag{7.152b}$$

Q_a and Q_b are now the respective charges on the Josephson junctions. Since the Josephson junctions are superconductor|insulator|superconductor structures, there is an insulating barrier in *both* arms of the SQUID. This creates a parallel *capacitance* in the arms of the circuit. In addition, there is also an effective *half-inductor* in each arm because the current is circulating around a *half*-loop. Normally, one loop is assumed for an inductor, but in the SQUID, instead, there are two half-loops. This is an important distinction. Depending on the conditions in the SQUID, there can be resistance, as well. Therefore, the dc SQUID device is generally comprised of three electronic elements which includes:

(1) capacitor (from junction barriers)
(2) inductor (from half-loop), and
(3) resistor (metallic wire/normal conditions)

For an electronic circuit *assuming all three elements are linear*, the corresponding voltage output equation for is given by

$$0 = L\frac{dI_j}{dt} + I_j R + \frac{Q_j}{C} - V_{in} \Rightarrow \frac{d^2 Q_j}{dt^2} + \frac{R}{L}\frac{dQ_j}{dt} + \frac{Q_j}{LC} = \frac{V_{in}}{LC} \tag{7.153}$$

The above equation is a linear ordinary differential equation with an exact solution for the junction charge $Q_j(t)$ given by

$$Q_j(t) = C V_{in}\left[1 - Ae^{-t/2\tau}\sin(\beta t + \varphi)\right] \tag{7.154}$$

where the solution parameters are given by

$$\tau = \frac{L}{R} \quad , \quad \alpha = \frac{R}{2L} \quad , \quad \beta = \frac{1}{2}\sqrt{\frac{R^2}{L^2} - \frac{4}{LC}} \tag{7.155a}$$

$$\varphi = \tan^{-1}\frac{\beta}{\alpha} \quad , \quad A = \frac{1}{\sin\varphi} \tag{7.155b}$$

Thus, even though a dc input is applied, we can see that the response of the charge and current in the circuit described by Eq. (7.153) is *harmonic* with damping/decay determined by the parameter $\tau = L/R$. Therefore, in the limit as $R \to 0$, it follows that $\tau \to \infty$ so the decay takes 'forever' to complete, *i.e.* the system oscillations have no damping/decay. Fig. 7.24 illustrates solutions based on Eq. (7.154) for different values of resistance R. It can be seen that as the resistance gets closer to zero, the current flowing in the circuit oscillates with less and less decay, approaching a purely harmonic response.

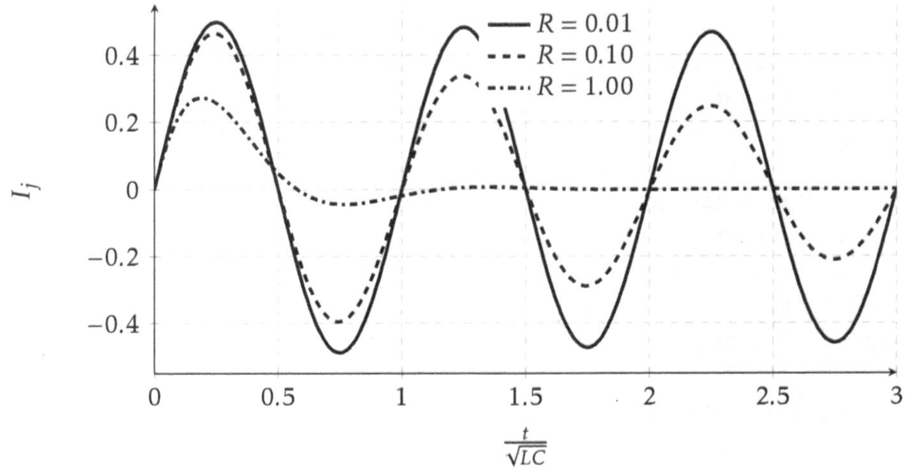

Figure 7.24: Current solutions of an RLC circuit with a dc voltage input (described by Eq. (7.153)), using $L = 1$, $C = 1$, and varying the resistance $R = 0.01, 0.1$, and 1.0. As the resistance decreases, the current becomes more purely harmonic, or non-damped.

When a SQUID transitions into the superconducting state, the system achieves the limiting case of $R = 0$, corresponding to a purely harmonic current/charge circuit between both arms of the SQUID. As long as both Josephson junctions are 'the same', then, the two currents I_a and I_b oscillate *in-phase*. Fig. 7.25 illustrates the current in both arms of the DC SQUID, solving the linear equations for the SQUID circuit illustrated in Fig. 7.23. The circuit has two parallel arms, each containing a capacitor, half-inductor and negligible resistance $i.e. R \approx 0$.

Perhaps one of the most subtle, yet, necessary conditions related to how a SQUID works can be understood by observing the SQUID arm currents I_a, I_b in Fig. 7.25. Notice that within a *half-cycle* of a current pulse, one current I_a or I_b always carries the top half of the *ac* pulse, while the other arm carries to bottom half of the *ac* pulse. Together, *they both contain a complete pulse in half the period.* For an ac current, ordinarily in a single loop, the emf drop around the entire loop would take the entire period of the pulse. In the SQUID, two complementary halves of the emf $\Delta V/2$ develop simultaneously, in only *half* the period of the pulse! This seemingly small electronic design modification happens to enable meeting the unique energetic requirements for a quantum state transition of a vortex of a superconductor.

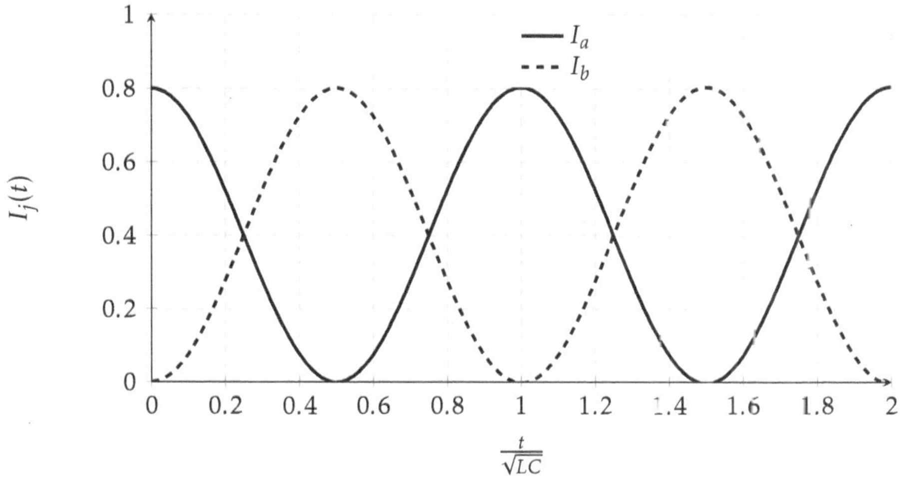

Figure 7.25: Current solutions of a linear squid circuit in superconducting conditions or $R \approx 0$, where $I = 0.4$.

7.12.1 Energy of Flux Quantum Transition

Let's determine how much energy V is created during a charge carrier's single round trip in the loop of a quantized vortex. In Subsec. 7.3.1, we found the charge carrier frequency ω_s around the flux field B to be

$$\omega_s = 2\pi f_s = \frac{eB}{m_0} \tag{7.156}$$

Then, using Eq. (7.150), we can also write

$$f_s = \frac{2e\Delta V_n}{h} \Rightarrow \Delta V_n = \frac{hf_s}{2e} = \frac{heB}{(2\pi)2em_0} = \frac{\hbar B}{2m_0} \tag{7.157}$$

Since $V = e\Delta V_n$, we then have

$$V = e\Delta V_n = \frac{hf_s}{2e} = \frac{\hbar eB}{2m_0} = \hbar\gamma B \tag{7.158}$$

Eq. (7.157) provides the amount of *energy for a conventional single inductance loop*. It also follows that if the same potential drop can be achieved in twice the frequency $2f_s$ or half the period, the amount of energy of the loop becomes

$$\Delta V = 2\hbar\gamma B \tag{7.159}$$

So, let's see how the energy of Eq. (7.158) and/or (7.159) compares to the energy it takes to excite a n–quantized vortex into an adjacent quantum state $n+1$. In Subsec. 7.3.1, we found the n–vortex energy to be

$$E_n = \hbar\omega\left(n+\frac{1}{2}\right) \tag{7.160}$$

Therefore, the energy of a transition from n to $n+1$ becomes

$$\Delta E_n = \hbar\omega_s\left(n+1+\frac{1}{2}\right)-\hbar\omega_s\left(n+\frac{1}{2}\right) \tag{7.161a}$$

$$= \hbar\omega_s\left(n+\frac{3}{2}\right)-\hbar\omega_s\left(n+\frac{1}{2}\right) \tag{7.161b}$$

$$= \hbar\omega_s\left(\frac{2}{2}\right) \tag{7.161c}$$

$$= \hbar\omega_s \tag{7.161d}$$

This result can also be written as

$$\Delta E_n = \hbar\omega_s = \frac{\hbar eB}{m_0} = \frac{2\hbar eB}{2m_0} = 2\hbar\gamma B \tag{7.162}$$

Comparing this result to Eq. (7.158), we conclude that

> R *Single conventional inductance loop*: a single conventional loop of induced emf does not create enough energy to excite or support a quantized vortex transition.

However, since Eq. (7.159) is identical to the energy for a half-period loop inductance,

> R *Two parallel half-period inductance loops*: The electric circuit design of two-halves of half-period inductance loops, as with a SQUID, creates exactly enough energy to excite a single-step quantized vortex transition.

Thus, the *SQUID* exploiting superconductors is an example of a device that achieves the emf or potential drop around a whole loop in half the period or twice the frequency $2f_s$. Because of this, the SQUID becomes capable of manipulations of the quantum states in superconductors through an external field.

7.13 Quantized Vortex Size Dependence On B

When a flux given by $\mathbf{B} \cdot \mathbf{A}_\Phi$ exists, where $\mathbf{A}_\Phi$ is the flux area, there are multiple ways the flux can change. In the case of a diamagnetic material, the magnetic moment cannot rotate in any direction away from the magnetic field. The vortices in a superconductor, for example, involve charges circulating around the magnetic field. Therefore, if the angle between $\mathbf{B}$ and $\mathbf{A}_\Phi$ is fixed, then the only other ways that the flux can change is via the field strength and the flux area A_Φ. The change is flux due to these parameters can be found by using the result for the quanized flux differential, which can be written as

$$d\Phi = \frac{h}{2e} = d(BA_\Phi) = dB \cdot A_\Phi + B \cdot dA_\Phi \tag{7.163}$$

If the flux area is $A_\Phi = \pi r^2$, then

$$dA_\Phi = 2\pi r dr$$

Substitution into Eq. (7.163), we have

$$\frac{h}{2e} = dB\pi r^2 + B(2\pi r dr) \tag{7.164}$$

Rearranging Eq. (7.164) and dividing by the flux $B\pi r^2$ gives

$$2\frac{dr}{r} = \frac{h}{2eB\pi r^2} - \frac{dB}{B} \tag{7.165}$$

$$= \frac{d\Phi}{\Phi} - \frac{dB}{B} \tag{7.166}$$

Using the quantized flux $\Phi_n = n\Phi_0$ of the vortex (*c.* Eq.(7.107)), Eq. (7.165) then becomes

$$2\frac{dr}{r} = \frac{d\Phi_n}{\Phi_n} - \frac{dB}{B} = \frac{dn}{n} - \frac{dB}{B} \tag{7.167}$$

or

$$\frac{dr}{r} = \frac{1}{2}\left(\frac{dn}{n} - \frac{dB}{B}\right) \tag{7.168}$$

Integrating both sides of Eq. (7.168), we have

$$\ln\left(\frac{r}{r_i}\right) = \frac{1}{2}\left[\ln\left(\frac{n}{n_i}\right) - \ln\left(\frac{B}{B_i}\right)\right] \tag{7.169}$$

$$= \frac{1}{2}\ln\left(\frac{n \cdot B_i}{n_i \cdot B}\right) \tag{7.170}$$

To get rid of the logs, we can take the exponential of both sides. This leads to the *quantized vortex radius dependence* given by

Quantized Vortex Radius Field Dependence

$$r_n = r_i\left(\frac{B_i}{n_i}\right)^{1/2}\left(\frac{n}{B}\right)^{1/2} \quad (n_i \geq 1, B > 0) \tag{7.171}$$

Eq. (7.171) gives the relation between a quantized vortex radius and the external flux field B. Fig. 7.26 illustrates the first 20 quantized radii lines as function of the external field.

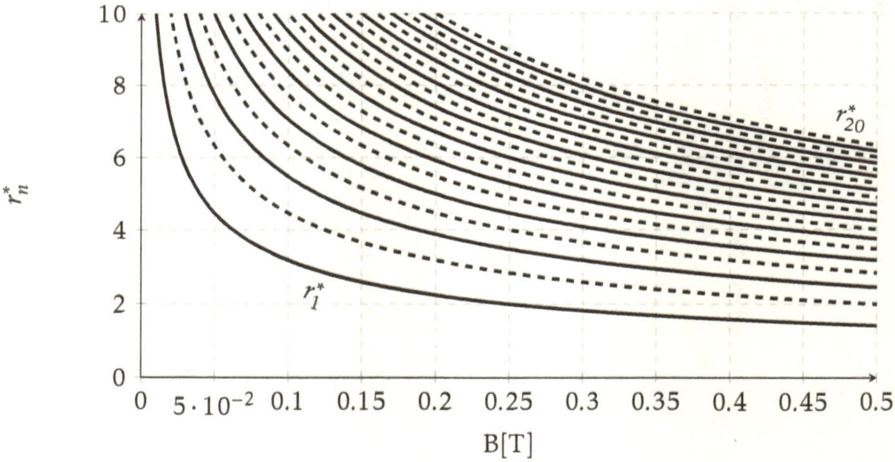

Figure 7.26: Quantized radial lines for a superconducting vortex. The y–axis is $r_n^* = \frac{r_n}{r_i}\frac{n_i}{B_i}^{1/2}$. In a SQUID, increasing the magnetic field corresponds to *horizontal* quantum jumps across the quantized radial lines.

In the SQUID device, for example, the *rigid* circuit design using superconductors restricts the vortex to an approximately *fixed* radius. As a result, changing the external field coupled to the vortex causes the quantum number n to cycle accordingly, so that quantum steps occur as the field changes.

7.14 Junction Supercurrent Dependence On Phase

Some of the above circuit analysis assumed linear elements. However, it is well-known that the *Josephson junction is not linear*. Fig. 7.27 illustrates a typical IV curve for a Josephson junction.

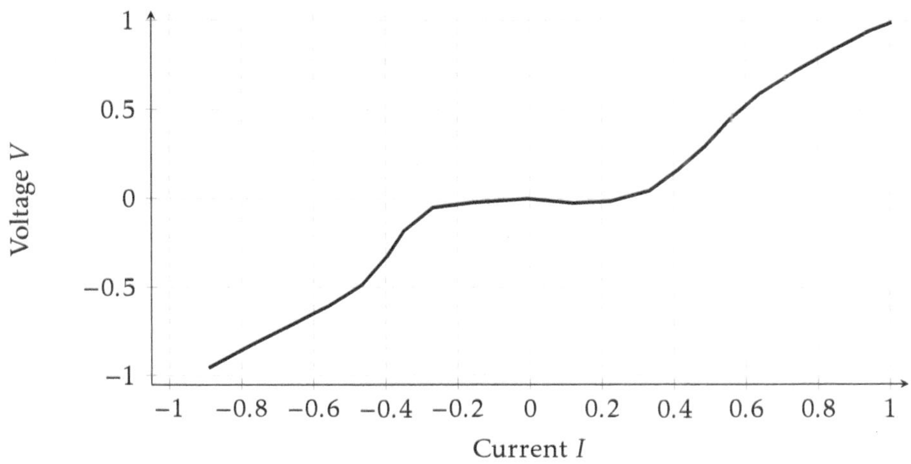

Figure 7.27: Example IV curve for a Josephson junction.

Nonlinearness does not change the fact that the system currents/voltages are harmonic/periodic. The shape of the periodic response, however, does deviate from a perfect sinusoid. If the DC SQUID were *linear*, the IV curve would be a straight line, and the device resistance R would be the slope of the line. As Fig. 7.27 illustrates, for the Josephson junction, the current-voltage relation is *not* a straight line. Thus, the effective device resistance depends on the current flowing through the device. However, notice that within a bounded range of current surrounding the origin, the curve has nearly zero voltage with *nonzero* current $I > 0$. The current value where the behavior transitions to ordinary linear behavior corresponds to the *critical current* of the superconducting junction $I_c \approx 0.3$.

If the device stays within the range $[-I_c, I_c]$, there are certain assumptions that can be made that enable finding the *current across the junction barrier*, specifically, in the superconducting state, as a function of the vortex phase θ. Since current flows without significant voltage across the junction, this is equivalent to $R \approx 0$. Therefore, in the superconducting state, the usual *scattering and reflections at the junction interface can be neglected*. As an insulating barrier, V_b is still relatively large because it includes the band gap between the valence and conduction bands. As an insulating barrier, we still have the condition $k_2^2 < 0$ so the exponential terms in Eq. (7.132) become $e^{\pm ik_2x} = e^{\pm i(ik_2')x} = e^{\mp k_2'x}$. Therefore, the complex eigenfunction becomes a real-valued exponential functions comprised of evanescent waves at the interfaces of the barrier. In this case, the wave function is given by

$$\psi(x) = C_1\cosh\left(k_b's_\varphi\right) + C_2\sinh\left(k_b's_\varphi\right) \tag{7.172}$$

$s_\varphi = r\varphi$ is the length variable along the φ dimension. k_b' is the wavenumber for the barrier of the junction where $k_b = ik_b'$ and $\hbar k_b^2/2m_0 < 0$. Therefore, $1/k_b'$ is the decay length of the evanescent wave in the barrier which facilitates *tunneling*. In this coordinate system, the origin is located at the center of the loop. Therefore, the junction interfaces are on symmetric sides of the origin. For the metallic superconductors of the junction, the wave functions are given by (*c*.Eq. (7.56))

$$\psi_{\text{superconductor}} = \frac{A_r}{\sqrt{2\pi}}e^{ik_\varphi r\varphi}$$

A_r is the vortex eigenfunction corresponding to the radial dimension r (*c*.Eq. 7.77). However, because of the azimuthal symmetry of the wave function, A_r is the same for both superconductor metals in the loop. Only φ varies in the loop. Then, denoting $\theta_i = k_\varphi r\varphi_i$, we have the following boundary conditions:

$$\frac{A_r}{\sqrt{2\pi}}e^{i\theta_1} = C_1\cosh\left(-k_b'a_\varphi\right) + C_2\sinh\left(-k_b'a_\varphi\right) \tag{7.173a}$$

$$\frac{A_r}{\sqrt{2\pi}}e^{i\theta_2} = C_1\cosh\left(+k_b'a_\varphi\right) + C_2\sinh\left(+k_b'a_\varphi\right) \tag{7.173b}$$

With two boundary conditions with two unknowns, we can solve for C_1 and C_2. Using the second equation to write C_1 in terms of C_2, we have

$$C_1 = \frac{\frac{A_r}{\sqrt{2\pi}}e^{i\theta_2} - C_2\sinh\left(+k_b'a_\varphi\right)}{\cosh\left(+k_b'a_\varphi\right)} \tag{7.174}$$

Substituting C_1 into Eq. (7.173a), then gives

$$\frac{A_r}{\sqrt{2\pi}}e^{i\theta_1} = \left[\frac{\frac{A_r}{\sqrt{2\pi}}e^{i\theta_2} - C_2\sinh\left(+k_b'a_\varphi\right)}{\cosh\left(+k_b'a_\varphi\right)}\right]\cosh\left(+k_b'a_\varphi\right) + C_2\sinh\left(-k_b'a_\varphi\right)$$

(7.175)

Canceling the cosh factor in the first term on the RHS, then rearranging to solve for C_2, we obtain

$$C_2 = \frac{A_r}{\sqrt{2\pi}} \frac{e^{i\theta_1} - e^{i\theta_2}}{\sinh\left(-k_b' a_\varphi\right) - \sinh\left(+k_b' a_\varphi\right)} \tag{7.176}$$

From the identity $\sinh x = (e^x - e^{-x})/2$, it follows that $\sinh{-x} - \sinh x = -2\sinh x$. Then, Eq. (7.176) can also be written as

$$C_2 = \frac{A_r}{\sqrt{2\pi}} \frac{e^{i\theta_2} - e^{i\theta_1}}{2\sinh\left(+k_b a_\varphi\right)} \tag{7.177}$$

Substituting Eq. (7.177) into Eq. (7.173b) gives

$$C_1 = \frac{A_r}{\sqrt{2\pi}} \frac{e^{i\theta_2} + e^{i\theta_1}}{2\cosh\left(+k_b a_\varphi\right)} \tag{7.178}$$

Having C_1 and C_2, we can proceed to evaluate the current density $\mathbf{J}_\varphi$. through the barrier of the Josephson junction. To evaluate $\mathbf{J}_\varphi$, we need $\psi\nabla\psi^*$ and $\psi^*\nabla\psi$. Since the current is around the loop, the gradient only includes the component along φ of a cylindrical coordinate system, where

$$\nabla\psi = \frac{1}{r}\frac{\partial\psi}{\partial\varphi} \tag{7.179}$$

The derivatives of the hyperbolic functions are given by

$$\frac{d}{dx}\sinh x = \cosh x \quad \text{and} \quad \frac{d}{dx}\cosh x = \sinh x$$

Using $s_\varphi = r\varphi$, evaluating the first term $\psi\nabla\psi^*$ leads to

$$\psi\frac{1}{r}\frac{\partial\psi^*}{\partial\varphi} = \left(C_1\cosh(k_b' s_\varphi) + C_2\sinh(k_b' s_\varphi)\right) \times \cdots$$

$$\frac{k_b' r}{r}\left(C_1^*\sinh(k_b' s_\varphi) + C_2^*\cosh(k_b' s_\varphi)\right)$$

$$= \underline{k_b' C_1 C_1^* \cosh(k_b' s_\varphi)\sinh(k_b' s_\varphi)} + k_b' C_1 C_2^* \cosh^2(k_b' s_\varphi) + \cdots$$

$$k_b' C_2 C_1^* \sinh^2(k_b' s_\varphi) + \underline{k_b' C_2 C_2^* \sinh(k_b' s_\varphi)\cosh(k_b' s_\varphi)} \tag{7.180}$$

Similarly, for the second term $\psi^* \nabla \psi$, we have

$$\psi^* \frac{1}{r} \frac{\partial \psi}{\partial \varphi} = \left(C_1^* \cosh(k_b' s_\varphi) + C_2^* \sinh(k_b' s_\varphi) \right) \times \cdots$$

$$\frac{k_b' r}{r} \left(C_1 \sinh(k_b' s_\varphi) + C_2 \cosh(k_b' s_\varphi) \right)$$

$$= \underline{k_b' C_1^* C_1 \cosh(k_b' s_\varphi) \sinh(k_b' s_\varphi)} + k_b' C_1^* C_2 \cosh^2(k_b' s_\varphi) + \cdots$$

$$k_b' C_2^* C_1 \sinh^2(k_b' s_\varphi) + \underline{k_b' C_2^* C_2 \sinh(k_b' s_\varphi) \cosh(k_b' s_\varphi)} \quad (7.181)$$

Since the *first* and *last* (underlined) terms in Eqs. (7.180) and (7.181) are identical, taking the difference between them gives

$$\psi \nabla \psi^* - \psi^* \nabla \psi = \cdots$$

$$k_b' (C_1 C_2^* - C_1^* C_2) \cosh^2(k_b' s_\varphi) + k_b' (C_1 C_2^* - C_2^* C_1) \sinh^2(k_b' s_\varphi) \quad (7.182)$$

We can use the hyperbolic function identity given by

$$\cosh^2 x = 1 + \sinh^2 x \quad (7.183)$$

With this substitution, all the coefficients of sinh cancel so that Eq. (7.182) simplifies to

$$\psi \nabla \psi^* - \psi^* \nabla \psi = k_b' (C_1 C_2^* - C_1^* C_2) \quad (7.184)$$

Substitution of C_1 and C_2 using Eqs. (7.177) and (7.178) into the first term $C_1 C_2^*$ of Eq. (7.184) leads to

$$C_1 C_2^* = \frac{A_r^2}{2\pi} \left[\frac{e^{i\theta_2} + e^{i\theta_1}}{2\cosh\left(+k_b' a_\varphi\right)} \right] \left[\frac{e^{-i\theta_2} - e^{-i\theta_1}}{2\sinh\left(+k_b' a_\varphi\right)} \right] \quad (7.185a)$$

$$= \frac{A_r^2}{2\pi} \frac{1 + e^{i(\theta_2 - \theta_1)} - e^{-i(\theta_2 - \theta_1)} - 1}{4\cosh\left(+k_b' a_\varphi\right) \sinh\left(+k_b' a_\varphi\right)} \quad (7.185b)$$

$$= \frac{A_r^2}{2\pi} \frac{e^{i(\theta_2 - \theta_1)} - e^{-i(\theta_2 - \theta_1)}}{4\cosh\left(+k_b' a_\varphi\right) \sinh\left(+k_b' a_\varphi\right)} \quad (7.185c)$$

Since the second term $C_1^* C_2$ is the complex conjugate of Eq. (7.185), $C_1 C_2^*$ becomes

$$C_1^* C_2 = \frac{A_r^2}{2\pi} \frac{e^{-i(\theta_2 - \theta_1)} - e^{i(\theta_2 - \theta_1)}}{4\cosh\left(+k_b' a_\varphi\right) \sinh\left(+k_b' a_\varphi\right)} \quad (7.186a)$$

$$= -\frac{A_r^2}{2\pi}\frac{e^{i(\theta_2-\theta_1)} - e^{-i(\theta_2-\theta_1)}}{4\cosh\left(+k_b' a_\varphi\right)\sinh\left(+k_b' a_\varphi\right)} \tag{7.186b}$$

Taking the difference $C_1 C_2^* - C_1^* C_2$, we have

$$C_1 C_2^* - C_1^* C_2 = \frac{A_r^2}{2\pi}\frac{e^{i(\theta_2-\theta_1)} - e^{-i(\theta_2-\theta_1)}}{2\cosh\left(+k_b' a_\varphi\right)\sinh\left(+k_b' a_\varphi\right)} \tag{7.187}$$

The numerator can be rewritten using the identity $e^{ix} - e^{-iz} = 2i\sin x$, so we have

$$\frac{C_1 C_2^* - C_1^* C_2}{\zeta} = \frac{iA_r^2}{2\pi}\frac{\sin(\theta_2-\theta_1)}{\cosh\left(+k_b' a_\varphi\right)\sinh\left(+k_b' a_\varphi\right)} \tag{7.188}$$

Using the above result, the current density $\mathbf{J}_s$ becomes

$$\mathbf{J}_\varphi = \frac{i\hbar k_b'}{2m_0}\left(\frac{iA_r^2}{2\pi}\frac{\sin(\theta_2-\theta_1)}{\cosh\left(+k_b' a_\varphi\right)\sinh\left(+k_b' a_\varphi\right)}\right) \tag{7.189}$$

$$= J_{\pi/2}\sin(\theta_2-\theta_1) \tag{7.190}$$

Therefore, we have the *current density across the Josephson junction* given by

Josephson Junction Current-Phase Relation

$$\mathbf{J}_\varphi = J_{\pi/2}\sin(\theta_2-\theta_1) \tag{7.191}$$

where

Josephson Junction Peak Current

$$J_{\pi/2} = -\frac{1}{4\pi m_0}\left(\frac{\hbar k_b' A_r^2}{\cosh\left(+k_b' a_\varphi\right)\sinh\left(+k_b' a_\varphi\right)}\right) \tag{7.192}$$

Fig. 7.28 plots the coefficient of Eq. (7.191), varying the barrier thickess ratio $k_b' a_\varphi$. .

Figure 7.28: $|J_{\pi/2}|$ as a function of the barrier thickness. The current via barrier tunneling increases with smaller barrier thickness.

The sine term in Eq. (7.191) can also be expressed alternatively by using the relation $\theta = k_\varphi r \varphi$. Denoting the current density coefficient as $J_{\pi/2}$, then, substituting into Eq. (7.191), we have

$$\widehat{\mathrm{J}}_\varphi = J_{\pi/2} \sin\left(k_\varphi r_n \varphi_2 - k_\varphi r_n \varphi_1\right) \tag{7.193}$$

$$= J_{\pi/2} \cdot \sin\left(\frac{-eBr_n}{\hbar} \cdot r_n \cdot (\varphi_2 - \varphi_1)\right) \tag{7.194}$$

$$= J_{\pi/2} \cdot \sin\left(\frac{-2eB\pi r_n^2}{h} \cdot (\varphi_2 - \varphi_1)\right) \tag{7.195}$$

$$= J_{\pi/2} \cdot \sin\left(\frac{-\Phi_n}{\Phi_0} \cdot (\varphi_2 - \varphi_1)\right) \tag{7.196}$$

Since φ_1 corresponds to left side of the circuit loop arm, while φ_2 corresponds to the right, a (+) positive *angle* corresponds to $\varphi_1 - \varphi_2$. Then, for the half-loop arm of a DC-SQUID, the angular span is $\varphi_1 - \varphi_2 = \pi$, so

Josephson Junction Current-Flux Relation

$$\widehat{\mathrm{J}}_\varphi = |J_{\pi/2}| \cdot \sin\left(\frac{|\Phi_n|}{\Phi_0} \cdot \pi\right) \tag{7.197}$$

Since $|\Phi_n|/\Phi_0 = n$, we have that

Ⓡ *DC SQUID π-periodicity*: The SQUID is periodic over integer multiples of π, rather than 2π.

The Josephson junction is one of several applications of superconductors. While it is still very early, there are efforts ongoing to make use of superconductors in a novel type of computer processor. We'll wrap up by discussing a topic related to this new potential application, known as a *quantum computer*.

7.15 Single Electron Transistor & Single Supercharge Box

A potentially new application of the Josephson junction, which exploits *superconductivity*, originated in the desire for highly sensitive detection of *charge* and charge variation with a sensitivity analogous to that of a superconducting quantum interference devices (SQUIDs). Earlier, we found the SQUID has the ultimate magnetic field resolution 10^{-15} Tesla. Consequently, they are the most sensitive magnetometers known today. Devices that sense charge are known as *electrometers*. The Josephson junction is the enabling device for the SQUID's sensitivity, however it exploits the diamagnetic property of superconductors, being proportional to a magnetic field up until the critical field B_c. But, can the Josephson junction somehow be used to detect charge in an analogously sensitive manner as it does the magnetic field? The *single electron transistor* (SET) turns out to be an offspring of such an inquiry of a sensitive electrometer based on superconductivity.

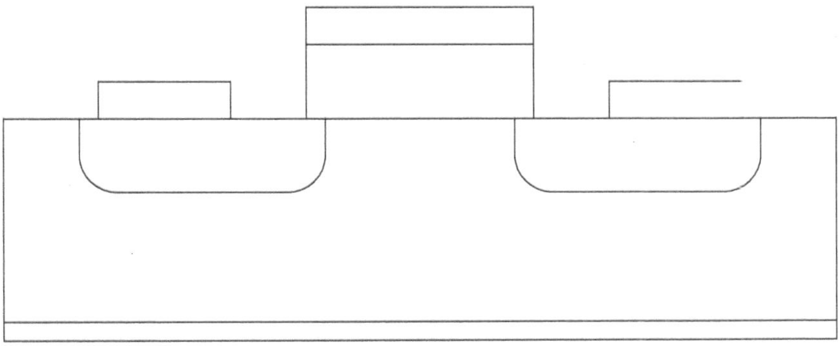

Like conventional transistors (see Fig. ??), the *SET* is also a three terminal device. Two of the three terminals are the *source* and *drain* terminals of the device, while the third terminal is known as *gate* terminal, which allows the application of a *gate voltage* V_g. The gate voltage is used to tune the amount of current between the source and drain terminals. So far, the description sounds like a conventional transistor. However, one of the key distinctions in the SET is the use of not one, but two Josephson junctions between the source and drain terminals. Moreover, between both Josephson junctions is an island of conduction material (or a quantum dot). This *conduction island* is metallic and patterned to a finite volume and surrounded by insulator. This enables a capacitant coupling to both junctions. With $V_g = 0$, the device behaves as an ordinary Josephson junction behaves (two in series). In Sec. **??**, we found that a superconductor requires a finite electrical energy input to exceed the energy gap Δ of the superconductor before supercurrent flows. Once current flows, there is negligible electrical loss owing to the superconducting state of the Josephson junction. By introducing a nonzero gate voltage $|V_g| > 0$, the energy gap can be modified such that current flows more readily *across the source-drain terminals*, particularly at smaller voltages across the source-drain terminals.

The utility of the metallic island in an SET is *essential* for the operation. When the metallic island, having a total gate capacitance C_g, is connected to a gate voltage V_g, the effect of shrinking the metallic island to a sufficiently small size is to reduce the capacitance of the combined structure so that $q_g = C_g \cdot V_g \approx e$. This can be satisfied by considering the relation for the capacitance given by

$$C_g = \frac{\epsilon_0 A_I}{d} \tag{7.198}$$

ϵ_0 is the permittivity in vacuum. A_I is the area of the island. If the device operation allows for a gate voltage of around $V_g = 1\text{mV}$, then the largest metallic island area A_I to get the charge stored on the metallic island to be $e \sim 1.6 \times 10^{-19}\text{C}$ can be estimated as

$$C_g V_g \approx e = \left(\frac{\epsilon_0 A_I}{d}\right) \times 10^{-3} \tag{7.199}$$

If the capacitant spacing d between the island and gate contact is $d \approx 1\text{nm}$, then solving for quantum dot area A_I gives

$$A_I \approx \frac{e \times 10^{-9}}{\epsilon_0 10^{-3}} = \frac{1.6 \times 10^{-19} \times 10^{-9}}{8.85 \times 10^{-12} 10^{-3}} \approx = 1.81 \times 10^{-14}\text{m}^2 \tag{7.200}$$

Taking the square root of this gives a measure of the size of an insulated metallic island that can be used to store a charge on the order of e.

Carrying this out, we find a size of around 135 nm can store a charge on the order of e. Moreover, the size requirement scales with the capacitant spacing square root $\sqrt{d}$. In other words, the device size s_d goes with

$$s_d \approx \frac{e \cdot d}{\epsilon \cdot V_g} \tag{7.201}$$

When the amount of the capacitant charge reaches this low a level ($\approx e$), it becomes possible for the device to be used to regulate the transfer of this amount of charge to the island, being essentially one charge at a time.

However, concomitant with these conditions is an exponential increase in the difficulty of a charge to transfer to the island because of the very sharp rise in electrostatic energy. This makes it *harder* for a new charge to join the island charge party. Consequently, a discrete transfer of electric charges to/from the island becomes pronounced. This additional electrical repulsive effect arising from the island capacitive properties is an example of an effect known as a *Coulomb blockade.*

Additionally, by choosing the ratio of A_I and d so that the capacitance C_g is small, it can also reduce the time constant τ for transferring charge, which is $\tau = R_g C_g$, where R_g is the resistance across the device. This not only enables the detection of e, but the detection of the transfer on relatively short time scales. Another advantage to the ability to detect charge transfer at higher frequencies (due to a shorter time-constant) is that it pushes the device operation into a region of lower $1/f$ noise. This type of noise, arising from thermal motion of background charges, is fundamental to solid state materials, and can limit resolution or sensitivity at lower device operating frequencies. When the device operates by exploiting these dynamical effects through use of a RF circuit added to the device, it is known as a *radio-frequency-single-electron-transistor* or *RF-SET*.

The basic device physics of a *single electron transistor* can be understood by considering a conventional transistor with a *quantum dot* placed in middle of the channel of the transistor between the source and drain. Normally, the metallic quantum dot contains a large number of electrons, however, *there is no net electrical charge on the quantum dot.* Thus, transferring a charge to the quantum dot through any sort of electrical action along the channel raises the energy of the quantum dot by an amount

$$\Delta E_{dot} = \frac{1}{2}\frac{Q^2}{C} \tag{7.202}$$

The charge for n transferred electrons is ne, so ΔE_{dot} becomes

$$\Delta E_{dot} = \frac{1}{2}\frac{n^2 e^2}{C} \tag{7.203}$$

e is the elementary charge of the electron. C is the capacitance, arising from the metal/semiconductor interfaces, which enables storage of charge on the surface of the quantum dot. This leads to an equivalent potential barrier at the quantum dot. Any conducting or tunneling electron must leap this barrier.

Per the above discussion, for a single electron to transfer to the island leads to an energy of $\frac{1}{2}\frac{e^2}{C}$. For two electrons $\frac{2e^2}{C}$, etc. Because the capacitance must be small to store small amounts of charge, a consequence is a large increase in energy ΔE_{dot}. This energy barrier which must be surmounted by electrons in the channel, is the Coulomb Blockade.

Additionally, there is resistance along the channel, and therefore, an effective RC circuit exists. RC circuits have time constant $\tau = RC$ dictating the time it takes the charge to store or discharge into the drain. It is therefore, better, for the resistance R to be as small as possible to ensure complete and orderly transfers useful for the device application.

An important point to keep in mind regarding the energy barrier of the quantum dot is that the barrier height must be sufficiently large compared to the mean electronic thermal energy, or

$$\Delta E >> k_B T \tag{7.204}$$

If this condition is not satisfied, then the stored electrons will be able to surmount the barrier via tunneling randomly, due to thermal excitation. This process is *decoherent*, or *random*, and is undesirable for controlling the charge storage.

The single charge of the quantum dot must satisfy the uncertainty principle, namely

$$\Delta E \Delta t \geq \frac{\hbar}{2} \tag{7.205}$$

Since the energy difference needed for a single electron is on the order of $\Delta E \approx e^2/2C$. And the transition time corresponding to this energy must is on the order of RC. Thus, we have the condition that

$$\Delta E \Delta t \approx \frac{e^2}{2C} \times RC \geq \frac{\hbar}{2} \Rightarrow R \geq \frac{\hbar}{e^2} \tag{7.206}$$

This lower bound on the resistance for a single electron, dictated by the uncertainty principle, is known as a *quanta of resistance* R_q equal to

Quanta of Resistance

$$R_q = \frac{\hbar}{e^2} \tag{7.207}$$

Substitution of the parameter values for $\hbar$ and e, the quanta of resistance takes a value of

$$R_q \approx 25.8k\Omega \tag{7.208}$$

The SET is currently an active area of research, based on the Josephson junction. The SET even has the potential to be used in another potential application based on superconductivity. We'll touch on this topic in the next section.

7.15.1 Challenges In Quantum Computing Using Superconductors

The field of *quantum computing* is currently premature, yet, is a very active area of technical research. Here, we'll lightly touch on the some of the balancing that needs to be achieved in order to build useful quantum computers based on superconductors. From earlier in this chapter, we know that surface currents are uniquely created in zero magnetic field. And a field is introduced, then increased, these circulating currents can transition from surface bound vortices to an array of ordered n-vortices.

The energies involved in superconductors have a contribution very similar in form to a *harmonic oscillator*, given by

$$E_r = \left(n + \frac{1}{2}\right)\hbar\omega_s \tag{7.209}$$

This term is the lowest order *quantized* contribution (others are higher order, and generally smaller, quantitatively). Thus, the quantized energies in a superconductor are dictated by this term, in particular.

All contemporary computers are designed to exploit a system of *binary states*, though quantum computers have the capacity to do more. In the first incarnations, however, they will likely not be different in this regard. For use in a quantum device, one of the challenges with the *as-is* harmonic oscillator behavior is that *all* the quantum states are separated by a single energy amount because

$$\Delta E_{10} = \left(1 + \frac{1}{2}\right)\hbar\omega_s - \left(0 + \frac{1}{2}\right)\hbar\omega_s = \left(\frac{3}{2} - \frac{1}{2}\right)\hbar\omega_s = \hbar\omega_s \tag{7.210a}$$

$$\Delta E_{21} = \left(2 + \frac{1}{2}\right)\hbar\omega_s - \left(1 + \frac{1}{2}\right)\hbar\omega_s = \left(\frac{5}{2} - \frac{3}{2}\right)\hbar\omega_s = \hbar\omega_s \tag{7.210b}$$

$$\Delta E_{32} = \left(3 + \frac{1}{2}\right)\hbar\omega_s - \left(2 + \frac{1}{2}\right)\hbar\omega_s = \left(\frac{7}{2} - \frac{5}{2}\right)\hbar\omega_s = \hbar\omega_s \tag{7.210c}$$

$$\vdots \tag{7.210d}$$

Transitioning to different quantum states requires excitation energy. For a processing or memory device, data is more stable if each quantum state requires a unique input to reach it. Because *DeltaE* is uniform, it makes a precise selection of states more challenging, especially, if you are not where you think you are, since the same amount of energy will always create a transition. This behavior can be distinguished from quantum transitional behavior of the hydrogen atom, for example, as all the energy transitions are unique. They are given by

$$\Delta E_n = -13.6\left(\frac{1}{n^2} - \frac{1}{(n+1)^2}\right) = -13.6\left(\frac{2n+1}{n^2(n+1)^2}\right)$$

The corresponding permissible energy transitions in the hydrogen atom are

$$\Delta E_{21} = -13.6\left(\frac{1}{1} - \frac{1}{2^2}\right) = -13.6 \times \frac{3}{4} \tag{7.211a}$$

$$\Delta E_{32} = -13.6\left(\frac{1}{4} - \frac{1}{9}\right) = -13.6 \times \frac{5}{36} \tag{7.211b}$$

$$\Delta E_{43} = -13.6\left(\frac{1}{9} - \frac{1}{16}\right) = -13.6 \times \frac{7}{144} \tag{7.211c}$$

$$\vdots \tag{7.211d}$$

Each transition is unique, and therefore, exciting this system with an energy that is very close to a particular transition frequency (provided the initial condition is correct), provides better control over quantum state transitions.

Thus, it is more desirable to have *nonuniformly spaced energy states,* as with an atom with bound states. This property is the reason for the unique atomic spectra. This property of nonuniform energy states, for a harmonic oscillator system, is known as *anharmonicity.* This makes a unique transition perfectly achievable with a given energy input. Changing the energy input, then, targets another transition, etc. The point is that one energy input targets a single transition, which effectively creates a more robust binary quantum system. In using superconductors with quantum computers, this is a key objective in the design.

As with the single-electron-transistor (SET), the temperature is also an important aspect. A typical frequency to excite transitions in a superconductor is around 10 GHz ($eB/m_0 = 10^{-19}/10^{-31}$). This has implications on a stable operating temperature because the mean kinetic energy of a atom is given by

$$E_k \approx \frac{k_B T}{2}$$

In order to avoid thermal energy *randomly* exciting quantum state transitions in a superconductor, the temperature must be low enough that the transition energy $\hbar\omega_{01}$, for example, satisfies

$$\hbar\omega_{01} >> \frac{k_B T}{2} \Rightarrow T << \frac{2\hbar\omega_{01}}{k_B}$$

If the transition frequency is $\omega_{01} \approx 10$ GHz ($\omega_{01} = 2\pi f$), substitution of $\hbar$ and the Boltzmann constant k_B gives

$$T \approx .96K$$

Thus, to exploit the physics of a Josephson junction/superconductor in a quantum computer, two anticipated challenges include

(1) designing atomic-like nonuniformly spaced quantum states and

(2) quantum state stability against thermal fluctuations causing random quantum transitions.

Lastly, there may still be a critical need to pay more attention to extending/improving the fidelity of the quantum physics describing these systems to enable more in-depth understanding of their physical properties, *etc.*, because quantum mechanics, as mentioned in the first section of the book, may still have blind-spots. Hopefully, some of the information in this book, and especially this chapter can help towards this end.

7.16 Chapter Summary

This chapter has focused, primarily, on the treatment of superconductors using the relativistic quantum mechanics. There is benefit in this approach because it has greater reach than any currently existing quantum mechanical treatment of these systems. The chapter began by discussing the relevant energy operators for superconductors, because they must be used, strictly, under appropriate conditions. This included higher order operators which prove to be indispensable to obtain an accurate description.

In the quantum energy operators, the role of both spin-orbit interaction, as well as other higher order terms were included. The SOI possesses the required symmetry properties as revealed by the BCS theory. The expectation values involving the SOI were also shown to lead to the well-known temperature-dependent behaviors including the discontinuity in resistance, as well as the electronic heat capacity. The exactly eigensolutions for superconductors where also worked out, leading to quantized vortices. The solutions allows the determination of the magnetic properties including the quantized flux and quantum flux, as well as the magnetic susceptibility properties. The relativistic treatment and results also lead to a clear explanation of the origins of Type I and Type II superconductors' magnetic responses.

The Josephson junction was also analyzed to explain their unique properties including voltage-frequency relation and the voltage-phase. This lead to a discussion of the superconducting-quantum-interference-device, which utilizes two junctions in parallel. In the SQUID, we found it to have a unique periodicity in πn, rather than $2\pi n$. The current relations of the SQUID were also derived.

Finally, the chapter was concluded with a light introduction and discussion of other applications using superconductors, such as single-electron-transistors and quantum computers.

Table 7.3 lists some of the key equations from this chapter.

Table 7.3: Chapter 7 Summary Equations.

Name	**Equation**
Azimuthal Wavenumber	$k_\varphi = -\frac{eBr}{\hbar}$
Orbital Frequency	$\omega_0 = -\frac{eB}{m_0}$
Azimuthal Solution	$\Phi(r,\varphi) = C_\varphi e^{-i\frac{e}{\hbar}Br^2\varphi}$
Energy Levels (r)	$E_r = \left(n+\frac{1}{2}\right)\hbar\omega_0$
Mean Binary Spin Potential	$\Delta_c(T) = \Delta_c(0)\tanh\left(\frac{\Delta_c(T)}{k_B T}\right)$
ρ-Discontinuity	$\frac{k_z(T)}{k_z^+} = \sqrt{\frac{V_{so}(0)}{E^+_{K,\text{cond}}}\tanh\left(\frac{V_{so}}{k_B T}\right)+1}$
C_{se}-Continuity	$C_{se} = \frac{V_{so}V_{so}(0)\text{sech}^2\left(\frac{V_{so}}{k_B T}\right)}{k_B T^2 - V_{so}(0)T\text{sech}^2\left(\frac{V_{so}}{k_B T}\right)}$
Flux Quantization Unit	$\Phi_0 = -\frac{h}{2e}$
Quantized Flux	$\Phi_n = n\Phi_0$
Corrected Magnetic Energy	$E_r = \frac{e\hbar B}{m_0}\left(n+\frac{1}{2}\right) - \frac{e^2\hbar^2B^2}{2m_0^3c^2}\left(n+\frac{1}{2}\right)^2$
Magnetic Moment per Carrier	$\mu_s = -\frac{\mu_0 e^2 H r^2}{2m_0}$
SQUID Loop Freq.	$f_s = \frac{2e\Delta V_n}{h}$
Quantized Radius	$r_n = r_i\left(\frac{B_i}{n_i}\right)^{1/2}\left(\frac{n}{B}\right)^{1/2}$
Jo-Junc Current-Phase	$\mathbf{J}_\varphi = -\frac{\hbar k_b' A_r^2}{4\pi m_0}\left(\frac{\sin(\theta_2-\theta_1)}{\cosh(+k_b' a_\varphi)\sinh(+k_b' a_\varphi)}\right)$
Jo-Junc Flux-Phase	$\widehat{\mathbf{J}}_\varphi = J_{\pi/2}\cdot\sin\left(\frac{\Phi_n}{\Phi_0}\cdot\pi\right)$

Table 7.3

Quanta of Resistance	$R_q = \frac{\hbar}{e^2}$

7.17 Chapter Problems

Problem 7.1 Using the relative permeability of a superconductor given by (c.7.120)

$$\mu_r = 1 - \frac{\langle n_s \rangle \mu_0 e^2 r^2}{2m_s}$$

estimate the maximum number of charge carriers per unit volume $\langle n_s \rangle$ in a tin superconducting cylinder of radii $r = 1.5 \times 10^{-3}$cm and $r = 5 \times 10^{-3}$cm, setting the unit charge and mass e, m_e to be that of the electron

Problem 7.2 Use the relative permeability from Prob. 7.1 to estimate the *relative* diamagnetic permeability of a superconductor with carrier density $\langle n_s \rangle = 10^{23}\text{m}^{-3}$, and a radius of 2×10^{-3}cm assuming the electron charge and mass.

Problem 7.3 Earlier in this chapter, we saw that the superconductor charge carriers have kinetic energy contributions given by

$$E_b = \frac{e\hbar B}{m_0}\left(n + \frac{1}{2}\right) - \frac{e^2\hbar^2 B^2}{2m_0^3 c^2}\left(n + \frac{1}{2}\right)^2$$

Calculate the ground state energy ($n = 0$) of a superconductor n–vortex in magnetic fields of .5,1 Tesla. Compare the resulting values to the result *without* the relativistic correction (*i.e.* without the 2nd term).

Problem 7.4 Using the coherence length given by

$$\nu_0 = \sqrt{\frac{\hbar^2}{m_0 E_0}}$$

Show that the ground state leads to

$$\nu_0 = \sqrt{\frac{2\hbar}{eB}}$$

Then, calculate ν_0 for an external field of 1.5Tesla.

Problem 7.5 Show that the orbital magnetic moment of the n-vortex core is given by

$$\mu_L = \frac{e^2 B r^2}{2m_0}$$

Hint: Use the definition of an orbital magnetic moment.

Problem 7.6 Use the solution for ψ_φ and the definition of the current density to show that the current density J_φ is given by

$$J_\varphi = v_\varphi |\psi_s|^2 = r\omega_0 |\psi_s|^2$$

where $\omega_0 = eB/m_0$.

Problem 7.7 For the Josephson Junction (JJ), we found the frequency-voltage relation given by (c.Eq7.149)

$$f_s = \frac{2e\Delta V_n}{h}$$

Using this relation, compute the voltage drop ΔV_n across a loop with a JJ in a magnetic field of $B = 0.5$Tesla. How will the voltage change with the magnetic field B?

Problem 7.8 Using the quantized flux in a superconductor (c.Eq. 7.107)

$$|\Phi_n| = n\Phi_0$$

calculate the first 3 flux states for $n = 1, 2$, and 3.

If the magnetic field creating the flux is increased for a fixed quantized vortex, how will the flux change (increase, decrease, or stay the same)?

Lastly, what conditions must be met to change the flux?

Problem 7.9 It was shown in the chapter that for a quantized vortex, the radius r_n is given by (c.Eq. 7.171)

$$r_n = r_i \left(\frac{B_i}{n_i}\right)^{1/2} \left(\frac{n}{B}\right)^{1/2} \quad (n_i \geq 1, B > 0)$$

Using the above relation, show that it leads to

$$\frac{\Phi_0}{\pi} = r_i^2 \frac{B_i}{n_i}$$

and

$$r_n = \frac{\Phi_0 n}{\pi B}$$

7.18 Suggested Readings & References

[1] K. Eason, *Quantum Mechanics With A SPIN On It*, Sym. Pub. (2019).

[2] A. A. Abrikosov. *On the magnetic properties of superconductors of the second group*. Soviet Physics JETP, **5**, 1774, (1957).

[3] U. Essmann and H. Trauble. *The direct observation of individual flux lines in type II superconductors*, Phys. Lett., **24a**, pp. 526, (1967).

[4] H. F. Hess *et. al.*, *Scanning-Tunneling-Microscope Observation of the Abrikosov Flux Lattice and the Density of States near and inside a Fluxoid*, **62**, pp. 214-, Phys. Rev. Lett. (1989).

[5] B. S. Deaver, Jr., and W. M. Fairbank. *Experimental Evidence For Quantized Flux In Superconducting Cylinders*. Phys. Rev. Lett., **7**, pp. 43-46, (1961).

CHAPTER 8

Historic Fundamentals Of Superconductors

Superconductivity is a phenomenon that appears in certain materials in conditions of relatively low temperature. When subjected to these conditions, superconductors demonstrate extraordinary characteristics, both *electrically* and *magnetically*. The physics of superconductors combines several concepts from areas including transport, spintronics, as well as statistical thermodynamics. They are all compulsory to understand the rich properties of superconductors. Superconductors are also examples of transport systems that exploit electronic holes[1], as well as a *spin-dependent* potential energy. But, there is more. To fully analyze the behavior of superconductors, we also need to consider the contribution of higher order energy operators which are derivable from the relativistic kinetic energy operator discussed in Chap. 2. It turns out that when these are brought together consistently, an insightful view into the physics of superconductors becomes possible. Before starting our analysis of superconductors using relativistic quantum mechanics, we review, in this chapter, the basics of some concepts relevant to superconductivity. Then, we'll be ready to discuss their origins, as best explained using a relativistic-quantum mechanical treatment.

[1]Discussed in Chap. 5

8.1 Earliest Theories Of Superconductivity

As mentioned, superconductors demonstrate unique characteristics both *electrically* and *magnetically*. Once a superconductor has transitioned into that state from the normal conduction state, the electrons of the superconductor take on especially unique characteristics. For example, magnetically, the electrons are responsible for nearly perfect *diamagnetic* behavior. However, in their magnetic response, the characteristics can be distinguished between two types. The only two flavors of superconductors that have been observed are known as *Type I* and *Type II*. Let's discuss what distinguishes these two types of superconductors.

8.1.1 Type I & Type II Superconductors

In earlier experiments carried out with single-element metallic superconductors (as opposed to combinations of different materials), one material having a more pronounced superconducting behavior was niobium (Nb). Later, insulated compounds were discovered that would demonstrate even larger critical temperature T_c, compared to any single element superconductor. The *critical temperature* is the temperature at which the material becomes superconducting. The demonstrated success with compounds indicates that superconducting properties *can* be tuned to have more favorable parameter values. This fact turns out to be important because it would be of extreme technological value to have materials that possess such properties at room temperature. Table 8.1 lists some superconducting compounds along with their transition temperatures T_C. With superconductors, there is a universal trend in the measured magnetization response to an external magnetic field B. We'll be deriving this trend from quantum mechanics in Sec. 7.8. Initially, the magnetization depends linearly on the external magnetic field, but only up until the so-called *critical field* B_c is reached. Then, the net magnetization takes on a nonlinear and unique response for $B > B_C$. An example of one type of superconductor magnetization response is illustrated in Fig. 8.1. In this case, the critical field corresponds to the amount of external field triggering a complete transition out of a purely superconductive state for $B > B_C$. There is another type of magnetic response that can be observed. The two types of superconductors are distinguished by their behavior beyond B_c, or for $B > B_C$. For *Type I*, as illustrated in Fig. 8.1, in the region $B > B_c$, the superconductor loses all of its net magnetization spontaneously. The first type ever observed by Kamerlingh Onnes in

Compound	T_c(K)
Nb_3Sn	18.1
Nb_3Ge	23.2
Nb_3Al	17.5
V_3Ga	16.5
V_3Si	17.1
La_3In	10.4
NbTi	10.0

Table 8.1: Transition temperatures and corresponding critical temperatures of some superconducting compounds.

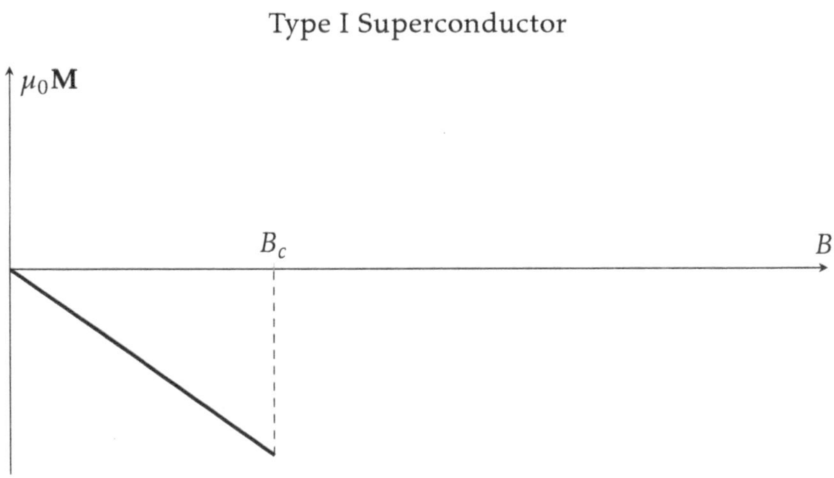

Figure 8.1: Magnetization dependence of the critical field B_c for a **Type I superconductor**. The slope of the curve is -1.

1911 was mercury (Hg) which is an example of a Type I superconductor.

However, with the appropriate combination of chemical elements, a second extended type of magnetic dependence on B has also been observed. Rather than the precipitous drop that takes place at the single-valued critical field B_c, in Type II, there is a more gradual exponential decay from a first critical field value B_{c1} down to zero at a second critical field B_{c2}, as illustrated in Fig. 8.2. . This behavior has significant implica-

Type II Superconductor

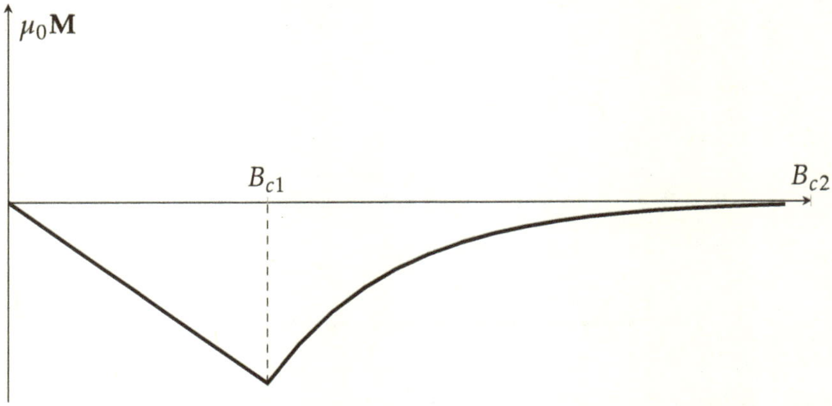

Figure 8.2: Magnetization dependence of the critical field B_c for a **Type II superconductor**. The behavior indicates two important magnetic field values, B_{c1} and B_{c2}.

tions because it leads to a superconductor that can carry larger currents without being triggered *fully* out of the superconducting state. For this reason, *type II* superconductors are often of more technological interest than *type I*. A list of some type II superconductors is given in Table 8.2. For Type II superconductors, for $B < B_{c1}$, the material is in a *pure superconducting state*. The region between $B = B_{c1}$ and $B = B_{c2}$, however, is a *mixed state* composed of normal + superconducting, going from pure superconducting at B_{c1} to pure mixed until B_{c2}, then to normal conductor for $B > B_{c2}$.

Compound	T_c[K]	B_{c2}[Tesla]
Nb_3Sn	18.5	24
Nb_3Ge	21.0	43.9
Nb_3Al	21.0	43.9
NbTi	10.5	12
NbZr	11	9

Table 8.2: Transition temperatures and critical fields of some **type II** superconductors. Because of their magnetic distinction, type II materials are capable of producing very large magnetic fields.

After the experimental observations of superconductivity began, it would not be long before some theoretical explanations would be attempted. Several notable physicists made serious attempts to offer both macroscopic and microscopic theories to explain observations with superconductors. The list of physicists who attempted theories for superconductivity includes Einstein, Bohr, Kronig, Landau, Bloch, Brillouin, Born, Feynman, and Heisenberg! This list serves as a testament to the level of difficulty of the problem of explaining the physics of superconductivity, at that time. When the ink finally settled onto paper, only a few had succeeded in offering theories for superconductivity that overwhelmingly agreed with the experimental data that was available, and persisted through the tests of time. One of them was worked out by a pair of brothers, Heinz (1907-1970) and Fritz London. Another valuable theoretical treatment was carried out by Ginzburg and Landau. Epecially important for Type II superconductors, another theoretical treatment was worked out by Abrikosov. And lastly, there was a theory developed by physics Bardeen, Cooper, and Schrieffer, known as the BCS theory. Each of these covers different aspects of the physics of superconductors. Here, we'll cover these important theories, starting with that of the London brothers.

8.1.2 London Equation Leads To Meissner Effect

Since superconductors have special electrical and magnetic properties, the London brothers thought to use the Maxwell equations to explain some of the physics of superconductors. Only two equations are needed to combine with the Maxwell equations and arrive at the results of the London brothers. Starting with the electrical current density flowing in the superconductor J_s, along with the Coulomb force equation, we have

$$\mathbf{J}_s = n_s q_s \mathbf{v}_s \tag{8.1a}$$

$$m_s \frac{d\mathbf{v}_s}{dt} = q_s \mathbf{E} \tag{8.1b}$$

n_s is the mean carrier density of superconducting charge carriers with physical units of inverse volume, *e.g.* m^{-3}. This is the classical analog to $|\psi(x,y,z)|^2$, have the same units, where $\psi(x,y,z)$ is the quantum mechanical wave function. q_s is the unit charge of the carrier. $\mathbf{v}_s$ is the carrier velocity, and $\mathbf{E}$ is the electric field in the superconductor. This pair of equations describes a *lossless* conductor, consistent with superconductivity. For electrical losses, a phenomenological dissipation term would have to be added to Eq. (8.1b). Using Eq. (8.1b) for $d\mathbf{v}_s/dt$, time-differentiation of Eq. (8.1a) leads to

$$\frac{\partial \mathbf{J}_s}{\partial t} = n_s q_s \frac{\partial \mathbf{v}_s}{\partial t} = \frac{n_s q_s^2}{m_s}\mathbf{E}$$

Applying the curl operator $\nabla\times$ to both sides gives

$$\nabla \times \frac{\partial \mathbf{J}_s}{dt} = \frac{n_s q_s^2}{m_s} \nabla \times \mathbf{E}$$

Changing the order of the operations on the LHS, substituting the Maxwell equation $\nabla \times \mathbf{E} = -\partial \mathbf{B}/\partial t$ on the RHS, and using the relation $\mathbf{B} = \nabla \times \mathbf{A}$, we have

$$\frac{\partial}{\partial t}(\nabla \times \mathbf{J}_s) = -\frac{n_s q_s^2}{m_s}\frac{\partial \mathbf{B}}{\partial t} = -\frac{n_s q_s^2}{m_s}\frac{\partial}{\partial t}(\nabla \times \mathbf{A})$$

Integration with respect to time leads to the so-called *London Equation*, given by[2]

London Equation

$$\mathbf{J}_s = -\frac{n_s q_s^2}{m_s}\mathbf{A} \tag{8.2}$$

Eq. (8.2) says that the superconducting current density $\mathbf{J}_s$ is proportional to the magnetic vector potential $\mathbf{A}$. Therefore, the result is only useful when a magnetic field is applied. This opens the door to the electromagnetic laws. Let us write Ampére's law in the form

$$\nabla \times \mathbf{B} = \mu \mathbf{J}_s$$

μ is the *magnetic permeability* where $\mathbf{B} = \mu\mathbf{H}$. Generally, μ has a time-dependence, but *not* a spatial one. Applying the curl operator $\nabla\times$ on both sides and using some of the relations above gives

$$\nabla \times \nabla \times \mathbf{B} = \mu \nabla \times \mathbf{J}_s = \mu \nabla \times \left(-\frac{n_s q_s^2}{m_s}\mathbf{A}\right)$$

[2]Strictly speaking, they are not equal, but within a constant of each other. This result assumes the constant to be null.

We can replace the LHS using the vector identity $\nabla\times\nabla\times\mathbf{B} = \nabla(\nabla\cdot\mathbf{B})-\nabla^2\mathbf{B}$. Since the electromagnetic equations already dictate that $\nabla\cdot\mathbf{B} = 0$, we have the following result:

$$-\nabla^2\mathbf{B} = -\left(\frac{\mu n_s q_s^2}{m_s}\mathbf{B}\right) \Rightarrow \nabla^2\mathbf{B} - k_L^2\mathbf{B} = 0$$

This is a well-known equation having a solution given by

$$\mathbf{B}(\mathbf{r}) = \mathbf{B}_0 e^{-\mathbf{k}_L\cdot\mathbf{r}} \tag{8.3}$$

$\mathbf{B}_0$ denotes the value of the magnetic induction at the boundary of the superconductor. Therefore, Eq. (8.3) describes a decaying induction field whose maximum is at the boundary, so that the field is essentially zero beyond a certain distance from the surface. The parameter k_L and length scale of the decay δ_L are related as

$$k_L^2 = \frac{1}{\delta_L^2} \Rightarrow \delta_L = \sqrt{\frac{m_s}{\mu n_s q_s^2}} \tag{8.4}$$

If we set $\mu = \mu_0$, it is known as the *London penetration depth* representing an upper-bound of the *magnetic induction penetration depth* into the superconductor. It is defined as

Maximum Magnetic Induction Penetration Depth

$$\delta_L = \sqrt{\frac{m_s}{\mu_0 n_s q_s^2}} \tag{8.5}$$

The upper-bound given by Eq. (8.68) can be estimated if we substitute reasonable values for the electron in vacuum where $n_s \approx 10^{23}$ m^{-3}. Using this value for n_s, we have[3]

$$\delta_L \approx \sqrt{\frac{m_0}{\mu_0 n_s e^2}}$$

$$= \sqrt{\frac{9.109\times10^{-31}}{4\pi10^{-7}\times10^{23}\times(1.602\times10^{-19})}}$$

[3]We will calculate this value for n_s in Sec. 7.8 after obtaining required quantum mechanical results.

$$= 16.805 \times 10^{-6}\text{m}(\mu m)$$

Fig. 8.3 illustrates the magnetic induction spatial dependence in a superconductor, demonstrating how the induction decays exponentially *from* the surface inward.

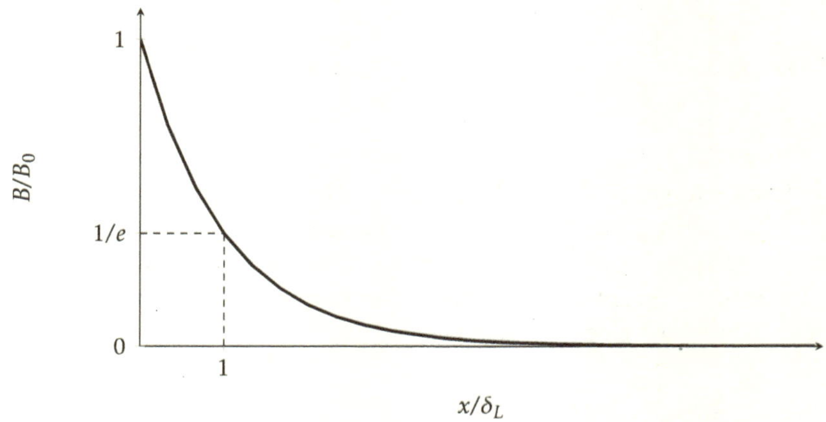

Figure 8.3: Penetration of magnetic induction into a superconductor with depth dimension along x. The location $x = 0$ represents the surface or boundary of the superconductor. At $x/\delta_L = 1$ the induction has reduced to $1/e \approx 1/2.718 = .37$ times the peak value on the boundary.

Unfortunately, in the London equation, we still have no information on conditions for how a superconductor reaches such a state. In fact, the London theory is very general in the way we have presented it. But we know that while in one set of conditions, the material is in a normal state, with the induction field normally penetrating more inside the material. However, changing the conditions, it suddenly gets expelled from the material when the superconducting state is reached. But, what distinguishes this as a theory for superconductors apart from the lossless assumption? The answer, which can only be addressed by quantum mechanics, lies in the behavior of the carrier concentration n_s which, as mentioned, is in proportion to $|\psi(x,y,z)|^2$.

One approach that *begins* to address this question was obtained using phenomenological thermodynamical parameters to better understand the behavior of n_s in the vicinity of the transition. Incidentally, the lowest order term in this theory turns out to be *consistent with quantum mechanics,* and this makes it very useful. It is known as the *Ginzburg-Landau theory,* introduced in the next section.

8.1.3 The Critical Magnetic Field Of A Superconductor

The expelling of the magnetic induction by a superconductor, as described from the London brothers result, implies that the magnetic induction *inside* the superconductor must be zero. This idea is illustrated in Fig. 8.4. The only way the magnetic induction can be zero in a

normal conductor $T > T_c$ superconductor $T < T_c$

Figure 8.4: Magnetic induction lines on (left) a normal nonmagnetic conductor vs. (right) a superconductor. In the superconductor, the magnetic induction lines are expelled out of the sample. This phenomenon is known as the *Meissner effect*.

material is if the magnetization **M** of the superconductor **M** aligns fully opposite to the direction of $\mathbf{B}_e$ so that we have

$$\mathbf{B} = \mathbf{B}_e + \mu_0 \mathbf{M} = 0 \Rightarrow \mu_0 \mathbf{M} = -\mathbf{B}_e \tag{8.6}$$

When the magnetization of a material responds to a magnetic field by pointing *opposite* to the direction of the external magnetic field, the material is said to be **diamagnetic**. This behavior in superconductors is not only *diagmagnetic*, but can be in the maximum possible form of it. This is known as **superdiamagnetism**, because for a diamagnetic material, the maximum relative magnetic susceptibility is given by

$$\chi_r = \frac{\mu_0 \mathbf{M}}{\mathbf{B}_e} = -\frac{\mu_0 \mathbf{B}_e}{\mu_0 \mathbf{B}_e} = -1 \tag{8.7}$$

This relates to the amount of electrical current I that can be passed through a superconductor. This is because the current I creates a magnetic field in the superconducting wire. In a circular wire, this field H_I on the surface of the wire is given by

$$B_I = \mu_0 H_I = \frac{\mu_0 I}{2\pi R} \tag{8.8}$$

For superconductors, there is greater sensitivity to the field at the surface because, as we'll find, this is where superconducting charge carriers tend to concentrate. As mentioned, the superconductor can be driven *out of the superconducting state* when this field exceeds a certain critical magnetic field. More generally, when any magnetic field exceeds a certain value, the superconductor can be driven out of the superconducting state. This magnetic field is known as the **critical magnetic field** B_c. The critical field depends on temperature. This is implied by the fact that without a field, there is a definite T_c as observed by Onnes, however, when the magnetic field is present at some temperature $T < T_c$, it can also be driven out of the superconducting state, i.e. another value of T_c is found with a magnetic field. Experimentally, for several materials including lead (Pb), mercury (Hg), tin (Sn), and indium (In), the B_c is approximately parabolic in temperature, and is often written in the following form:

$$B_c(T) = B_c(0)\left[1 - \frac{T^2}{T_c^2}\right] \tag{8.9}$$

T_c is the transition temperature at zero field. This function is illustrated in Fig. 8.5. Table 8.3 lists both the critical temperature at zero field T_c

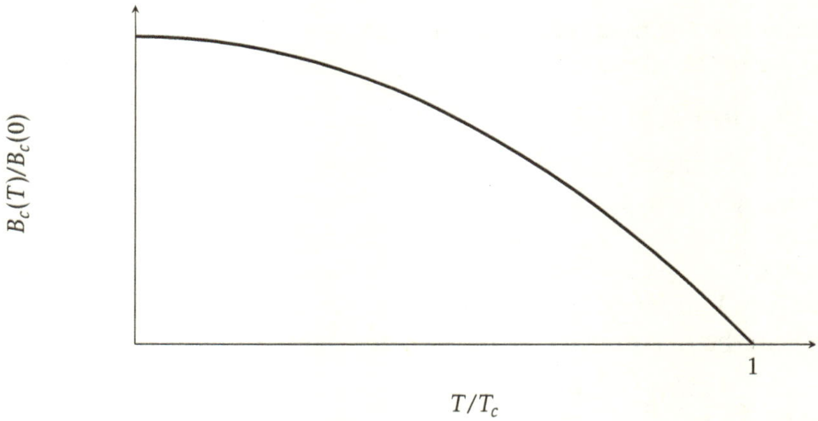

Figure 8.5: Parabolic temperature dependence of the critical field B_c for some single element superconductors.

along with the critical field at zero temperature $B_c(0)$. Since the *critical field* B_c is because of the electrical current flowing in the superconductor, there is also a corresponding average critical current density $J_c = I_c/A$.

Element	T_c[K]	$B_c(0)$[mT]
V	5.38	142.0
Nb	9.50	198.0
Tc	7.77	141.0
In	3.40	29.3
Sn	3.72	30.9
La	6.00	110.0
Ta	4.48	83.0
Hg	4.15	41.2
Pb	7.19	80.3

Table 8.3: Transition temperatures and critical fields of some selected superconducting elements.

Using Eq. (8.8), the critical current density is approximately given by

$$J_c = \frac{2\pi r B_c}{\mu_0 A} = \frac{2\pi r B_c}{\mu_0 \pi r^2} = \frac{2B_c}{\mu_0 r} \tag{8.10}$$

Thus, a superconductor has two or three important parameters that characterize it, or determine the ability to transition into/out of the superconducting state.

R *Superconducting state*: The superconducting state depends on two (or three) important parameters, given by the transition temperature T_c and the critical magnetic field B_c (also impacted by the critical current J_c).

With this understanding of the importance of the additional parameter given by the critical field, new opportunities to tune the superconducting state would lead to two distinct types of superconductors. We'll discuss these two types in the next section.

8.1.4 Ginzburg & Landau Theory

Prior to working on superconductivity, Lev Landau had already proposed a general theory of phase transitions. He took an interest in applying it to superconductors being that they are evidently going through a phase transition. Landau, together with Ginzburg would work to apply Landau's theory of phase transitions to provide more insight into the behavior of the probability density, which is a measure of the number of charge carriers per unit volume n_s. Since wave functions are *normalizable*, we can write

$$n_s = \psi_s^* \psi_s = |\psi_s|^2 \tag{8.11}$$

This parameter is positive-definite or $n_s \geq 0$. In the normal conduction state, we expect $n_s \approx 0$, while in the superconducting state, we expect $n_s > 0$. According to Landau's phase transition theory, such a parameter is ideal to use as a so-called *order parameter*. This is the first step in applying the phase transition theory, to identify an *order parameter*. The order parameter in the Landau-Ginzburg theory n_s, must generally possess the binary property where:

$$\begin{aligned} &(1) \quad \text{order parameter} = 0 \quad \text{for } T > T_c \\ &(2) \quad \text{order parameter} \neq 0 \quad \text{for } T < T_c \end{aligned}$$

T_c is the superconductor phase transition temperature. Any parameter that satisfies the above can be used as an order parameter. Since the wave function ψ_s generally depends on space, it follows that the order parameter does.

Let's now consider the *change in free energy* or free energy density, near the transition point. With an order parameter identified, the next step is to write the difference in the free energy density $\Delta f = f_s - f_n$ as an expansion with additional relevant energies associated with the phase transition. We'll include a kinetic energy contribution, as well as the added work done on the magnetization by the changing magnetic field. Since we are writing energy densities, the kinetic energy density in terms of the superconductor wave function ψ_s becomes $\psi_s^* \widehat{H}_K \psi_s$, where $\widehat{H}_K$ is the familiar kinetic energy operator. Including a magnetic energy, this leads to an energy density given by

$$f_s - f_n = \alpha |\psi_s|^2 + \frac{\beta}{2} |\psi_s|^4 + \psi_s^* \frac{(-i\hbar\nabla - q_s\mathbf{A})^2}{2m_s} \psi_s - \int \mathbf{M} \cdot d\mathbf{B} \tag{8.12}$$

We know that the last two terms are positive since they include the kinetic energy of the carrier and the contribution from an external magnetic field, which *can* be shown to be positive[4]. This minor detail leads to the fact that a function ψ_{min} that minimizes Eq. (8.12) also minimizes *only* the first two terms (since adding 2 positive terms will only make it larger). This can be used to relate the minimizing wave function ψ_{min} to the parameters α and β. Minimizing only the first two terms of Eq. (8.12) with respect to ψ_s^* then leads to

$$\psi_{\text{min}}(\alpha + \beta|\psi_{\text{min}}|^2) = 0$$

The above equation has two distinct solutions given by

$$\psi_{\text{min}} = 0 \quad \text{and} \quad |\psi_{\text{min}}|^2 = \frac{-\alpha}{\beta} \tag{8.13}$$

Since $|\psi_{\text{min}}|^2 \geq 0$, it must be that α and β *have opposite signs*. This makes sense for an expansion of this type with even powers. Since $|\psi_{\text{min}}|^{2q} \geq 0$, general convergence is only possible if the coefficients of the expansion have opposite sign. Otherwise, a nearly infinite expansion would tend to diverge. Therefore, for a favorable transition, $f_s - f_n$ should be negative, or the system has available to it, a new lower energy state corresponding to f_s. Therefore, the leading term which is taken to be the largest in amplitude, must be negative, while the next term is positive. Thus, $\alpha < 0$ and $\beta > 0$.

Minimizing *all* the terms of Eq. (8.12) w.r.t. ψ_s^* leads to the *Ginzburg-Landau equation* given by

Ginzburg-Landau Equation

$$\alpha\psi_s + \beta|\psi_s|^2\psi_s + \frac{(-i\hbar\nabla - q_s\mathbf{A})^2}{2m_s}\psi_s = 0 \tag{8.14}$$

Because second term on the LHS is third order in the order parameter $|\psi|$, the *Ginzburg-Landau equation* is generally nonlinear. However, it is useful to look at an approximate form of Eq. (8.14) because it more

[4]see Quantum Mechanics With A SPIN On It for a derivation, Chap. 15, Sec. 15.5.1 (Eason)

correctly reveals essential spatial information relating to $n_s = \psi_s^* \psi_s$. Neglecting the nonlinear term $\beta|\psi_s|^2\psi_s$ gives

$$\alpha\psi_s + \frac{(-i\hbar\nabla - q_s\mathbf{A})^2}{2m_s}\psi_s = 0 \Rightarrow \frac{(-i\hbar\nabla - q_s\mathbf{A})^2}{2m_s}\psi_s = -\alpha\psi_s = |\alpha|\psi_s \quad (8.15)$$

We should recognize Eq. (8.15) as a quantum mechanical eigenvalue equation for the kinetic energy operator, including a magnetic field. We have derived a similar solution for $\psi_s(x,y,x)$ given by Eq. (2.174).[5] The solution describes a quantized circulating vortex. Moreover, the solutions demonstrate that the highest probabilities for charge carrier location tends to the exterior edges of the vortex. This is, as we'll learn in the next chapter, the origin of the ideal *diamagnetism* in Type I superconductors. In the next section, we'll discuss, in detail, just how the two are related.

8.2 Cooper Pair Concept & Spin-Orbit Interaction

Since measuring a diamagnetic moment $\mu_0\mathbf{M}$ as a function of magnetic field B yields zero when $B = 0$, this indicates two possibilities:

(1) all the freed spin carriers are balanced so $S = 0$ when $B = 0$ or

(2) the electrons are not rotating/circulating so that $L = 0$, when $B = 0$.

In the first case, for the charge carriers contributing to $\mu_0\mathbf{M}$, there are equal numbers of spins up and spins down. In the second case, if the energy also depends on orbital angular momentum, then, a lack of orbit leads to $L = 0$, and there is no contributions to a diamagnetic moment, in particular. It turns out that there is quantum mechanical energy contribution with the property that if either (1) or (2) are true, then its energy contribution vanishes.

The Cooper problem begins to look at what kind of energy potential could lead to a transition characteristic in superconductors. In the most basic picture, we can consider this transition taking place between two different states, being a *normal conducting* and a *superconducting* state. Fig. 8.6 illustrates the two different states associated with this transition. *Before* an external electric field is applied, the electrons

[5] However, the specific solution for superconductors is worked out in Subsec. 7.3.1.

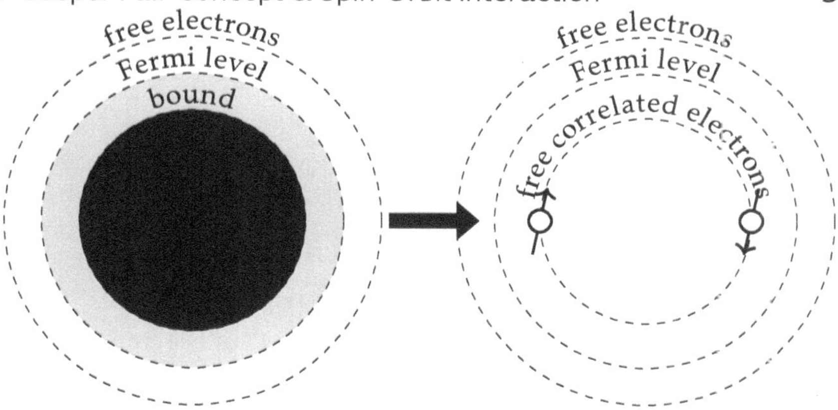

Figure 8.6: BCS picture of correlated electrons in a superconductor (left) originally in bounded states in finite temperature, prior to excitation by an external electric field; and (right) after excitation in a thermally activated potential turned on after cooling below a critical temperature T_c. The energy level for the free electrons on the right is lower than the bound state energy illustrated on the left.

ordinarily occupy bound states in their respective atoms (left side) below the Fermi edge, occupying the *valence band*. When an electric field is applied, the valence electrons are excited out of the bound state into the *conduction band*. In a superconductor, there is a catch. Ordinarily, an electron in the conduction band itinerates or hops to another atom finding its equivalent translated minimum energy state (or higher) in the lattice of atoms. In a superconductor, upon excitation, the electrons gain additional energy from a unique source, providing further excitation. This leads to a unique situation having electrons excited *beyond* what is provided by the electric field. It turns out that that the additional mechanism can be approximated as binary-symmetric in the carrier spin-states of the electron, particularly, when there is no magnetic field $B = 0$. The BCS theory begins to explore where such a picture of additional energy having such properties leads. So, let's zero-in on this picture a bit more.

8.2.1 Intro to the Cooper Problem

Let's start by stripping down a general Hamiltonian for an atom down to a simpler form. From there, we can take an initial look into how *superconductivity* might be possible, according to quantum mechanics. The simplest multi-electron problem we may consider has a *pair* of electrons. For two nuclei A and B and two electrons 1 and 2, the general atomic Hamiltonian can be written as

$$\widehat{H} = -\frac{\hbar^2}{2M}\left[\nabla_A^2 + \nabla_B^2\right]\psi_s - \frac{\hbar^2}{2m_s}\left[\nabla_1^2 + \nabla_2^2\right]\psi + \frac{e^2}{4\pi\epsilon_0}\left[-\frac{1}{d_{1A}} - \frac{1}{d_{1B}} - \frac{1}{d_{2A}} - \frac{1}{d_{2B}} + \frac{1}{r_{12}} + \frac{1}{R}\right]\psi_s \quad (8.16)$$

M is the sum of both nuclei masses, while m^* is the reduced mass obtained after expressing the system in the coordinates of the center of mass R and the electron relative position r_1 and r_2. In the *Born-Oppenheimer approximation*, the kinetic energy contributions from both nuclei are ignored. There is another justification for why it can be ignored in these conditions, namely, because we are primarily interested in free or conduction electrons, which are not bound to any nuclei. However, their respective attractions to the nuclei are still included. Then, Eq. (8.16) reduces to

$$\widehat{H} = -\frac{\hbar^2}{2m_s}\left[\nabla_1^2 + \nabla_2^2\right]\psi_s + \underbrace{\frac{Ze^2}{4\pi\epsilon_0}\left[-\frac{1}{d_{1A}} - \frac{1}{d_{2A}} + \frac{1}{r_{12}}\right]}_{\text{Electric Potential V}}\psi_s \quad (8.17)$$

We have generalized the atomic number beyond unity by including any number of nuclear protons Z. From Eq. (8.17), because of the combination of signs (electron-nuclei attraction + electron-electron repulsion), the last term in brackets can be positive or negative, depending on the *inter*distances. When substituting explicit forms for a multi-electron wave function, new interactions such as the *exchange integral* and the *Coulomb integral* emerge. With the exchange integral, for example, we found that bound electrons on different atoms with opposite spins can lower their energy by overlapping their respective wave functions. This gives rise to the well-known *covalent bond* for anti-aligned spins. However, ferromagnetism creates aligned spin systems. In general, since the Hamiltonian is the eigenoperator for the energy, and since the kinetic energy terms are positive, the energy can only be changed through some kind of combination of interactions between the electrons with themselves, and between the electrons with nuclei or atoms. So, let us write Eq. (8.17) more generally as

$$\widehat{H} = \left[-\frac{\hbar^2}{2m^*}\nabla_1^2 - \frac{\hbar^2}{2m^*}\nabla_2^2 - V_\sigma\right]\psi_s = E_s\psi_s \quad (8.18)$$

In the BCS picture, it is assumed that $V_\sigma > 0$, however $E_s \geq 0$ so that *the energy above the Fermi edge*. Otherwise, we'd have a bound state. Ignoring the interaction potential between the two electrons for the moment, Eq. (8.18) simplifies to

$$\widehat{H}_0 = \left[-\frac{\hbar^2}{2m^*}\nabla_1^2 - \frac{\hbar^2}{2m^*}\nabla_2^2 \right] \psi_k = 2\epsilon_k = E_k \psi_k \tag{8.19}$$

ψ_k is the two-electron eigenfunction without an interaction, and ϵ_k is the single electron kinetic energy. For this to be true, using $\widehat{p} = \hbar\widehat{k}$, it must be that

$$\frac{-\hbar^2 k_1^2}{2m} = \frac{-\hbar^2 k_2^2}{2m}$$

From this relation, we obtain conditions for the so-called *Cooper pair*, where we have

$$V(\mathbf{k}_1) = V(-\mathbf{k}_2) \quad \Leftrightarrow \quad V(\widehat{\sigma_1}) = V(-\widehat{\sigma_2}) \quad \text{B=0} \tag{8.20}$$

For these conditions, we can readily identify an energy operator that satisfies both conditions simultaneously, namely, the *spin-orbit interaction*. In Sec. 2.5, we found this interaction to be

$$\widehat{H}_{SO} = -\frac{\hbar}{4m_0^2c^2}\widehat{\boldsymbol{\sigma}} \cdot (\widehat{\mathbf{p}} + e\mathbf{A}) \times \nabla V = -\frac{\hbar}{4m_0^2c^2}\widehat{\boldsymbol{\sigma}} \cdot \widehat{\mathbf{p}} \times \nabla V \tag{8.21}$$

Without a magnetic field ($B = 0 \rightarrow A = 0$), Eq. (8.21) can be seen to satisfy Eq. (8.20). $\widehat{H}_{SO}$ is conserved for both spins because the pair $(+\mathbf{p}, +\boldsymbol{\sigma})$ has *equal* energy to $(-\mathbf{p}, -\boldsymbol{\sigma})$. Therefore

$$V(\widehat{\sigma_1}) = V(-\widehat{\sigma_2}) \tag{8.22}$$

These are hallmark conditions for a *Cooper pair*. The spin-orbit interactions can *change the energy of electrons* in such a way that spin-up and spin-down electrons are degenerate and correlated. Summarizing, we have the following for a *Cooper pair*:

Cooper pairs: They have equal energies for equal and opposite momentum, or for momenta 180° out of phase. Likewise, for their spin angular momentum. Two electrons moving in opposite directions, as well as two circling in a loop with 180° phase difference both satisfy the Cooper pair condition. But a magnetic field reduces the symmetry.

When a magnetic field is introduced, the symmetry property is reduced because $\widehat{\mathbf{P}} = \pm\widehat{\mathbf{p}} + e\mathbf{A}$. We know opposite charge and spin don't do the same things in a magnetic field. In addition to having 180° phase difference between the electron pair, another condition concerns the energetic conditions of electrons that are excited just beyond the Fermi level ϵ_F, to within a narrow band above it. This threshold energy is sometimes called the *cutoff energy* of $\epsilon_F + \hbar\omega_D$. Below the Fermi level, particularly when $T = 0$, the states are all occupied, and thus in a bound state. This changes for finite temperature, and we'll treat this case, as well. The Cooper-pair problem considers a potential under conditions consistent with observations with superconductors. With two states being central to our discussion, we can frame the two state system symmetric about some energy point of the system. Having these properties suggests we can use spin matrices to describe the properties.

8.3 BCS Energy & Spin Matrices

In the BCS theory, in addition to introducing the concept of Cooper pairs, the energy responsible for superconductivity is given certain properties to enable analyses. Let us consider some of those properties to understand how it leads to some well-known results in the BCS theory. Also, since superconductivity is treated rigorously with relativistic quantum mechanics in Chap. 7, this will be the extent of our objective here.

As mentioned, let the reference energy level be given by the *Fermi edge* ϵ_F. Relative to this point, there is additional kinetic energy as well as a potential energy having certain properties. By assuming two states based on the Cooper pair concept, it can be caste into an equivalent spin system with spin-up and spin-down electrons. If binary, then it can be formalized with a mathematical description using the 2×2 *principle spin matrices* $\widehat{\sigma}_x, \widehat{\sigma}_y$, and $\widehat{\sigma}_z$.

In considering the potential energy V, as an example, recall that the Hamiltonian $\widehat{H}_Z$ for the Zeeman energy (spin coupling to an external magnetic field) of the spin angular momentum is given by

$$\widehat{H}_Z = -\widehat{\mu} \cdot \mathbf{B} = \frac{g|\gamma|\hbar B}{2}\sigma_z = \frac{V_B}{2}\widehat{\sigma}_z$$

The factor of $1/2$ comes from the spin angular momentum relation $\widehat{S}_z = (\hbar/2)\widehat{\sigma}_z$. With spin matrices, we are describing *two* eigenenergy states the system can occupy. For a superconductor, the two states can correspond to the two electrons in the Cooper pair or the formation or non-formation of a Cooper pair, *etc.* As a symmetrical two-state system, lett us take the two eigenenergies E_λ of the system as

$$E_\lambda(\lambda = +1) = +\frac{V_B}{2} \tag{8.23a}$$

$$E_\lambda(\lambda = -1) = -\frac{V_B}{2} \tag{8.23b}$$

The splitting between the two states ΔE is just

$$\Delta E = E_{\lambda=-1} - E_{\lambda=+1} = +\frac{V_B}{2} - \left(-\frac{V_B}{2}\right) = 2 \times \frac{V_B}{2} = V_B \tag{8.24}$$

Since V_B is a splitting or energy difference, the BCS theory often denotes it using the symbol Δ. Therefore, an important attribute is that the states are *symmetric with respect to the middle of the splitting.* In the case of Zeeman energy, the symmetry point is the atomic electronic state without the external magnetic field coupling by the spin or orbital angular momentum. For a Cooper pair, the symmetry point is relative to the additional kinetic energy of the electrons in the Cooper pair, *beyond* the Fermi edge.

Let us take a *Cooper pair* to have two possible eigenstates:

$|1\rangle$ representing occupation of the Cooper state, and

$|0\rangle$ for *nonoccupation* of the Cooper pair state.

The ground state $|0\rangle$ corresponds to bound electrons having combined negative eigenenergies. Thus, it corresponding to no pair being formed. We want to construct a description of a *general transition* probability for a Cooper pair with momentum $\pm k_1, \pm\sigma_1$ to another state $\pm k_2, \pm\sigma_2$. We will denote such a transition as $k_1 \to k_2$.

Since the transition for the binary system uses spin matrices, let's look at some essential properties of spin matrices that are useful. For the principle spin matrices, each has corresponding eigenstates or 2×1 eigenvectors for each 2×2 matrix operator. The principal spin matrices $\widehat{\sigma_x}, \widehat{\sigma_y}$, and $\widehat{\sigma_z}$ are given by

$$\widehat{\sigma_x} = \begin{bmatrix} 0 & 1 \\ 1 & 0 \end{bmatrix} \quad \widehat{\sigma_y} = \begin{bmatrix} 0 & -i \\ i & 0 \end{bmatrix} \quad \text{and} \quad \widehat{\sigma_z} = \begin{bmatrix} 1 & 0 \\ 0 & -1 \end{bmatrix}$$

If $\widehat{\sigma_i}$ operates on its own eigenvector $|0\rangle$ or $|1\rangle$, the eigenvector remains unchanged. However, lets see what happens if we let $\widehat{\sigma_i}$ operate on the eigenvectors of the other two spin operators $\widehat{\sigma_j}$ and $\widehat{\sigma_k}$. The eigenvectors for $\widehat{\sigma_z}$ are given by

$$|1_z\rangle = \boldsymbol{\lambda}_{+1} = \begin{bmatrix}1\\0\end{bmatrix} \quad \text{and} \quad |0_z\rangle = \boldsymbol{\lambda}_{-1} = \begin{bmatrix}0\\1\end{bmatrix} \tag{8.25}$$

When we operate on these using $\widehat{\sigma_x}$, we have

$$\widehat{\sigma_x}|1_z\rangle = \begin{bmatrix}0 & 1\\1 & 0\end{bmatrix}\begin{bmatrix}1\\0\end{bmatrix} = \begin{bmatrix}0\\1\end{bmatrix} = |0_z\rangle$$

When operating on $|0_z\rangle$, we obtain

$$\widehat{\sigma_x}|0_z\rangle = \begin{bmatrix}0 & 1\\1 & 0\end{bmatrix}\begin{bmatrix}0\\1\end{bmatrix} = \begin{bmatrix}1\\0\end{bmatrix} = |1_z\rangle$$

Thus, when $\widehat{\sigma_x}$ operates on the eigenkets of $\widehat{\sigma_z}$, we find a *toggling of the energy states*.

Letting $\widehat{\sigma_y}$ operate on the eigenvectors of $\widehat{\sigma_z}$, we have

$$\widehat{\sigma_y}|1_z\rangle = \begin{bmatrix}0 & -i\\i & 0\end{bmatrix}\begin{bmatrix}1\\0\end{bmatrix} = \begin{bmatrix}0\\i\end{bmatrix} = i|0_z\rangle$$

Then, operating on $|0_z\rangle$ leads to

$$\widehat{\sigma_y}|0_z\rangle = \begin{bmatrix}0 & -i\\i & 0\end{bmatrix}\begin{bmatrix}0\\1\end{bmatrix} = \begin{bmatrix}-i\\0\end{bmatrix} = -i|1_z\rangle$$

We find that toggling is possible, however, only with $\widehat{\sigma_x}$. With $\widehat{\sigma_y}$, toggling includes a constant. It turns out that both together can be used to define a linear combination that is fully compliant with quantum mechanical systems. The chosen linear combinations form *two adjoint operators that create and annihilate the energy states* of a binary system in a way that well-known in quantum mechanical systems. If the eigenstates of the Cooper pair are chosen to be those of $\widehat{\sigma_z}$, we can define two general *unitary operators* $\widehat{\sigma^+}$ and $\widehat{\sigma^-}$ as

$$\widehat{\sigma^+} = \frac{1}{2}\left(\widehat{\sigma_x} + i\widehat{\sigma_y}\right) = \frac{1}{2}\left(\begin{bmatrix}0 & 1\\1 & 0\end{bmatrix} + i\begin{bmatrix}0 & -i\\i & 0\end{bmatrix}\right) = \begin{bmatrix}0 & 1\\0 & 0\end{bmatrix} \tag{8.26}$$

$$\widehat{\sigma^-} = \frac{1}{2}\left(\widehat{\sigma_x} - i\widehat{\sigma_y}\right) = \frac{1}{2}\left(\begin{bmatrix}0 & 1\\1 & 0\end{bmatrix} - i\begin{bmatrix}0 & -i\\i & 0\end{bmatrix}\right) = \begin{bmatrix}0 & 0\\1 & 0\end{bmatrix} \tag{8.27}$$

Eqs. (8.26) and (8.27) turn out to be *binary creation and annihilation operators*, where they are defined as

Binary Creation (+) And Annihilation (-) Operators

$$\widehat{\sigma}^+ = \frac{1}{2}\left(\widehat{\sigma}_x + i\widehat{\sigma}_y\right) \quad \text{and} \quad \widehat{\sigma}^- = \frac{1}{2}\left(\widehat{\sigma}_x - i\widehat{\sigma}_y\right) \tag{8.28}$$

Note that the two operators involved Eq. (8.28) *depend on the choice of system eigenkets*. When we apply the *binary creation operator* $\widehat{\sigma}^+$ to the eigenkets of $\widehat{\sigma}_z$, which we have chosen as the representation for the Cooper pair eigenstates (dropping the z subscript from here on out) of a two-state system, we obtain

$$\widehat{\sigma}^+|0\rangle = \frac{1}{2}\left(\widehat{\sigma}_x + i\widehat{\sigma}_y\right)|0\rangle = \begin{bmatrix} 0 & 1 \\ 0 & 0 \end{bmatrix}\begin{bmatrix} 0 \\ 1 \end{bmatrix} = \begin{bmatrix} 1 \\ 0 \end{bmatrix} = |1_z\rangle \tag{8.29}$$

However, when we apply $\widehat{\sigma}^+$ to $|1_z\rangle$, we get

$$\widehat{\sigma}^+|1\rangle = \begin{bmatrix} 0 & 1 \\ 0 & 0 \end{bmatrix}\begin{bmatrix} 1 \\ 0 \end{bmatrix} = \begin{bmatrix} 0 \\ 0 \end{bmatrix} \neq |0_z\rangle \tag{8.30}$$

Instead, one obtains the *nullket*. These are well-known characteristics of quantum mechanical *creation* and *annihilation* operators. They are examples of *ladder operators* which appears in the atomic orbital angular momentum problem, as well as the harmonic oscillator problem. In a quantum mechanical system, when the *maximum* energy state is reached in a finite population of possible quantum states, the creation operator should return the *nullket*, as there are no potential energy eigenstates above the maximum energy eigenstate. Likewise, for the annihilation operator with the ground state. For the annihilation operator, we have

$$\widehat{\sigma}^-|1\rangle = \frac{1}{2}\left(\widehat{\sigma}_x - i\widehat{\sigma}_y\right)|1\rangle = \begin{bmatrix} 0 & 0 \\ 1 & 0 \end{bmatrix}\begin{bmatrix} 1 \\ 0 \end{bmatrix} = \begin{bmatrix} 0 \\ 1 \end{bmatrix} = |0_z\rangle \tag{8.31}$$

However, when we apply $\widehat{\sigma}^-$ to the ground state $|0_z\rangle$, we get

$$\widehat{\sigma}^-|0\rangle = \begin{bmatrix} 0 & 0 \\ 1 & 0 \end{bmatrix}\begin{bmatrix} 0 \\ 1 \end{bmatrix} = \begin{bmatrix} 0 \\ 0 \end{bmatrix} = \text{null Ket} \tag{8.32}$$

Since $\widehat{\sigma}_x$ and $\widehat{\sigma}_y$ both have real eigenvalues, *i.e.* they are Hermitian operators, then the creation and annihilation operators satisfy the relation

$$\widehat{\sigma}^- = [\widehat{\sigma}^+]^T = [\widehat{\sigma}^+]^\dagger \quad \text{and} \quad \widehat{\sigma}^+ = [\widehat{\sigma}^-]^T = [\widehat{\sigma}^-]^\dagger \tag{8.33}$$

Both operators are respective transposes so the creation and annihilation operators are also *adjoints* of one another.

For a direct quantum state transition, $\widehat{\sigma}^+$ and/or $\widehat{\sigma}^-$ would suffice. However, it is more likely that Cooper pairs or superconducting state is reached from undergoing more than one energy transition. To allow for this, we can describe *a first transition* of the Cooper pair binary system from $\pm\mathbf{k}_1$ to $\pm\mathbf{k}_2$, say, from having kinetic energy $\pm\mathbf{k}_1$ to *not* occupying the Cooper pair state, then from the non-Cooper pair state into a different $\pm\mathbf{k}_2$ Cooper pair state. If we take the starting state for a transition to be $\pm\mathbf{k}_1$, the spin operator for transitioning to $\pm\mathbf{k}_2$ becomes the ladder operator sequence given by

$$\widehat{O}(k_1 \Rightarrow k_2) = \widehat{\sigma}^+_{k2}\widehat{\sigma}^-_{k1} \tag{8.34}$$

The first operator (right-most) annihilates the spin-pair state from $\pm\mathbf{k}_1$. Then, the second operator (left-most) creates the Cooper pair state $\pm\mathbf{k}_2$. Eq. (8.34) is an *indirect transition operator for a single Cooper pair* because it takes place as a sequence of two transitions. With this, the energy operator becomes

$$\widehat{\Delta H} = \widehat{\sigma}^+_{k2}\widehat{\sigma}^-_{k1} \tag{8.35}$$

If there are N Cooper pairs that can make such an indirect transition, then we can just sum all these pair transition energies as

$$\widehat{O}^N_c(k_1 \Rightarrow k_2) = \sum_{(k_1 \to k_2)} \widehat{\sigma}^+_{k2}\widehat{\sigma}^-_{k1} \tag{8.36}$$

Since this is all a description for the potential energy arising in the BCS theory, we can introduce the potential energy expectation value so we have

Cooper Pair Transition Energy Hamiltonian

$$\widehat{\Delta H}_c = \Delta_c \sum_{(k_1 \to k_2)} \widehat{\sigma}^+_{k2}\widehat{\sigma}^-_{k1} \tag{8.37}$$

Since the transitions are between different pair states, the transition energy corresponds to the splitting energy Δ_c. As a two-state system expressed using spin operators, the Hamiltonian given by Eq. (8.37) can also be expressed with the explicit eigenvectors for a *general* spin matrix. The general spinor for an electron spin matrix is given by

$$|s\rangle = \begin{bmatrix} \cos\frac{\theta}{2}e^{-i\frac{\varphi}{2}} \\ \sin\frac{\theta}{2}e^{i\frac{\varphi}{2}} \end{bmatrix} \tag{8.38}$$

With z as the reference axis, it dictates the value for $\theta = \pi/2$. Then, it follows that $\cos(\pi/4) = \sin(\pi/4) = 1/\sqrt{2}$. If we also shift the phase by $e^{+i\frac{\varphi}{2}}$, we have

$$|s\rangle = \frac{1}{\sqrt{2}}\begin{bmatrix} 1 \\ e^{i\varphi} \end{bmatrix}$$

Introducing probability amplitudes u and v to the state vector, we have

$$|s\rangle = \begin{bmatrix} u \\ ve^{i\varphi} \end{bmatrix} \tag{8.39}$$

u is the probability amplitudes for the non-occupation of the Cooper pair state, and v corresponds to that of occupation. This is the general form of a two-state system having a polar angle of $\theta = \pi/2$. The symmetric state probability amplitudes must satisfy the condition $u^2 + v^2 = 1$. By assuming the polar angle $\theta = \pi/2$, it is equivalent to the electrons having motion confined in the $x - y$ plane, with phase φ. By imposing the constraint that u, v are real, the phase φ is then relative to $\varphi_i = 0$. Using the definitions for the creation and annihilation operators from Eq. (8.28), the energy operator can be written as

$$\widehat{\Delta H}_c = \widehat{\sigma}^+_{k2}\widehat{\sigma}_{k1} = \left(\widehat{\sigma}_x + i\widehat{\sigma}_y\right)\left(\widehat{\sigma}'_x + i\widehat{\sigma}'_y\right) \tag{8.40}$$

$$= \widehat{\sigma}_x\widehat{\sigma}'_x + \widehat{\sigma}_y\widehat{\sigma}'_y + i\left(\widehat{\sigma}_y\widehat{\sigma}'_x - \widehat{\sigma}_x\widehat{\sigma}'_y\right) \tag{8.41}$$

Ultimately, the energy expectation value will be evaluated. As a first step, let us consider the operator

$$\widehat{\Delta H}_c = \widehat{\sigma}^+_{k2}\widehat{\sigma}_{k1} = \widehat{\sigma}_x\langle\widehat{\sigma}'_x\rangle + \widehat{\sigma}_y\langle\widehat{\sigma}'_y\rangle + i\left(\widehat{\sigma}_y\langle\widehat{\sigma}'_x\rangle - \widehat{\sigma}_x\langle\widehat{\sigma}'_y\rangle\right) \tag{8.42}$$

For $\widehat{\sigma}_x$, the expectation value is given by

$$\langle\widehat{\sigma}_x\rangle = \boldsymbol{\lambda}^+_{+1}\widehat{\sigma}_x\boldsymbol{\lambda}_{+1}$$

$$= \begin{bmatrix} \cos\frac{\theta}{2}e^{i\frac{\varphi}{2}} & \sin\frac{\theta}{2}e^{-i\frac{\varphi}{2}} \end{bmatrix}\begin{bmatrix} 0 & 1 \\ 1 & 0 \end{bmatrix}\begin{bmatrix} \cos\frac{\theta}{2}e^{-i\frac{\varphi}{2}} \\ \sin\frac{\theta}{2}e^{i\frac{\varphi}{2}} \end{bmatrix}$$

$$= \begin{bmatrix} \cos\frac{\theta}{2}e^{i\frac{\varphi}{2}} & \sin\frac{\theta}{2}e^{-i\frac{\varphi}{2}} \end{bmatrix} \begin{bmatrix} \sin\frac{\theta}{2}e^{i\frac{\varphi}{2}} \\ \cos\frac{\theta}{2}e^{-i\frac{\varphi}{2}} \end{bmatrix}$$

$$= \cos\frac{\theta}{2}\sin\frac{\theta}{2}e^{i\varphi} + \cos\frac{\theta}{2}\sin\frac{\theta}{2}e^{-i\varphi}$$

$$= \cos\frac{\theta}{2}\sin\frac{\theta}{2}\left(e^{i\varphi} + e^{-i\varphi}\right)$$

$$= \sin\theta\cos\varphi$$

Similarly, for $\langle\widehat{\sigma_y}\rangle$, we have

$$\langle\widehat{\sigma_y}\rangle = \boldsymbol{\lambda}_{+1}^{\dagger}\widehat{\sigma_y}\boldsymbol{\lambda}_{+1}$$

$$= \begin{bmatrix} \cos\frac{\theta}{2}e^{i\frac{\varphi}{2}} & \sin\frac{\theta}{2}e^{-i\frac{\varphi}{2}} \end{bmatrix} \begin{bmatrix} 0 & -i \\ i & 0 \end{bmatrix} \begin{bmatrix} \cos\frac{\theta}{2}e^{-i\frac{\varphi}{2}} \\ \sin\frac{\theta}{2}e^{i\frac{\varphi}{2}} \end{bmatrix}$$

$$= i\begin{bmatrix} \cos\frac{\theta}{2}e^{i\frac{\varphi}{2}} & \sin\frac{\theta}{2}e^{-i\frac{\varphi}{2}} \end{bmatrix} \begin{bmatrix} -\sin\frac{\theta}{2}e^{i\frac{\varphi}{2}} \\ \cos\frac{\theta}{2}e^{-i\frac{\varphi}{2}} \end{bmatrix}$$

$$= -i\cos\frac{\theta}{2}\sin\frac{\theta}{2}e^{i\varphi} + i\cos\frac{\theta}{2}\sin\frac{\theta}{2}e^{-i\varphi}$$

$$= i\cos\frac{\theta}{2}\sin\frac{i\theta}{2}\left(e^{i\varphi} - e^{-i\varphi}\right)$$

$$= \sin\theta\sin\varphi$$

Using $\theta = \pi/2$, we have

$$\langle\widehat{\sigma_x}\rangle = \cos\varphi, \quad \langle\widehat{\sigma_y}\rangle = \sin\varphi \tag{8.43}$$

Substitution of the expectation values into Eq. (8.42) gives

$$\widehat{\Delta H}_c = \widehat{\sigma_{k2}^{+}}\widehat{\sigma_{k1}} = \widehat{\sigma_x}\cos\varphi + \widehat{\sigma_y}\sin\varphi + i\left(\widehat{\sigma_y}\cos\varphi - \widehat{\sigma_x}\sin\varphi\right) \tag{8.44}$$

Now, we can substitute the principle spin matrices so that Eq. (8.44) becomes

$$\widehat{\sigma_{k2}^{+}}\widehat{\sigma_{k1}} = \widehat{\sigma_x}\cos\varphi + \widehat{\sigma_y}\sin\varphi + i\left(\widehat{\sigma_y}\cos\varphi - \widehat{\sigma_x}\sin\varphi\right) \tag{8.45a}$$

$$= \begin{bmatrix} 0 & 1 \\ 1 & 0 \end{bmatrix}\cos\varphi + \begin{bmatrix} 0 & -i \\ i & 0 \end{bmatrix}\sin\varphi + i\left(\begin{bmatrix} 0 & -i \\ i & 0 \end{bmatrix}\cos\varphi - \begin{bmatrix} 0 & 1 \\ 1 & 0 \end{bmatrix}\sin\varphi\right) \tag{8.45b}$$

$$= \begin{bmatrix} 0 & 2\cos\varphi - 2i\sin\varphi \\ 0 & 0 \end{bmatrix} \tag{8.45c}$$

$$= \begin{bmatrix} 0 & 2e^{-i\varphi} \\ 0 & 0 \end{bmatrix} \tag{8.45d}$$

The above result only applies to the potential energy contribution. As mentioned, there is also kinetic energy. To add symmetry to the kinetic energy contribution, we can choose the reference point to be the energy midway between the beginning Cooper pair state and the final state. Then, the kinetic energy becomes

$$\begin{bmatrix} +\xi & 0 \\ 0 & -\xi \end{bmatrix} \tag{8.46}$$

Then, combining the kinetic and potential energy matrices into a single one gives

$$\begin{bmatrix} +\xi & 2\Delta e^{-i\varphi} \\ 0 & -\xi \end{bmatrix} \tag{8.47}$$

Eq. (8.47) can be used to determine the eigenenergy, as well as the probability coefficients u and v. From here, the expectation value becomes

$$\begin{bmatrix} u & ve^{-i\varphi} \end{bmatrix} E_\lambda \begin{bmatrix} u \\ ve^{i\varphi} \end{bmatrix} = \langle \widehat{\Delta H_c} \rangle = \begin{bmatrix} u & ve^{-i\varphi} \end{bmatrix} \begin{bmatrix} +\xi & 2\Delta e^{-i\varphi} \\ 0 & -\xi \end{bmatrix} \begin{bmatrix} u \\ ve^{i\varphi} \end{bmatrix} \tag{8.48}$$

Evaluation leads to

$$\begin{bmatrix} u & ve^{-i\varphi} \end{bmatrix} \begin{bmatrix} u\xi + 2v\Delta \\ -\xi v e^{i\varphi} \end{bmatrix} = u^2\xi + 2\Delta uv - \xi v^2 = (u^2 + v^2)E_\lambda \tag{8.49}$$

As a quadratic in terms of the amplitudes, the eigenvalues can be determined from the following conditions

$$\frac{\partial}{\partial u} = 0 \Rightarrow 2\xi u + 2v\Delta = 2uE_\lambda \tag{8.50a}$$

$$\frac{\partial}{\partial v} = 0 \Rightarrow 2\Delta u - 2v\xi = 2vE_\lambda \tag{8.50b}$$

Notice that although $u^2 + v^2 = 1$, it was kept in the expression to evaluate Eq. (8.50). Using Eq. (8.50) to solve for u, we have

$$u = v\left[\frac{E_\lambda + \xi}{\Delta}\right] \tag{8.51}$$

Using Eq. (8.51) in the first of Eq. (8.50) then leads to

$$\xi v\left[\frac{E_\lambda + \xi}{\Delta}\right] + v\Delta = v\left[\frac{E_\lambda + \xi}{\Delta}\right]E_\lambda \tag{8.52}$$

v cancel on both sides of Eq. (8.52). Also, ξE_λ appears on both sides, so it also cancels. Then, we have the result

$$E_\lambda^2 = \xi^2 + \Delta^2 \tag{8.53}$$

or

Cooper Pair Transition Eigenenergy

$$E_\lambda = \pm\sqrt{\xi^2 + \Delta^2} \tag{8.54}$$

The probability amplitudes (starting with v) can be found using Eq. (8.49), with the relation $u^2 + v^2 = 1$. Using Eq. (8.51), this gives

$$u^2\xi + 2\Delta uv - \xi v^2 = E_\lambda \Rightarrow$$
$$v^2(E_\lambda + \xi)^2\xi + 2\Delta^2(E_\lambda + \xi)v^2 - v^2\Delta^2\xi = \Delta^2 E_\lambda \tag{8.55}$$

Solving for v^2, we have

$$v^2 = \frac{\Delta^2 E_\lambda}{(E_\lambda + \xi)^2\xi + 2\Delta^2(E_\lambda + \xi) - \xi\Delta^2} \tag{8.56}$$

Substitution of Eq. (8.54) into Eq. (8.56), we can also express Eq. (8.56) as a function of the dimensionless variable ξ/Δ, which leads to

$$v^2 = \frac{\sqrt{\frac{\xi^2}{\Delta^2} + 1}}{\left(\sqrt{\frac{\xi^2}{\Delta^2} + 1} + \frac{\xi}{\Delta}\right)^2 \frac{\xi}{\Delta} + 2\left(\sqrt{\frac{\xi^2}{\Delta^2} + 1} + \frac{\xi}{\Delta}\right) - \frac{\xi}{\Delta}} \tag{8.57}$$

Eq. (8.57) can be shown to be equal to the following:

Probability Of Cooper Pair Formation

$$v^2 = \frac{1}{2}\left(1 - \frac{\xi/\Delta}{\sqrt{1 + \frac{\xi}{\Delta}}}\right) \tag{8.58}$$

Similarly, the result for u^2 becomes

Probability Of Cooper Pair Nonformation

$$u^2 = \frac{1}{2}\left(1 + \frac{\xi/\Delta}{\sqrt{1+\frac{\xi}{\Delta}}}\right) \tag{8.59}$$

Eqs. (8.58) and (8.59) are plotted below. Fig. 8.7 illustrates that when

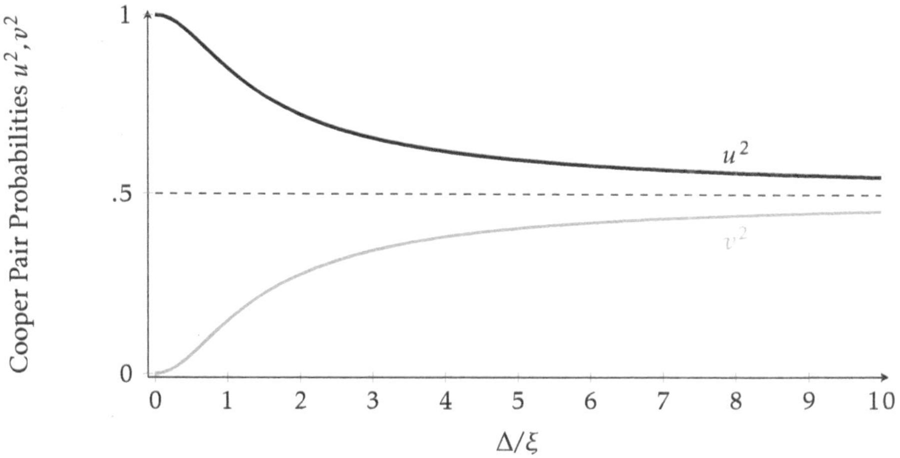

Figure 8.7: State transition probabilities for a Cooper pair as a function of the ratio of potential $\Delta = \Delta_c$ to symmetric kinetic energy ξ (Δ/ξ).

there is no potential energy Δ, *a Cooper pair cannot transition* or $v = 0$. Equivalently, the potential energy Δ enables the formation of a Cooper pair. The above is a relatively general treatment for a two-state system with particular symmetries. Only by assuming such properties, allowing for indirect transitions, results emerge that suggest why these special pairs exist in superconductors.

8.4 BCS Wave Function

Owing to the assumed symmetry of two potential eigenstates, the wave function discussed above has two eigenvectors. Then, the most general solution is a linear combination given by

$$\psi_p = u_p|0\rangle + v_p|1\rangle \quad \text{where} \quad u_p^2 + v_p^2 = 1 \tag{8.60}$$

u_p represents the *probability amplitude* of occupying the state $|0\rangle$, while v_p is that for the state $|1\rangle$. The second condition follows from the normalization condition of the general quantum state where we have

$$\begin{aligned}
1 &= \langle\psi_p|\psi_p\rangle \\
&= \big(\langle 0|u_p + \langle 1|v_p\big)\big(u_p|0\rangle + v_p|1\rangle\big) \\
&= u_p^2\langle 0|0\rangle + u_p v_p\langle 0|1\rangle + v_p u_p\langle 1|0\rangle + v_p^2\langle 1|1\rangle
\end{aligned}$$

Using the *orthogonality condition* for the eigenvectors, we have

$$\begin{aligned}
1 &= u_p^2\langle 0|0\rangle + u_p v_p\langle 0|1\rangle + v_p u_p\langle 1|0\rangle + v_p^2\langle 1|1\rangle \\
&= u_p^2\langle 0|0\rangle + v_p^2\langle 1|1\rangle
\end{aligned}$$

Then, using the *normalization condition* for the eigenvectors, we obtain the condition

$$1 = u_p^2 + v_p^2 \tag{8.61}$$

Eq. (8.61) also means that when this condition is satisfied, the generalized linear combination is normalized. The result can be used when we consider a further generalization of the wave function in Eq. (8.60). If there is no coupling between any of the Cooper pairs, the eigenstate can be written in the form of a product of each pair's transitional eigenstates. This defines the *BCS wave function* ϕ_{BCS} given by

$$\phi_{BCS} = \prod_{p=1}^{N_c} u_p|0\rangle + v_p|1\rangle \tag{8.62}$$

Using the above properties for a single Cooper pair, we can evaluate the normalization condition of the *BCS state*. This becomes

$$\langle\phi_{BCS}|\phi_{BCS}\rangle = \left\langle \prod_{p=1}^{N_c} \langle 0|u_p^\dagger + \langle 1|v_p^\dagger \prod_{q=1}^{N_c} u_q|0\rangle + v_q|1\rangle \right\rangle$$

The above represents an *integral of an inner-product of products*. However, for the left and right products, there are corresponding states which generally include spatial wave function components. For example, $p = 1, q = 1$, $p = 2, q = 2$, *etc.* are corresponding states. The spinor parts of the product effectively come out of the expectation value spatial integral. Under the integral remains only spatial parts of the wave function, which are also normalized. The spatial integral parts for each respective state, then, integrates to unity. Then, the product of products can be written as a single product given by

$$\langle\phi_{BCS}|\phi_{BCS}\rangle = \left\langle \prod_{p=q=1}^{N_c} \left(\langle 0|u_p^{\dagger} + \langle 1|v_p^{\dagger}\right)\left(u_q|0\rangle + v_q|1\rangle\right)\right\rangle \tag{8.63}$$

$$= \prod_{p=1}^{N_c}\left(|u_p|^2 + |v_p|^2\right) = \prod_{p=1}^{N_c} 1 = 1 \tag{8.64}$$

So the BCS wave function is *normalized* under the given assumptions. We can use Eq. (8.37) with Eq. (8.62) to evaluate the expectation value of the potential energy contribution. Substituting the BCS state, we have

$$\langle\Delta_c\rangle = \langle\phi_{BCS}|\widehat{\Delta H}_c|\phi_{BCS}\rangle \tag{8.65}$$

$$= \Delta_c \prod_{p=1}^{N_c} \langle 0|u_p^{\dagger} + \langle 1|v_p^{\dagger}\left(\sum_{(k_1\to k_2)} \widehat{\sigma_{k2}^{+}}\widehat{\sigma_{k1}^{-}}\right)\prod_{q=1}^{N_c} u_q|0\rangle + v_q|1\rangle \tag{8.66}$$

For each of the terms of the sum, the creation-annihilation operator only operates on one of the states in the product, while the remaining bras and corresponding kets come together to yield an inner product of unity. Therefore, $\langle\Delta_c\rangle$ becomes

$$\langle\Delta_c\rangle =$$

$$\Delta_c \prod_{p-} \langle 0|u_p^{\dagger} + \langle 1|v_p^{\dagger}\left(\sum_{(k_1\to k_2)} \left(\langle 0|u_{k2}^{\dagger} + \langle 1|v_{k2}^{\dagger}\right)\widehat{\sigma_{k2}^{+}}\widehat{\sigma_{k1}^{-}}\left(u_{k1}|0\rangle + v_{k1}|1\rangle\right)\right)\times$$

$$\prod_{q-} u_q|0\rangle + v_q|1\rangle = \Delta_c\left(\sum_{(k_1\to k_2)} \left(\langle 0|u_{k2}^{\dagger} + \langle 1|v_{k2}^{\dagger}\right)\widehat{\sigma_{k2}^{+}}\widehat{\sigma_{k1}^{-}}\left(u_{k1}|0\rangle + v_{k1}|1\rangle\right)\right) \tag{8.67}$$

$q-$ and $p-$ denotes that *the product has had one state removed*, and brought into the summation explicitly. Thus, the products have $N_c - 1$ terms for each contribution to the sum. We can evaluate this directly by using the property $\widehat{\sigma^{+}} = [\widehat{\sigma^{-}}]^{\dagger}$. This leads to

$$\langle\Delta_c\rangle = -\langle V_c\rangle\left(\sum_{(k_1\to k_2)} \left(\langle 0|u_{k2}^{\dagger} + \langle 1|v_{k2}^{\dagger}\right)\left[\widehat{\sigma_{k2}^{-}}\right]^{\dagger}\widehat{\sigma_{k1}^{-}}\left(u_{k1}|0\rangle + v_{k1}|1\rangle\right)\right)$$

$$= \Delta_c \left(\sum_{(k_1 \to k_2)} \left(0 \cdot u_{k2}^{\dagger} + \langle 0|v_{k2}^{\dagger}\right)(u_{k1} \cdot 0 + v_{k1}|0\rangle) \right)$$

$$= \Delta_c \sum_{(k_1 \to k_2)} v_{k2}^{\dagger} v_{k1}$$

The above result says that the energy contribution from each pair is proportional to the product of the probability amplitudes of k_2 and k_1. If either are zero, then all bets are off for a Cooper pair formation transitioning from k_1 to k_2. The quantity $v_{k2}^{\dagger} v_{k1}$, here, is proportional to the generalization of the expectation value given by $\tanh\Delta_c/k_B T$.

8.5 Chapter Summary

In this chapter, we have covered essential fundamental and early concepts in *superconductivity*. It is is a low temperature phenomenon which demonstrates a practical loss of resistivity below a critical temperature T_c. A superconductor is also an ideal *diamagnet*, demonstrating not just a negative susceptibility χ_c, but a maximum of it. We showed how the London equation suggests that any conducting material without electrical loss demonstrates ideal diamagnetism, through the fact that the magnetic induction vanishes inside the conductor. We also discussed the Ginzburg-Landau theory. The *linearized* form from this theory leads to a quantum mechanical eigenoperator equation which is relevant to superconductors. In fact, we solved that equation in Chap. 7.

Temperature effects were introduced by using symmetry properties similar to the forms of exchange interaction and spin-orbit coupling. From this, we considered a potential connected to a thermal bath at thermodynamic equilibrium. This implicitly assumes the potential energy is spin-dependent. Then, we examined the effects of temperature on parameters like the correlation energy, resistivity and heat capacity, all demonstrating discontinuities at the critical temperature T_c. Lastly, we discussed BCS energy and wave function.

Table 8.4 summarizes some of the key results from this chapter.

Table 8.4: Chapter 2 Summary Equations.

Name	Equation
London Equation	$\mathbf{J}_s = -\frac{n_s q_s^2}{m_s}\mathbf{A}$
B penetration depth	$\delta_L = \sqrt{\frac{m_s}{\mu_0 n_s q_s^2}}$
Ginzburg-Landau Equation	$\alpha\psi_s + \beta\lvert\psi_s\rvert^2\psi_s + \frac{(-i\hbar\nabla - q_s\mathbf{A})^2}{2m_s}\psi_s = 0$
Binary Creation Operator	$\widehat{\sigma}^{+} = \frac{1}{2}\left(\widehat{\sigma}_x + i\widehat{\sigma}_y\right)$
Binary Annihilation Operator	$\widehat{\sigma}^{-} = \frac{1}{2}\left(\widehat{\sigma}_x - i\widehat{\sigma}_y\right)$

8.6 Chapter Problems

Problem 8.1 Using the London equation (c.Eq. 8.2)

$$\mathbf{J}_s = -\frac{n_s q_s^2}{m_s}\mathbf{A}$$

and show that if $\mathbf{p}_B = e\mathbf{A}$ then

$$\mathbf{J}_s = nev_B$$

Problem 8.2 Using the London penetration depth

$$\delta_L = \sqrt{\frac{m_0}{\mu_0 n_s e^2}} \tag{8.68}$$

Estimate the induction field penetration depth for a magnetic induction field B in a superconductor having 10^{23}m^{-3}, 10^{24}m^{-3}, and 10^{25}m^{-3} electron charges per unit volume. How could introducing an external magnetic field change the penetration depth?

Problem 8.3 In the Ginzburg-Landau theory, the resulting equation was found to be (c.Eq. 8.14)

$$\alpha\psi_s + \beta|\psi_s|^2\psi_s + \frac{(-i\hbar\nabla - q_s\mathbf{A})^2}{2m_s}\psi_s = 0$$

What are the units of α and what does it approximate relating it to the pure quantum mechanical solution?

Problem 8.4 Let a magnetic energy in a superconductor have the form

$$V = -\mathbf{M} \cdot \mathbf{B}$$

If we consider the transition from the normal conducting state to the superconducting state, show that the change in energy or work by the field is given by

$$W = -\int \mathbf{M} \cdot d\mathbf{B} = +\frac{B^2}{2\mu}$$

Can this energy be negative where $W < 0$? If so, under what conditions?

What changes in work done if we add the higher order correction to the kinetic energy?

Problem 8.5 In Chap. ?? we saw that the magnetic vector potential A can be written as (c.Eq. 7.45)

$$\mathbf{A} = [0 \quad rB/2 \quad 0]^T$$

For magnetic fields of strength 0.3 and .6 Tesla in the superconductor of radius r_n = 1cm, with n_s = 10^{23}m^{-3} charge carriers, calculate the currents J_s.

Problem 8.6 In superconductors, we have leared there are two types known as Type I and II. If Type I corresponds to occupation of the ground state, then Type II should correspond to higher energy states. If this is the case, will the magnetic moments of Type II tend to be smaller, larger, or equal to those of Type I? Explain why? .

Problem 8.7 Using the magnetic field on the surface of a conductor (c.Eq. 8.8)

$$B = \frac{\mu_0 I}{2\pi R}$$

and the approximation for the critical magnetic field B_c given by (c.Eq. 8.9)

$$B_c(T) = B_c(0)\left[1 - \frac{T^2}{T_c^2}\right]$$

Calculate the corresponding critical current J_c for a superconductor having radius of 2cm and a critical magnetic field at zero temperature of $B_c(0) = 0.05$Tesla.

8.7 Suggested Readings & References

[1] K. Eason, *Quantum Mechanics With A SPIN On It*, Sym. Pub (2019).

[2] M. Holt *et. al.*, *Phonon dispersions in niobium determined by x-ray transmission scattering*, Phys. Rev. B **66**, 064303 (2002).

[3] P. K. Sharma *et. al.*, *Vibration Spectra and Debye Temperatures of Some Transition Metals*, Z. Naturforsch. **26 a**, pp.747-752 (1971).

[4] R. de Bruyn Ouboter, *Heike Kamerlingh Onnes's Discovery of Superconductivity*, Scientific American, pp.98-103 (1997).

[5] D. v Delft and P. Kes, *The Discovery Of Superconductivity*, Physics Today, pp.38-43 (2010).

[6] J. Bardeen, L. N. Cooper and J. R. Schreiffer, *Theory of Superconductivity*, Phys. Rev. B, Vol 8, Num 5, pp.1175-1204 (1957).

APPENDIX A

Appendix: Harmonic Oscillator

To do this, we will use *relativistic* rules and laws, and the reasons for this will become more clear as we get further and further into the text. Just note that any physics (non-relativistic) won't do, owing to the nature of the systems being described, being capable of describing any elementary particle, from a photon to an electron. We'll introduce the theory of relativity before we use it.

The traditional harmonic oscillator problem turns out to provide a template for other problems in quantum mechanics. For example, *cyclotron motion* of a elementary particle is described by a similar physics as the harmonic oscillator. The same is true of *superconductors,* though there is considerably more physics to that problem. Additionally, this problem has a well-known classical counterpart, which gives us an opportunity to make an intuitive comparison for consistency, *etc.* We would desire that, although they speak different languages, they still say the same thing! This notion of consistency between a classical system and a quantum one is known formally as the **Principle of Correspondence** (**PoC**) or the **Correspondence Principle**.

So, this makes the harmonic oscillator a very good starting place for us. With that said, let's first review some familiar results from the classical harmonic oscillator, then we'll solve the harmonic oscillator problem for a quantum mechanical system.

A.1 The Classical Harmonic Oscillator

Let's consider a classical harmonic oscillator having a mass m, connected to a spring of stiffness or force constant k_s, as illustrated in Fig. A.1. The

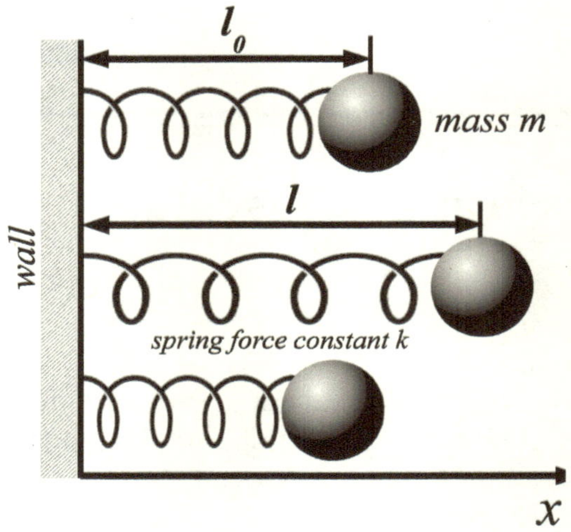

Figure A.1: A *classical* harmonic oscillator consisting of a mass m attached to a spring, showing different examples of displacement. There is an equilibrium position l_0 at which the system desires to be. Displacing the mass, attached to the spring, produces a force that counters the displacement $l - l_0$.

spring is only *reactive* and counters any displacement $l - l_0$ from equilibrium. Let's denote the displacement from *equilibrium* more tersely as $x = l - l_0$. In this way, the origin of our coordinate system corresponds to the equilibrium position, i.e. $x = 0 = l_0 - l_0$. This also allows us to write the change in potential energy $\Delta V = V - V(0)$ as simply V, since $V(0) = 0$. When the spring force F_s is taken to be *linear* in the displacement, the spring is said to follow *Hooke's law* given by

$$F_s = -k_s(l - l_0) = -k_s x \tag{A.1}$$

For a *conservative* system, the potential energy V exactly equates to the *possible* work done *by the spring* to bring the mass from $x \neq 0$ back to its equilibrium position $x = 0$. This is the negative of the work done *to* the spring. Then, ΔV is given by

$$\Delta V = V(x) = \int_0^x F_s dx = -\int_0^x k_s x dx = \frac{1}{2}k_s x^2 \tag{A.2}$$

Therefore, the classical potential energy for a *linear* spring system is

Linear Spring Potential Energy

$$V(x) = \frac{1}{2}k_s x^2 \tag{A.3}$$

Any system that is attached to a linear spring-like system possesses the potential energy given in Eq. (A.3). This is, in fact, the potential energy used in the harmonic oscillator, since its a mass attached to either a *literal* or *symbolic* spring.

What else is the linear force equation good for? Having the force F_s as an explicit function of x also allows us to express a classical ordinary differential equation (ODE) of motion (with x dependent on time t). Using $\sum F = F_k = ma$ leads to

$$-k_s x = m\frac{d^2x}{dt^2} \Rightarrow \frac{d^2x}{dt^2} + \frac{k_s}{m}x = 0 \tag{A.4}$$

Eq. (A.4) is a well-known differential equation having a general solution given by

$$x(t) = A\sin\left(\sqrt{\frac{k_s}{m}} \cdot t\right) + B\cos\left(\sqrt{\frac{k_s}{m}} \cdot t\right) \tag{A.5}$$

Note that the general solution is actually a linear combination of *two independent solutions* resulting from a second order equation. In other words, the sine and cosine functions, respectively, both satisfy Eq. (A.4), as well. We know that these functions are *periodic* in time, or *harmonic*, hence the name. This general solution can readily be plugged into Eq. (A.5) to verify that it is a solution. We can define an intuitive parameter ω, since it actually has units of radians per second or s^{-1} as the *natural frequency* of the system, as

Harmonic Oscillator Natural Frequency

$$\omega = \sqrt{\frac{k_s}{m}} \tag{A.6}$$

Combining these results, the potential energy V can also be written as

$$V = \frac{k_s x^2}{2} = \frac{m\omega^2 x^2}{2} \tag{A.7}$$

Since Eq. (A.2) is second order in time, it requires specification of both the initial displacement $x(t_0) = x_0$ as well as the initial velocity $dx/dt(t_0) = v(t_0) = v_0$. We then have

$$x(0) = A \cdot 0 + B \cdot 1 = x_0 \Rightarrow B = x_0$$

$$\frac{dx}{dt}(0) = A\omega \cdot 1 - B\omega \cdot 0 = 0$$

The result of applying both of these initial conditions to Eq. (A.5) results in $A = 0$, and the general solution simplifies to

Classical Harmonic Oscillator Solution

$$x(t) = x_0 \cos\omega t \tag{A.8}$$

This result means that whatever value is the initial displacement x_0, it determines the *amplitude* of the harmonic displacement oscillation, since it's determined by the coefficient of the cosine term. Fig. A.2 illustrates solutions for different initial conditions x_0 (top) and different spring stiffness constants k_s, which only changes the *natural frequency*. Changing the spring stiffness k_s has *no effect* on the amplitude of the oscillation. The *total* energy of the harmonic oscillator includes both the kinetic energy $K(t)$ and the potential energy $V(t)$ given by Eq. (A.7). Sticking with classical relations, the kinetic energy $K(t)$ is given by

$$K(t) = \frac{1}{2}m\left(\frac{dx}{dt}\right)^2 = \frac{m x_0^2 \omega^2}{2}\sin^2\omega t$$

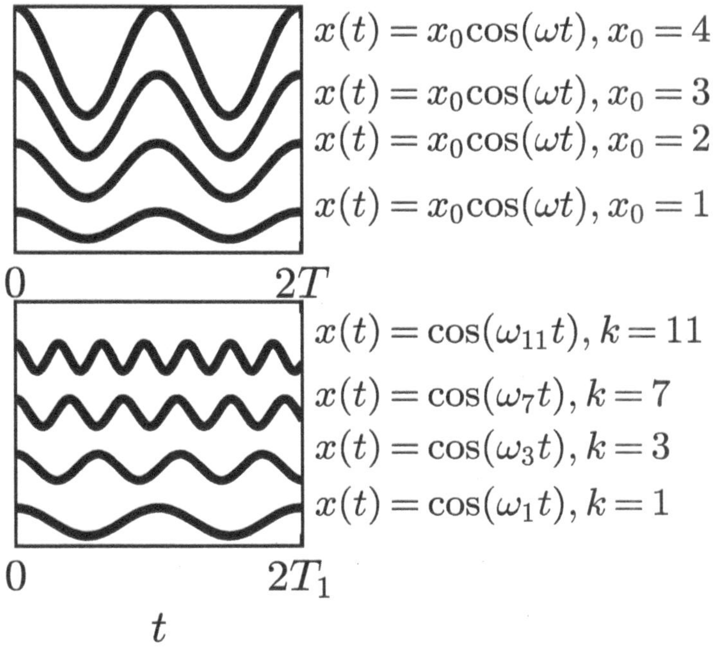

Figure A.2: Illustrations of the time variation of the displacement $x(t)$ where $T = 2\pi/omega$, of a classical oscillator, under (top) different initial displacements, x_0 and (bottom) different spring stiffnesses k.

The total energy $E = K + V$ then becomes

$$E = \frac{mx_0^2\omega^2}{2}\sin^2\omega t + \frac{k_s}{2}x_0^2\cos^2\omega t \tag{A.9}$$

Substitute of the spring stiffness k_s using Eq. (A.6), the total energy $E(t)$ becomes

$$E = \frac{mx_0^2\omega^2}{2}\left(\sin^2\omega t + \cos^2\omega t\right)$$

or

Conserved Harmonic Oscillator Energy

$$E = \frac{kx_0^2}{2} \tag{A.10}$$

We have used the fact that $\sin^2 \omega t + \cos^2 \omega t = 1$. Even though the mass is oscillating in time, the system's total energy is always *conserved*. Using Eqs. (A.6) and (A.10), it follows that the amplitude x_0 of the classical harmonic oscillator can be expressed as

Harmonic Oscillator Amplitude

$$x_0^2 = \frac{2E}{m\omega^2} \tag{A.11}$$

The results summarized above are well-established classical relations for a harmonic oscillator without any dissipation, or energy loss. For the quantum mechanical treatment of this same problem, we expect a consistent set of relations so we satisfy the *PoC*. So, let's find out what the Schrödinger equation has to say about the harmonic oscillator.

A.2 Quantum Mechanical Harmonic Oscillator

For the quantum mechanical harmonic oscillator problem, we solve the Schrödinger equation for a system with the potential energy V given by Eq. (A.2) or Eq. (A.7), which is second order in x. The Schrödinger equation in this case is spatially 1D, and is given by

$$i\hbar\frac{\partial\psi}{\partial t} = \left[-\frac{\hbar^2}{2m}\frac{\partial^2}{\partial x^2} + \frac{1}{2}m\omega^2 x^2\right]\psi \tag{A.12}$$

As we have seen in previous problems, Eq. (A.12) can also be solved using *separation of variables*. We, therefore, assume the wave function ψ can be expressed as $\psi = X(x)T(t)$. Substituting this form into Eq. (A.12), we have

$$i\hbar\frac{1}{T}\frac{dT}{dt} = -\frac{\hbar^2}{2m}\frac{1}{X}\frac{d^2X}{dx^2} + \frac{1}{2}m\omega^2 x^2 \tag{A.13}$$

Since the LHS depends solely on time t, while the RHS depends on x, both being independent variables, it follows that both sides must be equal to a constant given by the eigenenergy E_λ. This leads to the following two separated equations:

$$i\hbar\frac{1}{T}\frac{dT}{dt} = E_\lambda \tag{A.14a}$$

$$-\frac{\hbar^2}{2m}\frac{1}{X}\frac{d^2X}{dx^2} + \frac{1}{2}m\omega^2x^2 = E_\lambda \tag{A.14b}$$

We have seen Eq. (A.14a) before, for example, in the particle in a box (or infinitely deep potential well) problem in Chap. **??** (c. Eq. (**??**)). There, we found the solution for $T(t)$ to be given by

$$T(t) = T_0 e^{-i\frac{E_\lambda}{\hbar}(t-t_0)} \tag{A.15}$$

This gives the time-dependent part of the wave function $\psi = X(x)T(t)$. The determination of the spatial part of the wave function turns out to be a little more complicated, but, solvable nonetheless. In the next section, we'll utilize a particular method invented by mathematicians to determine $X(x)$.

A.3 Symmetry & Annihilation/Creation Operators

If we look closely at the Hamiltonian for the harmonic oscillator, we find an interesting and useful symmetry that can be exploited. Recall that the Schrödinger equation for the harmonic oscillator has the form

$$i\hbar\frac{\partial\psi}{\partial t} = \left[-\frac{\hbar^2}{2m}\frac{\partial^2}{\partial x^2} + \frac{1}{2}m\omega^2x^2\right]\psi \Rightarrow \widehat{H} = \frac{1}{2}m\omega^2x^2 - \frac{\hbar^2}{2m}\frac{\partial^2}{\partial x^2} \tag{A.16}$$

The Hamiltonian $\widehat{H}$ can be written in the following form:

$$\widehat{H} = \frac{1}{2m}\left[m^2\omega^2x^2 - \left(\hbar\frac{\partial}{\partial x}\right)^2\right] \tag{A.17}$$

We can readily identify a symmetry in Eq. (A.17) that allows an expression of lower order operators, as it resembles a scalar equation of the form

$$a^2 + b^2 = c_0(a+ib)(a-ib)$$

c is a scalar corresponding to $1/2m$. The real number a corresponds to $m\omega x$ in Eq. (A.17), and we can show a correspondence between the real number b and the momentum operator $-i\hbar\partial/\partial x$. Therefore, let us define the following operator

$$\widehat{A} = \frac{1}{\sqrt{2m}}[m\omega x + i\widehat{p}_x] \tag{A.18}$$

Since x is real-valued, and so is $\widehat{p}_x$ as a Hermitian operator, the first order operator $\widehat{A}$ necessarily has complex eigenvalues because of the purely imaginary momentum operator term. Moreover, because x and p_x change with time, the eigenvalues of the operator $\widehat{A}$ are generally time-dependent. For the other first order operator, we can define $\widehat{A}^{\dagger}$ as

$$\widehat{A}^{\dagger} = \frac{1}{\sqrt{2m}}[m\omega x - i\widehat{p}_x] \tag{A.19}$$

The eigenvalues of $\widehat{A}^{\dagger}$ are, therefore, complex conjugates of those of $\widehat{A}$. We know that any complex number multiplied by its complex conjugate is always a real number, guaranteeing the Hamiltonian remains Hermitian or having real eigenvalues. This is an important property resulting from the symmetry. Any operator $\widehat{A}^{\dagger}$ with complex eigenvalues that are complex conjugates of an operator $\widehat{A}$ is known as the **adjoint** of the operator $\widehat{A}$. The topic of *adjoint operators* is discussed in more detail in Chap. C. Let's express the Hamiltonian for the harmonic oscillator in terms of the operators $\widehat{A}$ and $\widehat{A}^{\dagger}$. This leads to

$$\widehat{A}\widehat{A}^{\dagger} = \frac{1}{\sqrt{2m}}[m\omega x + i\widehat{p}_x]\frac{1}{\sqrt{2m}}[m\omega x - i\widehat{p}_x]$$

$$= \frac{1}{2m}\left[m^2\omega^2x^2 + \widehat{p}_x^2 - im\omega\left(x\widehat{p}_x - \widehat{p}_x x\right)\right]$$

Using the definition for the square of the momentum operator, given by Eq. (**??**), we have

$$\widehat{A}\widehat{A}^{\dagger} = \frac{1}{2m}\left[m^2\omega^2x^2 - \hbar^2\frac{\partial^2}{\partial x^2} - im\omega\left(x\widehat{p}_x - \widehat{p}_x x\right)\right]$$

Comparing the above result with Eq. (A.17), we find that the first two terms by themselves equate to the Hamiltonian of the harmonic oscillator $\widehat{H}$, however, we also have an additional term in parentheses. Let's take a closer look at this third operator, which becomes

$$\left[x\left(-i\hbar\frac{\partial}{\partial x}\right) - \left(-i\hbar\frac{\partial}{\partial x}\right)x\right]\psi_n = x\left(-i\hbar\frac{\partial\psi_n}{\partial x}\right) - \left(-i\hbar\frac{\partial(x\psi_n)}{\partial x}\right)$$

$$= -ix\hbar\frac{\partial\psi_n}{\partial x} + ix\hbar\frac{\partial\psi_n}{\partial x} + i\hbar\psi_n$$

$$= i\hbar\psi_n$$

Hence, the operator $x\widehat{p}_x - x\widehat{p}_x$ turns out to be a constant, or has the eigenvalue given by

Position-Momentum Commutation Relation

$$x\widehat{p}_x - \widehat{p}_x x = i\hbar \qquad \text{(A.20)}$$

The result given in Eq. (A.20) illustrates an important property of quantum mechanical operators. Specifically, their *order* matters because Eq. (A.20) means that $x\widehat{p}_x \neq \widehat{p}_x x$ or $x\widehat{p}_x - \widehat{p}_x x \neq 0$. In quantum mechanics, this general rule must always be remembered when considering a sequence of operators. Because $x\widehat{p}_x - \widehat{p}_x x \neq 0$, x and $\widehat{p}_x$ are said *not* to **commute** with one another. We discuss the topic of *commutation* between operators in more detail in Chap. B. Using the above result, we have the following relation between $\widehat{AA^\dagger}$ and $\widehat{H}$:

$$\widehat{AA^\dagger} = \widehat{H} - \frac{1}{2m}[im\omega(i\hbar)]$$

$$= \widehat{H} + \frac{\hbar\omega}{2}$$

Thus, the Hamiltonian becomes

$$\widehat{H} = \widehat{AA^\dagger} - \frac{\hbar\omega}{2} \qquad \text{(A.21)}$$

If we apply the operator $\widehat{AA^\dagger}$ on the eigensolutions ψ_n, using the relation $\widehat{H}\psi_n = E_\lambda\psi_n$, we then have

$$\widehat{AA^\dagger}\psi_n = \left[\hbar\omega\left(n + \frac{1}{2}\right) + \frac{\hbar\omega}{2}\right]\psi_n$$

$$= (n+1)\hbar\omega\psi_n$$

Adding a constant of $\hbar\omega/2$ to both sides, we can write the following relation:

$$\left[\widehat{AA^\dagger} + \frac{\hbar\omega}{2}\right]\psi_n = \hbar\omega\left[(n+1) + \frac{1}{2}\right]\psi_n \qquad \text{(A.22)}$$

A similar analysis for $\widehat{A}^{\dagger}\widehat{A}$, instead, gives

$$\begin{aligned}\widehat{A}^{\dagger}\widehat{A} &= \frac{1}{2m}\left[m^2\omega^2x^2 - \hbar^2\frac{\partial^2}{\partial x^2} + im\omega\left(x\widehat{p}_x - \widehat{p}_x x\right)\right] \\ &= \widehat{H} - \frac{\hbar\omega}{2} \\ &= \hbar\omega\left(n + \frac{1}{2}\right) - \frac{\hbar\omega}{2} \\ &= \hbar\omega n\end{aligned}$$

In this case, subtracting a constant from both sides, we can write

$$\left[\widehat{A}^{\dagger}\widehat{A} - \frac{\hbar\omega}{2}\right]\psi = \hbar\omega\left[(n-1) + \frac{1}{2}\right]\psi \tag{A.23}$$

The above analysis shows that the operators $\widehat{A}\widehat{A}^{\dagger}$ and $\widehat{A}^{\dagger}\widehat{A}$ tend to shift the $n + 1/2$ term of the eigenvalues. Additionally, we find that changing the order of the operators leads to different results. In this case, changing the order affects the direction in which the integer n effectively changes. For the first case, there is a shift of $+\hbar\omega/2$. Adding and subtracting the constants we have chosen, lead to clean integers on the RHS.

It is more common to define these operators in dimensionless form as $\widehat{a} = \widehat{A}/\sqrt{\hbar\omega}$ and $\widehat{a}^{\dagger} = \widehat{A}^{\dagger}/\sqrt{\hbar\omega}$. In this dimensionless form, we have the following two first-order symmetric operators

$$\widehat{a} = \frac{1}{\sqrt{2m\hbar\omega}}\left[m\omega x + i\widehat{p}_x\right] \tag{A.24a}$$

$$\widehat{a}^{\dagger} = \frac{1}{\sqrt{2m\hbar\omega}}\left[m\omega x - i\widehat{p}_x\right] \tag{A.24b}$$

The dimensionless form leads to the following dimensionless eigenvalue equations, given by

$$\left[\widehat{a}\widehat{a}^{\dagger} + \frac{1}{2}\right]\psi = \left[(n+1) + \frac{1}{2}\right]\psi \tag{A.25a}$$

$$\left[\widehat{a}^{\dagger}\widehat{a} - \frac{1}{2}\right]\psi = \left[(n-1) + \frac{1}{2}\right]\psi \tag{A.25b}$$

Having defined the first order operators $\widehat{a}$ and $\widehat{a}^{\dagger}$, that relate directly to the second order Hamiltonian $\widehat{H}$ for the harmonic oscillator, we are in a position to examine new questions such as what these lower order operators do to an eigenstate $\psi_n(\zeta)$ of the harmonic oscillator. That is, what do we get when we compute $\widehat{a}\psi_n$? Substituting the operator $\widehat{p}_x = -i\hbar\partial/\partial x$, and using the definition of $\widehat{a}$, given by Eq. (A.24a), we have

$$\widehat{a}\psi_n = \frac{1}{\sqrt{2m\hbar\omega}}[m\omega x + i\widehat{p}_x]\psi_n$$

$$= \frac{1}{\sqrt{2}}\left[\sqrt{\frac{m\omega}{\hbar}}x\psi_n + \sqrt{\frac{\hbar}{m\omega}}\frac{\partial\psi_n}{\partial x}\right]$$

Using the earlier definition of $\alpha = \sqrt{m\omega/\hbar}$ and $\zeta = \alpha x$, the above becomes

$$\widehat{a}\psi_n = \frac{1}{\sqrt{2}}\left(\zeta\psi_n + \frac{\partial\psi_n}{\partial\zeta}\right) \tag{A.26}$$

We can evaluate these terms explicitly using the exact solution for ψ_n given by Eq. (3.86). First, for the term $\partial\psi_n/\partial\zeta$, we have

$$\frac{\partial\psi_n}{\partial\zeta} = C_n\left[-\zeta e^{-\frac{\zeta^2}{2}}H_n(\zeta) + e^{-\frac{\zeta^2}{2}}\frac{\partial H_n}{\partial\zeta}\right] \tag{A.27}$$

Now, we make use of another special property of Hermite polynomials known as a recurrence relation, namely

Hermite Polynomial Derivative Relation

$$\frac{dH_n}{d\zeta} = 2nH_{n-1}(\zeta) \tag{A.28}$$

Substituting this relation into Eq. (A.27) gives

$$\frac{\partial\psi_n}{\partial\zeta} = C_n\left[-\zeta e^{-\frac{\zeta^2}{2}}H_n(\zeta) + 2ne^{-\frac{\zeta^2}{2}}H_{n-1}(\zeta)\right] \tag{A.29}$$

Then, substituting this result into Eq. (A.37), we obtain

$$\widehat{a}\psi_n = \frac{C_n}{\sqrt{2}}\left(\zeta e^{-\frac{\zeta^2}{2}}H_n(\zeta) + -\zeta e^{-\frac{\zeta^2}{2}}H_n(\zeta) + 2ne^{-\frac{\zeta^2}{2}}H_{n-1}(\zeta)\right)$$

$$= \frac{2nC_n}{\sqrt{2}}e^{-\frac{\zeta^2}{2}}H_{n-1}(\zeta)$$

Using the definition for the normalization constant $C_n = \sqrt{\alpha/2^n n!\sqrt{\pi}}$, we can simplify further as follows:

$$\begin{aligned}
\widehat{a}\psi_n &= \sqrt{\frac{\alpha}{2^n n!}}\sqrt{2}ne^{-\frac{\zeta^2}{2}}H_{n-1}(\zeta) \\
&= \sqrt{n}\sqrt{\frac{2n\alpha}{2^n n!}}e^{-\frac{\zeta^2}{2}}H_{n-1}(\zeta) \\
&= \sqrt{n}\sqrt{\frac{\alpha}{2^{n-1}(n-1)!}}\sqrt{2}ne^{-\frac{\zeta^2}{2}}H_{n-1}(\zeta) \\
&= \sqrt{n}C_{n-1}e^{-\frac{\zeta^2}{2}}H_{n-1}(\zeta)
\end{aligned}$$

Therefore, we obtain the following result for $\widehat{a}\psi_n$:

Harmonic Oscillator Annihilation Operator Relation

$$\widehat{a}\psi_n = \sqrt{n}\psi_{n-1} \tag{A.30}$$

The first order operator $\widehat{a}$ takes any eigensolution ψ_n and returns a state proportional to the eigensolution ψ_{n-1}. Because of this property of effectively reducing the energy from n to $n-1$, $\widehat{a}$ earns the name of being the **annihilation operator**, and is also an example of a **lowering operator**.

A similar analysis can be done for $\widehat{a}^\dagger\psi_n(\zeta)$, which gives

$$\begin{aligned}
\widehat{a}^\dagger\psi_n &= \frac{1}{\sqrt{2m\hbar\omega}}[m\omega x - i\widehat{p}_x]\psi_n \\
&= \frac{1}{\sqrt{2}}\left[\sqrt{\frac{m\omega}{\hbar}}x\psi_n - \sqrt{\frac{\hbar}{m\omega}}\frac{\partial\psi_n}{\partial x}\right]
\end{aligned}$$

In terms of $\alpha = \sqrt{m\omega/\hbar}$ and the dimensionless parameter $\zeta = \alpha x$, we now have

$$\widehat{a}^\dagger\psi_n = \frac{1}{\sqrt{2}}\left(\zeta\psi_n - \frac{\partial\psi_n}{\partial\zeta}\right) \tag{A.31}$$

Then, after substitution of the result for $\partial\psi_n/\partial x$ given by Eq. (A.27), we get

$$\hat{a}^{\dagger}\psi_n = \frac{C_n}{\sqrt{2}}\left(2\zeta e^{-\frac{\zeta^2}{2}}H_n(\zeta) - 2ne^{-\frac{\zeta^2}{2}}H_{n-1}(\zeta)\right) \tag{A.32}$$

In the case of $\hat{a}^{\dagger}$, we make use of a different *recurrence relation* for Hermite polynomials, namely

Hermite Polynomial Recurrence Relation

$$H_{n+1}(\zeta) = 2\zeta H_n(\zeta) - 2nH_{n-1}(\zeta) \tag{A.33}$$

Substitution of this relation into Eq. (A.32) gives

$$\hat{a}^{\dagger}\psi_n = \frac{C_n}{\sqrt{2}}e^{-\frac{\zeta^2}{2}}H_{n+1}(\zeta) \tag{A.34}$$

Using the definition for the normalization constant C_n, we can simplify further as

$$\begin{aligned}\hat{a}^{\dagger}\psi_n &= \frac{\sqrt{n+1}\cdot C_n}{\sqrt{n+1}\sqrt{2}}e^{-\frac{\zeta^2}{2}}H_{n+1}(\zeta)\\ &= \sqrt{n+1}\sqrt{\frac{\alpha}{2^{n+1}(n+1)!}}e^{-\frac{\zeta^2}{2}}H_{n-1}(\zeta)\\ &= \sqrt{n+1}C_{n+1}e^{-\frac{\zeta^2}{2}}H_{n+1}(\zeta)\end{aligned}$$

Thus, we obtain the following result for $\hat{a}^{\dagger}\psi_n$:

Harmonic Oscillator Creation Operator Relation

$$\hat{a}^{\dagger}\psi_n = \sqrt{n+1}\psi_{n+1} \tag{A.35}$$

In this case, $\hat{a}^{\dagger}$ takes any eigensolution ψ_n and returns a state proportional to the eigensolution ψ_{n+1}, effectively increasing the energy from n to $n+1$. Hence, $\hat{a}^{\dagger}$ is known as the **creation operator**, and is an example of a **raising operator**. We can see that the unique properties that we have found for the eigensolutions of the harmonic oscillator are due, entirely, to the unique properties of the Hermite polynomials. It is straight forward to show, with Eqs. (A.30) and (A.35), that the following eigenvalues for $\widehat{a}\widehat{a}^{\dagger}$ and $\widehat{a}^{\dagger}\widehat{a}$ are given by

$$\widehat{a}\widehat{a}^{\dagger}\psi_{n-1} = n\psi_{n-1} \tag{A.36a}$$

$$\widehat{a}^{\dagger}\widehat{a}\psi_{n+1} = (n+1)\,\psi_{n+1} \tag{A.36b}$$

The first of these, given by Eq. (A.36a) also means that the eigenvalue of $\widehat{a}\widehat{a}^{\dagger}$ is the quantum number n.

A.4 Introduction To The Coherent State

One question that we have not yet answered is what are the general eigenvalues λ_a and eigenfunctions ψ_a of $\widehat{a}$ itself. Due to the form of $\widehat{a}$ and $\widehat{a}^{\dagger}$, we expect that the eigenvalues of $\widehat{a}^{\dagger}$ are complex conjugates of λ_a. This answer to this question defines the so-called **coherent state** of the harmonic oscillator. Thus, the *coherent state* is defined by the following eigenvalue problem:

Coherent State Definition

$$\widehat{a}\psi_a = \lambda_a\psi_a \tag{A.37}$$

Using the operator definition for $\widehat{a}$, Eq. (A.37) becomes

$$\frac{1}{\sqrt{2}}\left(\zeta\psi + \frac{\partial\psi}{\partial\zeta}\right) = \lambda_a\psi \Rightarrow \frac{\partial\psi}{\partial\zeta} = \left(\sqrt{2}\lambda_a - \zeta\right)\psi$$

The above result leads to the following equation relating differentials:

$$\frac{d\psi}{\psi} = \left(\sqrt{2}\lambda_a - \zeta\right)d\zeta \tag{A.38}$$

Generally, $\lambda_a(\zeta(t), p_x(t))$ is a function of ζ and the expected value for the momentum p_x. To get past this difficulty, let us only consider $\lambda_a(t=0) = \lambda$, or *the eigenvalue associated with the initial condition*. We can further assume that the initial condition for the momentum is $\widehat{p}_x = 0$. This simply means that we are starting from rest. Under these conditions for λ_a, we have

$$\lambda_a(t=0) = \frac{1}{\sqrt{2m\hbar\omega}}[m\omega x + i\widehat{p}_x]$$
$$= \frac{1}{\sqrt{2m\hbar\omega}} m\omega x$$
$$= \frac{\zeta_0}{\sqrt{2}}$$

We find that the initial condition of the eigenvalue λ_a for the coherent state is given by

Coherent State Eigenvalue Initial Condition

$$\lambda_a(0) = \frac{\zeta_0}{\sqrt{2}} \tag{A.39}$$

Using this initial eigenvalue, and integrating Eq. (A.38) leads to a solution corresponding to the initial condition, given by

$$\psi_a(\zeta,0) = C_n e^{-\frac{\zeta^2}{2}+\sqrt{2}\lambda_a\zeta}$$
$$= C_n e^{-\frac{\zeta^2}{2}+\zeta_0\zeta}$$

To facilitate the normalization of $\psi_a(\zeta,0)$, let us write the result for $\psi_a(\zeta)$ as

$$\psi_a(\zeta,0) = C_n e^{-\frac{1}{2}\left(\zeta^2-2\zeta_0\zeta+\zeta_0^2\right)} e^{\frac{\zeta_0^2}{2}}$$
$$= C_n e^{\frac{\zeta_0^2}{2}} e^{-\frac{1}{2}(\zeta-\zeta_0)^2}$$

We determine the normalization constant C_n or $|C_n|$ using the normalization condition, which gives

$$1 = \int_{-\infty}^{+\infty} \psi_a^\dagger(\zeta)\psi_a(\zeta)dx$$
$$= \frac{C_n^2}{\alpha} e^{\zeta_0^2} \int_{-\infty}^{+\infty} e^{-(\zeta-\zeta_0)^2} d\zeta$$

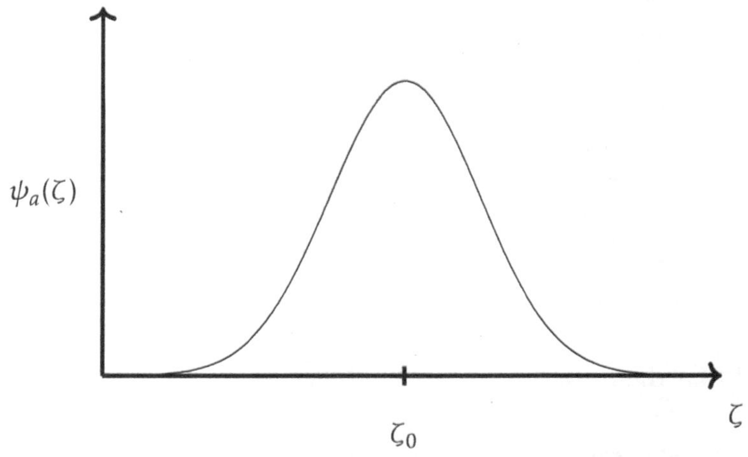

Figure A.3: The harmonic oscillator coherent state, $\psi_a(\zeta)$ is a Gaussian distribution centered at the initial dimensionless position ζ_0.

$$= C_n^2 \frac{\sqrt{\pi}}{\alpha} e^{\zeta_0^2}$$

Solving for C_n, we obtain

$$C_n = \sqrt{\frac{\alpha}{\sqrt{\pi}}} e^{-\frac{\zeta_0^2}{2}} \tag{A.40}$$

Substitution of this result into the above expression for $\psi_a(\zeta, 0)$ leads to our final form for the initial condition of the eigenstate of $\widehat{a}$, given by

Harmonic Oscillator Coherent State

$$\psi_a(\zeta, 0) = \sqrt{\frac{\alpha}{\sqrt{\pi}}} e^{-\frac{1}{2}(\zeta - \zeta_0)^2} \tag{A.41}$$

The coherent state given in Eq. (A.41) is illustrated in Fig. A.3.

$\psi_a(\zeta)$ is a Gaussian distribution centered at the initial dimensionless position $\zeta_0 = \alpha x_0$.

The coherent state of the harmonic oscillator is an important quantum state, and most of the direct evidence of this statement is given in Chap.

C, using a more concise notation known as *Bra-Ket* notation. There, the coherent state will be shown to be the most general form of the initial state of the oscillator demonstrating the following:

1. The *coherent state* has an identical eigenvalue (given by Eq. (A.39)) to the most general eigenstate for $\widehat{H}$, at time $t = 0$.

2. The expectation value of $x(t)$ using $\psi_a(\zeta)$ leads to the classical solution for the harmonic oscillator, therefore, establishing agreement with the *principle of correspondence.*

3. The first two conditions prove that $\psi_a(\zeta, 0) = \Psi(\zeta, 0)$, or is the initial state of the harmonic oscillator.

The second point is somewhat similar to our finding that *the most general solution* for a free elementary particle through a slit leads to the physical phenomenon of diffraction, as well as interference. We found this in the double slit problem. The most general form of the solution, in both cases, turns out to be necessary in order to be in agreement with the *principle of correspondence.* These points will be discussed further in Chap. C. Let's extend our results for the 1D harmonic oscillator to the domain of 3D.

A.5 Extending To 3D Harmonic Oscillator

Extending the harmonic oscillator to higher dimensions assumes the potential energy is dependent on all three spatial coordinates x, y, and z. Therefore, the potential energy $V(\mathbf{r})$ now has the form

$$V(\mathbf{r}) = \frac{1}{2}m\omega^2 r^2 = \frac{1}{2}m\omega^2(x^2 + y^2 + z^2) \tag{A.42}$$

The spatial part of the 3D Schrödinger equation is then given by

$$\widehat{H} = \nabla^2\psi + \frac{2m}{\hbar^2}\left[E_\lambda - V(\mathbf{r})\right]\psi = 0 \tag{A.43}$$

Substituting Eq. (A.42) into Eq. (A.43) and expanding, we have

$$\left(\frac{\partial^2\psi}{\partial x^2} + \frac{\partial^2\psi}{\partial y^2} + \frac{\partial^2\psi}{\partial z^2}\right) + \frac{2m}{\hbar^2}\left[E_\lambda - \frac{1}{2}m\omega^2(x^2 + y^2 + z^2)\right]\psi = 0 \tag{A.44}$$

Using the *method of separation of variables* again, we assume the wave function has the form

$$\psi(\mathbf{r}) = X(x)Y(y)Z(z) \tag{A.45}$$

Substitution of Eq. (A.45) into Eq. (A.44), then dividing by ψ leads to the following:

$$\left(\frac{1}{X}\frac{d^2X}{dx^2} - \frac{m^2\omega^2}{\hbar^2}x^2\right) + \left(\frac{1}{Y}\frac{d^2Y}{dy^2} - \frac{m^2\omega^2}{\hbar^2}y^2\right) + \left(\frac{1}{Z}\frac{d^2Z}{dz^2} - \frac{m^2\omega^2}{\hbar^2}z^2\right) = -\frac{2mE_\lambda}{\hbar^2} = -k^2 \tag{A.46}$$

We have an equation that sums a function of x, a function of y, and a function of z, and they sum to equal a constant. Since they are all functions of independent variables x, y, and z, a necessary condition is that each function, separately, must equal to a constant. Thus, we have

$$\frac{1}{X}\frac{d^2X}{dx^2} - \frac{m^2\omega^2}{\hbar^2}x^2 = -k_x^2 \tag{A.47a}$$

$$\frac{1}{Y}\frac{d^2Y}{dy^2} - \frac{m^2\omega^2}{\hbar^2}y^2 = -k_y^2 \tag{A.47b}$$

$$\frac{1}{Z}\frac{d^2Z}{dz^2} - \frac{m^2\omega^2}{\hbar^2}z^2 = -k_z^2 \tag{A.47c}$$

We can define k^2 in terms of three x, y, and z components where we have

$$k_x^2 + k_y^2 + k_z^2 = k^2 \tag{A.47d}$$

Eqs. (A.47a), (A.47b), and (A.47c) have the same form as what we have seen in the 1D harmonic oscillator problem, where the corresponding 1D equation was given by Eq. (A.14b). Since we have worked out the solution to that equation (using the method of Frobenius), given in Eq. (**??**), X, Y, and Z each take a corresponding form, based on the 1D solution. The 3D solution $\psi = X(x)Y(y)Z(z)T(t)$ becomes

$$\psi_{n_x,n_y,n_z} = C_{n_x}C_{n_y}C_{n_z}H_{n_x}(\zeta_1)H_{n_y}(\zeta_2)H_{n_z}(\zeta_3)e^{-\frac{1}{2}(\zeta_1^2+\zeta_2^2+\zeta_3^2)}e^{-i\frac{E_\lambda}{\hbar}(t-t_0)} \tag{A.48}$$

Eq. (A.48) can also be written as

3D Harmonic Oscillator Eigensolutions

$$\psi_{n_x,n_y,n_z} = C_{n_x,n_y,n_z} H_{n_x}(\alpha x) H_{n_y}(\alpha y) H_{n_z}(\alpha z) e^{-\frac{\alpha^2}{2}\mathbf{r}^2} e^{-i\frac{E_\lambda}{\hbar}(t-t_0)} \quad (A.49)$$

The parameter α is identical to the value in the 1D problem where $\alpha = \sqrt{m\omega/\hbar}$. To determine the normalization constant for the 3D harmonic oscillator, we use the same property of the Hermite polynomials, given by Eq. (3.87). In the case of the 3D oscillator, we end up with the product of three integrals, one in x, one in y, and one in z. Then, we can use Eq. (3.87) for each integral separately since they have identical form. This leads to

$$1 = \frac{C_{nx}^2}{\alpha}\left(2^{n_x} n_x! \sqrt{\pi}\right) \cdot \frac{C_{ny}^2}{\alpha}\left(2^{n_y} n_y! \sqrt{\pi}\right) \cdot \frac{C_{nz}^2}{\alpha}\left(2^{n_z} n_z! \sqrt{\pi}\right) \quad (A.50)$$

We can define the 3D normalization constant as

$$C_{n_x,n_y,n_z}^2 = C_{nx}^2 \cdot C_{ny}^2 \cdot C_{nz}^2 \quad (A.51)$$

Then, solving for C_{n_x,n_y,n_z}, we get

3D Harmonic Oscillator Normalization Constant

$$C_{n_x,n_y,n_z} = \sqrt{\frac{\alpha^3}{\pi^{3/2} \cdot 2^{n_x+n_y+n_z} \cdot n_x! n_y! n_z!}} \quad (A.52)$$

Also, using Eq. (A.47d), it follows that

$$k_x^2 + k_y^2 + k_z^2 = \frac{2mE_{\lambda x}}{\hbar^2} + \frac{2mE_{\lambda y}}{\hbar^2} + \frac{2mE_{\lambda z}}{\hbar^2} = \frac{2mE_\lambda}{\hbar^2} \quad (A.53)$$

Using the solution for the energy for the 1D problem, given in Eq. (3.75), the 3D energy solution becomes

$$E_\lambda = \left(n_x + \frac{1}{2}\right)\hbar\omega + \left(n_y + \frac{1}{2}\right)\hbar\omega + \left(n_z + \frac{1}{2}\right)\hbar\omega \quad (A.54)$$

Summing the terms of Eq. (A.54) and rearranging, the 3D eigenenergy $E_\lambda = E_{n_x,n_y,n_z}$ becomes

3D Quantum Mechanical Harmonic Oscillator Energy

$$E_{n_x,n_y,n_z} = \left(n_x + n_y + n_z + \frac{3}{2}\right)\hbar\omega \tag{A.55}$$

As in the case of the 1D harmonic oscillator problem, the lowest energy state or *ground state* is also nonzero, given by $(3/2)\hbar\omega$. We see that the wave function for the 3D harmonic oscillator is essentially the product of the 1D wave functions, while the energy is the sum. Having this pattern from $1D$ to $3D$, facilitates expressing the solutions for the $2D$ case, as well. Note that unlike in the 1D case, there is *degeneracy* in any oscillator in a dimension greater than 1D, because n_x+n_y and $n_x+n_y+n_z$ can be satisfied by several unique combinations, depending on the quantum numbers. For example, the energy E_λ for the quantum state (n_x, n_y, n_z) =(2,5,7) is the same as that for the quantum state (10,2,2), since both combinations sum to 14. Any other combination summing to 14 also has the same eigenenergy. The number of degenerate states grows rapidly as these integers grow. Although their energies are the same, their wave functions can differ with respect to space.

A.6 Chapter Summary

In this chapter, we have discussed the problem of the harmonic oscillator. The classical oscillator was first reviewed to provide a reference frame, to which the quantum mechanical results can be compared. Using the *Method of Frobenius*, we found the exact eigensolutions for the harmonic oscillator $\psi_n(\zeta,t)$, which involves the Hermite Polynomials $H_n(\zeta)$. The special properties of the Hermite polynomials were found to lead to a number of interesting characteristics for the eigensolutions ψ_n.

Then, exploiting symmetry in the Hamiltonian, we introduced the first order *annihilation* and *creation operators* for the oscillator and used them to define and solve for the *coherent state*. The coherent state is the most general initial condition for the harmonic oscillator, but we had to defer some additional details involving the coherent state, when this work can be done more concisely. Lastly, we extended the 1D result for the eigensolutions to 3D. Table A.1 lists some of the key results from this chapter.

Table A.1: Chapter A Summary Equations. $\zeta = \alpha \cdot x = \sqrt{m\omega/\hbar} \cdot x$; $\zeta_0 = \alpha \cdot x_0 = \sqrt{m\omega/\hbar} \cdot x_0$; m is mass, ω is the natural frequency, x_0 is the initial position.

Name	Equation
1D Oscillator Energy States	$E_\lambda = \left(n + \frac{1}{2}\right)\hbar\omega$
Eigenstates 1D harmonic oscillator	$\psi_n(\zeta, t) = C_n H_n(\zeta) e^{-\frac{1}{2}\zeta^2} e^{-i\frac{E_\lambda}{\hbar}(t-t_0)}$
1D Normalization Constant	$C_n = \sqrt{\frac{\alpha}{2^n n! \sqrt{\pi}}}$
Position-Momentum Commutation	$x\widehat{p}_x - \widehat{p}_x x = i\hbar$
Annihilation Operator	$\widehat{a} = \frac{1}{\sqrt{2m\hbar\omega}}[m\omega x + i\widehat{p}_x]$
Creation Operator	$\widehat{a} = \frac{1}{\sqrt{2m\hbar\omega}}[m\omega x - i\widehat{p}_x]$
Annihilation Operator Relation	$\widehat{a}\psi_n = \sqrt{n}\psi_{n-1}$
Creation Operator Relation	$\widehat{a}^{\dagger}\psi_n = \sqrt{n+1}\psi_{n+1}$
Oscillator Coherent State	$\psi_a(\zeta, 0) = \sqrt{\frac{\alpha}{\sqrt{\pi}}} e^{-\frac{1}{2}(\zeta - \zeta_0)^2}$
3D Oscillator Energy States	$E_\lambda = \left(n_x + n_y + n_z + \frac{3}{2}\right)\hbar\omega$
3D Normalization Constant	$C_{n_x,n_y,n_z} = \sqrt{\frac{\alpha^3}{(\pi^{3/2}) 2^{(n_x+n_y+n_z)} n_x! n_y! n_z!}}$

n	Hermite Polynomial H_n
0	$H_0(\zeta) = 1$
1	$H_1(\zeta) = 2\zeta$
2	$H_2(\zeta) = 4\zeta^2 - 2$
3	$H_3(\zeta) = 8\zeta^3 - 12\zeta$
4	$H_4(\zeta) = 16\zeta^4 - 48\zeta^2 + 12$
5	$H_5(\zeta) = 32\zeta^5 - 160\zeta^3 + 120\zeta$
6	$H_6(\zeta) = 64\zeta^6 - 480\zeta^4 + 720\zeta^2 - 120$
7	$H_7(\zeta) = 128\zeta^7 - 1344\zeta^5 + 3360\zeta^3 - 1680\zeta$
8	$H_8(\zeta) = 256\zeta^8 - 3584\zeta^6 + 13440\zeta^4 - 13440\zeta^2 + 1680$
9	$H_9(\zeta) = 512\zeta^9 - 9216\zeta^7 + 48384\zeta^5 - 80640\zeta^3 + 30240\zeta$
10	$H_{10}(\zeta) = 1024\zeta^{10} - 23040\zeta^8 + 161280\zeta^6 - 403200\zeta^4 + 302400\zeta^2 - 30240$
11	$H_{11}(\zeta) = 2048\zeta^{11} - 56320\zeta^9 + 506880\zeta^7 - 1774080\zeta^5 + 2217600\zeta^3 - 665280\zeta$
12	$H_{12}(\zeta) = 4096\zeta^{12} - 135168\zeta^1 0 + 1520640\zeta^8 - 7096320\zeta^6 + 13305600\zeta^4 - 7983360\zeta^2 + 665280$
13	$H_{13}(\zeta) = 8192\zeta^{13} - 319488\zeta^1 1 + 4392960\zeta^9 - 26357760\zeta^7 + 69189120\zeta^5 - 7983360\zeta^2 + 17297280\zeta$

Table A.2: List of First 14 Hermite Polynomials.

A.7 Chapter Problems

Problem A.1 The first two eigenstates for the harmonic oscillatcr can be written as

$$\psi_1 = \left(\frac{\alpha}{\pi}\right)^{1/4} e^{-\alpha x^2/2} \quad \text{and} \quad \psi_2 = \left(\frac{4\alpha^3}{\pi}\right)^{1/4} x e^{-\alpha x^2/2}$$

Show that these two states are orthogonal, or that

$$\int_{-\infty}^{\infty} \psi_1^* \psi_2 dx = 0$$

Problem A.2 Show that the harmonic oscillator solution $x(t) = A\sin\omega t + A\cos\omega t$, where $\omega = \sqrt{k/m}$, is also a solution to Newton's equation $F = m\dot{x}$ for the harmonic oscillator.

Problem A.3 Show that for the harmonic oscillator, the expectation of x is zero, and that the expectation $\langle x^2 \rangle$ for $n = 2$ is given by

$$\langle x^2 \rangle = \frac{5}{2} \frac{\hbar}{(mk)^{1/2}}$$

Problem A.4 Show that the expectation of p is zero, and the squared momentum $\langle p^2 \rangle$ for $n = 2$ is given by

$$\langle p^2 \rangle = \frac{5}{2} \hbar (mk)^{1/2}$$

Problem A.5 We know the Hamiltonian operator $\widehat{H}$ is given by

$$\widehat{H} = -\frac{\hbar^2}{2m} \nabla^2 + V(x, y, z)$$

Prove that for any eigensolution ψ of $\widehat{H}$, and for any linear operator $\widehat{O}$ that

$$\int_V \psi^* \left(\widehat{H}\widehat{O} - \widehat{O}\widehat{H}\right) \psi dV = 0$$

Problem A.6 Using the above result from the previous problem, define the linear operator $\widehat{O}$ to be

$$\widehat{O} = -i\hbar \left(x\frac{\partial}{\partial x} + y\frac{\partial}{\partial y} + z\frac{\partial}{\partial z} \right)$$

Show that

$$\widehat{H}\widehat{O} - \widehat{O}\widehat{H} = i\hbar\left(x\frac{\partial}{\partial x} + y\frac{\partial}{\partial y} + z\frac{\partial}{\partial z}\right) - 2i\hbar\widehat{K}$$

$\widehat{K}$ is the kinetic energy operator.

Problem A.7 The harmonic oscillator solution can be written as

$$x(t) = A\sin(\omega t + \phi)$$

Solving for t, one obtains

$$t = \frac{1}{\omega}\sin^{-1}\left(\frac{x}{A}\right) - \phi$$

Use both of these equations, differentiating the equation of t to prove that

$$\frac{dx}{dt} = \omega\sqrt{A^2 - x^2} = A\cos(\omega t + \phi)$$

Now, use the relation

$$p(x)dx = \frac{dx}{\pi\sqrt{A^2 - x^2}}$$

with the above to show that $p(x)$ satisfies the probability distribution condition given by

$$1 = \frac{\omega}{\pi}\int_0^{\pi/\omega} dt = \int p(x)dx$$

A.8 Suggested Readings & References

[1] A. Ghatak and S. Lokanathan, *Quantum Mechanics: Theory and Applications*, Springer-Science (2004)

APPENDIX B

Appendix: Orbital Angular Momentum

There are many applications today that exploit the electron's or the atom's net angular momentum. An example is found in magnetic materials, which are used in everything from refrigerator magnets, to microwaves, to providing levitation to the world's fastest trains. A more sophisticated example we have seen is in *spintronics*, *e.g.* spin-Hall effect. Because angular momentum is such an important concept in physical systems, we discuss the topic of *localized orbital angular momentum* and some related topics. Understanding angular momentum in quantum mechanics, even for a single electron, goes along way towards understanding the properties of a large number of electrons. We must also expect a consistency between a quantum mechanical system and the larger ensemble of a classical system. This is a fundamental concept in quantum mechanics known as the *Principle of Correspondence*[1]. In this appendix, we'll emphasize this principle, as well. Moreover, several of the operators used in the core of the book are developed here.

To introduce *quantum mechanical* orbital angular momentum, we will also classical and semi-classical orbital angular momentum concepts.

[1]The PoC specifies the expectation that a quantum mechanical system converges to its corresponding classical one upon taking the limit towards infinite quantum numbers, and/or infinite numbers of elementary particles.

For the quantum mechanical analogs, we must cover some of the historical mathematical tools and their origins to help find exact solutions to our problems involving orbital angular momentum. These tools include *Legendre polynomials* and *functions*. So, bear with us in introducing these topics as thoroughly as we can. However they are nontrivial, while providing incredible insight. Taking the time to learn them fortunately makes the quantum mechanical solutions for orbital angular momentum states easier to find. Dynamical transitions between orbital angular momentum quantum states is much more straightforward, but is delayed until Chap. , after discussing other prerequisite concepts. So, let us begin by reviewing some *classical* concepts from classical and semi-classical orbital angular momentum, then build from there.

B.1 Classical Orbital Angular Momentum Relations

In classical physics, recall the *non-relativistic* kinetic energy K (c.Eq.1.86) of any mass m with motion of speed $v = |\mathbf{v}|$, is given by

$$K = \frac{1}{2}m\mathbf{v}^2 = \frac{\mathbf{p}^2}{2m} \tag{B.1}$$

For a rotating mass on a perfectly circular (planar) path, the instantaneous linear speed is $v = r\omega$, where the angular velocity $\omega = d\theta/dt =$, as illustrated in Fig. B.1. The kinetic energy for a rotating body then becomes

$$K = \frac{1}{2}mv^2 = \frac{1}{2}mr^2\omega^2 = \frac{1}{2}I\omega^2 \tag{B.2}$$

If we define the *rotational inertia* parameter I as

$$I = mr^2$$

then, the kinetic energy takes the more concise form given by

$$K = \frac{1}{2}I\omega^2 \tag{B.3}$$

Notice that from Eqs. (B.2) and (B.3), we find analogous relationships between the **rotational inertia** I and **translational inertia** m. Likewise, there is a correspondence between the linear velocity v and the angular velocity ω. Using such relations, we can formally define a *classical* quantity known as the **angular momentum**, denoted by **L**, analogous to the linear momentum **p**. It is given by

$$L = I\omega \Leftrightarrow p = mv \quad \text{(B.4)}$$

Using Eq. (B.50), the kinetic energy K due to orbital motion becomes

$$K = \frac{1}{2}I\omega^2 = \frac{L^2}{2I} \quad \text{(B.5)}$$

Since rotational motion involves changes to the velocity *vector*, there is

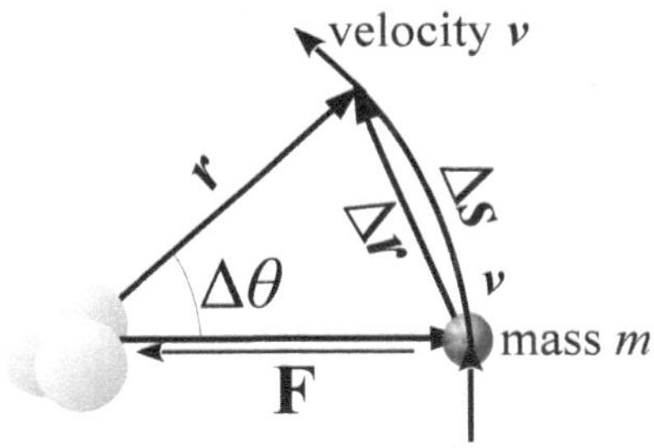

Figure B.1: Illustration of angular momentum L in circular motion. There is a correspondence between the translational degrees of freedom and rotational, summarized in Table B.1.

acceleration $\mathbf{a} = d\mathbf{v}/dt$. As a result, there must be some forces involved arising from a change in momentum. When a mass m moves in circular motion, it can be shown that there is a radial force **F** acting on the mass keeping it from going along a straight (tangential) trajectory. To see this, considering a mass m rotating in a circular orbit, noting that in taking the limits toward δr, δs, and $\delta\theta$ approaching zero, the change in the radial vector r approaches the change in the arc length s. Therefore, in this limit, $|dr| = |ds| = |rd\theta|$. Then, circular velocity is

$$\mathbf{v} = \frac{d\mathbf{s}}{dt} = \mathbf{r}\frac{d\theta}{dt} = \mathbf{r}\omega \tag{B.6}$$

The derivative of the velocity $\mathbf{v}$ can be a little non-intuitive because one might look at the equation above and conclude that since $|\mathbf{r}| = r$ and ω are constants, there will be no acceleration. However, *there will be a nonzero acceleration* due to a changing orientation of the velocity vector $\mathbf{v}$. Since we are dealing with vector quantities, defined by *both* magnitude and direction, for a vector *not* to change, it must be that neither the direction nor magnitude changes. In circular motion, though the magnitude of the position vector does *not* change, its direction does. Therefore, there is always a nonzero variation in $\mathbf{r}$ along the tangential direction. Thus, for the same reason, there is a change in $\mathbf{v}$ arising from its rotation. $\Delta\mathbf{v}$ is along the radial direction for a perfectly circular path. This idea is illustrated in Fig. B.2. The nonzero magnitude of this variation gives

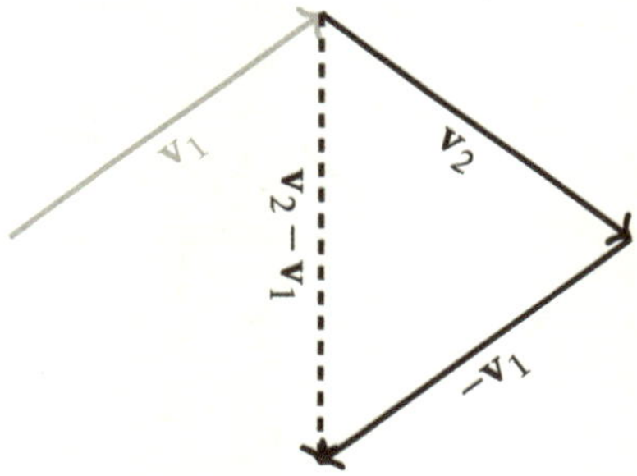

Figure B.2: Illustration of a vector change for two tangential velocity vectors $\mathbf{v}_1$ and $\mathbf{v}_2$. Note that a perfect radial direction follows from *perfect circular motion*. Any departure from perfect circular motion introduces additional vector components.

nonzero acceleration $\mathbf{a}$. For perfect circular motion, we have

$$\mathbf{a} = \frac{d\mathbf{v}}{dt} = \frac{d}{dt}(\mathbf{r}\omega) = \frac{d\mathbf{r}}{dt}\omega = \frac{d\mathbf{s}}{dt}\omega = \mathbf{v}\omega \tag{B.7}$$

The above can be used to obtain the resulting force responsible for the circular motion of mass m. We know from Newton's second law that

$$|\mathbf{F}| = \left|\frac{d\mathbf{p}}{dt}\right| = m|\mathbf{a}| = mv\omega = mr\omega^2 = \frac{mv^2}{r} \tag{B.8}$$

This force is known as a **centripetal force**, and it is present in any system with orbital angular momentum. Perhaps, this can be stated another way.

mv^2/r is not actually the force itself, but the resulting motion from the *centripetal force*, acting centrally and keeping the particle from ordinarily traveling along a straight tangential path.

With this, we can find an analog of $\mathbf{F} = m d\mathbf{v}/dt$ for a system with circular orbital angular momentum $\mathbf{L}$. Since $\mathbf{L} = \mathbf{r} \times \mathbf{p}$, taking the time derivative of $\mathbf{L}$, we get

$$\frac{d\mathbf{L}}{dt} = \frac{d\mathbf{r}}{dt} \times \mathbf{p} + \mathbf{r} \times \frac{d\mathbf{p}}{dt} \tag{B.9}$$

Recall that we have already established that $\mathbf{v} = d\mathbf{s}/dt = d\mathbf{r}/dt$, and since $\mathbf{p} = m\mathbf{v}$, the first term in Eq. (B.9) is zero because it involves the cross-product of two parallel vectors ($\mathbf{v} \times \mathbf{v}$). Therefore, we have the result

$$\frac{d\mathbf{L}}{dt} = \mathbf{r} \times \frac{d\mathbf{p}}{dt} = \mathbf{r} \times \mathbf{F} = \boldsymbol{\tau} \tag{B.10}$$

From this, we have that

$$\mathbf{F} = \frac{d\mathbf{p}}{dt} \Leftrightarrow \boldsymbol{\tau} = \frac{d\mathbf{L}}{dt} \tag{B.11}$$

In linear motion due to forces, the time-derivative of linear momentum $\mathbf{p}$ is the force $\mathbf{F}$. For a system in rotational motion, the time-derivative of angular momentum $\mathbf{L}$ is the *rotational force* or *torque* $\boldsymbol{\tau}$, causing rotational motion. The correspondences between linear and rotational motion discussed so far are summarized in Table B.1. Let's continue the discussion of angular momentum by taking the concept one step further in a specific application to the electron in circular motion in the presence of a magnetic field $\mathbf{B}$.

Translational	Rotational
inertia m	$I = mr^2$
velocity v	ω
momentum p	L
kinetic energy $\frac{1}{2}mv^2$	$\frac{1}{2}I\omega^2$
force $\mathbf{F} = \frac{d\mathbf{p}}{dt}$	$\tau = \frac{d\mathbf{L}}{dt}$

Table B.1: Correspondence between parameters for translational and rotational motion.

B.2 Canonical Problem Of The Magnetic Moment

Here, we define the **canonical** (or basic, standard) **problem** of the *magnetic moment* μ. It provides a uniquely proper framework for some important concepts we shall discuss. Of course, the first concept that must be defined is the *magnetic moment* itself. It has its origins in classical *magnetostatics* (of the Maxwell equations). In physical systems, the magnetic moment vector associated with an electron, for example, eventually aligns itself with any net magnetic fields, particularly, when *they are not initially aligned* with the magnetic moment. This is a result obtained from quantum mechanics, where the electron spin angular momentum has lower energy point away from the field, and thus, the magnetic moment aligns with it.

R *the magnetic moment*: the magnetic moment of an electron, or any negatively charged system, when coupled to a magnetic field, tends to align with the magnetic field, and so opposes the angular momentum vector, which anti-aligns with a magnetic field.

This statement subtly implies the importance of the canonical problem because it suggests the ideal *nonequilibrium initial condition*, which is a condition of *misalignment*. Therefore, when the magnetic moment is initially misaligned with **B**, it will naturally seek its energy minimum. Such a process must generally be described by a dynamical process. It is proper use of these details that will ultimately lead us to a Hamiltonian with dynamical capability. By using the *canonical problem of the magnetic moment*, we can not only lay the classical foundation for the dynamical behavior of a magnetic moment, denoted by $\boldsymbol{\mu}$, but quantum mechanics will also enable us to obtain exact dynamical descriptions when spin is introduced to a magnetic field **B**.

When a particle of charge q moves with velocity $v = |\mathbf{v}|$ in a uniform magnetic field **B**, the field exerts a force on the moving charge. This well-known empirical force is the *Lorentz force*, which can be derived in more than one way. For example, it can be obtained from Faraday's law of electromagnetics or from a relativistic correction to the electrostatic force $\mathbf{F}_e = q\mathbf{E}$. Both lead to the **Lorentz force** given by

$$\mathbf{F} = q\mathbf{v} \times \mathbf{B} \tag{B.12}$$

Based on Eq. (B.12), consider if instead of a point charge q, the charge is distributed uniformly along a circular loop, moving uniformly around the wire loop (e.g. an electrical current I_0), as illustrated in Fig. B.3, illustrating the *initial condition*, specifically, of the magnetic moment $\boldsymbol{\mu}$ subjected to a magnetic field. This condition defines the **canonical magnetic moment problem**, and we'll discuss the reasons for this, shortly, in more detail. With a distributed charge, we may speak of a *differential* Lorentz force $\delta\mathbf{F}$ acting on an elemental length of the circular wire Δs, possessing a *differential* charge δq. Taking the limit $\Delta\mathbf{s} \to d s$, $\delta\mathbf{F}$ becomes

$$\delta\mathbf{F} = \delta q \frac{d\mathbf{s}}{dt} \times \mathbf{B} = \frac{\delta q}{dt} d\mathbf{s} \times \mathbf{B} = I_0 d\mathbf{s} \times \mathbf{B} \tag{B.13}$$

I_0 is the strength of the loop current. Notice that in the extreme top and bottom portions of the loop, the currents move in opposite directions ($\pm d\mathbf{s}$) while the motion is also perpendicular to the same oriented magnetic field **B**, say along the $+z$ axis. At the same time, the components of current parallel/anti-parallel to **B** (on the left and right) do not contribute to the force at all because in this case $d\mathbf{s} \times \mathbf{B} = 0$. From this, we can deduce that there will be *two* Lorentz forces acting on the loop, being focused at the top and bottom, respectively. At the *top of the loop*, the force acts 'out of the diagram, while the *bottom force acts into of the diagram*. According the mechanics of any such forces, this constitutes a *magnetically-induced moment* or torque τ on the loop caused

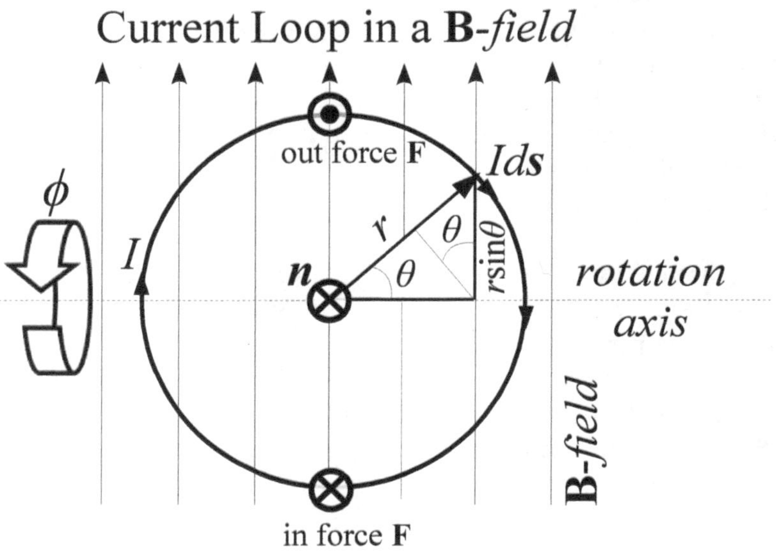

Figure B.3: Illustration of current I_0 circulation in a closed loop, in a uniform magnetic field **B**. Because the loop contains charged particles (in motion), the *Lorentz force* will begin to act on the loop, imparting a torque τ or moment on the loop.

by the two equal and opposite forces at the top and bottom of the loop. Consequently, this will tend to cause the loop to rotate with an angle ϕ, about the rotation axis. Since this force-pair contributes to the same rotational motion, we can write the differential torque $d\boldsymbol{\tau}$ given by the sum of both force-moments acting on top and bottom $d\mathbf{s}$ portions as

$$d\boldsymbol{\tau} = 2 \cdot \mathbf{r}\sin\theta \times \delta\mathbf{F} = 2\mathbf{r}\sin\theta \times I_0 d\mathbf{s} \times \mathbf{B} \tag{B.14}$$

$\mathbf{r}\sin\theta$ is the *moment arm* perpendicular to the axis of rotation, incidentally parallel to the magnetic field axis. Eq. (B.14) is a double cross product involving two distinct angles, namely θ (from the right-most product $ds \times B$) and the angle of the loop rotation ϕ about the rotation axis (from the other cross product). Using the relation $ds = rd\theta$, the differential torque $d\boldsymbol{\tau}$ becomes

$$d\boldsymbol{\tau} = 2\mathbf{r}\sin\theta \times I_0 \mathbf{r} d\theta \times \mathbf{B} = 2I_0 B r^2 \sin^2\theta \sin\phi d\theta \tag{B.15}$$

To obtain the *total* torque τ on the loop, we must integrate Eq. (B.15) with respect to θ from 0 to π (we only need to compute over one half the loop since we include the factor of 2 in the torque from two symmetric forces). This then gives

$$\tau = \int_{\theta=0}^{\theta=\pi} d\tau = 2I_0 Br^2 \sin\phi \int_0^{\pi} \sin^2\theta d\theta \tag{B.16}$$

The integral of $\sin^2\theta$ is given by $\theta/2 - \sin 2\theta/4$. Upon evaluating using the limits of 0 to π gives the result $\int \sin^2\theta d\theta = \pi/2$. Thus, Eq. (B.16) becomes

$$\tau = (\pi r^2) I_0 B \sin\phi \tag{B.17}$$

Since πr^2 is the orbital area A_o of the loop, we can also write Eq. (B.17) as

$$\tau = A_o I_0 B \sin\phi = I_0 (A_o \mathbf{n}) \times \mathbf{B} = I_0 \mathbf{A}_o \times \mathbf{B} \tag{B.18}$$

n is the unit vector normal to the orbital area A_o of the loop. Based on Eq. (B.18), we may now define an **orbital magnetic moment** μ_L, having units of Amps $\times$ m^2 as

Classical Orbital Magnetic Moment

$$\mu_L = I_0 \mathbf{A}_o \tag{B.19}$$

A magnetic moment is also known as a *orbital magnetic dipole*. Using Eq. (B.19), the total torque τ then becomes

Magnetostatic Torque On Orbital Magnetic Moment

$$\tau = \mu_L \times \mathbf{B} \tag{B.20}$$

Note that $\mathbf{A}_o = A_o \mathbf{n}$ is a vector quantity whose magnitude is given by the orbital area $A_o = \pi r^2$, while it points along the unit normal to the orbital area. The convention for determining this direction is to use the right-hand rule with the loop current I. Note that the final result here is also applicable to other closed loop shapes.

Since the current loop evidently has a torque acting on it due to an external magnetic field **B**, it begins to rotate and there is a *magnetic potential energy* V_M associated with the magnetic moment. This additional energy ΔE_0 added to the loop is given by $\int \tau d\phi$, which leads to

$$\Delta E = V_M = \int_{\phi_i}^{\phi_f} \tau d\phi = \int_{\phi_i}^{\phi_f} A_o I_0 B\sin\phi d\phi = -I_0 A_o B\cos\phi\Big|_{\pi/2}^{\phi} \quad \text{(B.21)}$$

Thus, we have the following result for the corresponding magnetic potential energy V_M:

Magnetic Potential Energy Of Magnetic Moment

$$V_M = -\boldsymbol{\mu}_L \cdot \mathbf{B} \quad \text{(B.22)}$$

Eq. (B.21) *uniquely* defines the *initial condition* for the *canonical magnetic moment problem* as

Defined Initial Condition Of Magnetic Moment

$$\phi_i = \frac{\pi}{2} \quad \text{(B.23)}$$

This initial condition, corresponding to Fig. B.3, turns out to be very important for the consideration of dynamical processes. As a *symmetry point,* it enables a crucial ability to treat the corresponding quantum mechanical to completeness where it becomes possible to obtain exact dynamical descriptions, even for nonlinear dynamical processes.

$\Delta E = V_M$ is known as the **magnetic potential energy** and it arises because charged particles in circular motion in a magnetic field **B** experience a torque. This particular concept of magnetic potential energy turns out to be widely used in the description of any magnetic moment $\boldsymbol{\mu}$. We shall not only have occasion to see these relations many times in later chapters, but we will also build on these ideas, later, in order to extend the contributions to the magnetic potential energy. Since the concept involves moving charged bodies orbiting in a loop, there must also be some angular momentum **L** associated with the loop. With both orbital angular momentum **L** and magnetic moment $\boldsymbol{\mu}$, we want to relate the two. This is done of the next section.

B.3 Orbital Magnetic Moment & Angular Momentum

Up to now, we have seen the classical magnetic moment definition $\boldsymbol{\mu}_L$, and we have also seen that the torque is given by the time-derivative of orbital angular momentum or $\boldsymbol{\tau} = d\mathbf{L}/dt$. Now, let us determine the explicit relationship between $\mathbf{L} = \mathbf{r}_o \times \mathbf{p}$ and $\boldsymbol{\mu}_L$. $\mathbf{r} = \mathbf{r}_o$ is the radius of the electron orbit. For a single electron orbiting in a circular loop in a unit time period T, the linear velocity v is given by

$$\mathbf{v} = v\mathbf{e}_\varphi = \frac{2\pi r}{T}\mathbf{e}_\varphi \Rightarrow \frac{1}{T} = \frac{v}{2\pi r} \tag{B.24}$$

$\mathbf{e}_\varphi$ is the unit vector along the loop tangential direction. Using Eq. (B.24), the current I for a single electron charge is

$$\mathbf{I} = -\frac{e}{T}\mathbf{e}_\varphi \Rightarrow I = -\frac{ev}{2\pi r_o} \tag{B.25}$$

$e > 0$ is the *elementary charge* for the proton and electron. Eq. (B.25) tells us that the conventional current I is opposite in direction to the motion of the electron, described by $\mathbf{v}$. Expressing the magnetic moment $\boldsymbol{\mu}_L = IA_o\mathbf{n}$, we then have

$$\mu_L = IA_o = -\frac{ev}{2\pi r_o} \cdot (\pi r_o^2) = -\frac{evr_o}{2}$$

For the more general case that $\mathbf{v}$ is *not perpendicular* to $\mathbf{r}_o$, we may write

$$\boldsymbol{\mu}_L = I\mathbf{A}_o = -\frac{e}{2}\mathbf{r}_o \times \mathbf{v} \tag{B.26}$$

Only the component perpendicular to $\mathbf{r}_o$ contributes to the magnetic moment. Therefore, the magnetic moment becomes

$$\boldsymbol{\mu}_L = -\frac{e}{2}\mathbf{r}_o \times \mathbf{v} = -\frac{e}{2m_e}\mathbf{r}_o \times m_e\mathbf{v} \tag{B.27}$$

Using the definition of $\mathbf{L} = \mathbf{r}_o \times \mathbf{p}$, we have the following relation between $\boldsymbol{\mu}_L$ and $\mathbf{L}$:

Orbital Magnetic Moment-Momentum Relation

$$\boldsymbol{\mu}_L = -\frac{e}{2m_e}\mathbf{L} \tag{B.28}$$

The constant of proportionality is typically denoted by γ, defined as

Single Electron Gyromagnetic Ratio

$$\gamma = -\frac{e}{2m_e} \tag{B.29}$$

It is known as the *single* **electron gyromagnetic ratio**. It is often defined as a positive quantity $\gamma = \frac{e}{2m_e}$, as well.

For an electron, in particular, we will show later in this chapter that the *quantum orbital angular momentum* is given by

$$L = |\mathbf{L}| = \sqrt{\ell(\ell+1)}\hbar$$

ℓ takes any *integer* value $\ell = 0, 1, 2, ...$ Since we know the magnetic moment $\boldsymbol{\mu}_L$ is proportional to $\boldsymbol{L}$, we should not be required to solve an additional eigenvalue equation for $\boldsymbol{\mu}_L$. In other words, because of this proportional relationship, knowing L is sufficient to know μ_L. Using Eq. (B.28), we may write the orbital magnetic moment as

$$\mu_L = |\boldsymbol{\mu}_L| = \sqrt{\ell(\ell+1)}\gamma\hbar \tag{B.30}$$

Just as $\sqrt{\ell(\ell+1)}$ can be regarded as counting how many $\hbar$ units are in the orbital angular momentum, we may identify the quantity $\gamma\hbar = e\hbar/2m_e$ as an analogous quantity for counting magnetic moment units. The constant $\gamma\hbar$ is known as the single electron **Bohr magneton** denoted by μ_B.

Single Electron Bohr Magneton

$$\mu_B = |\gamma|\hbar \tag{B.31}$$

From Eq. (B.30), μ_B can be regarded as a natural unit for counting magnetic moments, analogous to $\hbar$ for the orbital angular momentum L. Let's now examine how angular momentum and magnetic moment relate to an electron's potential energy.

These quantities relating to orbital angular momentum were originally introduced in the classical domain based on the electromagnetic equations. The classical frame provides a conceptual basis of their origins and some level of physical meaning. However, the classical discussion ends here. Next, we are going to take these concepts and apply them onto the domain of quantum mechanics. Thus, let us begin the quantum mechanics of orbital angular momentum, which is an *operator*.

B.4 Bohr's Atomic Hydrogen Model

Recall that in Chap. **??**, we discussed Rydberg's proposed *empirical* model that well-described the behavior of the hydrogen spectral lines, particularly, if only thermal excitation was used to excite the atoms. However, there still lacked any concrete insight as to the physical origin of these observations. Additionally, Rydberg's model turned out to be very limited if one wished to describe other atomic and molecular systems. Around 1913, this changed with the work of Danish physicist Neils Bohr (1885-1962). Bohr imagined the hydrogen atom as illustrated

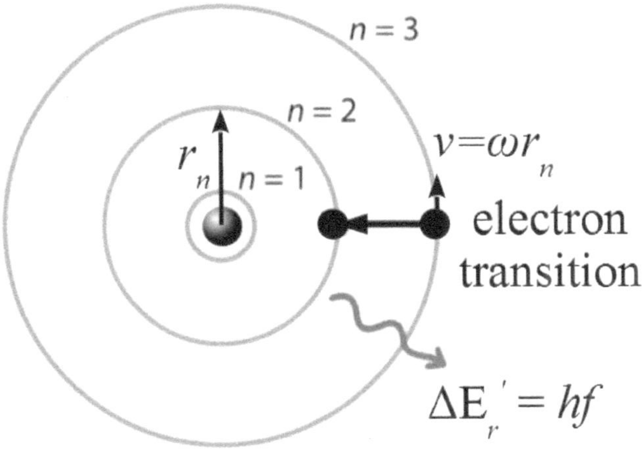

Figure B.4: Simplified diagram of a hydrogen atom, based on Bohr's model. Perfect circular discrete orbitals are assumed. Light is emitted only when transitioning from one orbital to another.

in Fig. B.4, assuming perfect circular discrete orbitals. At that time, the atomic model of the hydrogen atom was already believed to be a heavy nucleus with a light-weight electron whizzing around it. With this structure in mind, Bohr utilized a quanta assumption in line with what Planck and Einstein had already done. He assumed that in the hydrogen atom, the angular momentum L takes only discrete values. That is to say that the angular momentum L is given by

$$|\mathbf{L}| = L_n = rp = rmv = n\hbar \tag{B.32}$$

In (B.32), n is an integer value given by $1, 2, 3$, and so on which will end up as the principal quantum number appearing in the Rydberg formula. Considering hydrogen, which has a single electron, the Bohr model assumes the single electron is orbiting in planar loops around the nucleus, which contains one proton, but no neutron.

It was shown in Sec. B.1 that there exists a centripetal force for a mass in a steady circular orbit. For the electron, the Coulomb electric force balances this force. Thus, for a circular orbit, we can sum these forces to obtain the following (in SI units):

$$\sum \mathbf{F} = m\mathbf{a} \Rightarrow \frac{e^2}{4\pi\epsilon_0 r^2} = \frac{mv^2}{r} \tag{B.33}$$

After substituting the Bohr assumption for the angular momentum $L = L_{\dot{n}}$ from Eq. (B.32), we have a way to determine the orbital radius r using

$$\frac{e^2}{4\pi\epsilon_0 r^2} = \frac{n^2\hbar^2}{mr^3} \tag{B.34}$$

Solving for the radius r, we obtain

$$r = \frac{n^2\hbar^2 4\pi\epsilon_0}{me^2} = \frac{n^2h^2\epsilon_0}{m\pi e^2} \tag{B.35}$$

Eq. (B.35) reveals that a consequence of Bohr's assumption of the quanta of angular momentum L, is that the orbital radius r is also quantized because it is a function of the discrete integer n. We can find the total energy E for the hydrogen electron by summing the kinetic and potential energies as

$$E = K + V = \frac{L^2}{2mr^2} - \frac{e^2}{4\pi\epsilon_0 r} \tag{B.36}$$

This can also be written as

$$E = \frac{r^2v^2m^2}{2mr^2} - \frac{e^2}{4\pi\epsilon_0 r} = \frac{v^2m}{2} - \frac{e^2}{4\pi\epsilon_0 r} \tag{B.37}$$

By using the force balance equation in Eq. (B.33), the first term on the far right hand side can be written as

$$E = \frac{e^2}{8\pi\epsilon_0 r} - \frac{e^2}{4\pi\epsilon_0 r} = -\frac{e^2}{8\pi\epsilon_0 r}$$

The energy E is just half the Coulomb potential energy V, for any value of r. It follows that $|V|/K = 2$. This result turns out to be general, and is an example of more fundamental theorem known as the *Virial theorem*. After substituting the result for the orbital radius r, we find that the energy E is given by

Hydrogen Atom Discrete Energy Levels (Bohr Model)

$$E = -\frac{me^4}{8\epsilon_0^2 h^2}\frac{1}{n^2} \tag{B.38}$$

Eq. (B.38) reveals that the energy of the electron in the hydrogen atom is negative. This is the result of the fact that the attractive negative Coulomb's potential energy is the dominating energy which, consequently, keeps the electron *bound* to the hydrogen atom's nucleus. For this reason, negative energies always correspond to **bound energy states**. From his assumption utilizing the little known constant from Plank $\hbar$, Bohr found that the *bound energy states are quantized*, meaning that the available energies the electron can take are *not* arbitrary, but one of infinitely many discrete or quantum states. In the Bohr model, this is inherited from the quantized angular momentum $L = L_n$.

From Bohr's result, the smallest possible orbital radius r_0 for the electron in the hydrogen atom is not zero, but corresponds to setting the integer n to its lowest possible value, $n = 1$, which leads to

$$a_0 = \frac{h^2\epsilon_0}{m\pi e^2} = \frac{4\pi\hbar^2\epsilon_0}{me^2} = 5.2919 \times 10^{-11}\text{m} = 0.52919\text{Å} \tag{B.39}$$

Eq. (B.39) is known as the **Bohr radius** and is often used as a unit of measure for atomic distances. Bohr found that the spectral lines that were routinely measured for hydrogen were described well by *transitions* from one energy state to another. This lead to success in predicting the spectral lines of hydrogen, corresponding to changes in the quantum states of the electron, expressed by

Hydrogen Atom Discrete Energy Transitions (Bohr Model)

$$\Delta E = \frac{me^4}{8\epsilon_0^2 h^2}\left(\frac{1}{n_1^2} - \frac{1}{n_2^2}\right) \tag{B.40}$$

In Eq. (B.40), $n_2 > n_1$, which yields a positive change in energy during the transition, so it corresponds to when the electron descends from a higher energy state back to a lower one. Thus, the smaller n value also corresponds to a state closer to the nucleus, or to a more bound state. This state has a lower energy, or is more negative. This provides an interpretation of the observations that could be formulated by Bohr, that as the electron undergoes its transition back down to a lower state, it gives off photons, or quanta of light containing information about the energy levels of the electron. It also suggests that we may use Einstein and Planck's relationship for the light quanta emitted, which was known by 1911, to write the transition equation in terms of the wavelength λ

$$\Delta E(\lambda) = \hbar\omega = h\frac{c}{\lambda} = \frac{me^4}{8\epsilon_0^2 h^2}(\frac{1}{n_1^2} - \frac{1}{n_2^2}) \Rightarrow \frac{1}{\lambda} = \frac{me^4}{8c\epsilon_0^2 h^3}\left(\frac{1}{n_1^2} - \frac{1}{n_2^2}\right) \tag{B.41}$$

In terms of the frequency f, the energy $E(f)$ is then given by

$$f = \frac{me^4}{8\epsilon_0^2 h^3}\left(\frac{1}{n_1^2} - \frac{1}{n_2^2}\right) \tag{B.42}$$

Another impressive result from Bohr's model was that when this transition equation is expressed in the form of the empirical relation proposed by Rydberg (*c.* Eq. (**??**)), evaluation of the coefficient obtained by Bohr worked out to be

Rydberg Constant

$$R_d = \frac{me^4}{8c\epsilon_0^2 h^3} \approx 10965609 \quad \text{m}^{-1} \tag{B.43}$$

The value from Bohr's model turned out to be the same as the empirical parameter determined by Rydberg, applied to the hydrogen atom. This constant is known as the *Rydberg constant* R_d. Thus, Bohr had obtained a result that not only predicted the energy transitions of some hydrogen spectra, but he predicted the value of the Rydberg constant in terms of other known fundamental constants. After Planck's success with the blackbody radiation problem which lead to the discovery of h, and Einstein's success describing the photoelectric effect, which also requiring h, now Bohr's hydrogen atomic spectrum model also requires h. We were reaching a point where the idea of a quanta of energy was becoming an undeniable truth in these new kinds of physics problems.

Now that we have considered orbital angular momentum from the classical definitions, as well as from the earliest quantum mechanical model, we can now solve the hydrogen atom problem using the *non-relativistic* Schrödinger equation, where we'll find considerably more information reaching *beyond* only the amount of energy changes giving rise to the hydrogen atomic spectra.

B.5 Orbital Angular Momentum Operators $\widehat{L}$

In Eq. (2.127), several quantum mechanical operators were derived inside the relativistic quantum mechanical formulation, where the *kinetic energy operator* was given by

$$\widehat{H}_K = (\widehat{\mathbf{p}} + e\mathbf{A})^2/2m_0 \tag{B.44}$$

Later, in Sec. 2.6, we also showed that the operator in Eq. (B.44) *contains* another operator given by

$$\widehat{H}_L = \widehat{\mathbf{L}} \cdot \mathbf{B} \tag{B.45}$$

Recall that the vector components of the classical orbital angular momentum operator are given by

$$L_x = yp_z - zp_y \tag{B.46}$$

$$L_y = zp_x - xp_z \tag{B.47}$$

$$L_z = xp_y - yp_x \tag{B.48}$$

The quantum mechanical operators for the components of linear momentum are

$$\widehat{p}_x = -i\hbar\frac{\partial}{\partial x} \tag{B.49a}$$

$$\widehat{p}_y = -i\hbar\frac{\partial}{\partial y} \tag{B.49b}$$

$$\widehat{p}_z = -i\hbar\frac{\partial}{\partial z} \tag{B.49c}$$

Using the quantum mechanical operators for the linear momentum, the orbital angular momentum operators become

$$\widehat{L}_x = -i\hbar\left(y\frac{\partial}{\partial z} - z\frac{\partial}{\partial y}\right) \tag{B.50a}$$

$$\widehat{L}_y = -i\hbar\left(z\frac{\partial}{\partial x} - x\frac{\partial}{\partial z}\right) \tag{B.50b}$$

$$\widehat{L}_z = -i\hbar\left(x\frac{\partial}{\partial y} - y\frac{\partial}{\partial x}\right) \tag{B.50c}$$

This is identical to the results we obtained in Sec. 2.6. Like the energy operators, the angular momentum operators also operate on their corresponding *wave functions* ψ which satisfy their respective *eigenvalue* equation. Let us consider the angular momentum component $\widehat{L}_z$, for which the eigenvalue equation is defined as

$$\widehat{L}_z\psi = \lambda_z\hbar\psi \tag{B.51}$$

Substitution of Eq. (B.50c) gives

$$-i\hbar\left(x\frac{\partial\psi}{\partial y} - y\frac{\partial\psi}{\partial x}\right) = \lambda_z\hbar\psi \tag{B.52}$$

Because orbital angular momentum involves circular motion, it is convenient to transform the coordinate system from Cartesian (x,y,z) into polar coordinates (r,θ,φ), before solving the equations involving orbital angular momentum. Fig. B.5 illustrates the relationship between the two coordinate systems that we will utilize. From Fig. B.5, the relationships between the two coordinate systems are given by the following set of relations:

$$x = r\sin\theta\cos\varphi, \quad y = r\sin\theta\sin\varphi$$

$$z = r\cos\theta, \quad x^2 + y^2 + z^2 = r^2$$

The above relations can be used to obtain the following useful equations:

$$\frac{y}{x} = \tan\varphi, \quad \text{and} \quad \frac{x^2 + y^2}{z^2} = \tan^2\theta$$

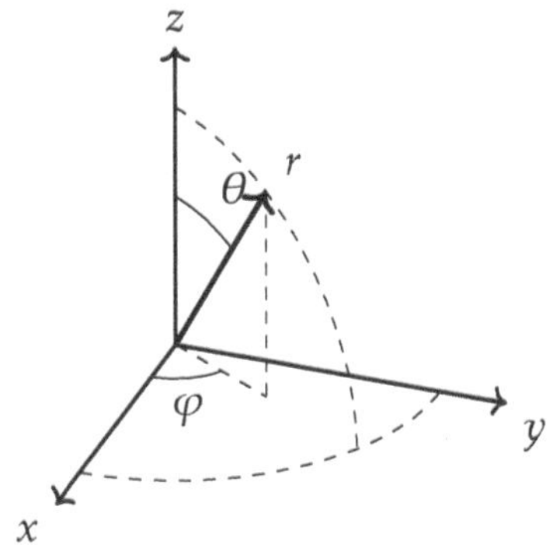

Figure B.5: Relations between the Cartesian and polar coordinates used in the orbital angular momentum quantum mechanical problem.

In order to render Eq. (B.52) into an explicit equation of (r,θ,φ), we make substitutions as follows:

$$x\frac{\partial\psi}{\partial y} = r\sin\theta\cos\varphi\left(\frac{\partial\psi}{\partial r}\frac{\partial r}{\partial y} + \frac{\partial\psi}{\partial\theta}\frac{\partial\theta}{\partial y} + \frac{\partial\psi}{\partial\varphi}\frac{\partial\varphi}{\partial y}\right) \tag{B.53a}$$

and

$$y\frac{\partial\psi}{\partial x} = r\sin\theta\sin\varphi\left(\frac{\partial\psi}{\partial r}\frac{\partial r}{\partial x} + \frac{\partial\psi}{\partial\theta}\frac{\partial\theta}{\partial x} + \frac{\partial\psi}{\partial\varphi}\frac{\partial\varphi}{\partial x}\right) \tag{B.53b}$$

From Eqs. (B.53), all terms involving x, y, and z must be replaced with a function of (r,θ,φ). This means the six terms $\partial r/\partial y$, $\partial\theta/\partial y$, $\partial\varphi/\partial y$, $\partial r/\partial x$, $\partial\theta/\partial x$, and $\partial\varphi/\partial x$ must be replaced in this way. Thus, the following substitutions can be used in substitution:

$$\frac{\partial r}{\partial x} = \frac{x}{r} = \sin\theta\cos\varphi, \quad \frac{\partial r}{\partial y} = \frac{y}{r} = \sin\theta\sin\varphi, \quad \frac{\partial\theta}{\partial x} = \frac{\cos\theta\cos\varphi}{r} \tag{B.54a}$$

$$\frac{\partial\theta}{\partial y} = \frac{\cos\theta\sin\varphi}{r}, \quad \frac{\partial\varphi}{\partial x} = -\frac{1}{r}\frac{\sin\varphi}{\sin\theta}, \quad \frac{\partial\varphi}{\partial y} = \frac{1}{r}\frac{\cos\varphi}{\sin\theta} \tag{B.54b}$$

After substitution of Eqs. (B.53) and Eq. (B.54) into Eq. (B.50), we obtain the result

$\widehat{L}_z$ Operator

$$\widehat{L}_z\psi = -i\hbar\frac{\partial\psi}{\partial\varphi} \tag{B.55}$$

Therefore, the *eigenvalue equation* $\widehat{L}_z = \lambda_z\hbar\psi$ for $\widehat{L}_z$ becomes

$\widehat{L}_z$ Eigenvalue Equation

$$-i\hbar\frac{\partial\psi}{\partial\varphi} = \lambda_z\hbar\psi \tag{B.56}$$

It is straightforward to check that the solution to the eigenvalue equation for $\widehat{L}_z$ is given by

$$\psi = C_n e^{i\lambda_z\varphi} \tag{B.57}$$

C_n is the normalization constant, independent of φ. The limits of φ range from 0 to 2π, and therefore, in order for the wave function ψ to be continuous at these limits, any number of complete revolutions must repeat itself at the limits. For this condition to be satisfied, it must be that

$$C_n e^{i\lambda_z\varphi} = C_n e^{i\lambda_z(\varphi+2\pi)} \tag{B.58}$$

This means that

$$e^{i2\pi\lambda_z} = 1 \tag{B.59}$$

Considering Euler's theorem $e^x = \cos x + i\sin x$, Eq. (B.59) tells us that $e^{i2\pi\lambda_z} = \cos(2\pi\lambda_z) + i\sin(2\pi\lambda_z) = 1$. But 1 being a real number, means it can only be true if $\sin(2\pi\lambda_z) = 0$. For this to be satisfied, λ_z can only be an integer $m = -2, -1, 0, 1, 2$, etc. The eigenfunction for L_z, denoted by $\psi = \Phi(\varphi)$, is given by

$$\Phi(\varphi) = C_n e^{im\varphi} \quad \text{where} \quad m = ...,\text{-}2,\text{-}1,0,1,2,...$$

The normalization constant C_n is found from the *normalization condition* $\int \Phi^*\Phi d\varphi = 1$, which leads to

$$1 = \int_0^{2\pi} C_n e^{im\varphi} C_n^* e^{-im\varphi} d\varphi = |C_n|^2 \int_0^{2\pi} d\varphi \Rightarrow |C_n| = \frac{1}{\sqrt{2\pi}}$$

Using the normalization condition prohibits an exact knowledge of C_n within a phase factor $e^{i\theta}$. This is because we solve for $|C_n|^2$. This sort of result is said to be *within a phase-factor*. Any complex number having the same amplitude, with arbitrary phase, still satisfies our normalization condition. The normalized eigenfunction Φ for the z component of the angular momentum is then given by

Normalized Eigenfunction Of $\widehat{L}_z$ Operator

$$\Phi(\varphi) = \frac{1}{\sqrt{2\pi}} e^{im\varphi} \qquad m = \cdots, -3, -2, -1, 0, 1, 2, 3, \cdots \tag{B.60}$$

It turns out that the eigenfunction for $\widehat{L}_z$ is not a function of r nor θ. This made the analysis simpler. However, the other two components of angular momentum are not strict functions of φ. In a similar manner as we've done for $\widehat{L}_z$, the other components of the orbital angular momentum can also be determined. Substitution of Eqs. (B.53) and Eq. (B.54) into Eq. (B.50), we obtain an operator $\widehat{L}_x$ given by

$\widehat{L}_x$ Operator

$$\widehat{L}_x = i\hbar\left(\sin\varphi\frac{\partial}{\partial\theta} + \cot\theta\cos\varphi\frac{\partial}{\partial\varphi}\right) \tag{B.61}$$

And for the $\widehat{L}_y$ component, we have

$\widehat{L}_y$ Operator

$$\widehat{L}_y = i\hbar\left(-\cos\varphi\frac{\partial}{\partial\theta} + \cot\theta\sin\varphi\frac{\partial}{\partial\varphi}\right) \tag{B.62}$$

Having $\widehat{L}_x, \widehat{L}_y$, and $\widehat{L}_z$ enables us to express any equation that depends on $\widehat{\mathbf{L}}$. Since kinetic energy involves the square of the momentum (i.e. $\widehat{p}^2$ or $\widehat{L}^2$), we can also find the operator for $\widehat{L}^2$. As we saw in Chap. **??** with the operator $\widehat{p}^2$, the square of an operator in quantum mechanics means that the operator is applied twice or consecutively to the object on the right of it. It does not mean multiplying something by itself, as it does with scalars. Therefore, $\widehat{L^2}$ is given by

$$\widehat{L^2} = \widehat{L_x^2} + \widehat{L_y^2} + \widehat{L_z^2} \tag{B.63}$$

From Eq. (B.56), the operator for L_z^2 is given by

$$\widehat{L_z^2}\psi = \widehat{L}_z(\widehat{L}_z\psi) = -i\hbar\frac{\partial}{\partial\varphi}\left(-i\hbar\frac{\partial\psi}{\partial\varphi}\right)$$

Therefore, we have the following eigenvalue equation $\widehat{L_z^2} = \lambda\psi$, which becomes

L_z^2 Operator Eigenvalue Equation

$$-\hbar^2\frac{\partial^2\psi}{\partial\varphi^2} = \lambda\hbar^2\psi \tag{B.64}$$

It is a matter of convenience to include the factor of $\hbar^2$, or not on the RHS of Eq. (B.64). Including it leads to dimensionless eiqenvalues that are pure integer functions. Eqs. (B.61) and (B.61) allow us to write $\widehat{L_x^2}$ and $\widehat{L_y^2}$ as

$$\widehat{L_x^2}\psi = -\hbar^2\left(\sin\varphi\frac{\partial}{\partial\theta} + \cot\theta\cos\varphi\frac{\partial}{\partial\varphi}\right)\left(\sin\varphi\frac{\partial}{\partial\theta} + \cot\theta\cos\varphi\frac{\partial}{\partial\varphi}\right)\psi \tag{B.65a}$$

$$\widehat{L_y^2}\psi = -\hbar^2\left(-\cos\varphi\frac{\partial}{\partial\theta} + \cot\theta\sin\varphi\frac{\partial}{\partial\varphi}\right)\left(-\cos\varphi\frac{\partial}{\partial\theta} + \cot\theta\sin\varphi\frac{\partial}{\partial\varphi}\right)\psi \tag{B.65b}$$

Eqs. (B.63), (B.65), (B.65b), and (B.64), can be used to write the operator $\widehat{L^2}$, which works out to be

$$\widehat{L^2}\psi = -\hbar^2\left[\frac{1}{\sin\theta}\frac{\partial}{\partial\theta}\left(\sin\theta\frac{\partial\psi}{\partial\theta}\right) + \frac{1}{\sin^2\theta}\frac{\partial^2\psi}{\partial\varphi^2}\right] \tag{B.66}$$

Eq. (B.66) defines the operator for $\widehat{L^2}$. Thus, we can solve the corresponding eigenvalue equation for $\widehat{L^2}$ by solving the eigenvalue equation given by

$$\widehat{L^2}\psi = -\hbar^2\left[\frac{1}{\sin\theta}\frac{\partial}{\partial\theta}\left(\sin\theta\frac{\partial\psi}{\partial\theta}\right) + \frac{1}{\sin^2\theta}\frac{\partial^2\psi}{\partial\varphi^2}\right] = \lambda\hbar^2\psi$$

or, after factoring $\hbar^2$, we have

$\widehat{L}^2$ Operator Eigenvalue Equation

$$L^2\psi = -\left[\frac{1}{\sin\theta}\frac{\partial}{\partial\theta}\left(\sin\theta\frac{\partial\psi}{\partial\theta}\right) + \frac{1}{\sin^2\theta}\frac{\partial^2\psi}{\partial\varphi^2}\right] = \lambda\psi \qquad \text{(B.67)}$$

We must remember, with this form, the resulting eigenvalues must be scaled by $\hbar^2$. Before we get to the solution of Eq. (B.67), which entails finding both the eigenvalues λ and the associated eigenfunctions or wave functions ψ, let us first discuss a relevant historical topic that provides some insight into the origin of the solutions for the $\widehat{L}^2$ operator, namely, the *Laplacian equation* and *Legendre polynomials*. This is the subject of the next section.

B.6 The Laplacian and Legendre Equations

It turns out that Eq. (B.67) has been seen before its appearance in the quantum mechanical orbital angular momentum operator $\widehat{L}^2$. The part if brackets is a part of the 3D Laplacian (∇^2) in spherical coordinates, given by

$$\nabla^2\phi = \frac{1}{r^2}\left[\frac{\partial}{\partial r}\left(r^2\frac{\partial}{\partial r}\right) + \frac{1}{\sin\theta}\frac{\partial}{\partial\theta}\left(\sin\theta\frac{\partial}{\partial\theta}\right) + \frac{1}{\sin^2\theta}\frac{\partial^2}{\partial\varphi^2}\right]\phi \qquad \text{(B.68)}$$

This relation generally describes a 3D potential ϕ (*e.g.* gravitational, electrical) of the form

$$\phi(r) = \frac{a}{|\mathbf{r} - \mathbf{r}_0|}$$

a is a constant independent of position. Whenever there is electrical *charge balance*, given by equal numbers of positive and negative charges so total charge $Q = 0$ (e.g. a dipole, quadrupole, octupole, etc.), anywhere in the distant space *outside* the body of charges, the Laplacian describes the potential by the equation $\nabla^2\phi = 0$. Since ϕ only depends on the distance between the position of interest ($\mathbf{r}$) and the source ($\mathbf{r}_0$), by choosing the origin of the coordinate system conveniently, the potential function can be transformed to the following form

$$\phi(\mathbf{r}) = \frac{a}{|\mathbf{r}-\mathbf{r}_0|} = \frac{1}{|r-r\cos\theta|} = \frac{1}{r}\frac{a}{|1-\cos\theta|} = f(r)g(\cos\theta) \qquad \text{(B.69)}$$

θ is the inclusive angle between the vectors $\mathbf{r}$ and $\mathbf{r}_0$. In mathematics, there is an infinite series known as the **geometric series** given by

$$g(\cos\theta) = g(\rho) = \sum_{n=0}^{\infty} a\rho^n = \frac{a}{1-\rho}$$

The name of this series originates from the early mathematical idea of using known geometrical shapes to obtain integrals or areas of mathematical functions such as parabolas. The idea was expressed as early as 250 B.C. by Greek mathematician Archimedes of Syracuse (287 B.C.-212 B.C.), who studied in Egypt. Each term in the series can be represented as a factor times the previous term. $g(\cos\theta)$ in Eq. (B.69) is in the form of a *geometric series*, with an interval of convergence $\rho = \cos\theta \in (-1,1)$. This implies that the potential function can be represented by an infinite series.

While investigating gravitational potential, French mathematician Adrien-Marie Legendre (1752-1833) found that such a potential can, indeed, be represented by a series of the form

$$\phi(r,\cos\theta) = 1 + rP_1(\cos\theta) + r^2P_2(\cos\theta) + r^3P_3(\cos\theta) + \ldots = \sum_{n=0}^{\infty} r^n P_n(\cos\theta) \qquad \text{(B.70)}$$

P_n is a polynomial of order n, where $n = 0,1,2,3,\ldots$, whose argument is $\cos\theta$. These polynomials are known as **Legendre polynomials**. It turns out that an equation describing only the polynomials P_n can be obtained by plugging the series form proposed by Legendre into Eq. (B.68), and carrying out some calculus. To do this, we'll consider the three parts of the Laplacian separately, then sum the results from these three parts together.

The radial part (the first term) of the Laplacian gives

$$\frac{\partial}{\partial r}\left(r^2\frac{\partial}{\partial r}\right)\sum_{n=0}^{\infty} r^n P_n(\cos\theta) = \sum_{n=0}^{\infty} n(n+1)r^n P_n(\cos\theta) \quad \text{(radial part)}$$

Note that since the Laplacian equation is equal to zero on the RHS, we may ignore the factor $1/r^2$, for now. The second term of the Laplacian then leads to

$$\frac{1}{\sin\theta}\frac{\partial}{\partial\theta}\left(\sin\theta\frac{\partial}{\partial\theta}\right)\sum_{n=0}^{\infty} r^n P_n(\cos\theta) =$$

$$-2\cos\theta\sum_{n=0}^{\infty} r^n \frac{dP_n}{d(\cos\theta)} + \sin^2\theta\sum_{n=0}^{\infty} r^n \frac{d^2P_n}{d(\cos\theta)^2}$$

Because the series is a function of r and θ only, and does not depend on φ, all derivatives w.r.t. φ are 0. Thus, the last term in the Laplacian is 0. Putting the results together, we have that the Laplacian $\nabla^2\phi$ is given by

$$\nabla^2\phi = \sum_{n=0}^{\infty} n(n+1)r^n P_n(\cos\theta) - 2\cos\theta\sum_{n=0}^{\infty} r^n \frac{dP_n}{d(\cos\theta)} +$$

$$\sin^2\theta\sum_{n=0}^{\infty} r^n \frac{d^2P_n}{d(\cos\theta)^2} = 0$$

Since the RHS is zero, and our equation must hold for any value of r, each coefficient of r^n in the series must also be zero. So we have that for all $n = 0, 1, 2, \ldots,$

$$n(n+1)P_n(\cos\theta) - 2\cos\theta\frac{dP_n}{d(\cos\theta)} + \sin^2\theta\frac{d^2P_n}{d(\cos\theta)^2} = 0 \tag{B.71}$$

Letting $x = \cos\theta$, Eq. (B.71) becomes

Legendre Differential Equation

$$(1-x^2)\frac{d^2P_n}{dx^2} - 2x\frac{dP_n}{dx} + n(n+1)P_n(x) = 0 \tag{B.72}$$

Eq. (B.72) is known as the **Legendre differential equation**, and it is the defining equation for the Legendre polynomials P_n. Therefore, the infinite series proposed by Legendre satisfies the Laplace equation if and only if P_n satisfies Eq. (B.72). The question remains as to what are the solutions for the Legendre polynomials P_n? One clever way to answer this question is to use a mathematical trick. Consider the function

$$f_n = (x^2-1)^n \tag{B.73}$$

We, first, want to generate a second order differential equation for f_n. Starting with differentiating Eq. (B.73) once, we obtain

$$\frac{df_n}{dx} = n(x^2-1)^{n-1}(2x)$$

Multiplying both sides by (x^2-1), and rearranging, we then have

$$(1-x^2)\frac{df_n}{dx} + 2xnf_n = 0$$

Differentiating once again for a second order differential equation, we obtain

$$(1-x^2)\frac{d^2f_n}{dx^2} + 2x(n-1)\frac{df_n}{dx} + 2nf_n = 0 \tag{B.74}$$

Though we have obtained a second order differential equation with f_n, we are not done yet. As a final step, let us continue to differentiate Eq. (B.74) n times. The result will be a differential equation for the $(n+2)^{nd}$ derivative of f_n. This is done using the generalized product rule for differentiation, also known as the **Leibniz rule**, which states that for the product of two n-times differentiable functions $y_1(x)$ and $y_2(x)$

Leibniz' Rule of Differentiation

$$\frac{d^n}{dx^n}(y_1 \cdot y_2) = \sum_{k=0}^{n} \frac{n!}{k!(n-k)!}\frac{d^k y_1}{dx^k}\frac{d^{n-k}y_2}{dx^{n-k}} \tag{B.75}$$

The formula may look complicated, but fortunately, it turns out to be fairly simple and straightforward. Let's apply the Liebniz' rule to Eq. (B.74), term by term. Applying Leíbniz' rule to the first term in Eq. (B.74), we obtain

$$\text{first term} \quad \rightarrow f_n^{n+2} - 2xnf_n^{n+1} - n(n-1)f_n^n$$

The superscripts $n+2$, $n+1$, and n denote the $n+2$, $n+1$, and n derivatives of f_n, respectively. Applying to the second term, we get

$$\text{second term} \quad \rightarrow 2x(n-1)f_n^{n+1} + 2n(n-1)f_n^n$$

And for the last term, we simply have

$$\text{third term} \quad \rightarrow 2xnf_n^n$$

Adding all three results together, we have

$$(1-x^2)f_n^{n+2} - 2xf_n^{n+1} + n(n+1)f_n^n = 0$$

We are getting close! Let us define another function $D_n(x)$ as the nth derivative of f_n or

$$D_n(x) \equiv \frac{d^n}{dx^n}(x^2-1)^n = f_n^n \tag{B.76}$$

Then, the differential equation for $D_n(x)$ becomes

$$(1-x^2)\frac{d^2D_n}{dx^2} - 2x\frac{dD_n}{dx} + n(n+1)D_n = 0 \tag{B.77}$$

Eqs. (B.77) and Eq. (B.72) are identical, which means that D_n satisfies the differential equation for P_n, and therefore, any function proportional to D_n is a solution to Eq. (B.72), or

$$P_n = C_0\frac{d^n}{dx^n}(x^2-1)^n \tag{B.78}$$

The convention for expressing Legendre polynomials P_n is to write P_n such that $P_n(1) = 1$. In evaluating d^nP_n/dx, it will be a sum of terms where all terms but one have a factor $x^2 - 1$. So, when $x = 1$, they all vanish and our condition becomes

$$C_0 n!2^n x^n + 0 = 1 \rightarrow C_0 = \frac{1}{2^n n!} \tag{B.79}$$

Therefore, the resulting Legendre polynomials that satisfy the condition $P_n(1) = 1$ are defined by

Rodrigues' Formula For Legendre Polynomials

$$P_n = \frac{1}{2^n n!}\frac{d^n}{dx^n}(x^2-1)^n \tag{B.80}$$

Eq. (B.80) is known as the **Rodrigues' formula**, first given in 1816 by the French mathematician Benjamin Olinde Rodrigues (1795-1851). It is the exact solution for the Legendre polynomials, and has been known for over 200 years! Table B.2 lists the first few Legendre polynomials.

Table B.2: First ten Legendre polynomials P_n. The Legendre polynomials arise as factors in the solution to the orbital angular momentum problem, and are part of the infinite series solution to the Laplace equation. They can be obtained directly and iteratively from the Rodriques formula.

Name	Equation
$n = 0$	$P_0(x) = 1$
$n = 1$	$P_1(x) = x$
$n = 2$	$P_2(x) = \frac{1}{2}(3x^2 - 1)$
$n = 3$	$P_3(x) = \frac{1}{2}(5x^3 - 3x)$
$n = 4$	$P_4(x) = \frac{1}{8}(35x^4 - 30x^2 + 3)$
$n = 5$	$P_5(x) = \frac{1}{8}(63x^5 - 70x^3 + 15x)$
$n = 6$	$P_6(x) = \frac{1}{16}(231x^6 - 315x^4 + 105x^2 - 5)$
$n = 7$	$P_7(x) = \frac{1}{16}(429x^7 - 693x^5 + 315x^3 - 35x)$
$n = 8$	$P_8(x) = \frac{1}{128}(6435x^8 - 12012x^6 + 6930x^4 - 1260x^2 + 35)$
$n = 9$	$P_9(x) = \frac{1}{128}(12155x^9 - 25740x^7 + 18018x^5 - 4620x^3 + 315x)$

A corresponding plot of the first ten Legendre polynomials is shown in Fig. B.6.

Figure B.6: First ten Legendre polynomials P_n, plotted over the range of interest $(-1, 1)$.

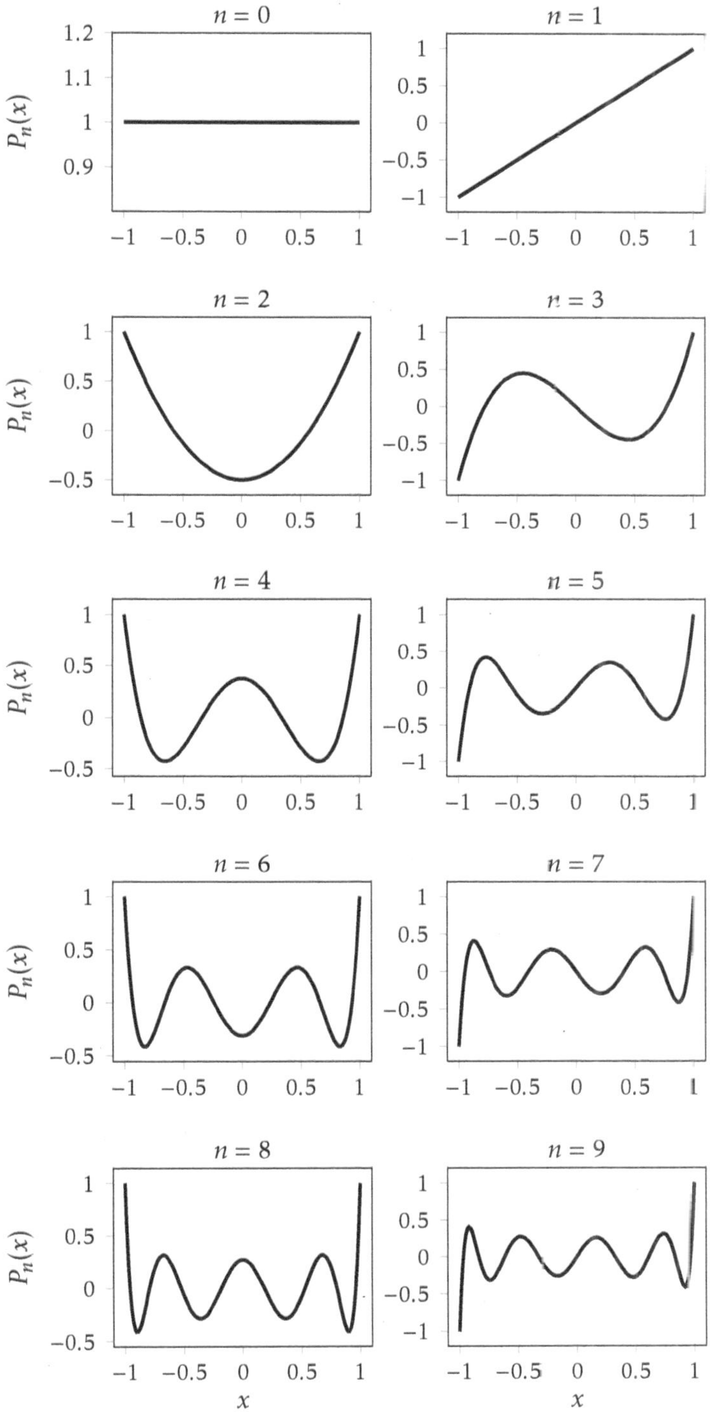
$n = 0$
$n = 1$
$n = 2$
$n = 3$
$n = 4$
$n = 5$
$n = 6$
$n = 7$
$n = 8$
$n = 9$
$P_n(x)$
x

Because of the symmetry of the potential function, consecutive values of n alternate between anti-symmetric and symmetric states. Let's extend these ideas a bit further to obtain a result that will be directly applicable to the orbital angular momentum problem.

B.7 Extending To Associated Legendre Functions

So we have discussed from whence the Legendre polynomials originate, and we have also seen that, through a clever choice of functions, we were able to use Leibniz' rule to obtain the exact solutions for P_n, the *Legendre polynomials*. Taking this idea further, lets see what happens when we apply Leibniz' rule to Eq. (B.67), differentiating m additional times. After carrying this out for all three terms as we did for the Legendre polynomials, then adding the results, we obtain

$$(1-x^2)\frac{d^2P_n^m}{dx^2} - 2x(m+1)\frac{dP_n^m}{dx} + (n(n+1) - m(m+1))\,P_n^m = 0 \quad \text{(B.81)}$$

P_n^m now represents the mth derivative of P_n. We now have a second order differential equation describing P_n^m. Using yet another mathematical trick, let us define a function Q by the following:

$$P_n^m = (1-x^2)^{-\frac{m}{2}}Q \quad \text{or} \quad Q = (1-x^2)^{\frac{m}{2}}P_n^m \quad \text{(B.82)}$$

After substitution of Eq. (B.82) into Eq. (B.81), multiplying the result by $(1-x^2)^{m/2}$, and after some algebra, we obtain the following second order differential equation for $Q(x)$.

$$(1-x^2)\frac{d^2Q}{dx^2} - 2x\frac{dQ}{dx} + \left[n(n+1) - \frac{m^2}{1-x^2}\right]Q = 0 \quad \text{(B.83)}$$

Using the definition for Q, given by Eq. (B.82), and using Rodrigues' formula for P_n, the function for Q, is given by

Associated Legendre Functions

$$A_n^m = \frac{C_n}{2^n n!}(1-x^2)^{\frac{m}{2}}\frac{d^{n+m}}{dx^{n+m}}(x^2-1)^n \quad \text{(B.84)}$$

Eq. (B.84) defines the **Associated Legendre Functions**, and they are closely *associated* with the Legendre polynomials P_n, however, their are significant differences. Just as with P_n, since $x = \cos\theta$, n, and m are all real numbers, A_n^m is a real-valued function. Among the differences between P_n and A_n^m, the *Associated Legendre functions* are only polynomials when m is an even integer. This condition leads to $(1-x^2)^{m/2}$ becoming a polynomial, multiplied by another polynomial. For our interests, the normalization condition is used to find the normalization constant C_n. To evaluate $\int (A_n^m)^2 dx = 1$, we can use a well-known mathematical property of associated Legendre functions, namely,

$$\int_{-1}^{1} A_n^m \cdot A_k^m dx = \frac{2(n+m)!}{(2n+1)(n-m)!}\delta_{n,k} \tag{B.85}$$

$\delta_{n,k}$ is the Kronecker delta function which is defined to be equal to 1 only if $n = k$, and 0 otherwise. Using this property, the normalization factor C_n for the Associated Legendre function is found to be

$$C_n = \sqrt{\frac{(2n+1)[(n-m)!]}{2(n+m)!}} \tag{B.86}$$

Then, we can write the explicit normalized associated Legendre function as

Normalized Associated Legendre Functions

$$A_n^m = \frac{1}{2^n n!}\sqrt{\frac{(2n+1)[(n-m)!]}{2(n+m)!}}(1-x^2)^{\frac{m}{2}}\frac{d^{n+m}}{dx^{n+m}}(x^2-1)^n \tag{B.87}$$

We should point out that we have departed from the usual convention in the notation for the associated Legendre functions. The common convention is P_n^m (rather than A_n^m, as we have used), but we intentionally avoid this notation here because we prefer to use P_n^m to denote the mth derivative of the Legendre polynomial P_n, which is also a polynomial. A_n^m, generally, is not a polynomial.

Eq. (B.84) implicitly dictates the allowed values of m, that ensures we always have a non-zero function. If these are to be useful as a basis set like Legendre polynomials, they cannot be zero-functions. They must be a constant (zeroth order) and/or higher order. A_n^m is the product of two terms, a explicit polynomial-*like* term and a pure polynomial derivative term. For A_n^m to result in a non-zero function, it is dictated by the derivative term on the RHS that the highest value m can take is that in which the $(n+m)^{th}$ derivative yields a non-zero constant, otherwise, the product vanishes. We must consider all possible values of m, being $m > 0$, $m < 0$, and $m = 0$. If $m > 0$, since $(x^2-1)^n$ is a polynomial of degree $2n$, we therefore have the condition that $n+m \leq 2n$ or $m \leq n$. If $m < 0$, $n-|m| \geq 0$, which gives $m \geq -n$ in order to ensure a zeroth derivative or higher. Thus, the general relation between the *degree n* and the *order m*, becomes

Relation Between Degree n and Order m Integers

$$|m| \leq n \tag{B.88}$$

So, $m = -n, -n+1, ..., -1, 0, 1, ..., n-1, n$, and any other values beyond this range lead to a trivial solution, or equals zero. When $m = 0$, we regain the Legendre polynomials P_n. As examples for $m = 1$, Table B.4 lists some Associated Legendre functions corresponding to $1 \leq n \leq 9$.

Table B.4: Associated Legendre functions A_n^m for $m = 1$. The Associated Legendre functions are factors in the solution to the orbital angular momentum eigen-value problem.

n	Some Associated Legendre Functions $A_n^1(x)$ (m=1,$n \leq 9$)
1	$A_1^1 = (1-x^2)^{1/2}$
2	$A_2^1 = 3x(1-x^2)^{1/2}$
3	$A_3^1 = \frac{1}{2}(15x^2-3)(1-x^2)^{1/2}$
4	$A_4^1 = \frac{1}{2^3}(140x^3-60x)(1-x^2)^{1/2}$
5	$A_5^1 = \frac{1}{2^3}(315x^4-210x^2+15)(1-x^2)^{1/2}$

Table B.4

6	$A_6^1 = \frac{1}{2^4}(1386x^5 - 1260x^3 + 210x)(1-x^2)^{1/2}$
7	$A_7^1 = \frac{1}{2^4}(3003x^6 - 3465x^4 + 945x^2 - 35)(1-x^2)^{1/2}$
8	$A_8^1 = \frac{1}{2^7}(51480x^7 - 72072x^5 + 27720x^3 - 2520x)(1-x^2)^{1/2}$
9	$A_9^1 = \frac{1}{2^7}(109395x^8 - 180180x^6 + 90090x^4 - 13860x^2 - 315)(1-x^2)^{1/2}$

Fig. B.7 plots the same associated Legendre functions. The reader should compare these curves to those of the Legendre polynomials in Fig. B.6

Figure B.7: First few normalized Associated Legendre functions $A_n^m(x)$ for $m = 1$ $(= Q_n^1)$, plotted over the range of interest $(-1, 1)$. The functions sequence through symmetric and anti-symmetric functions, corresponding to symmetric and anti-symmetric states. Associated Legendre functions tend to vanish at the limits of $x = \cos\theta$.

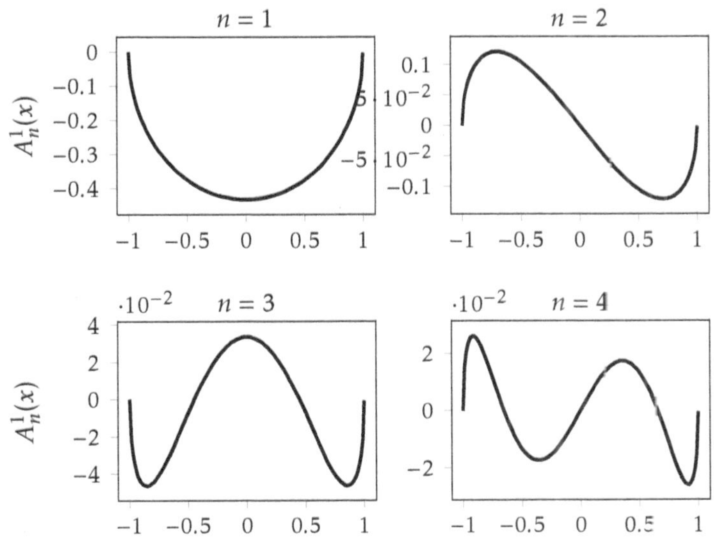

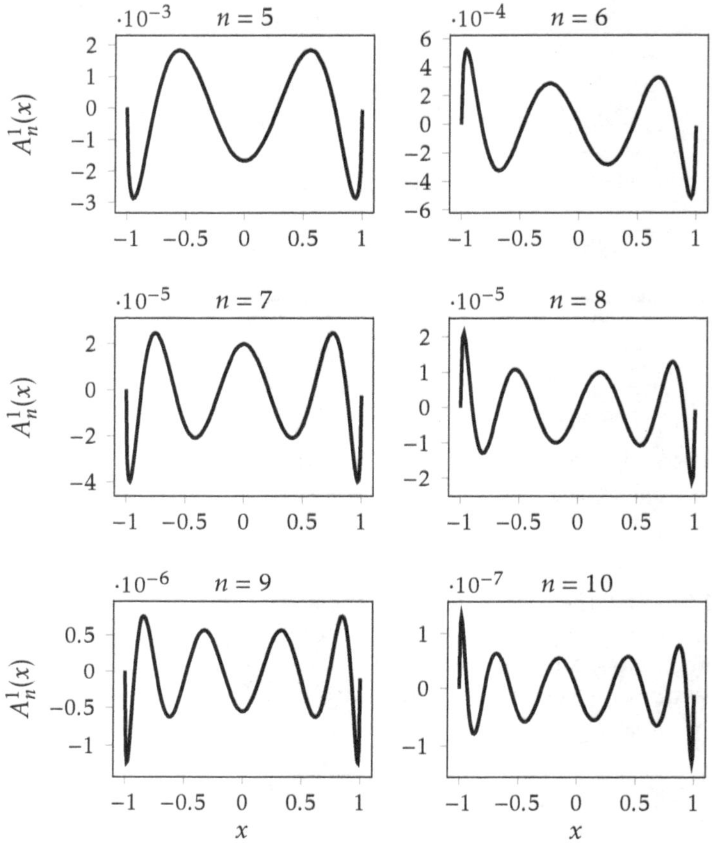

The functions sequence through symmetric and anti-symmetric functions with respect to the origin $x = 0$. We will see later that this characteristic will be inherited by the eigenfunctions for orbital angular momentum, corresponding to symmetric and anti-symmetric states. Unlike the Legendre polynomials, the Associated Legendre functions vanish at the limits of $x = \cos\theta$. The equations and solutions we have been dealing with have relevance to the orbital angular momentum problem we are considering in this chapter. We will see that the Associated Legendre functions emerge, once again, in solving Eq. (B.67).

B.8 Assoc. Legendre Functions & Orb. Angular Momentum

Now that we have discussed some background of the Legendre and Associated Legendre functions, let us return to the angular momentum problem we defined by Eq. (B.67). This equation is separable. We can always try this and if, after substitution of a separable function, we successfully obtain separated equations of independent variables that we can solve, then the method works. Thus, let us assume $\psi = Y_{\theta,\varphi} = F(\theta)\Phi(\varphi)$. Substitution into Eq. (B.67), multiplying by $\sin^2\theta$, then dividing by $Y_{\theta,\varphi}$, we obtain

$$\frac{\sin\theta}{F}\frac{d}{d\theta}\left(\sin\theta\frac{dF}{d\theta}\right)+\lambda\sin^2\theta=-\frac{1}{\Phi}\frac{d^2\Phi}{d\varphi^2} \tag{B.89}$$

The LHS is purely a function of θ while the RHS is purely a function of φ. As we have seen previously, when using the method of separation of variables, since both variables θ and φ are independent variables, it follows that both sides of Eq. (B.89) must be equal to a constant. Remember that we have dropped the $\hbar^2$. So, we set this constant equal to a perfect square m^2. When using the method of separation of variables to solve differential equations, we choose, for first order equations, equating to a constant to the first power, and for second order differential equations, equating to a constant to the second power, etc. So, we have

$$\frac{\sin\theta}{F}\frac{d}{d\theta}\left(\sin\theta\frac{dF}{d\theta}\right)+\lambda\sin^2\theta=m^2 \tag{B.90a}$$

$$-\frac{1}{\Phi}\frac{d^2\Phi}{d\varphi^2}=m^2 \tag{B.90b}$$

Eq. (B.90b) is the second order version of Eq. (B.56), describing L_z. The solution is given by Eq. (B.60). It can be shown that any function satisfying Eq. (B.56) must also satisfy Eq. (B.90b). This part of the equation for L^2 corresponds to contributions from L_z^2. We also know that m^2 is the square of an integer −2, −1, 0, 1, 2, 3, etc. We therefore only need to determine λ, as well as the function $F(\theta)$. Before we obtain the solution to Eq. (B.90a), let us transform the equation using

$$\cos\theta = x \rightarrow dx = -\sin\theta d\theta$$

Multiplying Eq. (B.90a) by $F/\sin^2\theta$, Eq. (B.90a) becomes

$$\frac{1}{\sin\theta}\frac{d}{d\theta}\left(\frac{\sin^2\theta}{\sin\theta}\frac{dF}{d\theta}\right)+\lambda F-\frac{m^2}{\sin^2\theta}F=0 \tag{B.91}$$

Making use of the transformation relations for x and dx, given above, we then have

$$\frac{d}{dx}\left[(1-x^2)\frac{dF}{dx}\right]+\lambda F-\frac{m^2}{1-x^2}=0 \tag{B.92a}$$

or

$$(1-x^2)\frac{d^2F}{dx^2}-2x\frac{dF}{dx}+\lambda F-\frac{m^2}{1-x^2}F=0 \tag{B.92b}$$

With this transformation, we need to solve for $F(x) = F(\cos\theta)$. Eqs. (B.92a) and (B.92b) must hold for all values of m. Let us consider, then, the simplest case of $m = 0$. In this case, Eq. (B.92b) becomes

$$(1-x^2)\frac{d^2F}{dx^2}-2x\frac{dF}{dx}+\lambda F=0 \tag{B.93}$$

Eq. (B.93) has the form of the *Legendre differential equation* given in Eq. (B.72). The only difference is that in Eq. (B.93), λ is in place of $n(n+1)$. If we can show that $\lambda = n(n+1)$, then we will have the complete solution, based on our analysis in the previous section, because then, F must be equal to the Associated Legendre function, given by Eq. (B.82).

To determine λ, Eq. (B.93) is sufficient. Our objective is to solve Eq. (B.93) for λ. The *Method of Frobenious* can be used, where we assume the solution takes the form of an infinite series. Thus

$$F = x^s\sum_{n=0}^{\infty}a_nx^n=\sum_{n=0}^{\infty}a_nx^{n+s} \tag{B.94}$$

The parameter s must be ≥ 0 in order to ensure a normalizable wave function at $x = 0$. Recall that our interval of convergence for x is $(-1,1)$. Substituting our infinite series solution into Eq. (B.93), we have

$$(1-x^2)\frac{d^2}{dx^2}\sum_{n=0}^{\infty}a_nx^{n+s}-2x\frac{d}{dx}\sum_{n=0}^{\infty}a_nx^{n+s}+\lambda\sum_{n=0}^{\infty}a_nx^{n+s}=0$$

After carrying out the derivatives of the series, and multiplying the factors, we obtain

$$\sum_{n=0}^{\infty}(n+s-1)(n+s)a_nx^{n+s-2}=$$

$$\sum_{n=0}^{\infty}[(n+s+1)(n+s)+2(n+s)-\lambda]a_n x^{n+s}$$

Multiplying by x^{2-s}, and rewriting the term on the RHS so that $(n+s)$ is factored, we may write

$$\sum_{n=0}^{\infty}(n+s-1)(n+s)a_n x^n = \sum_{n=0}^{\infty}[(n+s)(n+s+1)-\lambda]a_n x^{n+2} \tag{B.95}$$

We can make two observations from Eq. (B.95). First, because both sides are polynomials in x, and the equality must hold for any value of x, it necessitates that corresponding polynomial coefficients must be equal. Thus, x^0, x^1, x^2, etc. must have the same coefficients on both sides of Eq. (B.95). But, there is a difference in the corresponding exponents. The left side has an x^0 and x^1 term, while the right side's first exponent begins at x^2. This means that the coefficients for x^0 and x^1, corresponding to $n=0$ and $n=1$, respectively, must be equal to zero. This leads to the following *indicial equations* for this application of the method of Frobenius.

$$(s-1)s=0 \tag{B.96a}$$

$$(s+1)s=0 \tag{B.96b}$$

We have two second order equations that suggest three unique solutions $s=0$ (twice), $s=1$, and $s=-1$. However, the only allowable values for a bounded function are $s=0$ and $s=1$. $s=-1$ is not allowed since it leads to an infinite value at $x=0$ for the lowest order term in the series. This is a divergent series in the center of our interval $(-1,1)$. We therefore have the two solutions

$$s=0 \tag{B.97a}$$

$$s=1 \tag{B.97b}$$

When we evaluate the first few terms of Eq. (B.95), we get

$$s(s-1)a_0+s(s+1)a_1x+(s+1)(s+2)a_2x^2+(s+2)(s+3)a_3x^3+\ldots =$$
$$[s(s+1)-\lambda]a_0x^2+[(s+1)(s+2)-\lambda]a_1x^3+\ldots \tag{B.98}$$

Since the LHS coefficient index n is shifted by +2 relative to its corresponding power on the RHS, equating the coefficients leads to the following relation

$$(n+s+2)(n+s+1)a_{n+2}=[(n+s)(n+s+1)-\lambda]a_n$$

From this, we obtain a recurrence relation given by

Recurrence Relation For Orbital Angular Momentum

$$\frac{a_{n+2}}{a_n} = \frac{(n+s)(n+s+1)-\lambda}{(n+s+1)(n+s+2)} \tag{B.99}$$

Given our two solutions for s, we have the following

$$\frac{a_{n+2}}{a_n} = \frac{n(n+1)-\lambda}{(n+1)(n+2)} \quad \text{(s=0)} \tag{B.100}$$

$$\frac{a_{n+2}}{a_n} = \frac{(n+1)(n+2)-\lambda}{(n+2)(n+3)} \quad \text{(s=1)} \tag{B.101}$$

Evaluating the coefficients a_n in the asymptotic limit of $n \to \infty$, for a given λ, Eq. (B.99) leads to

$$\lim_{n\to\infty} \frac{a_{n+2}}{a_n} = \frac{n^2}{n^2} = 1$$

This tells us that for large n, the coefficients a_n approach a uniform single value. Let us denote this sufficiently large n by n^*. Then, the infinite series solution becomes

$$x^s \sum_{n=0}^{\infty} a_n x^n \to x^s(a_0 + a_1 x + ... + a_{n^*-1} x^{n^*-1}) + a x^s \sum_{n=n^*}^{\infty} x^n \tag{B.102}$$

In Eq. (B.102), we have uncovered a problem. For the solution to be valid for a well-behaved wave function, the limit of the function towards the boundaries of the problem, which are ± 1, must be zero, otherwise it is not normalizable. The series on the RHS is given by

$$\sum_{n=n^*}^{\infty} x^n = a_{n^*} x^{n^*} (1 + a_1^* x + a_2^* x^2 + ...) \tag{B.103}$$

This series diverges at ± 1 because convergence for any series of this form is only possible when $x < 1$. Therefore, the infinite series cannot be allowed as a solution for a wave function of the orbital angular momentum problem. The series, then, must be truncated to result in finite polynomials of order n, instead. Since the coefficients a_n are proportional to the recurrence relation, this is possible if and only if the recurrence relation vanishes at some finite value of n. For this, it must be that

$$\lambda = n(n+1) \quad \text{and} \quad \lambda = (n+1)(n+2) \tag{B.104}$$

Eqs. (B.104) has redundancy because according to Eq. (B.100) for $s = 0$, the possible values are $0 \cdot 1$, $1 \cdot 2$, $2 \cdot 3$,..., but using Eq. (B.101) for $s = 1$, the possible values are $1 \cdot 2$, $2 \cdot 3$, $3 \cdot 4$,... The case of $s = 0$, thus, covers all possible values of λ, being simply $n(n+1)$. The convention that we will use is to denote the integer n, which is associated with the eigenvalues of the orbital angular momentum operator $\widehat{L}$, as $n \to \ell$. With these results, we have achieved our objective in determining that the allowed values of λ, which are the eigenvalues of the total angular momentum L^2, are given by

Total Orbital Angular Momentum L^2 Eigenvalues

$$\lambda = \ell(\ell+1) \quad \ell = 0,1,2,... \tag{B.105}$$

Recall that our objective was to determine the values of λ, and to show that $\lambda = n(n+1) = \ell(\ell+1)$. We have therefore, also proven that Eq. (B.92b) becomes

$$(1-x^2)\frac{d^2 A_\ell^m}{dx^2} - 2x\frac{dA_\ell^m}{dx} + \ell(\ell+1)A_\ell^m - \frac{m^2}{1-x^2}A_\ell^m = 0 \tag{B.106}$$

In Eq. (B.106), we have replaced F with the associated Legendre function A_ℓ^m, which is the solution. Thus, we have obtained both $\Phi(\varphi)$ and $F(x = \cos\theta) = A_\ell^m(x)$, so we now have the complete solution for the total orbital angular momentum wave functions. The complete eigensolution for the L^2 operator becomes

Spherical Harmonics - Eigenfunctions of L^2 Operator

$$Y_\ell^m = C_n A_\ell^m e^{im\varphi} \tag{B.107}$$

The convention for these eigensolutions is to denote them by Y_ℓ^m. The normalization factor C_n is just the product of those for A_ℓ^m, given by Eq.

(B.86), and $\Phi(\varphi)$ determined to be $1/\sqrt{2\pi}$. So C_n becomes

$$C_n = \sqrt{\frac{(2n+1)(n-m)!}{4\pi(n+m)!}} \tag{B.108}$$

The eigenfunctions of the $\widehat{L^2}$ operator Y_ℓ^m, are known as **spherical harmonics** because they depend only on φ and θ. Given the effort involved in finding the eigenfunctions and corresponding eigenvalues for the orbital angular momentum operators, it should not be too surprising that the *spherical harmonics* are not trivial to visualize. It is helpful to see plots of the orbitals for various values of ℓ and m. But, before we get to that, let us discuss another convention used in expressing the atomic orbitals involving Y_ℓ^m.

We have seen that when we have a set of functions that are solutions to the Schrödinger equation, or eigensolutions, then it follows that any linear combination of them also satisfies the same equation. Thus, we can choose combinations of the spherical harmonics Y_ℓ^m that also satisfy the orbital angular momentum operators. For orbitals with $\ell = 2$, which are called d orbitals, some of the conventional linear combinations are

$$d_{z^2} = Y_2^0 = \left(\frac{5}{16\pi}\right)^{\frac{1}{2}}(3\cos^2\theta - 1) \tag{B.109a}$$

$$d_{xy} = \frac{1}{\sqrt{2}i}(Y_2^{+2} - Y_2^{-2}) = \left(\frac{5}{16\pi}\right)^{\frac{1}{2}}\sin^2\theta\sin 2\varphi \tag{B.109b}$$

$$d_{xz} = \frac{1}{\sqrt{2}}(Y_2^{-1} - Y_2^{+1}) = \left(\frac{5}{16\pi}\right)^{\frac{1}{2}}\sin\theta\cos\theta\sin\varphi \tag{B.109c}$$

$$d_{yz} = \frac{i}{\sqrt{2}}(Y_2^{-1} + Y_2^{+1}) = \left(\frac{5}{16\pi}\right)^{\frac{1}{2}}\sin\theta\cos\theta\cos\varphi \tag{B.109d}$$

$$d_{x^2-y^2} = \frac{1}{\sqrt{2}}(Y_2^{+2} + Y_2^{-2}) = \left(\frac{5}{16\pi}\right)^{\frac{1}{2}}\sin^2\theta\cos 2\varphi \tag{B.109e}$$

The d orbitals turn out to play an important role in magnetic materials. The logic behind the above convention will be clear upon visualizing these orbitals. With these linear combinations, the subscripts now tell us the orientation of the d orbitals in a Cartesian coordinate system. For example, the d_{xy} orbitals lie parallel to the xy plane. The d_{xz} is parallel to the xz plane, etc.

Below d ($\ell = 2$) orbitals are the orbitals corresponding to $\ell = 1$, called p orbitals and $\ell = 0$, called s orbitals. This transformation mimics the natural behavior of p orbits, as they align with the x, y, and z axes, respectively. Correspondingly, these are denoted by p_x, p_y, and p_z. All s orbitals are spherically symmetric. The lobular shapes of the orbitals

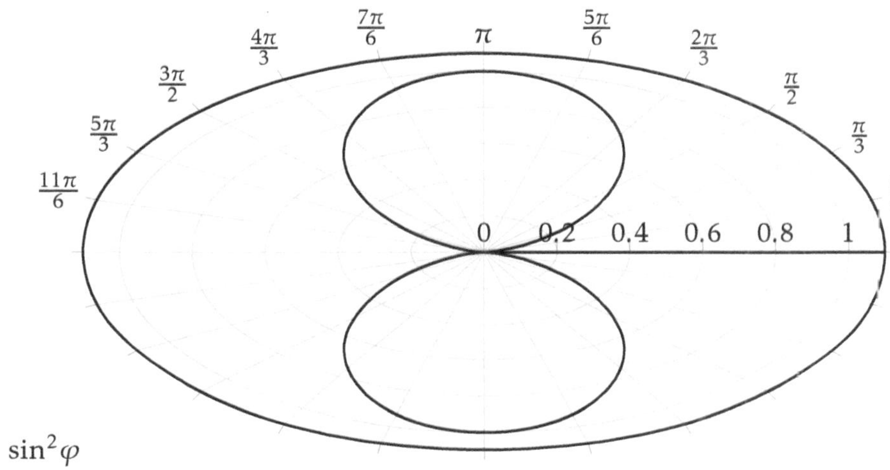

Figure B.8: 2D polar plots of $\sin^2\varphi$.

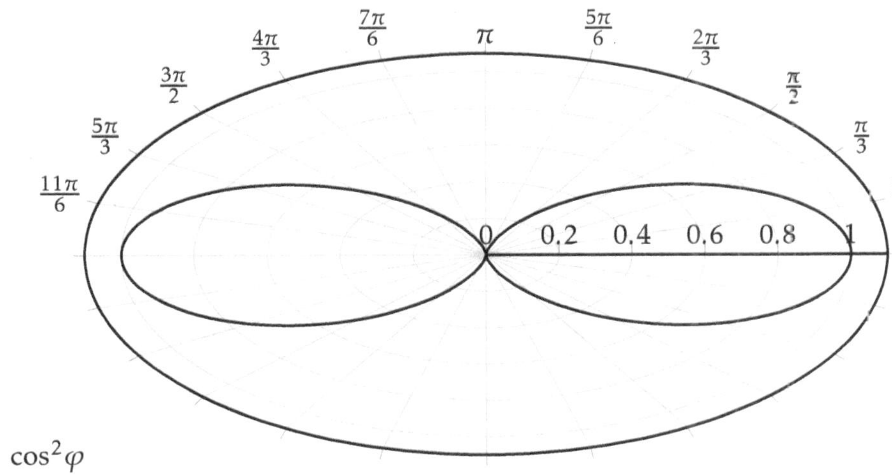

Figure B.9: 2D polar plot of $\cos^2\varphi$.

arise because of the trigonometric functions being plotted in a polar coordinate system. Our intuition tends to lie in Cartesian coordinates, in dealing with transformations between polar and Cartesian. However, in

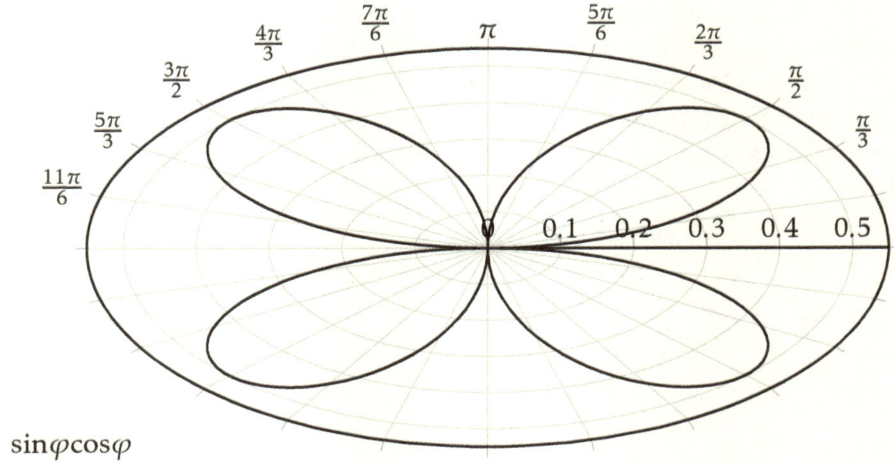

Figure B.10: 2D polar plots of $\sin\varphi\cos\varphi$. Ploting the functions in polar coordinates is the origin of orbital shapes, having an increasing number of lobes as the order increases.

polar coordinates, there is periodicity in the angle argument for sin and cos, and any higher order function of these terms. Hence, the resulting curves produce **polar lobes**.

Figs. B.8, B.9, and B.8 illustrate lobes in 2D polar plots corresponding to some of the components in the *d* orbitals. They illustrate the origin of orbital lobes, as well as the increasing number of lobes. Two lobes result from symmetric functions such as $\sin^2\varphi$ and $\cos^2\varphi$. The function $\cos^2\varphi$ is just a 90° rotation of $\sin^2\varphi$. And, when there is a $\sin\varphi \cdot \cos\varphi$ term, symmetry is broken, and two lobes become four since the product now has four zeros instead of two. As ℓ increases, more lobes generally can result. Fig. B.11 illustrates some 3D conventional orbitals, from *s* to *d*.

This completes the solution of the orbital angular momentum problem, known as *spherical harmonics* Y_ℓ^m. We started by introducing the quantum mechanical operators $\widehat{L}_x, \widehat{L}_y$, and $\widehat{L}_z$. This allowed us to define eigenvalue equations for both $\widehat{L}_z$ and $\widehat{L^2}$, and solve them exactly. Now, we'll turn our attention to an important topic in quantum mechanics that applies to any operator, but we'll introduce it using momentum operators. The concept is known as *commutation* and it's the subject of the next section.

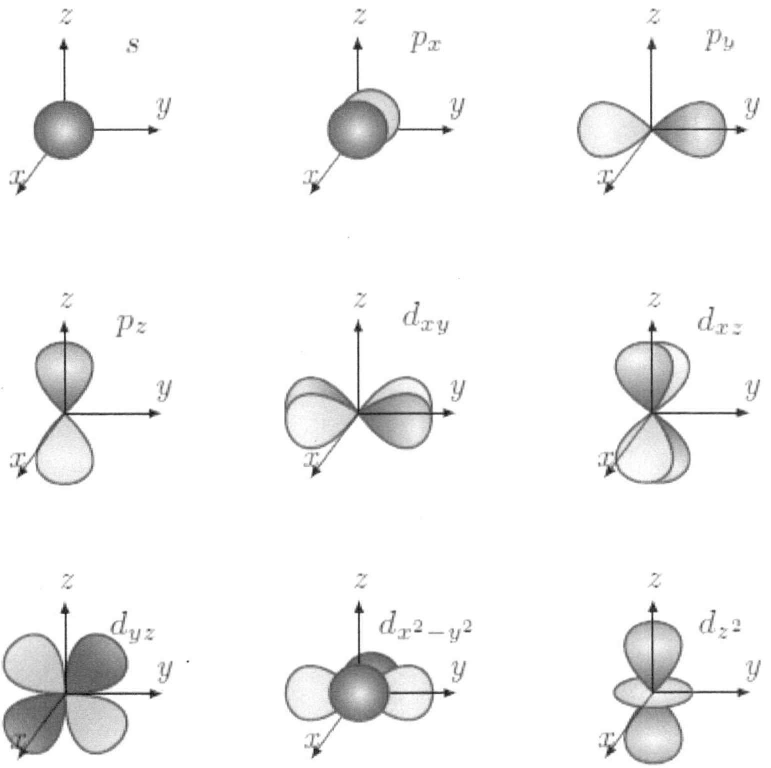

Figure B.11: Conventional single electron orbitals which are linear combinations of Y_ℓ^m.

B.9 Schrödinger's Hydrogen Atom Model

When Bohr treated the hydrogen atom, planar loops where assumed to express the angular momentum, simply as $L = mrv$. This is valid when p and r are orthogonal. This is not a necessary assumption for an atom, which lives in 3D. Additionally, Bohr treated only the electron energy to solve the problem. The nucleus was not considered at all, though this assumption turns out to be fine given the relatively large mass of the nucleus. But this approximation should be justified somehow. It turns out that with the Schrödinger equation, it is not necessary to ignore the nucleus. It is remarkable that although solving the hydrogen atom problem using quantum mechanics involves significantly more mathematics, we will see that it leads to results that are in complete agreement with Rydberg's empirical model, Bohr's early quantum model, and ultimately with even more general experimental data given by hydrogen's atomic spectra. So, let's work out the results from the Schrödinger equation.

Recall that the potential energy $V(\mathbf{r} = |\mathbf{r}_1 - \mathbf{r}_2|)$ of a system involving a single electron with position $\mathbf{r}_1$ and charge $q_1 = -e$, interacting with a nucleus having position $\mathbf{r}_2$ of charge $q_2 = Ze$, is given (in SI units) by

$$V(r) = -\frac{Ze^2}{4\pi\varepsilon_0|\mathbf{r}_1 - \mathbf{r}_2|} \tag{B.110}$$

This electrostatic potential describes any nuclear system of charge Ze having a single electron, like hydrogen, but also, singly-ionized helium (He^+), doubly-ionized lithium (Li^{++}), etc. Including the nucleus as well as the outer electron, the Hamiltonian eiqenvalue equation becomes

$$\widehat{H}\psi = -\frac{\hbar^2}{2m_1}\nabla^2\psi - \frac{\hbar^2}{2m_2}\nabla^2\psi + V\psi = E_\lambda\psi \tag{B.111}$$

Expanding Eq. (B.111), which contains two Laplacian operators, into its full 3D form gives

$$-\frac{\hbar^2}{2m_1}\left[\frac{\partial^2}{\partial x_1^2} + \frac{\partial^2}{\partial y_1^2} + \frac{\partial^2}{\partial z_1^2}\right]\psi - \frac{\hbar^2}{2m_2}\left[\frac{\partial^2}{\partial x_2^2} + \frac{\partial^2}{\partial y_2^2} + \frac{\partial^2}{\partial z_2^2}\right]\psi + V\psi = E_\lambda\psi \tag{B.112}$$

One difficulty with Eq. (B.112) is that it is *not separable*. Thus, we cannot use the method of separation of variables directly on this equation. This is because of the form of $V(|\mathbf{r}_1 - \mathbf{r}_2|)$. For separability, V should be either purely a function of $\mathbf{r}_1$, purely a function of $\mathbf{r}_2$, or a linear combination of any power of $\mathbf{r}_1$ and $\mathbf{r}_2$. However, since V is a function of $|\mathbf{r}_1 - \mathbf{r}_2|$ (a nonlinear function), we have this roadblock. Fortunately, it turns out that we can transform the terms of the Hamiltonian other than the potential to be an explicit function of $\mathbf{r}_1 - \mathbf{r}_2$, and another variable. So, we can trade the two vectors variables $[x_1, y_1, z_1]^T$ and $[x_2, y_2, z_2]^T$ we have for two transformed variables. This does the trick that will allow us to use the method of separation of variables. Let's begin the transformation of the Hamiltonian using the first variable suggested by the potential function, namely

$$\mathbf{r} = \mathbf{r}_1 - \mathbf{r}_2 = [x_1 - x_2, y_1 - y_2, z_1 - z_2]^T = [x, y, z]^T \tag{B.113}$$

For the second variable, we will use the **center of mass** of the hydrogen atom, denoted by **R**, defined as

$$\mathbf{R} = [X, Y, Z]^T \tag{B.114a}$$

The center of mass coordinates are defined as

$$X = \frac{m_1 x_1 + m_2 x_2}{m_1 + m_2} \tag{B.114b}$$

$$Y = \frac{m_1 y_1 + m_2 y_2}{m_1 + m_2} \tag{B.114c}$$

$$Z = \frac{m_1 z_1 + m_2 z_2}{m_1 + m_2} \tag{B.114d}$$

Since Eq. (B.112) involves second order derivative terms due to the Laplacian operator $\nabla^2\cdot$, we must replace all the derivative terms with their equivalent in terms of the relative position **r** and the center of mass **R**. We will build up this process to get $\partial^2\psi/\partial x_1^2$. The process goes the same for $\partial^2\psi/\partial x_2^2$ and $\partial^2\psi/\partial x_3^2$, etc. Because x_1 generally depends on both x and X, starting with the first derivative, we have

$$\frac{\partial\psi}{\partial x_1} = \frac{\partial\psi}{\partial x}\frac{\partial x}{\partial x_1} + \frac{\partial\psi}{\partial X}\frac{\partial X}{\partial x_1}$$

Using the definitions for r and R given in Eqs. (B.113)-(B.114d), the first derivative becomes

$$\frac{\partial\psi}{\partial x_1} = \frac{\partial\psi}{\partial x} + \frac{m_1}{m_1 + m_2}\frac{\partial\psi}{\partial X}$$

Differentiation for $\partial\psi/\partial x_1$ to obtain the second order derivative gives

$$\begin{aligned}\frac{\partial^2\psi}{\partial x_1^2} &= \frac{\partial}{\partial x_1}\left(\frac{\partial\psi}{\partial x}\right) + \frac{m_1}{m_1 + m_2}\frac{\partial}{\partial x_1}\left(\frac{\partial\psi}{\partial X}\right) \\ &= \frac{\partial^2\psi}{\partial x^2} + \frac{\partial^2\psi}{\partial x\partial X}\frac{\partial X}{\partial x_1} + \frac{m_1}{m_1 + m_2}\left(\frac{\partial^2\psi}{\partial x\partial X}\frac{\partial x}{\partial x_1} + \frac{\partial^2\psi}{\partial X^2}\frac{\partial X}{\partial x_1}\right)\end{aligned}$$

So, we have the following result for $\partial^2\psi/\partial x_1^2$:

$$\frac{\partial^2\psi}{\partial x_1^2} = \frac{\partial^2\psi}{\partial x^2} + \frac{2m_1}{m_1 + m_2}\frac{\partial^2\psi}{\partial x\partial X} + \left(\frac{m_1}{m_1 + m_2}\right)^2\frac{\partial^2\psi}{\partial X^2} \tag{B.115}$$

The final step is just multiplying Eq. (B.115) by $-\hbar^2/2m_1$, leading to the result

$$-\frac{\hbar^2}{2m_1}\frac{\partial^2\psi}{\partial x_1^2} = -\frac{\hbar^2}{2m_1}\frac{\partial^2\psi}{\partial x^2} - \frac{\hbar^2}{m_1 + m_2}\frac{\partial^2\psi}{\partial x\partial X} - \frac{\hbar^2 m_1}{2(m_1 + m_2)^2}\frac{\partial^2\psi}{\partial X^2} \tag{B.116}$$

Let us evaluate the corresponding term for the proton with mass m_2 and position x_2. After carrying out a similar process as above, the first derivative for x_2 becomes

$$\frac{\partial\psi}{\partial x_2} = -\frac{\partial\psi}{\partial x} + \frac{m_2}{m_1+m_2}\frac{\partial\psi}{\partial X}$$

Differentiation again w.r.t. x_2, and multiplying by $-\hbar^2/2m_2$ gives

$$-\frac{\hbar^2}{2m_2}\frac{\partial^2\psi}{\partial x_2^2} = -\frac{\hbar^2}{2m_2}\frac{\partial^2\psi}{\partial x^2} + \frac{\hbar^2}{m_1+m_2}\frac{\partial^2\psi}{\partial x\partial X} - \frac{\hbar^2 m_2}{2(m_1+m_2)^2}\frac{\partial^2\psi}{\partial X^2} \quad \text{(B.117)}$$

In the Hamiltonian given by Eq. (B.112), both these terms add together. The mixed partial derivative terms end up summing to zero since they have opposite signs. So, for the x- component of the summed Laplacian terms in the Hamiltonian, we have

$$-\frac{\hbar^2}{2m_1}\frac{\partial^2\psi}{\partial x_1^2} - \frac{\hbar^2}{2m_2}\frac{\partial^2\psi}{\partial x_2^2} = -\frac{\hbar^2}{2}\left(\frac{1}{m_1}+\frac{1}{m_2}\right)\frac{\partial^2\psi}{\partial x^2} - \frac{\hbar^2}{2}\left(\frac{1}{m_1+m_2}\right)\frac{\partial^2\psi}{\partial X^2} \quad \text{(B.118)}$$

Based on Eq. (B.118), we can define the following mass m^* for the combined system

Reduced Mass Definition of a Two-Body System

$$\frac{1}{m^*} = \frac{1}{m_1} + \frac{1}{m_2} \quad \text{(B.119)}$$

Thus, the masses of the proton nucleus and electron combine in the same way as electrical capacitors in series, since in this case

$$\frac{1}{C_{eq}} = \frac{1}{C_1} + \frac{1}{C_2}$$

Eq. (B.119) defines what is known as the **reduced mass** of a two-body system, and it is *always* smaller than the minimum of (m_1,m_2). From this definition, m^* is given by

$$m^* = \frac{m_1 m_2}{m_1+m_2}$$

It is straightforward to show that the reduced mass $m^* < \min(m_1, m_2)$. To see this, let m_1 be the smaller of the two. Then, it follows that

$$m^* = \frac{m_1 m_2}{m_1 + m_2} < \frac{m_1 m_2 + m_1^2}{m_1 + m_2} = m_1 \left(\frac{m_2 + m_1}{m_1 + m_2} \right) = m_1$$

This *reduced mass* is associated with the motion of the system, described by the kinetic energy. In the last term of Eq. (B.118), we see the mass $m_1 + m_2 = M$, which is the total mass of the system. Because all the terms look identical for x, y, and z, the process is the same for the other two components y_1, z_1 and y_2, z_2, respectively. So, our final result transforming the Hamiltonian becomes

$$-\frac{\hbar^2}{2m^*}\left(\frac{\partial^2\psi}{\partial x^2} + \frac{\partial^2\psi}{\partial y^2} + \frac{\partial^2\psi}{\partial z^2}\right) - \frac{\hbar^2}{2M}\left(\frac{\partial^2\psi}{\partial X^2} + \frac{\partial^2\psi}{\partial Y^2} + \frac{\partial^2\psi}{\partial Z^2}\right) - V\psi = E_\lambda \psi \tag{B.120}$$

Since $V = V(r = |\mathbf{r}_1 - \mathbf{r}_2|) = V(x, y, z)$, Eq. (B.120) *is* separable and therefore, we may use the *method of separation of variables* to solve it.

As we have done in the method of separation of variables, let $\psi = B(r)F(R)$. $B(r)$ describes the relative localized system behavior, corresponding to the bound states of the electron, while $F(\mathbf{R})$ describes the free motion of the center of mass $\mathbf{R}$ of the hydrogen atom. Substitution into Eq. (B.120), then dividing by $B \cdot F$ gives

$$-\frac{\hbar^2}{2m^*}\frac{1}{B(r)}\left(\frac{\partial^2 B}{\partial x^2} + \frac{\partial^2 B}{\partial y^2} + \frac{\partial^2 B}{\partial z^2}\right) + V(\mathbf{r}) - \frac{\hbar^2}{2M}\frac{1}{F(R)}\left(\frac{\partial^2 F}{\partial X^2} + \frac{\partial^2 F}{\partial Y^2} + \frac{\partial^2 F}{\partial Z^2}\right) = E_\lambda \tag{B.121}$$

The first part of Eq. (B.121) is purely a function of $\mathbf{r} = \mathbf{r}_1 - \mathbf{r}_2$, while the second part is purely a function of $\mathbf{R}$. Both of them add to equal the eigenvalue constant E_λ. Based on previous problems we have solved using the *method of separation of variables*, this is true if and only if both the function of r and R, respectively, equal to constants. So, we can write the following pair of equations:

$$-\frac{\hbar^2}{2m^*}\frac{1}{B(\mathbf{r})}\left(\frac{\partial^2 B}{\partial x^2} + \frac{\partial^2 B}{\partial y^2} + \frac{\partial^2 B}{\partial z^2}\right) + V(r) = E_r \tag{B.122a}$$

$$\frac{\hbar^2}{2M}\frac{1}{F(\mathbf{R})}\left(\frac{\partial^2 F}{\partial X^2} + \frac{\partial^2 F}{\partial Y^2} + \frac{\partial^2 F}{\partial Z^2}\right) = E_R \tag{B.122b}$$

The relationship between E_r, E_R, and E_λ becomes

$$E_r + E_R = E_\lambda \tag{B.123}$$

Let's focus on the first of these equations, describing $B(r)$ for the bound states. Eq. (B.122a) can be written as

$$\frac{\partial^2 B}{\partial x^2} + \frac{\partial^2 B}{\partial y^2} + \frac{\partial^2 B}{\partial z^2} + \frac{2m^*(E_r - V)}{\hbar^2} B = 0 \tag{B.124}$$

Since $\mathbf{r} = \mathbf{r}_1 - \mathbf{r}_2$, the *origin* of the relative electron's position r is the location of the proton-nucleus $\mathbf{r}_2$. Because of this, it is useful to express Eq. (B.124) in spherical coordinates (r, θ, φ). Rewriting the Laplacian in Cartesian coordinates in spherical coordinates gives

$$\frac{1}{r^2}\left[\frac{\partial}{\partial r}\left(r^2 \frac{\partial}{\partial r}\right) + \frac{1}{\sin\theta}\frac{\partial}{\partial \theta}\left(\sin\theta \frac{\partial}{\partial \theta}\right) + \frac{1}{\sin^2\theta}\frac{\partial^2}{\partial \varphi^2}\right] B + \frac{2m^*(E_r - V)}{\hbar^2} B = 0 \tag{B.125}$$

The second and third terms in the brackets, on the LHS, is the $\widehat{L^2}$ operator for the angular momentum, given by Eq. (B.66). Since Eq. (B.125) is obviously separable in r and the other two variables θ and φ, we may assume $B(r, \theta, \varphi)$ is given by

$$B(r, \theta, \varphi) = R(r) Y_{\ell,m}(\theta, \varphi) \tag{B.126}$$

$Y_{\ell,m}(\theta, \varphi)$ are the spherical harmonics we obtained in Chap. B, which are eigenfunctions of the $\widehat{L^2}$ and $\widehat{L_z}$ operators. If we assume this form and get all the way to a final solution successfully, then, it will mean that our assumption works! This is how we will approach the solution here. So, let's roll with it for now.

The spherical harmonics $Y_{\ell,m}$ are given by Eq. (B.107), with some examples in Table B.7. Substitution into Eq. (B.125), then dividing by $B = R \cdot Y_{\ell,m}$, we obtain the following *separated* differential equation:

$$\frac{1}{R}\frac{\partial}{\partial r}\left(r^2 \frac{\partial}{\partial r}\right) R - \frac{2m^* V r^2}{\hbar^2} + \frac{2m^* E_r r^2}{\hbar^2} + \frac{1}{Y_{\ell,m}}\left[\frac{1}{\sin\theta}\frac{\partial}{\partial \theta}\left(\sin\theta \frac{\partial}{\partial \theta}\right) + \frac{1}{\sin^2\theta}\frac{\partial^2}{\partial \varphi^2}\right] Y_{\ell,m} = 0 \tag{B.127}$$

Eq. (B.127) reveals to us that the method of separation of variables has worked, as we have transformed Eq. (B.125) to a sum of a function of r and a function of θ and φ, equal to a constant. This means that both functions separately also equal to constants, and therefore, the θ, φ part becomes equal to a constant. This then becomes identical to the $\widehat{L^2}$ operator. Then we may replace this part of the equation with the eigenvalue of the $\widehat{L^2}$ operator. After doing this, Eq. (B.127) becomes

$$\frac{1}{R}\frac{\partial}{\partial r}\left(r^2\frac{\partial}{\partial r}\right)R - \frac{2m^*Vr^2}{\hbar^2} + \frac{2m^*E_r r^2}{\hbar^2} - \frac{1}{Y_\ell^m}\left[\ell(\ell+1)\right]Y_\ell^m = 0 \quad \text{(B.128)}$$

Left-multiplying by $R(r)/r^2$, we have the following equation in terms of r:

Radial Part of Schrödinger Equation

$$\frac{1}{r^2}\frac{d}{dr}\left(r^2\frac{dR}{dr}\right) + \frac{2m^*}{\hbar^2}\left[E_r - V(r) - \frac{\ell(\ell+1)\hbar^2}{2m^*r^2}\right]R(r) = 0 \quad \text{(B.129)}$$

We have replaced the partial derivatives with total derivatives, since we now have a function of the single variable r. Eq.(B.129) is known as the *radial part of the Schrödinger equation.* In the next section, we shall determine the exact solution to this equation.

B.10 Schrödinger Equation Radial Solution

If we obtain an exact solution to the radial equation, we will have also obtained the solution $B(r,\theta,\varphi) = R(r)Y_{\ell,m}$, described by Eq. (B.125). To solve it, we'll further transform Eq. (B.129). Let's use the following transformation:

$$R(r) = \frac{U(r)}{r}$$

To complete the substitution, we will also need the derivatives of R in terms of U. The first derivative is given by

$$\frac{dR}{dr} = \frac{1}{r}\frac{dU}{dr} - \frac{U(r)}{r^2}$$

Then, multiplying by r^2, we have

$$r^2\frac{dR}{dr} = r\frac{dU}{dr} - U(r)$$

Differentiating the above result, the first order derivative terms cancel, and we have

$$\frac{1}{r^2}\frac{d}{dr}\left(r^2\frac{dR}{dr}\right)=\frac{1}{r}\frac{d^2U}{dr^2}$$

Substituting these results into Eq. (B.129), we have

$$\frac{1}{r}\frac{d^2U}{dr^2}+\frac{2m^*}{\hbar^2}\left[E_r-V(r)-\frac{\hbar^2\ell(\ell+1)}{2m^*r^2}\right]\frac{U(r)}{r}=0$$

Multiplying by r and replacing $V(r)$ with its definition using Eq. (B.110), we get the following corresponding differential equation for $U(r)$:

$$\frac{d^2U}{dr^2}+\left[\frac{2m^*E_r}{\hbar^2}+\frac{Zm^*e^2}{2\pi\hbar^2\varepsilon_0}\frac{1}{r}-\frac{\ell(\ell+1)}{r^2}\right]U(r)=0 \tag{B.130}$$

In addition to transforming the function R, we will also transform independent variable r to obtain a more convenient *normalized* form. This is done by introducing the dimensionless variable ζ defined by

$$\zeta=\alpha r\Rightarrow d\zeta=\alpha dr$$

We'll define α shortly. Then, the second derivative term becomes

$$\frac{d^2U}{dr^2}=\alpha^2\frac{d^2U}{d\zeta^2}$$

Eq. (B.130) then becomes

$$\alpha^2\frac{d^2U}{d\zeta^2}+\left[\frac{2m^*E_r}{\hbar^2}+\frac{Zm^*\alpha e^2}{2\pi\hbar^2\varepsilon_0}\frac{1}{\zeta}-\frac{\alpha^2\ell(\ell+1)}{\zeta^2}\right]U(\zeta)=0$$

Dividing by α^2, we have

$$\frac{d^2U}{d\zeta^2}+\left[\frac{2m^*E_r}{\hbar^2\alpha^2}+\frac{Zm^*e^2}{2\pi\hbar^2\varepsilon_0\alpha}\frac{1}{\zeta}-\frac{\alpha^2\ell(\ell+1)}{\zeta^2}\right]U(\zeta)=0 \tag{B.131}$$

We may define α^2 such that the first term in brackets becomes

$$\frac{2m^*E_r}{\hbar^2\alpha^2}=-\frac{1}{4}\Rightarrow\alpha^2=-\frac{8m^*E_r}{\hbar^2} \tag{B.132}$$

This choice recognizes *a priori*, that $E_r<0$, corresponding to bound states, and it ensures that $\alpha^2>0$. So, Eq. (B.131) may now be written as

$$\frac{d^2U}{d\zeta^2}+\left[-\frac{1}{4}+\frac{\beta}{\zeta}-\frac{\ell(\ell+1)}{\zeta^2}\right]U(\zeta)=0 \tag{B.133}$$

The parameter β is a constant given by

$$\beta \equiv \frac{Zm^*e^2}{2\pi\hbar^2\varepsilon_0\alpha} \tag{B.134}$$

With this form, let's determine the solution behavior in two asymptotic limits. First, for the asymptotic limit of $\zeta \to \infty$, in which case Eq. (B.133) becomes

$$\frac{d^2U}{d\zeta^2} - \frac{1}{4}U(\zeta) = 0 \tag{B.135}$$

The solution of Eq. (B.135) is given by

$$U_{\zeta\to\infty} = U_0 e^{-\frac{\zeta}{2}} = 0 \tag{B.136}$$

For the other asymptotic limit, when $\zeta \to 0$, Eq. (B.133) becomes

$$\frac{d^2U}{d\zeta^2} - \frac{\ell(\ell+1)}{\zeta^2}U(\zeta) = 0 \tag{B.137}$$

Eq. (B.137) can be solved assuming a solution of the form

$$U = \zeta^p \tag{B.138}$$

The second derivative is then given by

$$\frac{d^2U}{d\zeta^2} = p(p-1)\zeta^{a-2}$$

Plugging this into Eq. (B.137), and multiplying by ζ^2, we get

$$p(p-1)\zeta^p - \ell(\ell+1)\zeta^p = 0 \tag{B.139}$$

Since ζ is arbitrary and generally nonzero, it follows that

$$p^2 - p - \ell(\ell+1) = 0$$

The two roots of this quadratic equation are given by

$$p = \frac{1 \pm \sqrt{1+4\ell(\ell+1)}}{2} = \frac{1 \pm (2\ell+1)}{2} \to p = \ell+1, -\ell \tag{B.140}$$

For the problem we are considering, the negative exponent is not allowed for describing the wave function because it leads to a $1/\zeta^\ell$ factor, which does not approach zero in the limit of $\zeta \to 0$. This is an important point since this is the dominating term as ζ approaches zero. In this limit, this function must 'carry' the wave function to zero in order to be normalizable in spherical coordinates, just as the other asymptotic limit did for us when $\zeta \to \infty$. So, we have

$$p = \ell + 1 \quad \text{and} \quad U_{\zeta\to\infty}(\zeta) = U_0 \zeta^{\ell+1}$$

Incorporating both of asymptotic limits into the solution, let $U(\zeta) = U_{\zeta\to 0} \cdot U_{\zeta\to\infty} \cdot W(\zeta) = G \cdot W(\zeta)$. Since we know $G(\zeta)$ already, let us work to obtain an equation describing only $Y(\zeta)$. Then, if we solve for $W(\zeta)$,we will have the solution for $U = rR$, etc. Using this definition for U, the second derivative becomes

$$\frac{d^2U}{d\zeta^2} = W\frac{d^2G}{d\zeta^2} + 2\frac{dG}{d\zeta}\frac{dW}{d\zeta} + G\frac{d^2W}{d\zeta^2} \tag{B.141}$$

Since $G = \zeta^{\ell+1} \cdot e^{-\zeta/2}$, the first derivative of G is given by

$$\frac{dG}{d\zeta} = (\ell+1)\zeta^{\ell} e^{-\frac{\zeta}{2}} - \frac{1}{2}\zeta^{\ell+1} e^{-\frac{\zeta}{2}} = \left[\frac{\ell+1}{\zeta} - \frac{1}{2}\right] G(\zeta) \tag{B.142}$$

The second derivative then becomes

$$\begin{aligned}\frac{d^2G}{d\zeta^2} &= \frac{d}{d\zeta}\left[\left(\frac{\ell+1}{\zeta} - \frac{1}{2}\right)G(\zeta)\right] \\ &= \left(\frac{\ell+1}{\zeta} - \frac{1}{2}\right)\frac{dG}{d\zeta} - \left(\frac{\ell+1}{\zeta^2}\right)G(\zeta)\end{aligned}$$

Using the result for $dG/d\zeta$, we have

$$\frac{d^2G}{d\zeta^2} = -\left(\frac{\ell(\ell+1)}{\zeta^2} - \frac{\ell+1}{\zeta} + \frac{1}{4}\right)G(\zeta)$$

Substitution of these results into Eq. (B.141), then substituting that result into (B.133), and multiplying by ζ, we obtain the following differential equation describing $W(\zeta)$

Hydrogen's Confluent Hypergeometric Equation

$$\zeta\frac{d^2W}{d\zeta^2} + [2(\ell+1) - \zeta]\frac{dW}{d\zeta} - (\ell + 1 - \beta)W = 0 \tag{B.143}$$

Eq. (B.143) is another example of the *Confluent Hypergeometric Equation*, discussed in detail in Chap. A, obtained in the solution of the Schrödinger equation applied to the harmonic oscillator. The CHE is given by

$$x\frac{d^2y}{dx^2} + (c - x)\frac{dy}{dx} - ay(x) = 0$$

By inspection, we see that

$$c = 2(\ell + 1) \quad \text{and} \quad a = \ell + 1 - \beta \tag{B.144}$$

In solving the CHE in Chap. A, we obtained the solution by using the *method of Frobenius*, which assumes the solution has the form of an infinite series. However, we also found that in order for the solution to satisfy the normalization conditions of a localized wave function, the infinite series must be truncated to a finite polynomial. The infinite series we found is the *Confluent Hypergeometric function* ${}_1F_1(\zeta)$, and the truncated polynomials were scaled to obtain Hermite polynomials H_n. With this, we can deduce the solution for $W(\zeta)$. Moreover, recall that the truncation of the confluent hypergeometric function was achieved by imposing the important condition on the input parameter a, namely, that

$$a = \ell + 1 - \beta = -n_r \quad n_r = 0, 1, 2, \ldots$$

The parameter β must take values given by

$$\beta = n_r + \ell + 1 \tag{B.145}$$

Since n_r and ℓ both take values $0, 1, 2, \ldots$, we can simply define β as a **total quantum number** n, whose integer value is

$$n = n_r + \ell + 1 \quad \text{(total quantum number)} \tag{B.146}$$

In quantum mechanics, the *total quantum number* is also known as the **principal quantum number**, and takes the discrete values $1, 2, 3, \ldots$ (notice it cannot equal zero). Using the definition for β, we get

$$\beta = n = \frac{Zm^*e^2}{2\pi\hbar^2\varepsilon_0\alpha}$$

Squaring the result and substituting α^2 given by Eq. (B.132), then solving for $E_r = E_n$, we obtain the following for the allowed energy values:

Hydrogen Electron Bound Energy States (no magnetic field)

$$E_n = -\frac{Z^2 m^* e^4}{32\pi^2\hbar^2\varepsilon_0^2}\frac{1}{n^2} \tag{B.147}$$

In terms of the Planck constant h and the principal quantum number n, Eq. (B.147) can be written as

$$E_n = -\frac{Z^2 m^* e^4}{8h^2\varepsilon_0^2}\frac{1}{n^2} = -\frac{E_1}{n^2} \tag{B.148}$$

If we set $Z = 1$, the result is identical to Bohr's result given by Eq. (B.38). It is remarkable that after this relatively complicated analysis, the results have lead exactly to Bohr's result, for the case of the hydrogen atom $Z = 1$. E_1 represents the lowest possible energy where $E_1 \sim 13.6$ eV. Note that Bohr's model can also be generalized to obtain Eq. (B.148), as well. This difference is not arising from a quantum mechanical effect, but from assuming the nucleus in Bohr's model to have a charge of $Z = 1$. Schrödinger demonstrated this result in one of three papers he published in 1926, marking the introduction of his gift of the Schrödinger equation to the world, and the rest is history.

The energy difference ΔE_n, which is observed in the measurement of atomic and/or molecular spectra, is given by

Hydrogen Bound-State Energy Differences

$$\Delta E_n = \frac{Z^2 m^* e^4}{8h^2\varepsilon_0^2}\left(\frac{1}{n_1^2} - \frac{1}{n_2^2}\right) \tag{B.149}$$

In Eq. (B.149), $n_2 > n_1$, and this equation leads to the familiar hydrogen spectral series, and when $Z = 1$, gives rise to the series of Lyman, Balmer, Paschen, etc. We are now in a position to write down a complete wave function for the hydrogen atom. The complete solution for $U(\zeta)$ is given by

$$U(\zeta) = C_n \zeta^{\ell+1} e^{-\frac{\zeta}{2}} {}_1F_1(\zeta; a, c)$$

Parameters a and c are input parameters to the *Confluent Hypergeometric Function* described by Eq. (B.143). Because a takes the value of a negative integer, the infinite series is truncated. These truncated polynomials are proportional to Hermite polynomials H_n, however, since c also depends on the integer ℓ, we cannot substitute H_n for the CHF like we did in the harmonic oscillator problem where $c = 1/2$ was independent of any quantum number. Thus, we must keep this form to correctly account for all values of c. Then, with $\zeta = \alpha r$, $U(\zeta)$ becomes

$$U(r) = C_n \alpha_n^{\ell+1} r^{\ell+1} e^{-\frac{\alpha r}{2}} {}_1F_1(\zeta; a, c)$$

Since $R(r) = U(r)/r$, the **radial wave functions** for the hydrogen atom become

Hydrogen Bound-State Radial wave functions

$$R_{n,\ell}(r) = C_{n,\ell} (\alpha_n r)^\ell e^{-\frac{\alpha_n r}{2}} {}_1F_1(\alpha_n r; \ell + 1 - n, 2\ell + 2) \qquad \text{(B.150)}$$

$C_{n,\ell}$ is the normalization factor that must be determined from the normalization condition. This integral, in spherical coordinates, is given by

$$\int_0^\infty R^*_{n,\ell}(r) R_{n,\ell}(r) r^2 dr = 1 \qquad \text{(B.151)}$$

The normalization factor $C_{n,\ell}$ corresponding to $R_{n,\ell}$ works out to be

Radial Function Normalization Factor

$$C_{n,\ell} = \frac{\alpha_n^{3/2}}{(2\ell+1)!} \left[\frac{(n+1)!}{2n \cdot (n - \ell - 1)!} \right]^{1/2} \qquad \text{(B.152)}$$

$\alpha = \alpha_n$ is given by Eq. (B.132). Fig. B.12 illustrates some examples of the radial wave functions for $\ell = 0, \ell = 1$ and $\ell = 2$, with $n = 1 - 7$.

Fig. B.12 shows the radial wave functions. Keep in mind that the corresponding probability densities are given by $r^2 R^2_{n,\ell}$ since the 3D integration is over spherical coordinates. The left, middle and right columns correspond to $\ell = 0, 1, 2$, respectively. These functions also convolve with the spherical harmonics $Y_{\ell,m}$ to form the various s, p, and d orbitals, *etc.* These orbitals will be formally defined in Sec. **??**. As n and ℓ increase, the probability distributions, depending on $R_{n,\ell}(r)$, increasingly widen or broaden their spatial extent. Therefore, broadening is a characteristic of higher and higher energy states of the hydrogen atom, and it contributes to enlarging the effective atom or molecule size.

Figure B.12: Hydrogen radial wave functions $R_{n,\ell}$, for cases $\ell = 0$ (left column), $\ell = 1$ (middle column), and $\ell = 2$ (right column). r is normalized by the Bohr radius $a_0 \sim 0.52919$Å, given by Eq. (B.39).

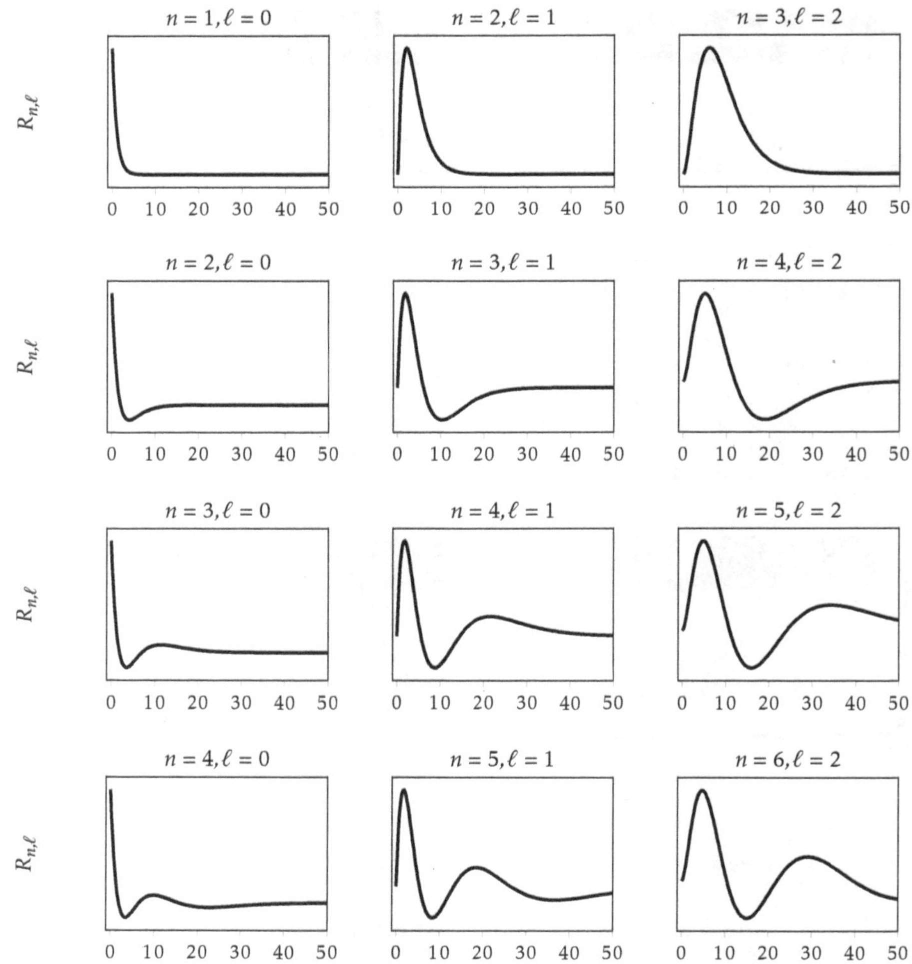

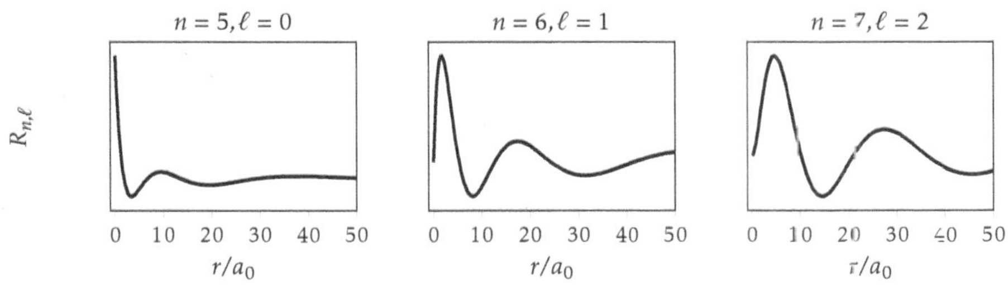

With $R_{n,\ell}(r)$, we can write the complete hydrogen atom wave functions for the bound states which are associated with bound-state energy E_r. The bound-state wave function, $B(r,\theta,\varphi) = R_{n,\ell}Y_{\ell,m}$ are given by

Hydrogen Bound-State wave functions

$$B_{\ell,m,n}(r,\theta,\varphi) = C_n\alpha_n^\ell r^\ell e^{-\frac{\alpha_n r}{2}}\,{}_1F_1(\alpha_n r;\ell+1-n,2\ell+2)Y_{\ell,m}(\theta,\varphi) \tag{B.153}$$

The normalization factor for B is the product of the normalization factors for the radial wave function, given by Eq. (B.152), and that of the spherical harmonics, given by Eq. (B.108). Therefore, $C_n = C_{n,\ell,m}$ becomes

$$C_{n,\ell,m} = \frac{\alpha^{3/2}}{(2\ell+1)!}\left[\frac{(2n+1)(n-m)!}{4\pi(n+m)!}\right]^{1/2}\left[\frac{(n+1)!}{2n\cdot(n-\ell-1)!}\right]^{1/2} \tag{B.154}$$

We now have the complete solution B for the bound states, and they are described by Eq. (B.125). The final part to the hydrogen atom solution from the Schrödinger equation is to obtain the solution for F, described by Eq. (B.122b). Since there is no potential energy term, it is identical to the free particle problem discussed in Sect. **??**, where $V = 0$. Eq. (B.122b) can be written as

$$\frac{\partial^2 F}{\partial X^2} + \frac{\partial^2 F}{\partial Y^2} + \frac{\partial^2 F}{\partial Z^2} + k^2 F = 0 \tag{B.155}$$

k^2 is the square of the wavenumber defined as

$$k^2 = \frac{2ME_R}{\hbar^2} \tag{B.156}$$

Eq. (B.155) is separable in X, Y, and Z, so F can also be written as $F = F_X F_Y F_Z$. If we substitute this into Eq. (B.155), then divide by F, we obtain

$$\frac{1}{F_X}\frac{\partial^2 F_X}{\partial X^2} + \frac{1}{F_Y}\frac{\partial^2 F_Y}{\partial Y^2} + \frac{1}{F_Z}\frac{\partial^2 F_Z}{\partial Z^2} + k^2 = 0 \tag{B.157}$$

Eq. (B.157) sums three independent functions of X, Y, and Z, equal to a constant. This is true if and only if the respective functions also equal to a constant equal to zero. We may also separate k^2 as

$$k^2 = k_x^2 + k_y^2 + k_z^2 \tag{B.158}$$

Then, Eq. (B.157) leads to the following three equations:

$$\frac{\partial^2 F_X}{\partial X^2} + k_x^2 F_X = 0, \quad \frac{\partial^2 F_Y}{\partial Y^2} + k_y^2 F_Y = 0, \quad \frac{\partial^2 F_Z}{\partial Z^2} + k_x^2 F_Z = 0 \tag{B.159}$$

Each of these are identical to Eq. (**??**), where we found the solution already. Using those results, the hydrogen center of mass solutions are the free-particle wave functions given by

$$F_X = F_{x0} e^{ik_x X}, \quad F_Y = F_{y0} e^{ik_y Y}, \quad F_Z = F_{z0} e^{ik_z Y} \tag{B.160}$$

Therefore, the total solution is given by

$$F = F_X F_Y F_Z = F_0 e^{i(k_x X + k_y Y + k_z Y)} = F_0 e^{i\mathbf{k}\cdot\mathbf{R}} \tag{B.161}$$

Using the de Broglie relation $p = \hbar k$, the solution becomes

Hydrogen Center-Of-Mass wave function

$$F(\mathbf{R}) = F_0 e^{\frac{i}{\hbar}\mathbf{p}\cdot\mathbf{R}} \tag{B.162}$$

Eq. (B.162) is the wave function for the hydrogen atom's center of mass, **R**. The complete hydrogen atom wave functions are given by the product of $F(\mathbf{R})$ and B, from Eq. (B.153).

While we are on the subject of orbital angular momentum, it's a good time to discuss some additional important quantum mechanical operators relating linear and angular momentum to magnetic fields. We begin by introducing the *commutation* operators as used in quantum mechanics.

B.11 Special Commutation Relations

In general, the order of operations on a function matters. Let's say I needed to give you directions to my house. Imagine if I told you to turn right, take ten steps, then turn left. This would certainly end up at a different location than if you turned left, took ten steps, then turned right, or reversed the order of operations. Similarly, with operators, there is no reason why changing their order should yield the same result. One linear operator that illustrates this well is matrices. Consider the following two matrices $\widehat{A}$ and $\widehat{B}$ operating on a vector v.

$$\widehat{A} = \begin{bmatrix} 1 & -1 \\ 3 & 0 \end{bmatrix} \quad , \quad \widehat{B} = \begin{bmatrix} 1 & 3 \\ 1 & -2 \end{bmatrix} \quad \text{and} \quad v = \begin{bmatrix} 1 \\ 1 \end{bmatrix}$$

If we let $\widehat{B}$ operate on v first, we see that $\widehat{A}\widehat{B}v$ gives

$$\begin{bmatrix} 1 & -1 \\ 3 & 0 \end{bmatrix}\begin{bmatrix} 1 & 3 \\ 1 & -2 \end{bmatrix}\begin{bmatrix} 1 \\ 1 \end{bmatrix} = \begin{bmatrix} 1 & -1 \\ 3 & 0 \end{bmatrix}\begin{bmatrix} 4 \\ -1 \end{bmatrix} = \begin{bmatrix} 5 \\ 12 \end{bmatrix}$$

However, reversing the order of operations, letting $\widehat{A}$ operate on v first, we see that $\widehat{B}\widehat{A}v$ gives

$$\begin{bmatrix} 1 & 3 \\ 1 & -2 \end{bmatrix}\begin{bmatrix} 1 & -1 \\ 3 & 0 \end{bmatrix}\begin{bmatrix} 1 \\ 1 \end{bmatrix} = \begin{bmatrix} 1 & 3 \\ 1 & -2 \end{bmatrix}\begin{bmatrix} 0 \\ 3 \end{bmatrix} = \begin{bmatrix} 9 \\ -6 \end{bmatrix} \neq \begin{bmatrix} 5 \\ 12 \end{bmatrix}$$

Two operators $\widehat{A}$ and $\widehat{B}$ are said to **commute**, if and only if the order of the operations does not matter. Thus, they *commute* when

$$\widehat{A}\widehat{B} = \widehat{B}\widehat{A} \tag{B.163a}$$

Alternatively, we can write

$$\widehat{A}\widehat{B} - \widehat{B}\widehat{A} = 0 \tag{B.163b}$$

In the notation above, the convention is to omit the object on which $\widehat{A}$ and $\widehat{B}$ operates (e.g. v in our example). If two operators commute, then operating with B first, then A, will yield the same result as operating with A first, then B. This leads to the definition of the **commutation operator**, which is defined as

Commutation Operator

$$[\widehat{A}, \widehat{B}] = \widehat{A}(\widehat{B}) - \widehat{B}(\widehat{A}) \tag{B.164}$$

Since quantum mechanics utilizes the notion of operators, the *commutation operator* turns out to be a useful concept that we can apply to linear and angular momentum. In the case of linear momentum, in the x direction, consider

$$[x, \widehat{p_x}] = x\widehat{p_x} - \widehat{p_x}x = x\left(-i\hbar\frac{\partial}{\partial x}\right) - \left(-i\hbar\frac{\partial}{\partial x}\right)x \tag{B.165}$$

Remember that an operator is always applied to a wave function sitting on the right of the operator. For example, applying Eq. (B.165) to a wave function ψ, we have

$$\left[x\left(-i\hbar\frac{\partial}{\partial x}\right) - \left(-i\hbar\frac{\partial}{\partial x}\right)x\right]\psi = x\left(-i\hbar\frac{\partial\psi}{\partial x}\right) - \left(-i\hbar\frac{\partial(x\psi)}{\partial x}\right) =$$

$$-ix\hbar\frac{\partial\psi}{\partial x} + ix\hbar\frac{\partial\psi}{\partial x} + i\hbar\psi = i\hbar\psi \neq 0$$

The commutation operator $[x, \widehat{p_x}]$ becomes

$$[x, \widehat{p_x}] = [x\widehat{p_x} - \widehat{p_x}x]\,\psi = i\hbar\psi \tag{B.166}$$

Since the result is not zero, it also means that x and $\widehat{p_x}$ *do not commute.* However, Eq. (B.166) is an eigenoperator equation for $[x, \widehat{p_x}]$. Though they do not commute, we are able to obtain its eigenvalues $\lambda = i\hbar$. Therefore, in many instances when we have the commutation operator $xp_x - p_x x$ involved, it may be replaced by the RHS of Eq. (B.166). In shorthand, we write the operator relation as

$$[x, \widehat{p_x}] = x\widehat{p_x} - \widehat{p_x}x = i\hbar$$

A similar result is found for y and z, where, for all components, we get

$$x\widehat{p_x} - \widehat{p_x}x = i\hbar \tag{B.167a}$$

$$y\widehat{p_y} - \widehat{p_y}y = i\hbar \tag{B.167b}$$

$$z\widehat{p_z} - \widehat{p_z}z = i\hbar \tag{B.167c}$$

These relations can be expressed more compactly, in terms of the commutation operator, as

$x, \widehat{p}$ Commutation Relation

$$[x_i, \widehat{p_i}] = i\hbar \tag{B.168}$$

In contrast, if we determine the commutation relation for different coordinates, such as

$$xp_y - p_y x = x\left(-i\hbar\frac{\partial}{\partial y}\right) - \left(-i\hbar\frac{\partial}{\partial y}\right)x$$

Operating on a wave function ψ, we have

$$\begin{aligned}\left[xp_y - p_y x\right]\psi &= \left[x\left(-i\hbar\frac{\partial}{\partial y}\right) - \left(-i\hbar\frac{\partial}{\partial y}\right)x\right]\psi \\ &= x\left(-i\hbar\frac{\partial\psi}{\partial y}\right) + i\hbar\left(\frac{\partial(x\psi)}{\partial y}\right) \\ &= -ix\hbar\frac{\partial\psi}{\partial y} + ix\hbar\frac{\partial\psi}{\partial y} = 0\end{aligned}$$

Since the result is zero, x and p_y commute. It is straightforward to show this result extends to all components where the coordinate of $\mathbf{r} = [x, y, z]^T$ is different from that of $\widehat{\mathbf{p}} = [\widehat{p_x}, \widehat{p_y}, \widehat{p_z}]^T$. So, we have that

$x, \widehat{p}$ Commutation Relation

$$\left[x_j, \widehat{p_i}\right] = 0 \quad (j \neq i) \tag{B.169}$$

Let us extend these commutation relation ideas to angular momentum $\widehat{\mathbf{L}} = [\widehat{L_x}, \widehat{L_y}, \widehat{L_z}]^T$. Since we know the angular momentum operators for L_x, L_y, and L_z, we can determine the commutation relations for the angular momentum operators *with each other*, using Eq. (B.170). First, let's evaluate

$$\widehat{L_y}\widehat{L_z} - \widehat{L_z}\widehat{L_y} \tag{B.170}$$

The components of the orbital angular momentum are given by

$$\widehat{L_x} = -i\hbar\left(y\frac{\partial}{\partial z} - z\frac{\partial}{\partial y}\right) \tag{B.171a}$$

$$\widehat{L}_y = -i\hbar\left(z\frac{\partial}{\partial x} - x\frac{\partial}{\partial z}\right) \tag{B.171b}$$

$$\widehat{L}_z = -i\hbar\left(x\frac{\partial}{\partial y} - y\frac{\partial}{\partial x}\right) \tag{B.171c}$$

Therefore, Eq. (B.170) becomes

$$\widehat{L}_y\widehat{L}_z - \widehat{L}_z\widehat{L}_y = -\hbar^2\left(z\frac{\partial}{\partial x} - x\frac{\partial}{\partial z}\right)\left(x\frac{\partial}{\partial y} - y\frac{\partial}{\partial x}\right) + \hbar^2\left(x\frac{\partial}{\partial y} - y\frac{\partial}{\partial x}\right)\left(z\frac{\partial}{\partial x} - x\frac{\partial}{\partial z}\right) \tag{B.172}$$

This expands to the following:

$$\widehat{L}_y\widehat{L}_z - \widehat{L}_z\widehat{L}_y = -\hbar^2\left[z\frac{\partial}{\partial x}\left(x\frac{\partial}{\partial y}\right) - z\frac{\partial}{\partial x}\left(y\frac{\partial}{\partial x}\right) - x\frac{\partial}{\partial z}\left(x\frac{\partial}{\partial y}\right) + x\frac{\partial}{\partial z}\left(y\frac{\partial}{\partial x}\right)\right] + \hbar^2\left[x\frac{\partial}{\partial y}\left(z\frac{\partial}{\partial x}\right) - x\frac{\partial}{\partial y}\left(x\frac{\partial}{\partial z}\right) - y\frac{\partial}{\partial x}\left(z\frac{\partial}{\partial x}\right) + y\frac{\partial}{\partial x}\left(x\frac{\partial}{\partial z}\right)\right] \tag{B.173}$$

Though it may appear to be complicated with so many terms, most of them cancel, and we have the result

$$\widehat{L}_y\widehat{L}_z - \widehat{L}_z\widehat{L}_y = \hbar^2\left[y\frac{\partial}{\partial z} - z\frac{\partial}{\partial y}\right] \tag{B.174}$$

We can also use the operator $\widehat{L}_x$ on the RHS, given in Eq. (B.171a), to write Eq. (B.174) as

$$\widehat{L}_y\widehat{L}_z - \widehat{L}_z\widehat{L}_y = i\hbar\widehat{L}_x \tag{B.175}$$

Evaluating the commutation rules based on Eq. (B.170) for the other two pairs $\widehat{L}_x\widehat{L}_z$ and $\widehat{L}_x\widehat{L}_y$, we find the following commutation rules for the orbital angular momentum.

$$\widehat{L}_y L_z - \widehat{L}_z\widehat{L}_y = i\hbar\widehat{L}_x \tag{B.176a}$$

$$\widehat{L}_z\widehat{L}_x - \widehat{L}_x\widehat{L}_z = i\hbar\widehat{L}_y \tag{B.176b}$$

$$\widehat{L}_x\widehat{L}_y - \widehat{L}_y\widehat{L}_x = i\hbar\widehat{L}_z \tag{B.176c}$$

This is written more compactly as

Orbital Angular Momentum Commutation Relation

$$\left[\widehat{L}_i, \widehat{L}_j\right] = i\hbar\widehat{L}_k \tag{B.177}$$

Because the LHS is also the vector cross-product, it is alternatively written as $\widehat{\mathbf{L}} \times \widehat{\mathbf{L}} = i\hbar\widehat{\mathbf{L}}$. Another set of commutation relations we will consider involves the magnetic vector potential $\mathbf{A}$. From the Maxwell equations, the magnetic field $\mathbf{B}$ can alternatively be expressed in terms of $\mathbf{A}$, where $\mathbf{B} = \nabla \times \mathbf{A}$. Let us then define an **extended momentum operator** $\widehat{\mathbf{P}}$ by

Extended Momentum Operator

$$\widehat{\mathbf{P}} \equiv \widehat{\mathbf{p}} + e\mathbf{A} \tag{B.178}$$

e is the elementary charge. We are going to consider two commutation relations involving $\widehat{\mathbf{P}}$. First, lets consider

$$\left[\widehat{P}_i, \widehat{P}_j\right] = \widehat{P}_i\widehat{P}_j - \widehat{P}_j\widehat{P}_i \tag{B.179}$$

i and j denote the dimension index $1 \rightarrow x$, $2 \rightarrow y$, and $3 \rightarrow z$. In evaluating this operator, let us start with x and y where we have

$$[\widehat{P}_x, \widehat{P}_y] = \left(\widehat{p}_x + eA_x)(\widehat{p}_y + eA_y) - (\widehat{p}_y + eA_y)(\widehat{p}_x + eA_x\right) \tag{B.180}$$

Using the definition for the momentum operator $\widehat{p}_i = -i\hbar\partial/\partial x_i$, the commutation operator in Eq. (B.180) becomes

$$[\widehat{P}_x, \widehat{P}_y]\psi = \left[(\widehat{p}_x + eA_x)\left(\widehat{p}_y + eA_y\right) - \left(\widehat{p}_y + eA_y\right)(\widehat{p}_x + eA_x)\right]\psi =$$
$$\left[\left(-i\hbar\frac{\partial}{\partial x} + eA_x\right)\left(-i\hbar\frac{\partial}{\partial y} + eA_y\right) - \left(-i\hbar\frac{\partial}{\partial y} + eA_y\right)\left(-i\hbar\frac{\partial}{\partial x} + eA_x\right)\right]\psi \tag{B.181}$$

The products work out to four terms, and the commutation operation becomes

$$[\widehat{P}_x, \widehat{P}_y]\psi = \left[-\hbar^2 \frac{\partial}{\partial x}\frac{\partial}{\partial y} - i\hbar e \frac{\partial}{\partial x}\left(A_y\right) - i\hbar e A_x \frac{\partial}{\partial y} + e^2 A_x A_y\right]\psi -$$

$$\left[-\hbar^2 \frac{\partial}{\partial y}\frac{\partial}{\partial x} - i\hbar e \frac{\partial}{\partial y}(A_x) - i\hbar e A_y \frac{\partial}{\partial x} + e^2 A_y A_x\right]\psi \quad \text{(B.182)}$$

Each terms operate on the wave function ψ. The last term involves all scalars, thus, it is only multiplication. Since multiplication of scalars is commutative, we can use the following relation:

$$e^2 A_x A_y \psi = e^2 A_y A_x \psi$$

Therefore, this term vanishes. Additionally, the first term involves mixed partials, for which the order also does not matter. Thus, we can use the following relation

$$-\hbar^2 \frac{\partial}{\partial x}\frac{\partial}{\partial y}\psi = -\hbar^2 \frac{\partial^2 \psi}{\partial x \partial y} = -\hbar^2 \frac{\partial^2 \psi}{\partial y \partial x}$$

From this, the first terms vanishes, as well. We now have left

$$[\widehat{P}_x, \widehat{P}_y]\psi = i\hbar e\left[-\frac{\partial}{\partial x}\left(A_y \psi\right) - A_x \frac{\partial \psi}{\partial y}\right] -$$

$$i\hbar e\left[-\frac{\partial}{\partial y}(A_x \psi) - A_y \frac{\partial \psi}{\partial x}\right] \quad \text{(B.183)}$$

The derivatives of the products must be expanded *using the product rule* for differentiation, which leads to

$$[\widehat{P}_x, \widehat{P}_y]\psi = i\hbar e\left[-\frac{\partial A_y}{\partial x}\psi - A_y \frac{\partial \psi}{\partial x} - A_x \frac{\partial \psi}{\partial y}\right] -$$

$$i\hbar e\left[-\frac{\partial A_x}{\partial y}\psi - i\hbar A_x \frac{\partial \psi}{\partial y} - A_y \frac{\partial \psi}{\partial x}\right] \quad \text{(B.184)}$$

The terms involving derivatives of the wave function ψ all cancel, leaving the result

$$[\widehat{P}_x, \widehat{P}_y]\psi = -i\hbar e \frac{\partial A_y}{\partial x}\psi + i\hbar e \frac{\partial A_x}{\partial y}\psi = -i\hbar e\left(\frac{\partial A_y}{\partial x} - \frac{\partial A_x}{\partial y}\right)\psi \quad \text{(B.185)}$$

The term in parentheses is just the z component of the magnetic field $\mathbf{B} = \nabla \times \mathbf{A}$, so we have

$$[\widehat{P}_x, \widehat{P}_y]\psi = -i\hbar e B_z \psi \quad \text{(B.186a)}$$

The result is similar for the other two pairs of momentum operators, where we get

$$\left[\widehat{P_y}, \widehat{P_z}\right]\psi = -i\hbar e B_x \psi \tag{B.186b}$$

$$\left[\widehat{P_z}, \widehat{P_x}\right]\psi = -i\hbar e B_y \psi \tag{B.186c}$$

The reader should verify that if we take $[\widehat{P_i}, \widehat{P_i}]\psi$, the result would vanish meaning that they commute. Thus, the indexes must be distinct for the result to be nonzero. More compactly, we may write this above results as

Extended Momentum Commutation Relation

$$\left[\widehat{P_i}, \widehat{P_j}\right] = -i\hbar e \varepsilon_{ijk} B_k \tag{B.187}$$

Therefore, we have that the commutation of two *extended momentum* components is a scalar operator, given by Eq. (B.187). ε_{ijk} is known as the *Levi-Civita symbol* defined as

$$\varepsilon_{ijk} = \begin{cases} +1 & \text{if } ijk = (1,2,3),(2,3,1), \text{or} \quad (3,1,2) \\ -1 & \text{if } ijk = (3,2,1),(1,3,2), \text{or} \quad (2,1,3) \\ 0 & \text{if } i = j, j = k, \text{ or } i = k \end{cases}$$

This notation maintains use of a right-handed coordinate system. We can use these results to determine more commutation relations of higher order in $\widehat{P_i}$. For example, let us determine

$$\left[\widehat{P_i^2}, \widehat{P_j}\right] = \widehat{P_i^2}\widehat{P_j} - \widehat{P_j}\widehat{P_i^2} \tag{B.188}$$

This will be used to obtain our end goal, which is to determine the following:

$$\left[\widehat{P^2}, \widehat{P_j}\right] = \left[\widehat{P_x^2} + \widehat{P_y^2} + \widehat{P_z^2}, \widehat{P_j}\right]$$

Let's start by considering the component $\widehat{P_y}$ with $\widehat{P_x^2}$, where we first have

$$\begin{aligned} [\widehat{P_x^2}, \widehat{P_y}] &= \widehat{P_x^2}\widehat{P_y} - \widehat{P_y}\widehat{P_x^2} \\ &= \widehat{P_x}\widehat{P_x}\widehat{P_y} - \widehat{P_y}\widehat{P_x}\widehat{P_x} \end{aligned}$$

$$= \widehat{P_x}\left(-ie\hbar B_z + \widehat{P_y}\widehat{P_x}\right) - \widehat{P_y}\widehat{P_x}\widehat{P_x} \quad \text{(using Eq. (B.187))}$$

$$= \widehat{P_x}(-ie\hbar B_z) + \left(\widehat{P_x}\widehat{P_y} - \widehat{P_y}\widehat{P_x}\right)\widehat{P_x}$$

$$= \widehat{P_x}(-ie\hbar B_z) + (-ie\hbar B_z)\widehat{P_x} \quad \text{(again using Eq. (B.187))}$$

$$= \widehat{P_x}(-2ie\hbar B_z) = (-2ie\hbar B_z)\widehat{P_x}$$

The last step of commuting follows since the term in parentheses is scalar. Carrying out a similar analysis for $[\widehat{P_z^2}, \widehat{P_y}]$, we have

$$[\widehat{P_z^2}, \widehat{P_y}] = \widehat{P_z^2}\widehat{P_y} - \widehat{P_y}\widehat{P_z^2}$$

$$= \widehat{P_z}\widehat{P_z}\widehat{P_y} - \widehat{P_y}\widehat{P_z}\widehat{P_z}$$

$$= \widehat{P_z}\left(ie\hbar B_x + \widehat{P_y}\widehat{P_z}\right) - \widehat{P_y}\widehat{P_z}\widehat{P_z} \quad \text{(using Eq. (B.187))}$$

$$= \widehat{P_z}(ie\hbar B_x) + \left(\widehat{P_z}\widehat{P_y} - \widehat{P_y}\widehat{P_z}\right)\widehat{P_z}$$

$$= \widehat{P_z}(ie\hbar B_x) + (ie\hbar B_x)\widehat{P_z} \quad \text{(again using Eq. (B.187))}$$

$$= \widehat{P_z}(2ie\hbar B_x) = (2ie\hbar B_x)\widehat{P_z}$$

Again, the last step follows because the term in parentheses is a scalar. For the third case of $[\widehat{P_y^2}, \widehat{P_y}]$, which turns out to be simple because it is the only case that, exploiting Eq. (B.187), we may write

$$[\widehat{P_y^2}, \widehat{P_y}] = \widehat{P_y}[\widehat{P_y}, \widehat{P_y}] = 0$$

Therefore, summing the results for $\widehat{P_y}$, it follows that

$$[\widehat{P_x^2} + \widehat{P_y^2} + \widehat{P_z^2}, \widehat{P_y}] = 2ie\hbar\left(B_x\widehat{P_z} - B_z\widehat{P_x}\right) \tag{B.189a}$$

The reader should confirm that carrying out the same analysis for $\widehat{P_x}$ and $\widehat{P_z}$, the result becomes

$$[\widehat{P_x^2} + \widehat{P_y^2} + \widehat{P_z^2}, \widehat{P_x}] = 2ie\hbar\left(B_z\widehat{P_y} - B_y\widehat{P_z}\right) \tag{B.189b}$$

$$[\widehat{P_x^2} + \widehat{P_y^2} + \widehat{P_z^2}, \widehat{P_z}] = 2ie\hbar\left(B_y\widehat{P_x} - B_x\widehat{P_y}\right) \tag{B.189c}$$

These results can be expressed more compactly as

Commutation Relation of Square Extended Momentum

$$[\widehat{\mathbf{P}}^2, \widehat{\mathbf{P}}] = 2ie\hbar\widehat{\mathbf{P}} \times \mathbf{B} \tag{B.190}$$

Defining the extended momentum $\widehat{\mathbf{P}}$ in this way turns out to be *necessary* in quantum mechanics because it renders the Schrödinger and Dirac equations to be capable of describing charged particles in both electric and magnetic fields *that are independent of time*. This will shown fully in the next section, where we will use the results above to show that by defining the extended momentum $\widehat{\mathbf{P}} = \widehat{\mathbf{p}} + e\mathbf{A}$, one obtains an equation of motion consistent with Newton's laws, including the response to the *Lorentz Force*, arising from an electric and magnetic field.

B.12 Extended Momentum Leads to Lorentz Force

In Chap. B, we introduced the extended momentum operator $\widehat{\mathbf{P}}$ given by

$$\widehat{\mathbf{P}} = \widehat{\mathbf{p}} + e\mathbf{A}$$

$\mathbf{A}$ is commonly known as the *magnetic vector potential*, whose vector curl is the magnetic field $\mathbf{B}$. Mathematically, we say that $\mathbf{B} = \nabla \times \mathbf{A}$. From the above, $\mathbf{A}$ can also be interpreted as a *magnetic momentum per unit charge*. We had to postpone the full justification for this definition until Eq. (4.22) could be derived. Now, we can provide the justification for the extended definition of the momentum. For the linear momentum in the absence of a magnetic field, the Hamiltonian $\widehat{H}$ of a system has been given by

$$\widehat{H}\psi = \left[\frac{\widehat{\mathbf{p}}^2}{2m} + V\right]\psi \Rightarrow \widehat{H} = \frac{\widehat{\mathbf{p}}^2}{2m} + V$$

Let us now replace $\widehat{\mathbf{p}}$ with the extended momentum $\widehat{\mathbf{P}}$, then apply Eq. (4.22). Thus, we have

$$\widehat{H}' = \frac{\widehat{\mathbf{P}}^2}{2m} + V = \frac{(\widehat{\mathbf{p}} + e\mathbf{A})^2}{2m} + V \tag{B.191}$$

If we assume that the magnetic vector potential $\mathbf{A}$ *is independent of time,* then the extended momentum operator has no dependence on time. It follows that

$$\left\langle \frac{\partial \widehat{\mathbf{P}}}{\partial t} \right\rangle = 0 \tag{B.192}$$

Using Eq. (4.22) to compute the time derivative of the expectation value of $\widehat{\mathbf{P}}$, we have

$$\frac{d\langle \widehat{\mathbf{P}} \rangle}{dt} = \frac{i}{\hbar} \left\langle \left[\widehat{H}', \widehat{\mathbf{P}} \right] \right\rangle \tag{B.193}$$

In what follows, we will omit the expectation value notation for convenience, so that we may write

$$\begin{aligned}
\frac{d\widehat{\mathbf{P}}}{dt} &= \frac{i}{\hbar} \left[\widehat{H}', \widehat{\mathbf{P}} \right] \\
&= \frac{i}{\hbar} \left[\left(\frac{\widehat{\mathbf{P}}^2}{2m} + V \right) \widehat{\mathbf{P}} - \widehat{\mathbf{P}} \left(\frac{\widehat{\mathbf{P}}^2}{2m} + V \right) \right] \\
&= \frac{i}{\hbar} \left[\frac{1}{2m} \left(\widehat{\mathbf{P}}^2 \widehat{\mathbf{P}} - \widehat{\mathbf{P}} \widehat{\mathbf{P}}^2 \right) + \left(V\widehat{\mathbf{P}} - \widehat{\mathbf{P}} V \right) \right]
\end{aligned}$$

The first term in the last line has the form of the operator given by Eq. (B.190). Using that result, we have

$$\frac{d\widehat{\mathbf{P}}}{dt} = \frac{i}{\hbar} \left[\frac{1}{2m} \left(\widehat{\mathbf{P}}^2 \widehat{\mathbf{P}} - \widehat{\mathbf{P}} \widehat{\mathbf{P}}^2 \right) + \left(V\widehat{\mathbf{P}} - \widehat{\mathbf{P}} V \right) \right] \tag{B.194a}$$

$$= \frac{i}{\hbar} \left[\frac{1}{2m} \left(2ie\hbar \widehat{\mathbf{P}} \times \mathbf{B} \right) + \left(V\widehat{\mathbf{P}} - \widehat{\mathbf{P}} V \right) \right] \tag{B.194b}$$

Let's now consider the potential term more closely. It can be written as follows:

$$V\widehat{\mathbf{P}} - \widehat{\mathbf{P}} V = V(\widehat{\mathbf{p}} + e\mathbf{A}) - (\widehat{\mathbf{p}} + e\mathbf{A})V \tag{B.195a}$$

$$= -i\hbar V \nabla + eV\mathbf{A} + (i\hbar \nabla)V - e\mathbf{A}V \tag{B.195b}$$

$$= -i\hbar V \nabla + i\hbar V \nabla + i\hbar \nabla V \tag{B.195c}$$

$$= i\hbar \nabla V \tag{B.195d}$$

Using this result in Eq. (B.194b), we have

$$\frac{d\widehat{\mathbf{P}}}{dt} = -\frac{e}{m} \widehat{\mathbf{P}} \times \mathbf{B} - \nabla V = -e\widehat{\mathbf{v}} \times \mathbf{B} - \nabla V \tag{B.196}$$

From Newton's second law, we know that $\mathbf{F} = d\langle \mathbf{P} \rangle / dt$. Thus, we have the general result

General Electromagnetostatic Force

$$\mathbf{F} = -e\langle \widehat{\mathbf{v}} \times \mathbf{B} \rangle - \nabla V \tag{B.197}$$

$\widehat{\mathbf{v}}$ is the velocity operator given by $\widehat{\mathbf{v}} = \widehat{\mathbf{P}}/m$. We recognize the first term as the *Lorentz force* $\mathbf{F}_L$ due to a magnetic field acting on a moving charge, while the second term is the electrostatic force $\mathbf{F}_e = e(\mathbf{F}_e/e) = e\mathbf{E}$ due to the electric field $\mathbf{E}$. This is the motivation behind the definition of the *extended momentum* operator $\widehat{\mathbf{P}}$. It provides a more complete description of a charged particle in an electric and magnetic field.

This definition of $\widehat{\mathbf{P}}$ extends quantum mechanics to a level of consistency with magnetostatics. It is straightforward to show that this definition of $\widehat{\mathbf{P}}$ also follows from the Maxwell equations. First, recall that in the full electromagnetics of the Maxwell equations, the most general electric field $\mathbf{E}$ is given by

$$\mathbf{E} = -\nabla V_e - \frac{\partial \mathbf{A}}{\partial t}$$

V_e is the electric potential. But, $\mathbf{E}$ is just the force per unit charge $\mathbf{F}_\epsilon/q$. Thus, for an electron, the force becomes

$$\mathbf{F}_e = q\mathbf{E} = -q\nabla V_e - q\frac{\partial \mathbf{A}}{\partial t}$$

If we take $\mathbf{F} = md\mathbf{v}/dt$, we have

$$m\frac{d\mathbf{v}}{dt} = -q\nabla V_e - q\frac{\partial \mathbf{A}}{\partial t} \tag{B.198a}$$

$$= -\nabla V - q\frac{\partial \mathbf{A}}{\partial t} \tag{B.198b}$$

V is the *conservative* electric potential energy. The first term corresponds to motion along the direction of the electrical potential energy variation, while the second term corresponds to motion along the direction of the change in $\mathbf{A}$. The ordinary momentum operator $\widehat{\mathbf{p}}$, used in quantum mechanics, corresponds to the first term. The second term is the new term. Eq. (B.198b), then, equating the most general force $\mathbf{F}_e$ leads to

$$\frac{d\mathbf{P}}{dt} = \frac{d\mathbf{p}}{dt} - \frac{\partial(q\mathbf{A})}{\partial t} \Rightarrow \mathbf{P} = \mathbf{p} + e\mathbf{A}$$

e is the elementary charge relating to the electron charge as $q = -e$. Although the above analysis is general, note that in quantum mechanics, it is always assumed that there is no time dependence for $\mathbf{A}$ (which means no time dependence for $\widehat{\mathbf{P}}$). However, time integration of the above leads to our result $\mathbf{P} = \mathbf{p} + e\mathbf{A}$, which must be satisfied everywhere in space, since $\mathbf{A}$ varies with spatial coordinates. Thus, we see that quantum mechanics and the Maxwell equations are placed on the same footing, in this regard, by defining the extended momentum $\widehat{\mathbf{P}}$.

Finally, we discuss a pair of special operators based on the angular momentum operators. They are the *lowering* and *raising operators*, and are the subject of the next section.

B.13 Angular Momentum Ladder Operators

The orbital angular momentum operators and eigenfunctions can be determined in one fell swoop, by working out the solution to the eigenvalue equations, which also dictate all the possible values of the quantum numbers. We found earlier that the solution to the eigenequation for $\widehat{L^2}$ is given by

$$Y_{\ell,m}(\theta,\varphi) = A_\ell^m(\theta)e^{im\varphi} \tag{B.199}$$

$A_\ell^m(\theta)$ is the *associated Legendre function* we derived earlier, given by Eq. (B.84). ℓ is the quantum number for the orbital angular momentum, while m is the magnetic quantum number associated with $\widehat{L}_z$, which manifests in the splitting by a magnetic field. Using the operators for $\widehat{L}_x$ given by Eq. (B.61) and $\widehat{L}_y$, given by Eq. (B.62), let us define a new operator $\widehat{L}_-$ as

Orbital Angular Momentum Lowering Operator Definition

$$\widehat{L}_- \equiv \widehat{L}_x - i\widehat{L}_y \tag{B.200}$$

Eq. (B.200) is known as the **orbital angular momentum lowering operator.** We'll find out shortly why it bears this name. After substituting $\widehat{L}_x$ and $\widehat{L}_y$ from Eqs. (B.61) and (B.61), into the definition for $\widehat{L}_-$, we have

$$\widehat{L}_x - i\widehat{L}_y = i\hbar\left(\sin\varphi\frac{\partial}{\partial\theta} + \cot\theta\cos\varphi\frac{\partial}{\partial\varphi}\right) + \hbar\left(-\cos\varphi\frac{\partial}{\partial\theta} + \cot\theta\sin\varphi\frac{\partial}{\partial\varphi}\right)$$

$$= \hbar(i\sin\varphi - \cos\varphi)\frac{\partial}{\partial\theta} + \hbar(i\cos\varphi + \sin\varphi)\cot\theta\frac{\partial}{\partial\varphi}$$

$$= \hbar\left(-e^{-i\varphi}\frac{\partial}{\partial\theta} + ie^{-i\varphi}\cot\theta\frac{\partial}{\partial\varphi}\right)$$

We have made use of the relations $\cos\varphi - i\sin\varphi = e^{-i\varphi}$ and $\sin\varphi + i\cos\varphi = ie^{-i\varphi}$. Thus, the *orbital angular momentum lowering operator* defined above, is given by

Lowering Operator $\widehat{L}_-$

$$\widehat{L}_- = \hbar e^{-i\varphi}\left[-\frac{\partial}{\partial\theta} + i\cot\theta\frac{\partial}{\partial\varphi}\right] \tag{B.201}$$

We know the eigenfunction $Y_{\ell,m}$ we obtained for the $\widehat{L^2}$ operator, so if we take the lowest possible energy state of orbital angular momentum, for a given value of ℓ, or

$$Y_{\ell,m=-\ell}(\theta,\varphi) = A_\ell^{-\ell}e^{-i\ell\varphi}$$

Let us explore what happens when we apply the lowering operator $\widehat{L}_-$ to $Y_{\ell,m=-\ell}$. Evaluating this leads to

$$\widehat{L}_- Y_{\ell,-\ell} = \hbar e^{-i\varphi}\left[-\frac{\partial}{\partial\theta} + i\cot\theta\frac{\partial}{\partial\varphi}\right]Y_{\ell,-\ell}$$

$$= \hbar e^{-i\varphi}\left[-\frac{\partial}{\partial\theta} + i\cot\theta\frac{\partial}{\partial\varphi}\right]A_\ell^{-\ell}e^{-i\ell\varphi}$$

$$= \hbar e^{-i\varphi}\left[-\frac{\partial A_\ell^{-\ell}}{\partial\theta}e^{-i\ell\varphi} + i(-i\ell)\cot\theta e^{-i\ell\varphi}A_\ell^{-\ell}\right]$$

$$= \hbar e^{-i(\ell+1)\varphi}\left[-\frac{\partial A_\ell^{-\ell}}{\partial\theta} + \ell\cot\theta A_\ell^{-\ell}\right]$$

In the above result, we have two terms involved in $\widehat{L}_- Y_{\ell,-\ell}$, with the exponential coefficient and the term in brackets, which is purely a function of θ. For this special case, our concern is with the term in brackets. We will show that this term is identically zero, which means that when the lowering operator acts on a minimum energy state, it returns null. To see this, let us first express the *associated Legendre function* $A_\ell^{-\ell}$ as follows:

$$\begin{aligned}
A_\ell^{-\ell} &= C_\ell\left(1-x^2\right)^{\frac{-\ell}{2}}\frac{d^{\ell+(-\ell)}}{dx}\left(x^2-1\right)^\ell \\
&= C_\ell\left(1-x^2\right)^{\frac{-\ell}{2}}\left(x^2-1\right)^\ell \\
&= (-1)^\ell C_\ell\left(1-x^2\right)^{\frac{\ell}{2}}
\end{aligned}$$

Also, substituting $x=\cos\theta$, evaluation of $dA_\ell^{-\ell}/d\theta$ gives

$$\begin{aligned}
\frac{dA_\ell^{-\ell}}{d\theta} &= \frac{dA_\ell^{-\ell}}{dx}\frac{dx}{d\theta} \\
&= (-1)^\ell\frac{\ell}{2}C_\ell(-\sin\theta)(-2x)\left[\left(1-x^2\right)^{\frac{\ell}{2}-1}\right] \\
&= (-1)^\ell\frac{\ell}{2}C_\ell(-\sin\theta)(-2x)\left[\left(1-x^2\right)^{\frac{\ell}{2}}(1-x^2)^{-1}\right]
\end{aligned}$$

Then, substituting $1-x^2=\sin^2\theta$ leads to

$$\begin{aligned}
\frac{dA_\ell^{-\ell}}{d\theta} &= (-1)^\ell \ell C_\ell(-\sin\theta)(-\cos\theta)\left[\left(1-x^2\right)^{\frac{\ell}{2}}\frac{1}{\sin^2\theta}\right] \\
&= \ell(-1)^\ell C_\ell\frac{\cos\theta}{\sin\theta}\left(1-x^2\right)^{\frac{\ell}{2}} \\
&= \ell\cot\theta A_\ell^{-\ell}
\end{aligned}$$

From this result, we can see that the following statement is true:

$$\widehat{L}_- Y_{\ell,-\ell} = \hbar e^{-i(\ell+1)\varphi}\left[-\frac{\partial A_\ell^{-\ell}}{\partial\theta}+\ell\cot\theta A_\ell^{-\ell}\right]=0$$

This result, being only dependent on $A_\ell^{-\ell}$ gives an important condition for the lowest energy state $A_\ell^{-\ell}$, given by

Lowest-Level Associated Legendre Function Condition

$$-\frac{\partial A_\ell^{-\ell}}{\partial\theta} + \ell\cot\theta A_\ell^{-\ell} = 0 \tag{B.202}$$

This is a property of the associated Legendre function. Moreover, this relation can be used to derive an alternative form for the lowest energy function $A_\ell^{-\ell}$. It we take the relation obtained above, which can be written as

$$\frac{\partial A_\ell^{-\ell}}{\partial\theta} = \ell\cdot\cot\theta A_\ell^{-\ell} = \ell\cdot\frac{\cos\theta}{\sin\theta}\cdot A_\ell^{-\ell}$$

If we let $u = \sin\theta \rightarrow du = \cos\theta d\theta$. Substitution of these relations into the above gives

$$\frac{dA_\ell^{-\ell}}{A_\ell^{-\ell}} = \ell\frac{du}{u} \Rightarrow \ln A_\ell^{-\ell} = \ell\ln(\sin\theta) + \text{constant} = \ln\left(\sin^\ell\theta\right) + \text{constant}$$

From this, we have a result for $A_\ell^{-\ell}$ given by

Lowest-Level Associated Legendre Function $A_\ell^{-\ell}$

$$A_\ell^{-\ell} = C_n\sin^\ell\theta \tag{B.203}$$

As usual, C_n must be determined by the normalization condition. The evaluation of this condition in spherical coordinates gives

$$\begin{aligned}
1 &= \int_0^\pi |Y_\ell^{-\ell}|^2\sin\theta d\theta d\varphi \\
&= C_n^2\int(\sin^\ell\theta)e^{+i\ell\varphi}(\sin^\ell\theta)e^{-i\ell\varphi}\sin\theta d\theta d\varphi \\
&= 2\pi C_n^2\int(\sin^{2\ell+1}\theta)d\theta
\end{aligned}$$

The factor of 2π follows from evaluating the part of the double integral w.r.t. φ. To evaluate the resulting single variable integral in θ, we can use a well-known identity involving integrals of powers of sin functions, known as a *trigonometric reduction formula*. It states that

Trigonometric Reduction Formula

$$\int \sin^n(x)dx = -\frac{1}{n}\sin^{n-1}(x)\cos(x) + \frac{n-1}{n}\int \sin^{n-2}(x)dx \quad \text{(B.204)}$$

The first term on the RHS becomes zero because we are evaluating with limits from $x = \theta \in (0, \pi)$, so the sine term vanishes. So, we can continuously evaluate the simplified integral down to the last evaluation, with a couple of little tricks, to obtain C_n. Please follow the patterns closely. This process goes as follows:

$$\begin{aligned}
1 &= 2\pi C_n^2 \int_0^\pi \sin^{2\ell+1}\theta d\theta \\
&= 2\pi C_n^2 \frac{2\ell}{2\ell+1} \int_0^\pi \sin^{2\ell-1}\theta d\theta \\
&= 2\pi C_n^2 \frac{2\ell(2\ell-2)}{(2\ell+1)(2\ell-1)} \int_0^\pi \sin^{2\ell-3}\theta d\theta \\
&\vdots \\
&= 2\pi C_n^2 \frac{2\ell(2\ell-2)(2\ell-4)...(2\ell-2\ell+2)}{(2\ell+1)(2\ell-1)...(2\ell-2\ell+1)} \int_0^\pi \sin d\theta \\
&= 2\pi C_n^2 \frac{2^\ell \cdot \ell(\ell-1)...1}{(2\ell+1)(2\ell-1)...(2\ell-2\ell+1)} \times 2 \\
&= 2\pi C_n^2 2^\ell \ell! \frac{1}{(2\ell+1)(2\ell-1)...(2\ell-2\ell+1)} \times 2 \\
&= 2\pi C_n^2 2^\ell \ell! \frac{2\ell(2\ell-2)...(2\ell-2\ell+2)}{(2\ell+1)2\ell(2\ell-1)...1} \times 2 \\
&= 2\pi C_n^2 \left(2^\ell \ell!\right)^2 \frac{2}{(2\ell+1)!}
\end{aligned}$$

From this, we obtain the result

$$1 = 2\pi C_n^2 \left(2^\ell \ell!\right)^2 \frac{2}{(2\ell+1)!}$$

Solving for C_n, we have

Normalization Factor For $Y_\ell^{-\ell}(\theta,\varphi)$

$$C_n = \frac{1}{2^\ell \ell!}\sqrt{\frac{(2\ell+1)!}{4\pi}} \tag{B.205}$$

We can define another operator as the *adjoint* operator of $\widehat{L}_-$, denoted by $\widehat{L}_+$ and it is defined as

Orbital Angular Momentum Raising Operator Definition $\widehat{L}_+$

$$\widehat{L}_+ = \widehat{L}_x + i\widehat{L}_y \tag{B.206}$$

By this definition, the eigenvalues of $\widehat{L}_+$ are complex conjugates to those of $\widehat{L}_-$, and we say the operators are *adjoint operators* of one another. This operator is known as the **orbital angular momentum raising operator** $\widehat{L}_+$. We can find the explicit operator by substituting $\widehat{L}_x$ and $\widehat{L}_y$ into the definition for the $\widehat{L}_+$, where we obtain

$$\begin{aligned}
\widehat{L}_x + i\widehat{L}_y &= i\hbar\left(\sin\varphi\frac{\partial}{\partial\theta} + \cot\theta\cos\varphi\frac{\partial}{\partial\varphi}\right) - \hbar\left(-\cos\varphi\frac{\partial}{\partial\theta} + \cot\theta\sin\varphi\frac{\partial}{\partial\varphi}\right)\\
&= \hbar(i\sin\varphi + \cos\varphi)\frac{\partial}{\partial\theta} + \hbar(i\cos\varphi - \sin\varphi)\cot\theta\frac{\partial}{\partial\varphi}\\
&= \hbar\left(e^{i\varphi}\frac{\partial}{\partial\theta} + ie^{i\varphi}\cot\theta\frac{\partial}{\partial\varphi}\right)
\end{aligned}$$

Thus, for $\widehat{L}_+$, we have the following operator:

Orbital Angular Momentum Raising Operator $\widehat{L}_+$

$$\widehat{L}_+ = \hbar e^{i\varphi}\left[\frac{\partial}{\partial\theta} + i\cot\theta\frac{\partial}{\partial\varphi}\right] \tag{B.207}$$

The *raising operator* $\widehat{L}_+$ is analogous to the lowering operator in its utility, but with the highest energy state. With the lowering operator $\widehat{L}_-$, we found that if we consider the lowest energy state of the spherical harmonics, given by $Y_\ell^{-\ell}$, that $\widehat{L}_- Y_\ell^{-\ell} = 0$. Well, if we now consider the highest energy state Y_ℓ^ℓ for a given ℓ, let us see what happens when $\widehat{L}_+$ operates on this state. We then have

$$\begin{aligned}
\widehat{L}_+ Y_{\ell,\ell} &= \hbar e^{i\varphi}\left[\frac{\partial}{\partial\theta} + i\cot\theta\frac{\partial}{\partial\varphi}\right] Y_{\ell,\ell} \\
&= \hbar e^{i\varphi}\left[\frac{\partial}{\partial\theta} + i\cot\theta\frac{\partial}{\partial\varphi}\right] A_\ell^\ell e^{i\ell\varphi} \\
&= \hbar e^{i\varphi}\left[\frac{\partial A_\ell^\ell}{\partial\theta} e^{i\ell\varphi} + i(i\ell)\cot\theta e^{i\ell\varphi} A_\ell^\ell\right] \\
&= \hbar e^{i\varphi} e^{i\ell\varphi}\left[\frac{\partial A_\ell^\ell}{\partial\theta} - \ell\cot\theta A_\ell^\ell\right]
\end{aligned}$$

Our concern is with the term in brackets. Similar to the case of $\widehat{L}_- Y_\ell^{-\ell}$, because of this term, $\widehat{L}_+ Y_\ell^\ell$ turns out to be identically zero. Let us consider the *associated Legendre function* A_ℓ^ℓ given by

$$A_\ell^\ell = C_\ell\left(1 - x^2\right)^{\frac{\ell}{2}} \frac{d^{2\ell}}{dx^{2\ell}}\left(x^2 - 1\right)^\ell$$

In the derivative term, we are now considering the highest order derivative possible before this term vanishes. Each derivative taken knocks the exponent down to a coefficient, while the exponent lessens by one. From this pattern, one finds this term becomes the constant given by

$$\frac{d^{2\ell}}{dx^{2\ell}}\left[\left(x^2 - 1\right)^\ell\right] = (2\ell)! \tag{B.208}$$

Then, we can write A_ℓ^ℓ as

$$A_\ell^\ell = C_\ell(2\ell)!\left(1 - x^2\right)^{\frac{\ell}{2}} = D_\ell\left(1 - x^2\right)^{\frac{\ell}{2}} \tag{B.209}$$

Evaluating the terms involved in $L_+ Y_\ell^\ell$ using Eq. (B.209), we have

$$\begin{aligned}
\frac{dA_\ell^\ell}{d\theta} &= \frac{dx}{d\theta}\frac{dA_\ell^\ell}{dx} \\
&= D_\ell(-\sin\theta)(-2x)\left[\frac{\ell}{2}(1 - x^2)^{\frac{\ell}{2}-1}\right]
\end{aligned}$$

$$= D_\ell(-\sin\theta)(-2x)\left[\frac{\ell}{2}(1-x^2)^{\frac{\ell}{2}}(1-x^2)^{-1}\right]$$

Again, substituting $1 - x^2 = \sin^2\theta$ leads to

$$\frac{dA_\ell^\ell}{d\theta} = D_\ell\,(-2\cos\theta)\,(-\sin\theta)\left[\frac{\ell}{2}\left(1-x^2\right)^{\frac{\ell}{2}}\frac{1}{\sin^2\theta}\right]$$

$$= \ell D_\ell \frac{\cos\theta}{\sin\theta}\left(1-x^2\right)^{\frac{\ell}{2}}$$

$$= \ell\cot\theta A_\ell^\ell$$

It readily follows from this result, that

$$\widehat{L}_+ Y_{\ell,\ell} = \hbar C_n e^{i(\ell+1)\varphi}\left[\frac{\partial A_\ell^\ell}{\partial\theta} - \ell\cot\theta A_\ell^\ell\right] = 0$$

This gives a relation that must be satisfied by A_ℓ^ℓ, namely

Highest-Level Associated Legendre Function Condition

$$\frac{\partial A_\ell^\ell}{\partial\theta} - \ell\cot\theta A_\ell^\ell = 0 \qquad \text{(B.210)}$$

Let's now find the eigenvalues of $\widehat{L}_-$ and $\widehat{L}_+$ in relation to those of $\widehat{L}$ and $\widehat{L}_z$. The definitions of $\widehat{L}_\pm$ lead to the fact that they both have complex eigenvalues λ_+, λ_-, respectively, and this leads to the following relations:

$$\widehat{L}_- Y_\ell^m = \lambda_- Y_\ell^m$$

$$\widehat{L}_+ Y_\ell^m = \lambda_+ Y_\ell^m$$

$$\widehat{L}_-\widehat{L}_+ Y_\ell^m = \lambda_-\lambda_+ Y_\ell^m = |\lambda_+|^2 Y_\ell^m = |\lambda_-|^2 Y_\ell^m$$

Next, let us consider the commutation relation, which we will need, between $\widehat{L}_z$ and $\widehat{L}_+$ given by

$$\left[\widehat{L}_z,\widehat{L}_+\right] = \widehat{L}_z(\widehat{L}_x + i\widehat{L}_y) - (\widehat{L}_x + i\widehat{L}_y)\widehat{L}_z$$

$$= \left(\widehat{L}_z\widehat{L}_x - \widehat{L}_x\widehat{L}_z\right) + i\left(\widehat{L}_z\widehat{L}_y - \widehat{L}_y\widehat{L}_z\right)$$

$$= i\hbar\widehat{L}_y + i(-i\hbar\widehat{L}_x)$$

$$= \hbar \widehat{L}_+$$

In the above, we have used the commutation relations $\widehat{\mathbf{L}} \times \widehat{\mathbf{L}} = i\hbar \widehat{\mathbf{L}}$. This leads to the following equation for $\widehat{L}_+$:

$$\widehat{L}_z \widehat{L}_+ Y_\ell^m = \widehat{L}_+ \widehat{L}_z Y_\ell^m + \left[\widehat{L}_z, \widehat{L}_+\right] Y_\ell^m$$

$$\begin{aligned}
\widehat{L}_z \widehat{L}_+ Y_\ell^m &= \widehat{L}_+ \widehat{L}_z Y_\ell^m + \hbar \widehat{L}_+ Y_\ell^m \\
&= m\hbar \widehat{L}_+ Y_\ell^m + \hbar \widehat{L}_+ Y_\ell^m \\
&= \hbar(m+1) \widehat{L}_+ Y_\ell^m
\end{aligned}$$

Based on the eigensolutions we obtained for $\widehat{L}_z \Phi = \hbar m \Phi$, which also leads to

$$\widehat{L}_z Y_\ell^{m+1} = \hbar(m+1) Y_\ell^{m+1}$$

Since all eigenstates of $\widehat{L}_z$ are distinct orthogonal states, these two states must be proportional to one another, such that

$$\widehat{L}_+ Y_\ell^m = \alpha Y_\ell^{m+1} \tag{B.211}$$

A similar conclusion follows for $\widehat{L}_-$, because in this case, the commutation relation gives

$$\begin{aligned}
\left[\widehat{L}_z, \widehat{L}_-\right] &= \widehat{L}_z(\widehat{L}_x - i\widehat{L}_y) - (\widehat{L}_x - i\widehat{L}_y)\widehat{L}_z \\
&= \widehat{L}_z \widehat{L}_x - i\widehat{L}_z \widehat{L}_y - \widehat{L}_x \widehat{L}_z + i\widehat{L}_y \widehat{L}_z \\
&= i\hbar \widehat{L}_y + i(i\widehat{L}_x) \\
&= -\hbar \widehat{L}_-
\end{aligned}$$

This leads to the following equation for $\widehat{L}_- Y_\ell^m$:

$$\begin{aligned}
\widehat{L}_z \widehat{L}_- Y_\ell^m &= \widehat{L}_- \widehat{L}_z Y_\ell^m + \left[\widehat{L}_z, \widehat{L}_-\right] Y_\ell^m \\
&= \widehat{L}_- \widehat{L}_z Y_\ell^m - \hbar \widehat{L}_- Y_\ell^m \\
&= m\hbar \widehat{L}_- Y_\ell^m - \hbar \widehat{L}_- Y_\ell^m \\
&= \hbar(m-1) \widehat{L}_- Y_\ell^m
\end{aligned}$$

Therefore, analogous to $\widehat{L}_+$, we find the eigenvalue equation for $\widehat{L}_-$ to be given by

$$\widehat{L}_- Y_\ell^m = \beta Y_\ell^{m-1} \tag{B.212}$$

The above relations allow us to determine eigenvalues α and β in terms of ℓ and m, as follows:

$$Y_\ell^m \widehat{L}_+^\dagger \widehat{L}_+ Y_\ell^m = Y_\ell^{m+1} |\alpha|^2 \hbar^2 Y_\ell^{m+1} \tag{B.213}$$

$$Y_\ell^m \widehat{L}_-^\dagger \widehat{L}_- Y_\ell^m = Y_\ell^{m-1} |\beta|^2 \hbar^2 Y_\ell^{m-1} \tag{B.214}$$

But $L_+^\dagger = \widehat{L}_-$, and $L_-^\dagger = \widehat{L}_+$, since both operators are *adjoint operators* of one another. Then, we have

$$\begin{aligned}
\widehat{L}_+ \widehat{L}_- Y_\ell^m &= \left(\widehat{L}_x + i\widehat{L}_y\right)\left(\widehat{L}_x - i\widehat{L}_y\right) Y_\ell^m \\
&= \left[\widehat{L}_x \widehat{L}_x + i(\widehat{L}_y \widehat{L}_x - \widehat{L}_x \widehat{L}_y) + \widehat{L}_y \widehat{L}_y\right] Y_\ell^m \\
&= \left[\widehat{L}^2 - \widehat{L}_z^2 + i\left(-i\hbar \widehat{L}_z\right)\right] Y_\ell^m \\
&= \left(\ell(\ell+1)\hbar^2 - m^2\hbar^2 + m\hbar^2\right) Y_\ell^m \\
&= (\ell+m)(\ell-m+1)\hbar^2 Y_\ell^m
\end{aligned}$$

This result allows us to determine $|\beta|^2$. By substituting the result into Eqs. (B.213) and (B.215), we have

$$(\ell+m)(\ell-m+1)\left(Y_\ell^m\right)^2 = |\beta|^2 \left(Y_\ell^{m+1}\right)^2 \tag{B.215}$$

Since the wave functions are normalized, integration of Eq. (B.215) gives the result

$$|\beta|^2 = (\ell+m)(\ell-m+1) \tag{B.216}$$

Similarly, for $|\alpha|^2$, we obtain

$$\begin{aligned}
\widehat{L}_- \widehat{L}_+ A_\ell^m &= \left(\widehat{L}_x - i\widehat{L}_y\right)\left(\widehat{L}_x + i\widehat{L}_y\right) A_\ell^m \\
&= \left[\widehat{L}_x \widehat{L}_x + i(\widehat{L}_x \widehat{L}_y - \widehat{L}_y \widehat{L}_x) + \widehat{L}_y \widehat{L}_y\right] A_\ell^m \\
&= \left[\widehat{L}^2 - \widehat{L}_z^2 + i\left(i\hbar \widehat{L}_z\right)\right] A_\ell^m \\
&= \left(\ell(\ell+1)\hbar^2 - m^2\hbar^2 - m\hbar^2\right) A_\ell^m
\end{aligned}$$

$$= (\ell - m)(\ell + m + 1)A_\ell^m$$

Substitution of this result into Eq. (B.213), integrating it, we have the following results for $|\beta|$ and $|\alpha|$ given by

$$|\alpha| = \sqrt{(\ell - m)(\ell + m + 1)} \tag{B.217a}$$

$$|\beta| = \sqrt{(\ell + m)(\ell - m + 1)} \tag{B.217b}$$

Finally, we can state two important relations for the lowering and raising operators $\widehat{L}_-, \widehat{L}_+$, defining the two operator eigenvalue equations given by

Angular Momentum Raising Eigenoperator Equations

$$L_+ Y_\ell^m = \sqrt{(\ell - m)(\ell + m + 1)}Y_\ell^{m+1} \tag{B.218}$$

and

Angular Momentum Lowering Eigenoperator Equation

$$L_- Y_\ell^m = \sqrt{(\ell + m)(\ell - m + 1)}Y_\ell^{m-1} \tag{B.219}$$

To demonstrate the utility of the above relations using the lowering and raising operators, let's do a few examples. The idea is that given any state $Y_\ell^{-\ell}$, one can use the *raising operator* $\widehat{L}_+$ as many of 2ℓ times to obtain higher level states Y_ℓ^{m+1}. And likewise, one can use the *lowering operator* $\widehat{L}_-$ as many of 2ℓ times, on Y_ℓ^ℓ, to obtain lower level states Y_ℓ^{m-1}. This is a utility of these kinds of operators. Next, we'll consider an example to illustrate how these operators can be used.

Let's consider the lowest energy state Y_1^{-1} (for $\ell = 1$). Using Eq. (B.204), we have

$$Y_1^{-1} = C_n \sin^1\theta e^{-i\varphi} = \sqrt{\frac{3}{8\pi}}\sin\theta e^{-i\varphi}$$

Using the operator equation for $\widehat{L}_+$ given in Eq. (B.218), we have

$$L_+ Y_1^{-1} = \sqrt{(\ell - m)(\ell + m + 1)}Y_\ell^0 = \sqrt{2 \cdot 1}Y_\ell^0$$

Substituting the operator for $\widehat{L}+$, the LHS then becomes

$$L_+Y_1^{-1} = \hbar e^{i\varphi}\left(\frac{\partial}{\partial\theta} + i\cot\theta\frac{\partial}{\partial\varphi}\right)Y_1^{-1}$$

$$= \sqrt{\frac{3}{8\pi}}\hbar e^{i\varphi}\left(\frac{\partial}{\partial\theta} + i\cot\theta\frac{\partial}{\partial\varphi}\right)\sin\theta e^{-i\varphi}$$

$$= \sqrt{\frac{3}{8\pi}}\hbar\left[\cos\theta + i\cot\theta(-i\sin\theta)\right]$$

$$= \sqrt{\frac{3}{8\pi}}\hbar(2\cos\theta) = \sqrt{2}Y_1^0$$

From this, we have the following result for Y_1^0:

$$Y_1^0(\cos\theta) = \frac{2}{\sqrt{2}}\sqrt{\frac{3}{8\pi}}\hbar\cos\theta = \sqrt{\frac{3}{4\pi}}\hbar\cos\theta \quad \text{(B.220)}$$

This process could continue, providing a way to obtain all the higher energy states for the orbital $\ell = 1$. These orbitals are examples of p orbitals, defined by the quantum number $\ell = 1$. This demonstrates a powerful utility of the raising (or lowering) operator, particularly, if only the ground state is known.

B.14 Chapter Summary

To examine orbital angular momentum in quantum mechanics, we first reviewed the concept in the classical domain. From there, we took the definitions and applied the quantum mechanical operators for $\widehat{\mathbf{p}}$, to obtain the orbital angular momentum operators. We then solved for the eigenvalues for $\widehat{L_z^2}$, given by m^2 and for $\widehat{L^2}$, given by $\ell(\ell+1)$. We also obtained the corresponding wave functions given by the spherical harmonics $Y_{\ell,m}$. In obtaining the wave functions, we also learned that the solutions to the orbital angular momentum problem were enabled by the earlier work of Legendre and others, in analyzing gravitational potential expansions. It all came together ultimately to provide a path to help solve the angular momentum problem exactly.

We then introduced the commutation operator and relevant relations involving $\widehat{\mathbf{p}}$ and $\widehat{\mathbf{L}}$. This allowed us to introduce the lowering and raising operators for the orbital angular momentum, and demonstrate their utility with the spherical harmonics. This completed our objectives with the orbital angular momentum problem. Table B.7 lists some of the key results for the total orbital angular momentum.

Table B.7: Chapter B Summary Equations.

Name	Equation
L_z eigen-function	$\Phi(\varphi) = \frac{1}{\sqrt{2\pi}} e^{im\varphi}$
L_z eigenvalues	$m = ..., -2, -1, 0, 1, 2, ...$
L^2 eigenvalues	$\lambda = \ell(\ell + 1)$
L^2 quantum numbers	$\ell = 0, 1, 2, 3, ...$
Relation between m and ℓ	$\|m\| \leq \ell$
Legendre polynomials P_ℓ	$P_\ell = \frac{1}{2^\ell \ell!} \frac{d^\ell}{dx^\ell} (x^2 - 1)^\ell$
Associated Legendre functions	$A_\ell^m = \frac{1}{2^\ell \ell!} (1 - x^2)^{\frac{m}{2}} \frac{d^{\ell+m}}{dx^{\ell+m}} (x^2 - 1)^\ell$
L^2 eigen-function	$Y_{\ell,m}(\theta, \varphi) = A_\ell^m (x = \cos\theta) e^{im\varphi}$
Lowering Operator	$\widehat{L}_- = \widehat{L}_x - i\widehat{L}_y$
Raising Operator	$\widehat{L}_+ = \widehat{L}_x + i\widehat{L}_y$

Table B.8 summarizes and illustrates more examples of spherical harmonic functions $Y_{\ell,m}$ in 3D. Note that only positive m values are shown, however, keep in mind that the corresponding $-m$ cases are only rotations of their $+m$ counterparts.

Table B.8: Real part of the spherical harmonics Y_ℓ^m (projected into cartesian coordinates), indicating both quantum numbers ℓ and m, the Associated Legendre Polynomial factor, and the azimuthal factor from L_z.

ℓ	m	$A_\ell^m(\cos\theta)$	$\Phi(\varphi)$	Real Part$[Y_{\ell,m}]$
0	0	$\frac{1}{(4\pi)^{1/2}}$	1	
1	0	$\left(\frac{3}{4\pi}\right)^{1/2}\cos\theta$	1	
1	1	$-\left(\frac{3}{8\pi}\right)^{1/2}\sin\theta$	$e^{i\varphi}$	
2	0	$\left(\frac{5}{16\pi}\right)^{1/2}(3\cos^2\theta - 1)$	1	
2	1	$-\left(\frac{15}{8\pi}\right)^{1/2}\cos\theta\sin\theta$	$e^{i\varphi}$	
2	2	$\left(\frac{15}{32\pi}\right)^{1/2}\sin^2\theta$	$e^{i2\varphi}$	
3	0	$\left(\frac{7}{4\pi}\right)^{1/2}(5\cos^2\theta - 3)\cos\theta$	1	
3	1	$-\left(\frac{7}{48\pi}\right)^{1/2}(1 - 5\cos^2\theta)\sin\theta$	$e^{i\varphi}$	
3	2	$\left(\frac{7}{480\pi}\right)^{1/2}\sin^2\theta\cos\theta$	$e^{i2\varphi}$	
3	3	$-\left(\frac{7}{2880\pi}\right)^{1/2}\sin^3\theta$	$e^{i3\varphi}$	

Table B.8

4	0	$\left(\frac{9}{4\pi}\right)^{1/2}(35\cos^4\theta - 30\cos^2\theta + 3)$	1	

B.15 Chapter Problems

Problem B.1 Show that the two associated Legendre functions A_1^1 and A_1^0 are normal where

$$\int_0^{\pi} \left(A_1^0(\theta)\right)^* A_1^1(\theta)d\theta$$

Use Table B.8, which lists these functions.

Problem B.2 Show that the function

$$u(r) = \frac{1}{r}$$

is a solution to the equation given by

$$\frac{1}{r^2}\frac{\partial}{\partial r}\left(r^2\frac{\partial u}{\partial r}\right) = 0$$

Substitute $u(r)$ into this equation.

Problem B.3 Earlier, we used what was called the Trigonometric Recurrence Relation to obtain the normalization constant. Use integration by parts, or

$$\int udv = uv - \int vdu$$

And prove that

$$\int \sin^n(\theta)d\theta = -\sin^{n-1}\theta\cos\theta + n - 1\int \sin^{n-2}\theta\cos\theta d\theta$$

Hint: Start with

$$\int \sin^n\theta d\theta = \int \sin^{n-1}\theta\sin\theta d\theta = \int udv$$

Problem B.4 Let $u(\varphi,\theta) = Y_1^1$. Show that u satisfies the following equation

$$\frac{1}{r^2\sin\theta}\frac{\partial}{\partial\theta}\left(\sin\theta\frac{\partial u}{\partial\theta}\right)+\frac{1}{r^2\sin^2\theta}\frac{\partial^2 u}{\partial\varphi^2}=-\frac{2}{r^2}u$$

We are using only the Laplacian terms dependent on φ and θ, since u is independent of r.

Problem B.5 In Eq. (B.220), we used the raising operator $\widehat{L}_+$ to obtain $Y_1^0(\cos\theta) = \sqrt{\frac{3}{4\pi}}\hbar\cos\theta$. Use this result along with the raising operator to show that $Y_1^1 = -\sqrt{\frac{3}{8\pi}}\sin\theta e^{i\phi}$.

Problem B.6 Prove that

$$\left[\widehat{L}^2,\widehat{\mathbf{L}}_z\right]=0$$

Discuss the properties of the Hamiltonian that lead to this result.

Problem B.7 The lowest spherical harmonic state for $\ell = 2$ is

$$Y_2^{-2}(\varphi,\theta)=\sqrt{\frac{15}{32\pi}}\sin^2\theta e^{-2i\varphi}$$

Use the raising ladder operator $\widehat{L}_+$ to show that

$$Y_2^{-1}(\varphi,\theta)=\sqrt{\frac{15}{8\pi}}\sin\theta\cos\theta e^{-i\varphi}$$

Appendix C

Appendix: BRA & KET Algebra

In this chapter, we introduce a popular and concise notation used in quantum mechanics. It was originally introduced by P. A. M. Dirac. It is both powerful and more concise than conventional calculus notation, and for this reason and more, is regarded as an *elegant* notation. It provides an alternative language or notation to express all the operations and eigenstates that we have seen throughout the text. Though it's more concise, it still allows great depth in analyzing problems, as we'll show. This notation has come to be known as *Bra-Ket Algebra,* named for its use of bracket components $\langle$ and $\rangle$. We'll begin with some preliminaries and definitions before using Bra-Ket Algebra.

C.1 Bra-Ket Preliminaries

From any eigenoperator equation, such as the Schrödinger equation, one obtains the solutions known as eigenfunctions ψ, along with their associated eigenvalues. For bound states, it also leads to a given set of quantum numbers and the resulting eigenvalues and eigenfunctions define the eigenstates of the system. The system may be in one of them, or some combination of them. Measurements of the observable states can serve to determine this combination, explicitly. In the Bra-Ket notation, each eigenstate A is represented by a **Ket-vector** $|\cdot\rangle$, expressed as

$$\psi_A : \text{state}A \quad \Rightarrow |A\rangle$$

The name *Ket* is obtained from the fact that *Ket* is the right side of Bra-Ket (as in bracket). That's right, its that simple. Hence, a *Ket-vector points to the right*. Note that it represents a *generalized eigenvector* like the 2×1 spinor or Dirac's 4-state vector, etc. The Ket-vector operates in a linear space, which means that any Ket can also be expressed as a linear combination of other properly chosen *Ket* vectors. For example, consider any two *Ket* vectors $|A\rangle$ and $|B\rangle$ such that their sum is

$$|A\rangle + |B\rangle = |C\rangle$$

Then, $|C\rangle$ is also a *Ket* vector living in the same mathematical space. More generally, the following state is a *Ket* vector in a linear space:

$$c_1|A\rangle + c_2|B\rangle = |D\rangle$$

Coefficients c_1 and c_2 are scalars and the property of a *linear space* is that $|D\rangle$ is also a Ket vector. Additionally, for any Ket vector $|A\rangle$, there also exists a *bra vector* $\langle A|$. In this case, *Bra* is the left side of the word *Bra-Ket*, hence, the *bra vector points to the left*. Given the Bra and Ket vectors, we can write the *inner product* or *scalar product* of any two states using Bra and Ket vectors as

$$\text{inner product} \Rightarrow \langle B|A\rangle$$

Having an *inner product*, we can then define *orthogonal states*. Two states are said to be *orthogonal* if

$$\langle B|A\rangle = 0 \Rightarrow |B\rangle \perp |A\rangle$$

A state $|A\rangle$ is also said to be *normalized* if

$$\langle A|A\rangle = 1 \quad \text{(normalization condition)} \tag{C.1}$$

Eq. (C.1) is the usual *normalization condition*, which we have discussed in the previous chapters. As a scalar product is generally complex. Then, it must also have a complex conjugate. Using Bra-Ket notation, the complex conjugate, denoted by $\overline{\langle A|B\rangle}$ is given by

Complex-Conjugate Using Bra-Ket Notation

$$\overline{\langle A|B\rangle} = \langle B|A\rangle$$

It follows from the complex conjugate definition, that if $|B\rangle = |A\rangle$, then

$$\overline{\langle A|A\rangle} = \langle A|A\rangle \Rightarrow \langle A|A\rangle \text{ is real}$$

For a nontrivial state $|B\rangle$, if $\langle A|B\rangle = 0$ for any Bra vector $\langle A|$, then $|B\rangle$ must be the **Null Ket vector**. Let's consider some explicit examples of *Bras* and *Kets*.

Consider the following *Ket* vectors:

$$|A\rangle = \begin{bmatrix} a \\ b \end{bmatrix} \quad \text{and} \quad |B\rangle = \begin{bmatrix} c \\ d \end{bmatrix}$$

a, b, c, and d are all scalars, and thus, both Kets are 2×1 vectors. Their corresponding Bra vectors, are then given by

$$\langle A| = \begin{bmatrix} a^* & b^* \end{bmatrix} \quad \text{and} \quad \langle B| = \begin{bmatrix} c^* & d^* \end{bmatrix}$$

Here a^*, b^*, c^*, and d^* are complex conjugates of a, b, c, and d, respectively. Thus, for a Ket vector, the Bra vector is the *transpose of the complex conjugates*.

Adding these two Kets (2×1 vectors) leads to another Ket (2×1 vectors), as we have

$$|A\rangle + |B\rangle = \begin{bmatrix} a + c \\ b + d \end{bmatrix}$$

Likewise, adding the Bra vectors, which are 1×2 vectors, we have

$$\langle A| + \langle B| = \begin{bmatrix} a^* + c^* & b^* + d^* \end{bmatrix}$$

Computing $\langle A|A\rangle$ becomes

$$\langle A|A\rangle = \begin{bmatrix} a^* & b^* \end{bmatrix} \begin{bmatrix} a \\ b \end{bmatrix} = a^*a + b^*b = |a|^2 + |b|^2 \geq 0$$

Note that because $|a| \geq 0$ and $|b| \geq 0$, then $\langle A|A\rangle = 0$ *if and only if* $|a| = |b| = 0$. For this reason, the inner product can only vanish if one of the states is the *Null Ket*. Otherwise, we have $\langle A|A\rangle > 0$.

Let us now take $a = b = 1$. Then

$$\langle A|A\rangle = 1 + 1 = 2$$

To normalize the Ket $|A\rangle$, we would have to scale $|A\rangle$ by $1/\sqrt{2}$, because then we have

$$\langle A|A\rangle = \frac{1}{\sqrt{2}}\begin{bmatrix} a^* & b^* \end{bmatrix}\frac{1}{\sqrt{2}}\begin{bmatrix} a \\ b \end{bmatrix} \doteq \frac{1}{2}(1+1) = 1$$

This defines a *normalized Ket* $|A'\rangle$ given by

$$|A'\rangle = \frac{1}{\sqrt{2}}\begin{bmatrix} 1 \\ 1 \end{bmatrix}$$

$\langle A'|A'\rangle = 1$ satisfies the normalization condition in Eq. (C.1). Let the normalized Ket vector $|B'\rangle$ be given by

$$|B'\rangle = \frac{1}{\sqrt{2}}\begin{bmatrix} 1 \\ -1 \end{bmatrix}$$

Then $\langle B'|A'\rangle$ becomes

$$\langle B'|A'\rangle = \frac{1}{\sqrt{2}}\begin{bmatrix} 1 & -1 \end{bmatrix}\frac{1}{\sqrt{2}}\begin{bmatrix} 1 \\ 1 \end{bmatrix} = \frac{1}{2}(1-1) = 0$$

In this case, these two vectors are said to be *orthogonal* to each other. This is easy to visualize in this example, because $|A'\rangle$ is a 45° vector, in the first quadrant of a $2D$ plane, while $|B'\rangle$ lies at −45° in the fourth quadrant of a $2D$ plane, i.e. they are 90° apart. This is illustrated in Fig. C.1. Let us also take a similar set of vectors, involving complex numbers.

$$|C'\rangle = \frac{1}{\sqrt{2}}\begin{bmatrix} 1 \\ -i \end{bmatrix} \quad \text{and} \quad |D'\rangle = \frac{1}{\sqrt{2}}\begin{bmatrix} 1 \\ +i \end{bmatrix}$$

Testing for orthogonality, we have

$$\langle D'|C'\rangle = \frac{1}{2}\begin{bmatrix} 1 & -i \end{bmatrix}\cdot\begin{bmatrix} 1 \\ -i \end{bmatrix} = \frac{1}{2}[1\cdot 1 + -i\cdot(-i)] = 0$$

This set of vectors is analogous to $|A'\rangle$ and $|B'\rangle$. They form vectors that are 90° apart in the *complex plane*, as shown in Fig. C.2. Note that, in this case, the Bra vector of $|D'\rangle$ and the Ket vector $|C'\rangle$ contain the same elements, however, they do not equate, because a Bra can never equal a

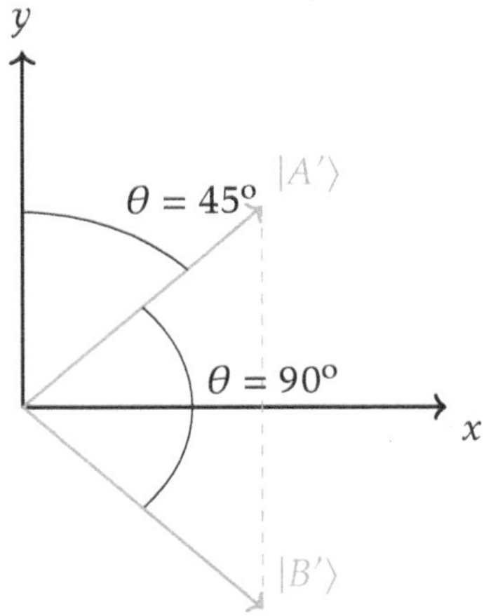

Figure C.1: Illustration of orthogonal states $|A'\rangle$, represented by Kets. In the state space, the two states form a 2D coordinate system that can describe any state that is a combination of the two eigenstates. These are the conditions for eigenstates which are orthogonal states.

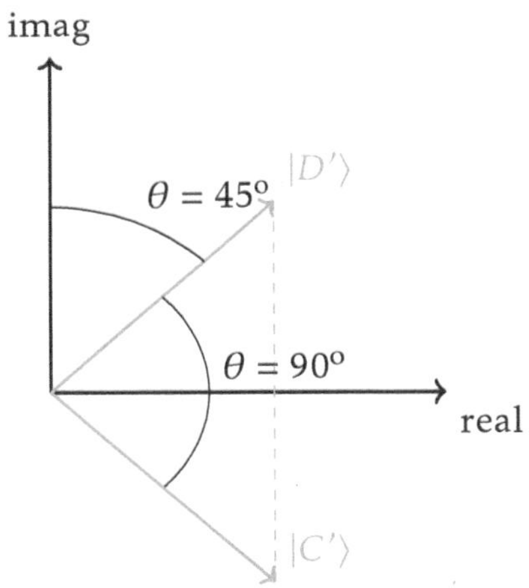

Figure C.2: Illustration of orthogonal states $|A'\rangle$, represented by Kets. In the state space, the two states form a 2D coordinate system that can describe any state that is a combination of the two eigenstates. These are the conditions for eigenstates obtained, which are orthogonal states.

Ket. They are defined as distinct constructs. For this reason, the vector space of Bras and Kets is said to be a **dual vector space**.

All the examples above contain *fixed* numbers, or constants. They do *not* depend on a variable. These concepts, however, extend to variable elements, as well. An important difference that arises when the Ket vector depends on space **r** is that the inner product then integrates over all of space. For example, let us consider the following states

$$|E\rangle = \psi(x) \quad \text{and} \quad |F'\rangle = \beta(x)$$

Then, the inner product $\langle E|F'\rangle$ becomes

$$\langle E|F'\rangle \equiv \int_{x=-\infty}^{+\infty} \psi^*(x)\beta(x)dx$$

Then, the *orthogonality* condition becomes

$$\langle E|F'\rangle \equiv \int_{x=-\infty}^{+\infty} \psi^*(x)\beta(x)dx = 0$$

Thus, any two states satisfying this condition are said to be orthogonal. This is restating what we have already learned prior to introducing Bra and Ket algebra. However, Bra-Ket notation provides an alternative, more concise notation for expressing this condition. Likewise, the *normalization condition* becomes

$$\langle E|E\rangle \equiv \int_{x=-\infty}^{+\infty} \psi^*(x)\psi(x)dx = 1$$

The concepts extend to 3D, where

$$\langle E|F'\rangle \equiv \int_{z=-\infty}^{+\infty}\int_{y=-\infty}^{+\infty}\int_{x=-\infty}^{+\infty} \psi^*(x,y,z)\beta(x,y,z)dxdydz = 0$$

and

$$\langle E|E\rangle \equiv \int_{z=-\infty}^{+\infty}\int_{y=-\infty}^{+\infty}\int_{x=-\infty}^{+\infty} \psi^*(x,y,z)\psi(x,y,z)dxdydz = 1$$

The conciseness of the Bra-Ket notation is more evident, the higher the dimension or the more degrees of freedom a system has. Let us now discuss the next step of applying operators to Bras and Kets.

C.2 Operators With Bras and Kets

In Chap. ??, we introduced the idea of a Hermitian operator. Now, we will use Bra-Ket algebra as a tool to explore properties of operators, including Hermitian or self-adjoint operators. Let's consider a self-adjoint operator where $\overline{\Omega} = \Omega$. And let the eigenoperator equation satisfied by Ω be

$$\Omega|A_n\rangle = \lambda_n|A_n\rangle \tag{C.2}$$

For now, we do not assume anything about λ_n. However, our goal is to show that if $\overline{\Omega} = \Omega$, then it follows that λ_n *must be real*. First, let $|C_n\rangle = \Omega|A_n\rangle$. Our results in the previous sections revealed that this leads to $\langle C_n| = \langle A_n|\overline{\Omega}$. Since Ω is self-adjoint, then we also have

$$\langle A_n|\overline{\Omega} = \langle A_n|\Omega = \lambda_n^*\langle A_n| \tag{C.3}$$

In Eq. (C.3), we have used the eigenoperator equation. Then, right-multiplying by $|A_n\rangle$, we have

$$\langle A_n|\Omega|A_n\rangle = \lambda_n^*\langle A_n|A_n\rangle \tag{C.4}$$

However, if we left-multiply Eq. (C.2) by $\langle A_n|$, we get the following:

$$\langle A_n|\Omega|A_n\rangle = \langle A_n|\lambda_n|A_n\rangle = \lambda_n\langle A_n|A_n\rangle \tag{C.5}$$

But, Eq. (C.4) and Eq. (C.5) are the same, both giving $\langle A_n|\Omega|A_n\rangle$. Then, it follows that

$$\lambda_n^*\langle A_n|A_n\rangle = \lambda_n\langle A_n|A_n\rangle$$

If we assume that $|A_n\rangle$ is not equal to the *null Ket* (this is always assumed for any eigenstate), then it follows that

$$\lambda_n^* = \lambda_n \tag{C.6}$$

The only numbers for which Eq. (C.6) is true are real numbers. Hence, λ_n *is real*. So, *a self-adjoint or Hermitian operator always has real eigenvalues*. This is why it is also called a *real operator*. Recall that we proved this first in Chap. **??**, using conventional mathematical notation. Here, the same result has been obtained using Bra-Ket algebra.

C.2.1 Orthogonal Eigenstates

Let us go further into some consequences for a *self-adjoint operator* using Bra-Ket notation. Consider two distinct eigenstates m and n defined by

$$\Omega|A_n\rangle = \lambda_n|A_n\rangle \tag{C.7a}$$

$$\Omega|A_m\rangle = \lambda_m|A_m\rangle \tag{C.7b}$$

Since Ω is a self-adjoint operator, then both λ_n and λ_m are real, and $\lambda_n \neq \lambda_m$. Now, let $|C_n\rangle = \Omega|A_n\rangle$. Multiplying Eq. (C.7b) by $\langle A_n|$, we have

$$\langle A_n|\Omega|A_m\rangle = \langle A_n|\lambda_m|A_m\rangle = \lambda_m\langle A_n|A_m\rangle \tag{C.8}$$

Additionally, from the definition of $|C_n\rangle$, it follows that

$$\langle C_n| = \langle A_n|\overline{\Omega} = \langle A_n|\Omega = \lambda_n\langle A_n| \tag{C.9}$$

Right-multiplying Eq. (C.9) by $|A_m\rangle$ leads to

$$\langle C_n|A_m\rangle = \lambda_n\langle A_n|A_m\rangle \tag{C.10}$$

Subtracting Eq. (C.10) from Eq. (C.8), we have

$$(\lambda_m - \lambda_n)\langle A_n|A_m\rangle = 0 \tag{C.11}$$

Since we assume $\lambda_m \neq \lambda_n$, then we have the result

$$\langle A_n|A_m\rangle = 0 \tag{C.12}$$

This is the definition of **orthogonal eigenstates**. Thus, for any two distinct states of a self-adjoint operator having distinct eigenvalues, they are orthogonal eigenstates.

C.3 Linear Harmonic Oscillator Revisited

To illustrate more utility of Bra-Ket algebra and symmetry in operators, we will revisit the problem of the linear harmonic oscillator (LHO), discussed in Chap. A. The LHO problem can be solved using Bra-Ket algebra. However, before we do this, lets revisit two useful anti-symmetric operators, originally introduced by P. A. M. Dirac. Recall that the Dirac equation, discussed in Chap. 2, also recognized symmetry in the relativistic energy $\overline{E}$, and exploited the symmetry to devise a first order operator for $\overline{E}$. This is given by Eqs. (**??**) and (**??**). This result illustrated how an operator can mirror the symmetry in the eigenvalues.

Recall an earlier result we found with $[x,p] = xp - px$, and given by Eq. (B.166), where we determined

$$[x,\widehat{p}_x] = x\widehat{p}_x - \widehat{p}_x x = i\hbar \tag{C.13}$$

x and $\widehat{p}_x$ *do not commute*. In our first encounter with the LHO problem (1D), we found the Hamiltonian $\widehat{H}$ to be

$$\widehat{H} = \frac{\widehat{p}^2}{2m^*} + \frac{1}{2}m^*\omega^2 x^2 \tag{C.14}$$

In Eq. (C.14), m^* is the reduced mass, ω is the oscillator frequency. Just as we observed in the relativistic energy, which lead to the Dirac equation, the Hamiltonian in Eq. (C.14) also possesses symmetry of the form $a^2 + b^2 = (a + ib)(a - ib)$. Let us define the following *two* 1D anti-symmetric operators:

$$\widehat{a} = \frac{m^*\omega x + ip}{\sqrt{2m^*\hbar\omega}} \quad \text{and} \quad \widehat{a}^\dagger = \frac{m^*\omega x - ip}{\sqrt{2m^*\hbar\omega}} \tag{C.15}$$

Note that $\widehat{a}^\dagger$ represents the *adjoint* of $\widehat{a}$, and they are *not equal*. Thus, $\widehat{a}$ is an example of an operator that is *not self-adjoint*. Consider the operator $\hbar\omega\widehat{a}\left(\widehat{a}^\dagger\right) = \hbar\omega\widehat{a}\widehat{a}^\dagger$, given by

$$\begin{aligned}
\hbar\omega\widehat{a}\overline{\widehat{a}} &= \left[\frac{m^*\omega x + ip}{\sqrt{2m^*\hbar\omega}}\right]\left[\frac{m^*\omega x - ip}{\sqrt{2m^*\hbar\omega}}\right] \\
&= \frac{1}{2}m^*\omega^2 x^2 + \frac{p^2}{2m^*} - i\frac{\omega}{2}(xp_x - p_x x) \\
&= \frac{1}{2}m^*\omega^2 x^2 + \frac{p^2}{2m^*} + \frac{\hbar\omega}{2} = \widehat{H} + \frac{\hbar\omega}{2}
\end{aligned}$$

In obtaining the above result, we have used Eq. (C.13). It follows that $\widehat{H}$, the Hamiltonian of the LHO, is also given by

$$\widehat{H} = \hbar\omega\left(\widehat{a}\widehat{a}^\dagger - \frac{1}{2}\right) \tag{C.16}$$

Similarly, for $\hbar\omega\widehat{a}^\dagger\widehat{a}$, we have

$$\begin{aligned}
\hbar\omega\widehat{a}^\dagger\widehat{a} &= \left[\frac{m^*\omega x - ip}{\sqrt{2m^*\hbar\omega}}\right]\left[\frac{m^*\omega x + ip}{\sqrt{2m^*\hbar\omega}}\right] \\
&= \frac{1}{2}m^*\omega^2 x^2 + \frac{p^2}{2m^*} + i\frac{\omega}{2}(xp_x - p_x x) \\
&= \frac{1}{2}m^*\omega^2 x^2 + \frac{p^2}{2m^*} - \frac{\hbar\omega}{2} = \widehat{H} - \frac{\hbar\omega}{2}
\end{aligned}$$

From this, we obtain

$$\widehat{H} = \hbar\omega\left(\widehat{a}^{\dagger}\widehat{a} + \frac{1}{2}\right) \tag{C.17}$$

This result leads to expectation value for the number of states n, since the we know for the harmonic oscillator, the energy eigenvalue is $\hbar\omega(n+1/2)$. Thus, we have $\langle n\rangle$ given by

Expectation Value Of Number Of States

$$\langle n\rangle = \widehat{a}^{\dagger}\widehat{a} \tag{C.18}$$

Then, right-multiplying $\hbar\omega\widehat{a}\widehat{a}^{\dagger}$ by $\widehat{a}$, and left-multiplying $\hbar\omega\widehat{a}^{\dagger}\widehat{a}$ by $\widehat{a}$ must lead to equal quantities where we have

$$\hbar\omega\widehat{a}\widehat{a}^{\dagger}\widehat{a} = \widehat{H}\widehat{a} + \frac{\hbar\omega}{2}\widehat{a}$$

$$\hbar\omega\widehat{a}\widehat{a}^{\dagger}\widehat{a} = \widehat{a}\widehat{H} - \frac{\hbar\omega}{2}\widehat{a}$$

Setting these two equations equal leads to

$$\widehat{H}\widehat{a} = \widehat{a}\widehat{H} - \hbar\omega\widehat{a} \tag{C.19}$$

Applying Eq. (C.19) to any *eigenket* (or eigenstate) of $\widehat{H}$ where $\widehat{H}|A\rangle = \lambda|A\rangle$, we have

$$\widehat{H}\widehat{a}|A\rangle = \widehat{a}\widehat{H}|A\rangle - \hbar\omega\widehat{a}|A\rangle = \lambda\,(\widehat{a}|A\rangle) - \hbar\omega(\widehat{a}|A\rangle) \tag{C.20}$$

Eq. (C.20) ultimately leads to two important consequences. To explore the first of these, if we define Ket $|B\rangle = \widehat{a}|A\rangle$, we then have that

$$\widehat{H}|B\rangle = \lambda|B\rangle - \hbar\omega|B\rangle = (\lambda - \hbar\omega)\,|B\rangle \tag{C.21}$$

This process can be continued. Let us now apply Eq. (C.19) to $|B\rangle$. So we define $|C\rangle = a|B\rangle$. We then obtain

$$\begin{aligned}\widehat{H}\widehat{a}|B\rangle = \widehat{H}|C\rangle &= \widehat{a}\widehat{H}|B\rangle - \hbar\omega\widehat{a}|B\rangle \\ &= (\lambda - \hbar\omega)\widehat{a}|B\rangle - \hbar\omega\widehat{a}|B\rangle = (\lambda - 2\hbar\omega)\,|C\rangle\end{aligned}$$

Note that we have used Eq. (C.21) to substitute for $\widehat{H}|B\rangle$. Continuing this systematic process, we obtain a generalized result given by

$$\widehat{H}\widehat{a}^n|A\rangle = (\lambda - n\hbar\omega)|A\rangle \tag{C.22}$$

Note that $|A\rangle \equiv \widehat{a}^0|A\rangle$, $|B\rangle = \widehat{a}|A\rangle$, $|C\rangle = \widehat{a}|B\rangle = \widehat{a}^2|A\rangle$, *etc.* It is important to point out here that $|A\rangle$ is a *non-null eigenket* of the LHO Hamiltonian $\widehat{H}$.

The second important consequence we will obtain, arising from Eq. (C.20), can be worked out as follows: First, for $|B\rangle = \widehat{a}|A\rangle$, let us write

$$\hbar\omega\langle B|B\rangle = \hbar\omega\langle A|\widehat{a}^\dagger\widehat{a}|A\rangle = \langle A|\hbar\omega\widehat{a}^\dagger\widehat{a}|A\rangle \tag{C.23}$$

$$= \langle A|\widehat{H} - \frac{\hbar\omega}{2}|A\rangle \tag{C.24}$$

$$= (\lambda - \frac{\hbar\omega}{2})\langle A|A\rangle \tag{C.25}$$

In the above, we have used our earlier result $\widehat{H} - \frac{\hbar\omega}{2}|A\rangle = (\lambda - \frac{\hbar\omega}{2})|A\rangle$. Since we take $|A\rangle$ as a non-null *eigenket* of $\widehat{H}$, then $\langle A|A\rangle > 0$. The LHS is semi-positive definite ($\hbar\omega\langle B|B\rangle \geq 0$), which leads to the result that the eigenvalues of $\widehat{H}$ satisfy the condition

$$\lambda \geq \frac{\hbar\omega}{2} \tag{C.26}$$

Now, let's continue the process with $|C\rangle = \widehat{a}|B\rangle$:

$$\hbar\omega\langle C|C\rangle = \hbar\omega\langle B|\widehat{a}^\dagger\widehat{a}|B\rangle = \langle B|\hbar\omega\widehat{a}^\dagger\widehat{a}|B\rangle$$

$$= \langle B|\widehat{H} - \frac{\hbar\omega}{2}|B\rangle$$

$$= \langle B|H|B\rangle - \frac{\hbar\omega}{2}|B\rangle$$

$$= (\lambda - \hbar\omega)\langle A|A\rangle - \frac{\hbar\omega}{2}\langle B|B\rangle$$

$$= (\lambda - \hbar\omega)\langle A|A\rangle - \frac{1}{2}(\lambda - \frac{1}{2}\hbar\omega)\langle A|A\rangle$$

Again, because the LHS is semi-positive definite (≥ 0), and $\langle A|A\rangle > 0$, we have the result:

$$\frac{\lambda}{2} \geq \frac{3}{4}\hbar\omega \Rightarrow \lambda \geq \frac{3}{2}\hbar\omega \tag{C.27}$$

Again, for $|D\rangle = \hat{a}|C\rangle$, we have

$$\begin{aligned}\hbar\omega\langle D|D\rangle = \hbar\omega\langle C|\hat{a}^{\dagger}\hat{a}|C\rangle &= \langle C|\hbar\omega\hat{a}^{\dagger}\hat{a}|C\rangle \\ &= \langle C|\hat{H} - \frac{\hbar\omega}{2}|C\rangle \\ &= \langle C|H|C\rangle - \frac{\hbar\omega}{2}|C\rangle \\ &= (\lambda - 2\hbar\omega)\langle A|A\rangle - \frac{\hbar\omega}{2}\langle C|C\rangle \\ &= (\lambda - 2\hbar\omega)\langle A|A\rangle - \frac{1}{2}(\lambda - \frac{3}{2}\hbar\omega)\langle A|A\rangle\end{aligned}$$

Hopefully, the pattern is becoming clear. Continuing this process for any Ket $\hat{a}^n|A\rangle$, we obtain the general result

$$\lambda \geq \left(n + \frac{1}{2}\right)\hbar\omega \tag{C.28}$$

With the inequality given in Eq. (C.28), equality holds *if and only if* the LHS is identically zero, or if and only if

$$\langle B|B\rangle = \langle C|C\rangle = \langle A|\bar{\hat{a}}^{n}\hat{a}^{n}|A\rangle = 0 \tag{C.29}$$

Thus, to show that $\hat{a}|A\rangle = 0$, which also means that $\hat{a}|B\rangle = \hat{a}|C\rangle = \hat{a}|D\rangle ... =$ 0, proves the equality implied by Eq. (C.28). So, let us show that $\hat{a}|A\rangle$ must, in fact, be the null Ket, even though $|A\rangle$ is not. Recall that

$$\hbar\omega\hat{a}\hat{a}^{\dagger} = \lambda + \frac{\hbar\omega}{2} \Rightarrow \hbar\omega\hat{a}\hat{a}^{\dagger}\hat{a} = \left(\lambda + \frac{\hbar\omega}{2}\right)\hat{a}$$

Recall the following result we obtained earlier:

$$\hbar\omega\hat{a}\hat{a}^{\dagger}\hat{a} = \hat{a}\hat{H} - \frac{\hbar\omega}{2}\hat{a} \Rightarrow \left(\lambda - \frac{\hbar\omega}{2}\right)\hat{a}$$

Therefore, for any eigenket $|A\rangle$, of $\hat{H}$, it follows that

$$\left(\lambda - \frac{\hbar\omega}{2}\right)\hat{a}|A\rangle = \left(\lambda + \frac{\hbar\omega}{2}\right)\hat{a}|A\rangle \Rightarrow \hbar\omega\hat{a}|A\rangle = 0$$

Since $\hbar\omega \neq 0$, we have the result

$$\hat{a}|A\rangle = 0 \tag{C.30}$$

This result is interesting, considering that $\hat{a}$ is a *non-null operator*, and $|A\rangle$ is a non-null *eigenket*, yet, we have $\widehat{a}|A\rangle = 0$. In these conditions, we say that $|A\rangle$ is in the **null-space** of the operator $\widehat{a}$. Therefore, since $|B\rangle = \widehat{a}|A\rangle = 0$, then $|C\rangle = \widehat{a}|B\rangle = 0$, $|D\rangle = \widehat{a}|C\rangle = 0$, etc. The results above then lead to

$$\hbar\omega\langle B|B\rangle = 0 = (\lambda - \frac{\hbar\omega}{2})\langle A|A\rangle$$

$$\hbar\omega\langle C|C\rangle = 0 = (\lambda - \frac{3}{2}\hbar\omega)\langle A|A\rangle$$

$$\hbar\omega\langle D|D\rangle = 0 = (\lambda - \frac{5}{2}\hbar\omega)\langle A|A\rangle$$

$$\vdots$$

Thus, the LHS vanishes, and because $\langle A|A\rangle > 0$, we have the following result for λ:

$$\lambda = \frac{\hbar\omega}{2}$$

$$\lambda = \frac{3}{2}\hbar\omega$$

$$\lambda = \frac{5}{2}\hbar\omega$$

$$\vdots$$

This result for λ can be expressed, generally, as

Linear Harmonic Oscillator Eigenvalues

$$\lambda = \left(n + \frac{1}{2}\right)\hbar\omega \tag{C.31}$$

Eq. (C.31) is identical to Eq. (3.75), obtained in Chap. A, however we have obtained this result utilizing *Bra-Ket algebra*. The approach that has been taken here is also known as the **Dirac Method**. From this, the lowest energy state, or *ground state*, is given from the condition

$$\widehat{a}|A\rangle = 0 \quad \text{(ground state condition)} \tag{C.32}$$

This can be used to determine the ground state wave function.

C.4 Annihilation and Creation Operators Revisited

The operators $\widehat{a}$ and $Bra-Ketalgebra$ have special properties based on the analysis we have conducted. We wish to describe them here using the Bra-Ket notation. Let's begin by recalling an earlier result where we obtained

$$\widehat{H}\widehat{a}|A\rangle = \lambda\,(\widehat{a}|A\rangle) - \hbar\omega\,(\widehat{a}|A\rangle) = (\lambda - \hbar\omega)\widehat{a}|A\rangle \tag{C.33}$$

$|A\rangle$ is the *eigenket* or eigenstate of the harmonic oscillator $\widehat{H}$. For the harmonic oscillator problem, we have

$$\widehat{H}|A\rangle = \lambda|A\rangle = (n + \frac{1}{2})\hbar\omega|A\rangle = (n + \frac{1}{2})\hbar\omega|n\rangle$$

Representing $|A\rangle$ by $|n\rangle$ reminds us that each eigenket is associated with a specific integer n. It follows that $\widehat{a}$ has a direct impact on this integer, and therefore, an impact on the eigenstate because

$$\widehat{H}\widehat{a}|n\rangle = ((n + \frac{1}{2})\hbar\omega - \hbar\omega)\widehat{a}|n\rangle$$

$$= (n - \frac{1}{2})\hbar\omega\widehat{a}|n\rangle = \left([n-1] + \frac{1}{2}\right)\hbar\omega\widehat{a}|n\rangle$$

This means that $\widehat{a}$ operating on eigenket $|n\rangle$ leads to a eigenket that is proportional to the eigenket $|n-1\rangle$, or

$$\widehat{a}|n\rangle = d_n|n-1\rangle$$

Because of this property, $\widehat{a}$ is also known as an **annihilation operator** or **destruction operator**. In order to find the proportionality constant d_n, let Ket $B = \widehat{a}|n\rangle$. Then, we know from Eq. (C.23) that

$$\hbar\omega\langle B|B\rangle = \langle n|\hbar\omega\bar{\widehat{a}}\widehat{a}|n\rangle = \langle n|\widehat{H}|n\rangle - \frac{1}{2}\hbar\omega\langle n|n\rangle$$

$$= \left(n + \frac{1}{2}\right)\hbar\omega - \frac{1}{2}\hbar\omega$$

$$= n\hbar\omega$$

We can also use this relation:

$$\hbar\omega\langle B|B\rangle = \hbar\omega|d_n|^2\langle n-1|n-1\rangle = \hbar\omega|d_n|^2$$

We have taken the eigenkets of $\widehat{H}$ to be normalized. From this, we have

$$\hbar\omega|d_n|^2 = n\hbar\omega \Rightarrow d_n = \sqrt{n} \tag{C.34}$$

Note that since we only have $|d_n|^2$ to be real, d_n is within a phase factor (it can be complex, for example).

A similar analysis can be done for $\widehat{a}^\dagger$, starting with the relation

$$\widehat{H}\widehat{a}^\dagger|A\rangle = \lambda(\widehat{a}^\dagger|A\rangle) + \hbar\omega(\widehat{a}^\dagger|A\rangle) = (\lambda + \hbar\omega)\widehat{a}^\dagger|A\rangle \tag{C.35}$$

It follows that

$$\begin{aligned}\widehat{H}\widehat{a}^\dagger|n\rangle &= ((n + \frac{1}{2})\hbar\omega + \hbar\omega)\widehat{a}^\dagger|n\rangle \\ &= (n + \frac{3}{2})\hbar\omega\widehat{a}^\dagger|n\rangle = \left([n+1] + \frac{1}{2}\right)\hbar\omega\widehat{a}^\dagger|n\rangle\end{aligned}$$

Therefore, $\widehat{a}^\dagger$ operating on eigenket $|n\rangle$ leads to a eigenket that is proportional to the eigenket $|n+1\rangle$, or

$$\widehat{a}^\dagger|n\rangle = c_n|n+1\rangle$$

And because of this unique property, $\widehat{a}^\dagger$ is also known as a **creation operator** or **ladder operator**. We can also determine c_n by letting Ket $B = \widehat{a}^\dagger|n\rangle$. Then we have

$$\begin{aligned}\hbar\omega\langle B|B\rangle = \langle n|\hbar\omega\widehat{a}^\dagger\widehat{a}|n\rangle &= \langle n|\widehat{H}|n\rangle + \frac{1}{2}\hbar\omega\langle n|n\rangle \\ &= \left(n + \frac{1}{2}\right)\hbar\omega + \frac{1}{2}\hbar\omega \\ &= (n+1)\hbar\omega\end{aligned}$$

And since

$$\hbar\omega\langle B|B\rangle = \hbar\omega|c_n|^2\langle n+1|n+1\rangle = \hbar\omega|c_n|^2$$

We have taken the eigenkets of $\widehat{H}$ to be normalized. From this, we have

$$\hbar\omega|c_n|^2 = (n+1)\hbar\omega \Rightarrow c_n = \sqrt{n+1} \tag{C.36}$$

Thus, we have that operators $\widehat{a}$ and $\widehat{a}^\dagger$ are, alternatively, given by

Annihilation and Creation Operators

$$\widehat{a}|n\rangle = \sqrt{n}|n-1\rangle \quad \text{and} \quad \widehat{a}^{\dagger}|n\rangle = \sqrt{n+1}|n+1\rangle \tag{C.37}$$

This form of the operators is useful in calculating expectation quantities. Recall that our previous definitions for determining expectation values of x, x^2, and $\widehat{p}$ are by evaluating

$$\langle x\rangle = \frac{\int \psi^* x\psi dx}{\int \psi^*\psi dx} = \int \psi^* x\psi dx \tag{C.38a}$$

$$\langle x^2\rangle = \frac{\int \psi^* x^2\psi dx}{\int \psi^*\psi dx} = \int \psi^* x^2\psi dx \tag{C.38b}$$

$$\langle \widehat{p}\rangle = \frac{\int \psi^* \widehat{p}\psi dx}{\int \psi^*\psi dx} = \int \psi^* \left[-i\hbar\frac{\partial}{\partial x}\right]\psi dx \tag{C.38c}$$

The last form assumes the wave functions to be normalized so that $\int \psi^*\psi dx = 1$. Now that we have the annihilation and creation operators, these quantities can be computed with them, rather than using the relations in Eqs. (C.38a) - (C.38c). We do not even require the wave functions explicitly. Using the operators $\widehat{a}$ and $\overline{a}$, we can relate them to x as follows:

$$\widehat{a}+\widehat{a}^{\dagger} = \frac{m^*\omega x + ip}{\sqrt{2m^*\hbar\omega}} + \frac{m^*\omega x - ip}{\sqrt{2m^*\hbar\omega}} = \frac{2m^*\omega}{\sqrt{2m^*\hbar\omega}}x = \sqrt{\frac{2m^*\omega}{\hbar}}x$$

Solving for x gives

$$x = \sqrt{\frac{\hbar}{2m^*\omega}}\left(\widehat{a}+\widehat{a}^{\dagger}\right) \tag{C.39}$$

Eq. (C.39) can be used, along with Eq. (C.37), to compute the expectation value of x or $\langle x\rangle$ as

$$\langle x\rangle = \langle n|x|n\rangle = \sqrt{\frac{\hbar}{2m^*\omega}}(\langle n|\widehat{a}|n\rangle + \langle n|\widehat{a}^{\dagger}|n\rangle)$$

$$= \sqrt{\frac{\hbar}{2m^*\omega}}(\langle n|\sqrt{n}|n-1\rangle + \langle n|\sqrt{n+1}|n+1\rangle)$$

$$= \sqrt{\frac{\hbar}{2m^*\omega}}(\sqrt{n}\langle n|n-1\rangle + \sqrt{n+1}\langle n|n+1\rangle)$$

$$= 0$$

We have used the fact that the eigenstates are *orthogonal*, which is the case for any Hermitian Hamiltonian, as we have for the harmonic oscillator. Carrying out a similar calculation for $\langle x^2\rangle$, we have

$$\langle x^2\rangle = \langle n|x^2|n\rangle = \frac{\hbar}{2m^*\omega}\langle n|(\widehat{a} + orthogonal)(\widehat{a} + \widehat{a}^{\dagger})|n\rangle$$

$$= \frac{\hbar}{2m^*\omega}\langle n|(\widehat{a}\cdot\widehat{a} + \widehat{a}\cdot\widehat{a}^{\dagger} + \widehat{a}^{\dagger}\cdot\widehat{a} + \widehat{a}^{\dagger}\cdot\widehat{a}^{\dagger})|n\rangle$$

$$= \frac{\hbar}{2m^*\omega}(\langle n|\widehat{a}\cdot\widehat{a}|n\rangle + \langle n|\widehat{a}\cdot\widehat{a}^{\dagger}|n\rangle + \langle n|\widehat{a}^{\dagger}\cdot\widehat{a}|n\rangle + \langle n|\widehat{a}^{\dagger}\cdot\widehat{a}^{\dagger}|n\rangle)$$

We have the sum of four terms. Let's consider each term one by one, starting with $\widehat{a}\cdot\widehat{a}$ and $\widehat{a}^{\dagger}\cdot\widehat{a}^{\dagger}$, where we have

$$\langle n|\widehat{a}\cdot\widehat{a}|n\rangle = \langle n|\widehat{a}\sqrt{n}|n-1\rangle$$

$$= \sqrt{n}\langle n|\widehat{a}|n-1\rangle = \sqrt{n(n-1)}\langle n|n-2\rangle$$

$$= 0$$

Similarly, for $\widehat{a}^{\dagger}\cdot\widehat{a}^{\dagger}$, we get

$$\langle n|\widehat{a}^{\dagger}\cdot\widehat{a}^{\dagger}|n\rangle = \langle n|\widehat{a}^{\dagger}\sqrt{n+1}|n+1\rangle$$

$$= \sqrt{n+1}\langle n|\widehat{a}|n+1\rangle = \sqrt{(n+1)(n+2)}\langle n|n+2\rangle$$

$$= 0$$

For $\widehat{a}\cdot\widehat{a}^{\dagger}$ and $\widehat{a}^{\dagger}\cdot\widehat{a}$, we obtain *nonzero* results, as we have

$$\langle n|\widehat{a}\cdot\widehat{a}^{\dagger}|n\rangle = \langle n|\widehat{a}\sqrt{n+1}|n+1\rangle$$

$$= \sqrt{n+1}\langle n|\widehat{a}|n+1\rangle = (n+1)\langle n|n\rangle$$

$$= n+1$$

And for $\widehat{a}^{\dagger}\cdot\widehat{a}$, we have

$$\langle n|\widehat{a}^{\dagger}\cdot\widehat{a}|n\rangle = \langle n|\widehat{a}^{\dagger}\sqrt{n}|n-1\rangle$$

$$= \sqrt{n}\langle n|\widehat{a}^{\dagger}|n-1\rangle = n\langle n|n\rangle$$

$$= n$$

Putting the results together, we obtain $\langle x^2 \rangle$ to be given by

$$\langle x^2 \rangle = \frac{\hbar}{2m^*\omega}[0 + n + (n+1) + 0] = \frac{\hbar}{2m^*\omega}(2n+1) \tag{C.40}$$

$$= \frac{\hbar}{m^*\omega}\left(n + \frac{1}{2}\right) \tag{C.41}$$

If we also determine $\langle \widehat{p} \rangle$ and $\langle \widehat{p}^2 \rangle$, as we have done for $\langle x \rangle$ and $\langle x^2 \rangle$, the uncertainties Δx and Δp can both be determined for the harmonic oscillator. A relationship between $\widehat{p}$ and $\widehat{a}$, etc. can be found taking the difference between $\widehat{a}$ and $\widehat{a}^\dagger$, where we have

$$\widehat{a} - \widehat{a}^\dagger = \frac{m^*\omega x + ip}{\sqrt{2m^*\hbar\omega}} - \frac{m^*\omega x - ip}{\sqrt{2m^*\hbar\omega}} = i\frac{2}{\sqrt{2m^*\hbar\omega}}\widehat{p} = i\sqrt{\frac{2}{m^*\hbar\omega}}\widehat{p}$$

Solving for $\widehat{p}$ gives

$$\widehat{p} = -i\sqrt{\frac{m^*\hbar\omega}{2}}(\widehat{a} - \widehat{a}^\dagger) \tag{C.42}$$

Using this result to compute $\langle \widehat{p} \rangle$, we have

$$\langle \widehat{p} \rangle = \langle n|\widehat{p}|n\rangle = -i\sqrt{\frac{m^*\hbar\omega}{2}}(\langle n|\widehat{a}|n\rangle - \langle n|\widehat{a}^\dagger|n\rangle)$$

$$= -i\sqrt{\frac{m^*\hbar\omega}{2}}(\langle n|\sqrt{n}|n-1\rangle + \langle n|\sqrt{n+1}|n+1\rangle)$$

$$= -i\sqrt{\frac{m^*\hbar\omega}{2}}(\sqrt{n}\langle n|n-1\rangle - \sqrt{n+1}\langle n|n+1\rangle)$$

$$= 0$$

Carrying out the same calculation for $\langle \widehat{p}^2 \rangle$, we have

$$\langle \widehat{p}^2 \rangle = \langle n|\widehat{p}^2|n\rangle = -\frac{m^*\hbar\omega}{2}\langle n|\left(\widehat{a} - \widehat{a}^\dagger\right)\left(\widehat{a} - \widehat{a}^\dagger\right)|n\rangle$$

$$= -\frac{m^*\hbar\omega}{2}\langle n|(\widehat{a}\cdot\widehat{a} - \widehat{a}\cdot\widehat{a}^\dagger - \widehat{a}^\dagger\cdot\widehat{a} + \widehat{a}^\dagger\cdot\widehat{a}^\dagger)|n\rangle$$

$$= -\frac{m^*\hbar\omega}{2}(\langle n|\widehat{a}\cdot\widehat{a}|n\rangle - \langle n|\widehat{a}\cdot\widehat{a}^\dagger|n\rangle - \langle n|\widehat{a}^\dagger\cdot\widehat{a}|n\rangle + \langle n|\widehat{a}^\dagger\cdot\widehat{a}^\dagger|n\rangle)$$

This contains the same four terms we had in the determination of $\langle x^2 \rangle$. We determined these separately above, so we can use the same results to compute the above, where we have for $\langle \widehat{p} \rangle$

$$\langle \widehat{p}^2 \rangle = -\frac{m^* \hbar \omega}{2}(0 - n - (n+1) + 0)$$

$$= \frac{m^* \hbar \omega}{2}(2n+1) = m^* \hbar \omega \left(n + \frac{1}{2}\right)$$

Having these four quantities is sufficient to determine the uncertainty Δx as

$$\Delta x = \sqrt{\langle x^2 \rangle - \langle x \rangle^2} = \sqrt{\langle x^2 \rangle} = \sqrt{\frac{\hbar}{m^* \omega}\left(n + \frac{1}{2}\right)}$$

And for $\Delta \widehat{p}$, we now obtain

$$\Delta \widehat{p} = \sqrt{\langle \widehat{p}^2 \rangle - \langle \widehat{p} \rangle^2} = \sqrt{\langle x^2 \rangle} = \sqrt{m^* \hbar \omega \left(n + \frac{1}{2}\right)}$$

This allows the uncertainty $\Delta x \Delta \widehat{p}$, for the harmonic oscillator to be determined (using Bra-Ket algebra), yielding the result

Harmonic Oscillator Uncertainty Principle

$$\Delta x \Delta \widehat{p} = \hbar \left(n + \frac{1}{2}\right) \tag{C.43}$$

Thus, Eq. (C.43) gives a result in accordance with the general *uncertainty principle* derived in Chap.??. There, we found that $\Delta x \Delta \widehat{p} \geq \hbar$. Since we have identified the relation defining the ground state, as well as the raising/lowering operators, next we'll make use of these operators to obtain additional higher order states using Bra-Ket algebra.

C.5 Ground And Above-Ground States of LHO

In this section, we look at how to obtain the explicit wave functions $|n\rangle = \psi_n(x)$ using the operators $\widehat{a}$ and/or $\widehat{a}^\dagger$. This connects the Bra-Ket algebra directly to the classical mathematical notation used throughout the text. We begin with the ground state of the harmonic oscillator. Let the ground-state be denoted by $\psi_0 = |0\rangle =$. Eq.(C.30) is the condition leading to the ground state, defined by

Ground State Condition

$$\widehat{a}|0\rangle = 0 \tag{C.44}$$

This condition means that

$$\widehat{a}|0\rangle = \widehat{a}\psi_0 = \frac{1}{\sqrt{2m^*\hbar\omega}}[m^*\omega x + ip]\psi_0 = 0 \tag{C.45}$$

Since the RHS equals zero, we can ignore the constant factor. Then, using the momentum operator for $\widehat{p}$, Eq. (C.45) becomes

$$\left[m^*\omega x + i\left(-i\hbar\frac{d}{dx}\right)\right]\psi_0 = 0$$

This leads to

$$\frac{1}{\psi_0}\frac{d\psi_0}{dx} = \frac{m^*\omega}{\hbar}x \Rightarrow \frac{d\psi_0}{\psi_0} = \frac{m^*\omega}{\hbar}xdx \tag{C.46}$$

Upon integration of Eq. (C.46), we have

$$\int\frac{d\psi_0}{\psi_0} = -\frac{m^*\omega}{\hbar}\int xdx \Rightarrow \log\psi_0 = -\frac{m^*\omega}{2\hbar}x^2 + c_i \tag{C.47}$$

c_i is an integration constant. Taking the exponential of both sides, we have the solution given by

$$\psi_0 = C_0 e^{-\frac{m^*\omega}{2\hbar}x^2} = C_0 e^{-\frac{x^2}{2\sigma^2}} \tag{C.48}$$

Introducing a dimensionless quantity $\xi = x/\sigma$, we can write the wave function $\psi_0 = |0\rangle$ as

Ground State For LHO

$$|0\rangle = C_0 H_0 e^{-\frac{1}{2}\xi^2} \tag{C.49}$$

H_0 denotes the zeroth order Hermite polynomial. The latter form uses the simple fact that $H_0 = 1$, however, we'll see that this pattern generalizes further. Eq. (C.49) is identical to the ground state obtained in Chap. A, for the linear harmonic oscillator. Thus, the condition given by Eq. (C.44) defines the ground-state wave function $|0\rangle$. The normalization constant C_0 is found from the normalization condition $\langle 0|0\rangle = \int \psi_0^* \psi_0 dx = 1$, which gives

$$1 = \int \psi_0^* \psi_0 dx = C_0^2 \int_{-\infty}^{+\infty} e^{-\frac{x^2}{2\sigma^2}} e^{-\frac{x^2}{2\sigma^2}} dx \tag{C.50}$$

$$= C_0^2 \int_{-\infty}^{+\infty} e^{-\frac{x^2}{\sigma^2}} dx \tag{C.51}$$

$$= C_0^2 \left(\sigma\sqrt{\pi}\right) = C_0^2 \sqrt{\frac{\pi\hbar}{m^*\omega}} \tag{C.52}$$

The normalization constant C_0 is, therefore, given by

$$C_0 = \left[\frac{m^*\omega}{\pi\hbar}\right]^{\frac{1}{4}} \tag{C.53}$$

Having the ground state $|0\rangle$, along with the creation/annihilation operators in terms of $\widehat{a}$ enables the determination of above-ground states. This is done using the creation operator, where we have

$$\begin{aligned}\widehat{a}^{\dagger}|n\rangle = \sqrt{n+1}|n+1\rangle \Rightarrow \psi_{n+1} &= \frac{1}{\sqrt{n+1}}\widehat{a}^{\dagger}\psi_n \\ &= \frac{1}{\sqrt{n+1}}\left[\frac{m^*\omega x - i\widehat{p}}{\sqrt{2m^*\hbar\omega}}\right]\psi_n \\ &= \frac{1}{\sqrt{2(n+1)}}\left[\sqrt{\frac{m^*\omega}{\hbar}}x - i\frac{1}{\sqrt{m^*\hbar\omega}}\widehat{p}\right]\psi_n\end{aligned}$$

Replacing the momentum operator $\widehat{p} = -i\hbar d/dx$, leads to

$$\begin{aligned}\psi_{n+1} &= \frac{1}{\sqrt{2(n+1)}}\left[\sqrt{\frac{m^*\omega}{\hbar}}x - \sqrt{\frac{\hbar}{m^*\omega}}\frac{d}{dx}\right]\psi_n \\ &= \frac{1}{\sqrt{2(n+1)}}\left[\frac{x}{\sigma} - \sigma\frac{d}{dx}\right]\psi_n\end{aligned}$$

In terms of the dimensionless quantity $\xi = x/\sigma$, we can write the above relation as

Above-Ground State Iterative Equation For LHO

$$\psi_{n+1}(\xi) = \frac{1}{\sqrt{2(n+1)}}\left[\xi - \frac{d}{d\xi}\right]\psi_n \quad \text{n=0,1,2,...} \tag{C.54}$$

Having the ground state and the relation given by Eq. (C.54) allows us to determine any state $n+1$, using it's previous state n. For example, starting with the ground state $|0\rangle = \psi_0$ to obtain $|1\rangle = \psi_1$, we have

$$\begin{aligned}
|1\rangle = \psi_1(\xi) &= \frac{1}{\sqrt{2(0+1)}}\left[\xi - \frac{d}{d\xi}\right]\psi_0 \\
&= \frac{1}{\sqrt{2}}\left[\xi - \frac{d}{d\xi}\right]C_0 e^{-\frac{1}{2}\xi^2} \\
&= \frac{C_0}{\sqrt{2}}[\xi + \xi]e^{-\frac{1}{2}\xi^2} \\
&= C_1\,[2\xi]e^{-\frac{1}{2}\xi^2}
\end{aligned}$$

Referring to Table A.2, which lists the Hermite polynomials H_n, we find that $H_1(\xi) = 2\xi$, so that $|1\rangle$ can be written as

$$|1\rangle = C_1 H_1(\xi) e^{-\frac{1}{2}\xi^2} \tag{C.55}$$

We can write the normalization constant as

$$C_1 = \frac{C_0}{\sqrt{2 \cdot 1}} \tag{C.56}$$

Evaluating $|2\rangle = \psi_2(\xi)$, we have

$$\begin{aligned}
|2\rangle &= \frac{1}{\sqrt{2(1+1)}}\left[\xi - \frac{d}{d\xi}\right]\psi_1 \\
&= \frac{1}{2}\left[\xi - \frac{d}{d\xi}\right]C_1\,[2\xi]e^{-\frac{1}{2}\xi^2} \\
&= \frac{C_1}{2}\left[(2\xi^2) - \left(2 - 2\xi^2\right)\right]e^{-\frac{1}{2}\xi^2} \\
&= C_2\,[4\xi - 2]e^{-\frac{1}{2}\xi^2}
\end{aligned}$$

The term in brackets is the 2nd order Hermite polynomial (see Table A.2), so that $|2\rangle$ is given by

$$|2\rangle = C_2 H_2(\xi) e^{-\frac{1}{2}\xi^2} \tag{C.57}$$

and

$$C_2 = \frac{C_1}{\sqrt{2\cdot 2}} = \frac{C_0}{\sqrt{2\cdot 1}}\frac{1}{\sqrt{2\cdot 2}} = \frac{C_0}{\sqrt{2!2^2}} \tag{C.58}$$

As the last example, evaluation of $|3\rangle = \psi_3$ gives

$$\begin{aligned}
|3\rangle &= \frac{1}{\sqrt{2(2+1)}}\left[\xi - \frac{d}{d\xi}\right]\psi_2 \\
&= \frac{C_2}{\sqrt{2\cdot 3}}\left[\left(4\xi^3 - 2\xi\right) - \left(8\xi + [4\xi^2 - 2](-\xi)\right)\right] e^{-\frac{1}{2}\xi^2} \\
&= \frac{C_2}{\sqrt{2\cdot 3}}\left[8\xi^3 - 12\xi\right] e^{-\frac{1}{2}\xi^2} \\
&= C_3 H_3(\xi) e^{-\frac{1}{2}\xi^2}
\end{aligned}$$

where

$$C_3 = \frac{C_2}{\sqrt{2\cdot 3}} = \frac{C_0}{\sqrt{2!2^2}}\frac{1}{\sqrt{3\cdot 2}} = \frac{C_0}{\sqrt{3!2^3}}$$

Hopefully, we have established a clear pattern of the wave functions $|n+1\rangle = \psi_{n+1}$. The general form of the solution is given by

1D Linear Harmonic Oscillator Eigenkets

$$|n+1\rangle = C_{n+1}\cdot H_{n+1}(\xi)\cdot e^{-\frac{1}{2}\xi^2} \tag{C.59}$$

H_{n+1} are *Hermite polynomials* (c. Chap. A). This result is identical to our earlier result given by Eq. (C.48). The normalization constant C_n is given by

1D Harmonic Oscillator Normalization Factor

$$C_n = \frac{C_0}{\sqrt{n! \cdot 2^n}} \tag{C.60}$$

C_0 is the *ground state normalization factor* given by Eq. (C.53). Thus, the *Dirac method* has also lead us to the same eigenstates or eigenkets obtained by solving the Schrödinger equation directly. However, this was done *using two first order operators, rather than one second order operator.* We can go on to express the most general form of the solution. This is the objective of the following section.

C.6 Coherent State Of The Oscillator Revisited

Therefore, the relationship between the eigenvalue β and the initial amplitude x_0 is given by

$$x_0 = 2|\beta|\sqrt{\frac{\hbar}{2m^*\omega}} \tag{C.61}$$

$$= |\beta|\sqrt{\frac{2\hbar}{m^*\omega}} \tag{C.62}$$

Thus, we have the classical result if the eigenvalue β is given by

Eigenvalue Of Initial State

$$|\beta| = x_0\sqrt{\frac{m^*\omega}{2\hbar}} \tag{C.63}$$

The eigenvalue given by Eq. (C.63) is identical to the eigenvalue we obtained for the coherent state, derived in Chap. A, and given by Eq. (A.39). Therefore, *the eigenvalues are the same for coherent state and initial condition of the most general state.* Thus, the two functions are one and the same. Also, $|\beta|^2$ becomes

$$x_0^2 = 4|\beta|^2\frac{\hbar}{2m^*\omega} = 2|\beta|^2\frac{\hbar}{m^*\omega} \Rightarrow |\beta|^2 = \frac{1}{2}x_0^2\frac{m^*\omega}{\hbar} \tag{C.64}$$

A similar analysis can also be carried out for $\langle p \rangle$ using, instead, the relation

$$\widehat{p} = i\sqrt{\frac{m^*\hbar\omega}{2}}(\widehat{a}^\dagger - \widehat{a})$$

We leave this as an exercise for the reader. This completes the chapter on *Bra-Ket algebra*. In the next section, we'll summarize this chapter's content.

C.7 Chapter Summary

In this chapter, we have introduced a popular notation for quantum mechanics, known as the *Bra-Ket algebra*. The notation was originally introduced by Dirac. We have linked the classical notation to the Bra-Ket algebra to demonstrate how it can also be used to analyze quantum mechanical problems. It not only provides an improved degree of conciseness, but it proves to have far reaching capabilities in analysis, as well.

We revisited the harmonic oscillator problem to demonstrate the use of the Bra-Ket algebra to solve this problem. The annihilation and creation operators were revisited, and used to obtain the solutions and expectation values for the harmonic oscillator. Table C.1 summarizes some of the key formulas from this chapter.

Table C.1: Chapter C Summary Equations.

Name	**Equation**
Dirac symmetry operator I	$\widehat{a} = \frac{m^*\omega x + ip}{\sqrt{2m^*\hbar\omega}}$
Dirac symmetry operator II	$\widehat{a}^\dagger = \frac{m^*\omega x - ip}{\sqrt{2m^*\hbar\omega}}$
Oscillator Eigenenergies	$E_\lambda = \left(n + \frac{1}{2}\right)\hbar\omega$
Annihilation operator	$\widehat{a}\|n\rangle = \sqrt{n}\|n-1\rangle$

Table C.1

Creation operator	$\widehat{a}^{\dagger}\|n\rangle = \sqrt{n+1}\|n+1\rangle$
Oscillator Uncertainty	$\Delta x \Delta \widehat{p} = \hbar\left(n + \frac{1}{2}\right)$
Above-Ground State	$\|n+1\rangle = \frac{1}{\sqrt{2(n+1)}}\left[\xi - \frac{d}{d\xi}\right]\|n\rangle$
Coherent State	$\|C\rangle = e^{-\frac{\|\beta\|^2}{2}} \sum_{n=0}^{\infty} \frac{\beta^n}{\sqrt{n!}}\|n\rangle$
General LHO State	$\|\Psi(x,t)\rangle = e^{-\frac{\|\beta\|^2}{2}} \sum_{n=0}^{\infty} \frac{\beta^n}{\sqrt{n!}}\|n\rangle e^{-i\left(n+\frac{1}{2}\right)\omega t}$
Coherent State Eigenvalue	$\|\beta\| = x_0\sqrt{\frac{m^*\omega}{2\hbar}}$

C.8 Chapter Problems

Problem C.1 The first order operators introduced by Dirac were given by

$$\widehat{a} = \frac{m^*\omega x + ip}{\sqrt{2m^*\hbar\omega}} \quad \text{and} \quad \widehat{a}^{\dagger} = \frac{m^*\omega x - ip}{\sqrt{2m^*\hbar\omega}}$$

In the Bra-Ket notation, we found

$$\widehat{a}|n\rangle = \sqrt{n}|n-1\rangle \quad \text{and} \quad \widehat{a}^{\dagger}|n\rangle = \sqrt{n+1}|n+1\rangle$$

Use these two relations strictly in Bra-Ket form to prove that

$$\widehat{a}^{\dagger}\widehat{a}|n\rangle = n|n\rangle$$

Problem C.2 Use the above eigenrelations for $\widehat{a}$ and $\widehat{a}^{\dagger}$ to show that reversing the order of the operators gives

$$\widehat{a}\widehat{a}^{\dagger}|n\rangle = (n+1)|n\rangle$$

Problem C.3 Use Eq. (C.54) to show that the harmonic oscillator state $|2\rangle$ can be expressed in terms of the ground state $|0\rangle$ as

$$|2\rangle = \frac{1}{2\sqrt{2}}\left[\frac{d^2|0\rangle}{d\xi^2} - (1+2\xi) + \xi^2|0\rangle\right]$$

Problem C.4 Use the relation

$$\widehat{p} = i\sqrt{\frac{m^*\hbar\omega}{2}}(\overline{a} - \widehat{a})$$

to show that $\langle \widehat{p}^4 \rangle$ is given by

$$\langle \widehat{p}^4 \rangle = (m^*\hbar\omega)^2 \left(n^2 + n + \frac{1}{4}\right)$$

Start with the following relation and use the properties of the cperators $\widehat{a}$ and $\widehat{a}^+$.

$$\langle n| \left(\frac{m^*\hbar\omega}{2}\right)^2 \left(\widehat{a}^+ - \widehat{a}\right)^4 |n\rangle$$

Problem C.5 With eigenvalues of the form found for $\langle \widehat{p}^4 \rangle$, discuss the conditions of one-to-one-ness with quantum number n for a value of $\langle \widehat{p}^4 \rangle$.?

Problem C.6 Use the relation

$$\widehat{p} = i\sqrt{\frac{m^*\hbar\omega}{2}}(\overline{a} - \widehat{a})$$

along with the general time-dependent harmonic oscillator state to show that $\langle p \rangle(t)$ is given by

$$\langle \widehat{p} \rangle(t) = -m^*\omega x_0 \sin\omega t$$

Index

D

E

F

G

H

I

J

K

L

M

N

O

P

www.ingramcontent.com/pod-product-compliance
Lightning Source LLC
LaVergne TN
LVHW091534200926
848483LV00002B/395
* 9 7 8 0 5 7 8 5 7 8 4 5 3 *